# Medicinal and Aromatic Plants of the World

Volume 3

**Series Editor**
Ákos Máthé
University of West Hungary
Faculty of Agriculture and Food Science
Mosonmagyaróvár, Hungary

More information about this series at http://www.springer.com/series/11192

Mohamed Neffati • Hanen Najjaa
Ákos Máthé
Editors

# Medicinal and Aromatic Plants of the World - Africa Volume 3

Springer

*Editors*
Mohamed Neffati
Arid Zone Research Institute
Médenine, Tunisia

Ákos Máthé
Faculty of Agriculture and Food Science
University of West Hungary
Mosonmagyaróvár, Hungary

Hanen Najjaa
Min of Agriculture, Range Ecology Laboratory
Arid Zone Research Institute
Médenine, Tunisia

ISSN 2352-6831 ISSN 2352-684X (electronic)
Medicinal and Aromatic Plants of the World
ISBN 978-94-024-1119-5 ISBN 978-94-024-1120-1 (eBook)
DOI 10.1007/978-94-024-1120-1

Library of Congress Control Number: 2017944445



Printed on acid-free paper

This Springer imprint is published by Springer Nature
The registered company is Springer Science+Business Media B.V.
The registered company address is: Van Godewijckstraat 30, 3311 GX Dordrecht, The Netherlands

# Foreword

The utilization of medicinal and aromatic plants (MAPs) dates back to the beginnings of mankind. This fact is increasingly true in the case of Africa, a continent that is considered to be the cradle of mankind.

The rich and diverse flora of Africa has been basis/home for the evolution of African traditional medicine. This holistic discipline incorporates the use of natural plant resources (in the form of indigenous herbalism) as well as African spirituality. In the course of millennia, through the ages, the use of MAPs has also seen extreme development, so that to date, the most sophisticated medicines are already available on the African continent as well. Nonetheless, healing with medicinal plants has retained its central place in the life of the masses. Beyond traditions, this can be attributed to the fact that synthetic medicines are still simply not affordable for many.

Scientific progress has brought about a progress also in the use of traditional medicinal plants. This means that the medicinal and aromatic plants historically used by the practitioners of traditional African medicine (diviners, midwives, herbalists) are being gradually subjected to scientific investigations. Modern pharmacy wants to isolate and utilize their active principles in the form of modern preparation forms that – in turn – can be rendered accessible to consumers already on a global scale.

As a result of this briefly (schematically) described process, the MAP resources – formerly used only by the local people – must come up to cover global demands. Obviously, this is not possible without farther protection, sustainable management and, as an ultimate option, cultivation.

In view of the intensive R+D activities into African MAPs, the special literature abounds in publications. Although even the present pace of research is not considered as sufficient by many, from handbooks to clinical research papers, the number of both comprehensive and specific publications related to MAPs is imposing.

Volume 3 distinguishes itself from most available volumes on African MAPs by avoiding the semblance of completeness. Squeezing all knowledge into one volume would be absolutely impossible.

Starting out from an overview of the status of MAPs in Africa, two introductory chapters offer most comprehensive, valuable reviews of most important African MAPs, as well as their traditional uses. The chapters to follow are compilations of up-to-date information on selected characteristic African MAP species (e.g. African cherry, bird plum), geographic regions (East and North Africa, Mauritius, sub-Saharan Africa, Morocco, Nigeria, Tunisia) or MAPs growing under special ecological conditions (arid and desert zones) and in cultivation, in Nigeria. In most contributions the utilization, botanical, phytochemical as well as pharmacological properties/traditional medicinal uses, of relevant species will be discussed.

It is unhidden hope of the author's collective that the volume will serve as a useful and updated complement to what has already been achieved in summarizing and publishing basic knowledge on African medicinal and aromatic plants.

Médenine, Tunisia Mohamed Neffati
Mosonmagyaróvár, Hungary Ákos Máthé
Médenine, Tunisia Hanen Najjaa

# Contents

# Introduction to Medicinal and Aromatic Plants in Africa

**Ákos Máthé, Mohamed Neffati, and Hanen Najjaa**

**Abstract** Natural substances of plant origin, have been used by man from the very early times of civilizations. Traditional Medicine (TM), a heritage of Africa, is still servicing approx. 80% of the population of this huge continent comprizing 55 countries, 800 languages and an unmeasurable wealth of plant diversity. The continent with a varied flora of an estimated 68,000 plant species is home to about 35,000 endemic species (Cunningham A, African medicinal plants. Setting priorities at the interface between conservation. UNESCO, Paris, 1993), growing in seven centers of endemism. Africa's not yet fully explored medicinal plant resources frequently come into the forefront of interest. The recent upsurge is witnessed by the fact that ca. 60% of all recent publications on African medicinal plants, appeared during the last decade. There is a tremendous potential for developing new crops and new products. Research into MAPs seem to offer a nearly unlimited scope for exploration. The ongoing screening for the biological activity of MAPs is also broad and covers several important health related aspects including antimalarial and other antiprotozoal products, anti-proliferative potential, anti-inflammatory and analgesic activities, anti-diabetes activity antioxidant activity and activity on the reproductive cardiovascular and central nervous systems, etc. Despite of this marked recent increase, Africa is lagging behind Europe and Asia in terms of the number of products that have been commercialized and the percentage of flora utilized for international trade. The growing demand on resources, however, has resulted, in some cases, the overexploitation, the occasional local disappearance of favored sources of botanicals and have reduced the diversity of species. By elaborating international and national standards of sustainability, these dangers should be either eased or eradicated. Domestication of wild-crafted species can also contribute to preserve the genetic resources and to reduce the sometimes imminent threats of extinction. The recent initiative to develop national and ultimately an African Herbal

Á. Máthé (✉)
Faculty of Agriculture and Food Science, University of West Hungary, Mosonmagyaróvár, Hungary
e-mail: acatius2005@gmail.com

M. Neffati • H. Najjaa
Arid Zone Research Institute, 4119 Médenine, Tunisia
e-mail: neffati.mohamed@ira.rnrt.tn; hanen.najjaa@yahoo.fr



M. Neffati et al. (eds.), *Medicinal and Aromatic Plants of the World - Africa Volume 3*, Medicinal and Aromatic Plants of the World,
DOI 10.1007/978-94-024-1120-1_1

Pharmacopoeia is likely to mark an important milestone in meeting quality, safety and efficacy standards both in African and international perspectives.

**Keywords** Medicinal and aromatic plants • Traditional medicine • African flora • Biodiversity hotspots • Sustainable management • Cultivation • Pharmacopoeia

## 1 Introduction

Africa is the world's second largest continent after Asia, both in terms of area and population. In 2010, its entire population was estimated at 1010.3 million people (FAO 2012). The African continent comprises ca. 55 countries, 800 languages, 3000 dialects, and as such it is a veritable treasure of genetic resources including medicinal plants (Simon et al. 2007). More than half of Africa's population makes a living from agriculture (FAO 2012). Medicinal plants are still widely used in the healthcare systems of Africa, particularly by the African population.

According to Mander and Le Breton (2005), alone in southern Africa, there are up to 100 million traditional-remedy consumers and ca. 500,000 traditional healers. Some 700,000 tons of plant material is consumed annually with an estimated value of as much as 150 million US dollars per annum.

Medicinal plants are often a basic requirement for the treatment of certain conditions irrespective of education and income levels of the patient (Cocks and Dold 2000). The use of traditional medicines is not confined to rural, low-income groups, but also prevails in urban areas.

The trade in traditional medicines (TM) forms part of a multimillion-Rand 'hidden economy' in southern Africa (Cunningham 1989). It is stimulated by high population growth rates, rapid urbanization and the important cultural value placed on traditional medicines, so that this trade is now greater than at any time in the past (Mander et al. 2006).

The African-continent has an important role as a source of medicinal and aromatic plants for the world's food, drug, herb, dietary supplement, and cosmetics markets. Although, in changing supplies, numerous African raw plant materials still have an established, strong international market presence. People around the world enjoy Africa's manifold "culinary contributions" including the peanut (*Arachis hypogaea*, Fabaceae), yam (*Dioscorea* spp., Dioscoreaceae), watermelon (*Citrullus lanatus*, Cucurbitaceae), okra (*Abelmoschus esculentus*, Malvaceae), as well as numerous other foods and flavors like coffee (*Coffea* spp., Rubiaceae) (Blumenthal 2011).

This wealth of plant species has been subject to a multitude of scientific investigations that also reveal the versatility of international contributions Africa's. As such, the number of references on "African MAPs" in the scientific database SCOPUS exceeds 25,000, in the period 2010–2015 and it seems to be farther increasing at a steady pace (e.g. *ca.* 9330 entries, in the current year of 2016).

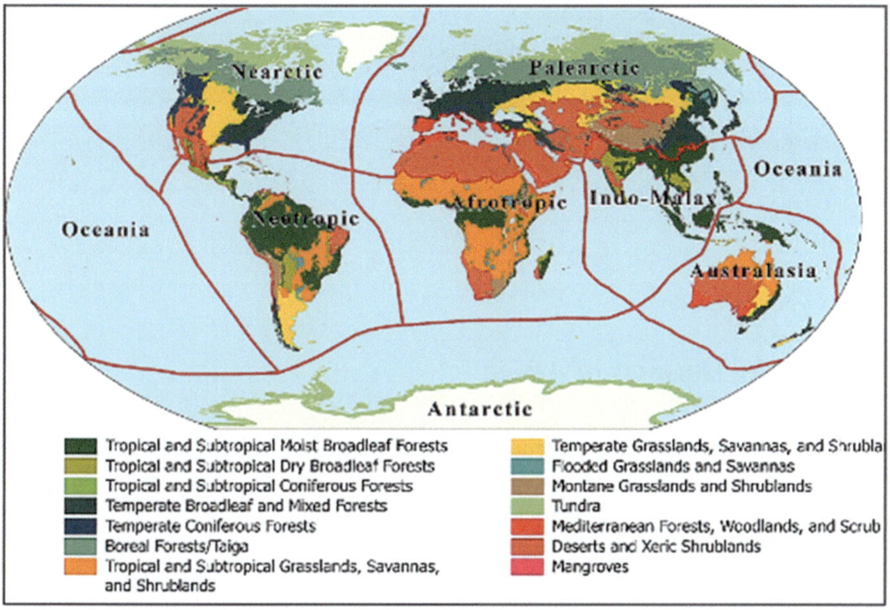


**Fig. 1** Terrestrial ecoregions of the world: 14 biomes and eight biogeographic realms (Olson et al. 2001)

Remarkably, despite of the relevant dynamic increase in the number of publications on indigenous medicinal plants and dietary supplements, Africa is lagging behind Europe and Asia. Similarly, a relatively low percentage of the African flora has been utilized for international trade (Van Wyk 2015). So, Africa remains to be a continent with nearly unlimited potentials for research and development in the exploration of new MAP crops and new products.

## *Africa as a Botanical Continent*

According to the World Geographical Scheme for Recording Plant Distributions (WGSRPD) (Brummitt and Pando 2001), which provides definitions and codes for recording plant distributions, Africa is a **botanical continent** that corresponds closely to the usual geographical definition. It excludes the Sinai Peninsula, which is placed in the region of Western Asia. To the west, it includes islands grouped as Macaronesia, comprising the Azores, the Canary Islands, and the Cape Verde Islands. To the east, it includes Madagascar and other Indian Ocean islands out as far as the island of Rodrigues (Udvardy 1975) and (Brummitt et al. 2001) (Fig. 1).

**Table 1** Biodiversity hotspots in Africa

| | Plant species | Endemic species |
|---|---|---|
| Hotspot | *Number* | |
| Eastern Arc and Coastal Forests of Tanzania/Kenya | 4000 | 1500 |
| West African Forests | 9000 | 2250 |
| Cape Floristic Region | 8200 | 5682 |
| Succulent Karoo | 4849 | 1940 |
| Madagascar | 12,000 | 8704 |

After Myers et al. (2000)

## *Biodiversity Hotspots of Africa*

The concept of "**hotspots**" has been elaborated to denote those natural areas that are characterized by exceptional concentrations of species with high levels of endemism and are experiencing unusually rapid rates of depletion (Myers 1988) and (Klopper et al. 2002). On a worldwide scale, 44% of all vascular plant species are confined to 25 recognized hotspots. This amounts to merely 1.4% of the Earth's terrestrial surface. Four out of the 25 hotspots can be found on the African continent and an additional one is Madagascar (Table 1).

According to Conservation International (Anonymous 2005) biodiversity hotspots, the Earth's biologically richest places with high numbers of endemic species, face extreme threats and have already lost at least 70% of their original vegetation.

The overexploitation of Africa's natural resources has also frequently lead to the destruction of both natural populations and vast stretches of unique natural areas. As habitat loss is often accompanied by the extinction of equally unique organisms and ecosystems, ultimately, the rich and unique biological heritage of the African continent can be at stake. In the lack of exact and comprehensive documentation of both extent and threatened species, this process can be regarded as greatly threatening (Klopper et al. 2002).

## *Flora of Africa*

The African continent has a unique diversity of geographic and climatic factors. It has a similarly exceptionally rich and varied flora with an estimated 68,000 plant species, of which about 35,000 are known to be endemic (Cunningham 1993). Exceptionally rich in endemic species (8500 species, each) are the Guineo-Congolian and Cape biogeographic units. The latter is known as the Cape Floristic Region and encompasses five of the South Africa's 12 endemic plant families. The Cape region has predominantly Mediterranean type of vegetation (Van Wyk et al. 2009). On the other hand, the Guineo-Congolian region can be ranked to the African

**Table 2** Seven centers of endemism in Africa

| Biogeographic unit | | Area 1000 $km^2$ | Plants | |
|---|---|---|---|---|
| | | | No. of species | % |
| 1. | Guineo-Congolian | 2815 | 8000 | 80 |
| 2. | Zambesian | 3939 | 8500 | 54 |
| 3. | Sudanian | 3565 | 2750 | 33 |
| 4. | Somali-Masai | 1990 | 2500 | 50 |
| 5. | Cape | 90 | 8500 | 80 |
| 6. | Karoo-Namib | 629 | 3500 | 50 |
| 7. | Afro-montane | 647 | 3000 | 75 |

After Cunningham (1993)

rain forests. Lowest frequency of endemism has been explored in the Somali-Masai floristic region, a mostly arid semi-desert grassland and scrubland (White 1983). The main geographical units of plant endemisms, according to Cunningham (1993) are summarized in Table 2.

## *Conservation and Sustainable Management of MAP Resources in Africa*

The sustainable management of traditional medicinal plant resources has become an important issue, not only because of the value of the natural flora as a potential source of new drugs (Elujoba et al. 2005), but also due to population growth accompanied by the significant reliance, in many African countries, on the use of traditional medicinal plants for health. The indiscriminate harvesting of spontaneous flora, including forests has already resulted in severe losses in habitats and genetic diversity. Additional sources of threat are forest clearance for agriculture, afforestation of montane grasslands, uncontrolled burning and livestock grazing. Frequently, it is the exclusion from core conservation areas that affects medicinal plant supplies adversely. The effect of competing resource uses (e.g.: as timber logging, commercial harvesting for export and extraction of pharmaceuticals, and use for building materials and fuel) is similar.

In brief, a growing demand for fewer resources has resulted, in some cases, the local disappearance of favored and effective sources of botanicals that have lead to reduced species diversity. In this process, the popular, slow growing or slow to propagate species or species with specific habitat requirements and a limited distribution are most vulnerable (Cunningham 1993).

In 2010, over 2000 plant species, about 1000 species of mammals and nearly 2000 species of fish were threatened, in Africa, whereas globally, in the same year, the relevant numbers were: more than 10,000 plant species, about 6000 species of fish and over 3000 species of mammals (FAO 2012).

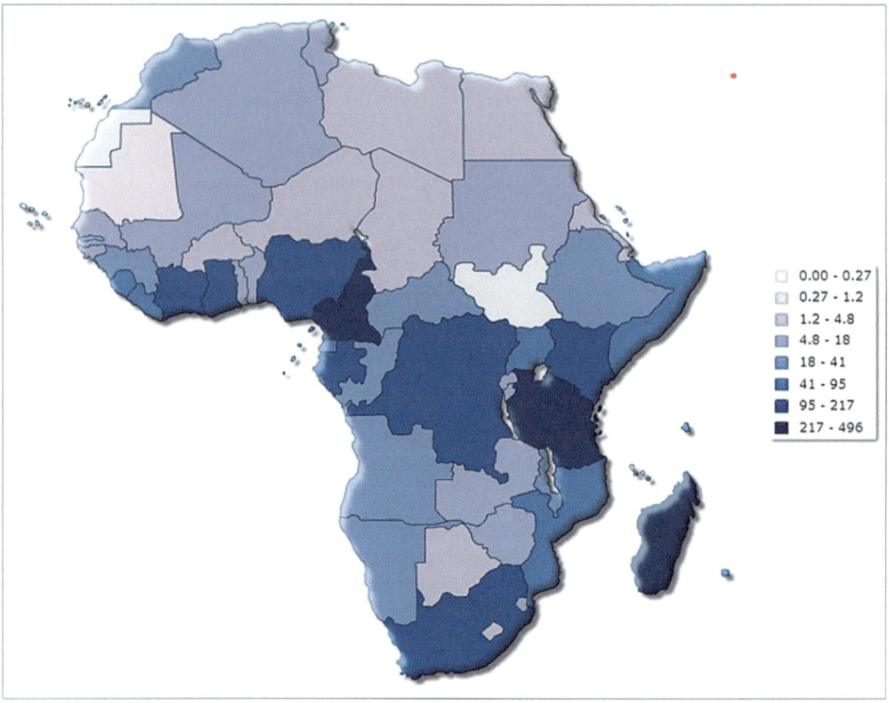


**Fig. 2** Endangered plant species in Africa (IndexMundi 2016)

A country-wise statistic of **endangered species** reveals that the African country with the highest value of 496 species, is Tanzania and the country with the lowest value of 0, is Mauritania (IndexMundi 2016) (Fig. 2).

As a part of natural diversity, the above described trends in the overexploitation of natural resources, have already lead to substantial losses in African MAP resources. The importance of medicinal plants in primary health care, both in self-medication and in national health services was first recognized by participants of the WHO/IUCN/WWF International Consultation on Conservation of Medicinal Plants, held in Chiang Mai, 21–26 March 1988 (Van Seters 1997).

Alarmed by the consequences of loss of plant diversity around the world, the so called Chiang Mai Declaration was issued under the motto: "Saving Lives by Saving Plants". The Declaration affirms the importance of medicinal plants and calls on the United Nations, its agencies and Member States, as well as other international organizations, to take action for the **conservation and sustainable use of medicinal plants** (Máthé 2015).

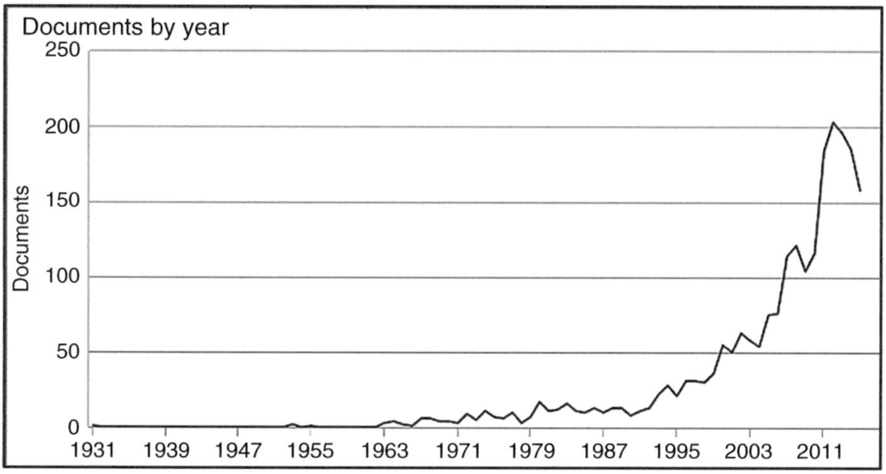


**Fig. 3** Number of SCOPUS documents of the search "medicinal plants Africa" for the period 1930–2015

## *Saving Plants that Save Lives: Effect of Chiang Mai Declaration (1988)*

The belated growth in international awareness about the declining supply capacity of MAPs, the over-harvesting of natural resources, the destructive harvesting practices accompanied by habitat loss, forest degradation, etc. are apparent, also on the African continent.

It seems that the Chiang Mai Declaration (1988) calling for coordinated international efforts to safeguard the natural populations of plant biodiversity has had a spurring effect on the study of African medicinal plants, too.

Our survey of the relevant special literature of the decades following the Declaration have revealed the emergence of new terms previously not used by the scientific community. E.g. the term "exploitation of MAPs" did not figure in the SCOPUS databases of special literature, until 1998 (Fig. 3).

Pharmacology, toxicology and pharmaceutics (788), Agricultural and Biological Sciences (634), as well as Medicine (601) have been the most popular scientific areas on African MAPS with entries mostly from South Africa (752), Nigeria (160), United States (139), Cameroon (105) and the United Kingdom (100).

As regards biodiversity and conservation of medicinal and aromatic plants in Africa, some important contributions have opened up a comprehensive insight into both problems and possibilities (Okigbo and Mmeka 2006; Makunga et al. 2008; Rukangira 2001; Dzoyem and Tshikalange 2013), etc. These authors also emphasize the importance of saving Africa's medicinal plant resources from extinction and call for intensive management as well as conservation, more research and increased level of public awareness about this vanishing African heritage. Effective

**Table 3** Spice and herb export (in US $) of 10 major exporting countries

| | Exported value (US $) | | | |
|---|---|---|---|---|
| Exporters | 2011 | 2012 | 2013 | 2014 |
| World | 8,066,358 | 5,611,580 | 4,847,018 | 5,793,174 |
| Africa Aggregation | 534,619 | 487,354 | 464,563 | 512,673 |
| Madagascar | 180,362 | 185,462 | 135,84 | 247,714 |
| South Africa | 64,444 | 64,429 | 66,741 | 70,229 |
| Tanzania (United Republic of) | 31,912 | 3817 | 43,283 | 45,627 |
| Comoros | 17,628 | 10,348 | 7836 | 33,57 |
| Nigeria | 42,285 | 88,537 | 9863 | 33,134 |
| Morocco | 17,925 | 17,158 | 22,744 | 28,487 |
| Ethiopia | 38,897 | 31,377 | 31,238 | 2257 |
| Kenya | 11,146 | | 12,976 | 674 |
| Mauritius | 1538 | 4172 | 1731 | 5065 |
| Uganda | 922 | 5581 | 529 | 382 |
| Tunisia | 31,961 | 2878 | 306 | 3636 |
| Egypt | 33,144 | 27,336 | | |

After ITC Trade statistics (2016)

conservation strategies are needed especially in the following four main areas: *in-situ* and *ex-situ* conservation, education and research.

Ultimately, the future of medicinal and aromatic plants in Africa seem to depend on the ability to resolve the current conflicts between conservation and resource use, as well as a shift towards more resource based agriculture that is already being challenged by the globalization of economics (Okigbo et al. 2008).

## *Commercial Potential of African MAPs*

Africa is a continent that not only possesses a vast treasure of medicinal plants but also produces some exclusive materials for the world market.

A more recent statistics, as available from the International Trade Center (Market Analysis and Research n.d.), indicates that in terms of sales values Madagascar, South Africa and Tanzania were the three main exporters countries (Table 3). These countries preserved their leading place also in the four consecutive years of analysis. In these years, the exports showed an increasing trend, with the exception of Madagascar, where a set-back was experienced, in the year 2013. As a contrast, the aggregate spice and herbs exports from Africa have seen a slight drop back, which has been the case also with the World figures (Market Analysis and Research n.d.; Profound Advisers in Development 2014; Dzoyem et al. 2013).

In the lack of exact assessments, the importance of this special economic botanical group of plants has been demonstrated in an indirect way by van Wyk (2015). His method was to evaluate the relevant search results in SCOPUS (Elsevier), the largest abstract and citation database of peer-reviewed scientific journals, books and conference proceedings. On the basis of a "strongly focused inventory of species" that was meant to cover most of the medicinal plant raw materials traded internationally (i.e.: exported from African countries) some informative observations/conclusions have been made.

It has been established that more than some 5400 plant species are used in traditional medicine (TM), in Africa. Less than 10% of these have been commercially developed to some extent. Africa can be regarded as home to more than 80 valuable commercial species that are regularly traded in international markets. These include phytomedicines (e.g. *Harpagophytum procumbens* and *Pelargonium sidoides*), functional foods (e.g. *Adansonia digitata* and *Hibiscus sabdariffa*) and sources of pure chemical entities (e.g. caffeine from *Coffea arabica* and yohimbine from *Pausinystalia johimbe*) (Van Wyk 2015).

## *Chemical and Pharmacological Potential of African MAPs*

Several African species have been found to contain active principles that can have a great potential in the fight against several global health issues. Examples are manifold. African medicinal plants of the Families Moraceae, Guttiferae, Fabaceae, etc. have been found to offer good alternative treatment against cancer which has been recognized as a critical health problem in Africa, too (Kuete 2013).

In terms of pharmacological activities, African MAPs are also manifold. They have been widely studied according to their following activities: Antimalarial and Other Antiprotozoal Products, Anti-proliferative Potential, Anti-inflammatory and Analgesic Activities, Anti-diabetes Activity, Antioxidant Activity and activity on the Reproductive Cardiovascular and Central Nervous Systems (Atawodi 2005; Fennell et al. 2004; Juliani et al. 2007; Amoa Onguéné et al. 2013; Van Wyk 2011; McGaw et al. 2014; Kuete et al. 2013; Tekwu et al. 2013).

## *Cultivation of MAPs in Africa: A New Dimension in Agriculture*

To-date, wild-crafting (collection) is still the main ways of harvesting medicinal plants in Africa. However, as earlier discussed, this has a significant detrimental impact on both the ecosystem and the survival of plant species.

Cultivation of highly valued medicinal plants is therefore expected to create new dimensions in the field of agriculture also for the developing African nations. Elaborate food-crop production technologies and techniques can be deployed in cultivation programs that could ultimately contribute to improving also the medicinal value of plant species.

According to Wiersum and Shackleton (2005) there exist a wide array of available plant production systems and different degrees of (semi)-domestication of wild plants.

It should be also noted that frequently the domestication process is inferred in context with the modification of morphological characteristics and genetic make-up of plants, as well as the "artificialization" of the biophysical environment in which the plant is growing (Wiersum et al. 2006). This concept has been also interpreted as to include the "acculturalization" of a plant to a given social management environment (Wiersum 1997). According to this interpretation, the stimulation process involves a socio-technical approach in which attention is given to the interaction between technical, cultural and institutional dimensions of domestication, i.e.: technology development is not a value-free process, but rather a location-specific development process in which cultivation practices are developed on the basis of both technical options, socio-cultural conditions and specific livelihood conditions of the intended cultivators.

This, in view of the development needs of underprivileged groups of people, means that the technologies to be elaborated should be adjusted to the specific social values and livelihood assets of the intended target groups.

## *Cultivation in Farming Systems vs. Sustainable Exploitation of MAPs*

In Africa, much attention has been given to the possibility of combining biodiversity conservation and poverty alleviation.

In an overview of medicinal commercial plant harvesting and its relevance to medicinal plant conservation and the self-sufficiency of traditional medical practitioners, Cunningham (1997) emphasized that generally, the most vulnerable species are the popular, slow reproducing species with specific habitat requirements and limited distribution. In most African countries, the sustainable use of barks, roots or whole plants used as herbal medicines is, in theory, possible, still due to the high level of input of resources (e.g.: money and manpower required for intensive management of slow-growing species) in multiple-species systems is likely to be rather limited,. It seems, therefore, necessary to secure alternative supply sources for popular, high conservation priority species outside of the core conservation areas. In the core conservation areas, however, the slow-growing species constitute a priority for both *ex-situ* conservation and strict protection. As for some species high prices can be achieved, they make potential new crop plants, either for agroforestry systems (e.g. *Warburgia salutaris, Garcinia kola, G. afzellii, G. epunctata*) or agricultural production (e.g. *Siphonochilus aethiopicus*).

This trend has received a strong impetus, especially, since the Johannesburg Earth Summit, in 2002. According to Wiersum et al. (2006), in this regards, two options merit special attention: i.e. the stimulation of sustainable exploitation of wild growing plant resources (e.g. in natural forests) or the cultivation in farming systems.

The way people use and manage wild plant resources will significantly influence the sustainability of their livelihoods and the conservation of diversity (Canter et al. 2005).

## *The Example of* Prunus africana

The bark of African cherry (*Prunus africana* (Hook. f.) Kalm.) has been used, for the last 35 years, in the treatment of benign prostatic hyperplasia and other disorders (Stewart 2003). The bark used for the preparation of Pygeum (bark extract used in the treatments) takes its origin entirely from wild-collection. Major exporting countries include Cameroon, Madagascar, Equatorial Guinea, and Kenya. Global exports of dried bark in 2000 have been estimated at 1350–1525 tons per annum, with a peak of 3225 tons, in 1997.

The natural occurrence of *Prunus africana* ranges from east to west and from central to southern Africa, posing several, different threats to the species. According to (Jimu 2011) *P. africana* populations in west (Cameroon and Equatorial Guinea) and east (Kenya, DRC, Uganda), as well as the island of Madagascar are threatened mainly by commercial harvesting, whereas in southern Africa the main threat is caused by habitat degradation/fragmentation accompanied by invasive alien species and traditional wild-harvesting for medicinal purposes (The latter are also common in west and east Africa). Climate change is likely to affect the species throughout its range of area, especially in the lack of measures to ensure the adaptation of this species. All these factors call for **country or region specific conservation strategies**.

Since *P. africana* occurs in ecologically important habitats, i.e. the afromontane forests that are rich in endemic, endangered and rare fauna and flora, the focus should be on the conservation of these ecosystems.

Sustainable harvesting should be promoted together with other conservation strategies (e.g. *in-situ*, *circa-situ* and *ex-situ* conservation): the ultimate aim should be to avoid/reduce both habitat destruction and destructive wildfires.

Since 1995, *Prunus africana* has been included in CITES (the Convention on International Trade in Endangered Species of Wild Fauna and Flora) Appendix II as an endangered species (Cunningham 1997).

As a result, various project initiatives have been launched to elaborate alternatives to wild-collection while also meeting future market demands: these include conservation practices, enrichment plantings, small- and large-scale production, and protection of genetic resources. There is hope that the status of this species will shift from the verge of transition of an exclusively wild-collected species to that of a cultivated medicinal tree (Stewart 2003).

## *Organic Production of MAPs in Africa*

Growing international organic markets seem to stimulate organic production also in Africa. Several non-governmental organizations (NGOs) and development partners have designed and supported projects aimed at increasing farmers' incomes by tapping into the opportunities offered by organic production. Many farmers produce organically by default, making it relatively easy to convert to true organic agricultural practices. Smallholders, however, often do not have access to international markets.

As a rule, already available certifications, e.g. Kilimohai (East African Organic Products Standard) used by them regionally, are not yet approved at the international markets. The East African organic products standard by TOAM (Tanzania Organic Agriculture Movement) is mostly used for self-assessment by producers, declarations of conformity in the marketplace, certification by certification bodies in the region, or other kinds of verification. If the standard is used for the purposes of third-party certification, inspection and certification should be carried out in accordance with international norms, such as ISO Guide 65 or the IFOAM Accreditation Criteria. If adherence to the standard is verified through other mechanisms, those mechanisms shall adhere to the principles of competency, integrity and transparency (TOAM 2016). Generally, most individual buyers are still not able to meet the quality and quantity requirements of international buyers, however, there are positive examples, e.g. Tanzania Organic Agriculture Movement (TOAM) coordinates and promotes the development of the organic farming sector among smallholder producers, in order to realize sustainable livelihoods in Tanzania (Barrett et al. 2001; Bakewell-Stone et al. 2008; Profound Advisers in Development 2014).

## *MAPs in Traditional Medicine: A Heritage in Africa*

African medicinal plant resources are abundant. A recent volume by Iwu (Iwu 2014) provides a comprehensive review of more than 2000 species of plants deployed in indigenous African medicine, with references from over 1100 publications.

Despite of their vast natural resources, most African countries are still struggling to provide basic amenities for their people. Health facilities are deplorable and most of the population continues to rely on traditional remedies derived from medicinal plants. In consequence, **Traditional African Medicine** (TAM) as a socio-economic and socio-cultural heritage, is still servicing over 80% of the populations in Africa (Elujoba et al. 2005).

As a, quasi, international ramification (impact) of African TAM, quite a number of African medicinal herbs have become increasingly popular, recently as dietary supplements, food supplements, natural health products, therapeutic goods, traditional medicines, and even conventional drugs (obviously, only in harmony with the regulatory system of the relevant countries). These herbs, according to Blumenthal

(2011) include the controversial diet aid hoodia (*Hoodia gordonii*, Asclepiadaceae), the antioxidant red tea rooibos (*Aspalathus linearis*, Fabaceae), the anti-malarial cryptolepis (*Cryptolepis sanguinolenta*, Asclepiadaceae), the 5-hydroxy-L-tryptophan-containing griffonia (*Griffonia simplicifolia*, Fabaceae), anti-inflammatory and analgesic devil's claw (*Harpagophytum procumbens*, Pedialiaceae), sausage tree (*Kigelia africana*, Bignoniaceae), the anti-tonsillitis and anti-bronchitis umckaloabo (*Pelargonium sidoides*, Geraniaceae), the difficult-to-sustain prostate aid pygeum (*Prunus africana*, Rosaceae; syn. *Pygeum africanum*), the anti-depressive and mood-enhancing sceletium (*Sceletium tortuosum*, Aizoaceae), and many others.

The current upsurge of interest in herbal medicines all over the world had lead to the unregulated exploitation of Africa's bio-resources, is a threat to the extinction of plant species, while the traditional healers themselves are hopelessly being observed to advance in age and gradually disappearing from the surface of the earth. Therefore, traditional medicine knowledge should be learned, acquired, documented and used for the benefit of both African people and the entire mankind (Elujoba et al. 2005).

## *African Herbal Pharmacopoeia and Research to Promote the Medicinal Plant Sector*

With the increased popularity of medicinal and aromatic plants, an inevitable need has emerged for enhanced quality control of raw materials, including tests for botanical identity and potential – accidental or intentional – adulterants. The present volume is expected to contribute to the medicinal plant sector in African countries by offering complementary, comprehensive information on a wide range of medicinal and aromatic plants.

There have been various resolutions adopted by the World Health Assembly and Regional Committee for Africa that call upon Member States, to develop herbal pharmacopoeias and to develop/apply scientific criteria and methods for proof of safety and efficacy of medicinal plant products. However, only few countries have developed national herbal pharmacopoeias. The number of plant species with sufficient information on the scientific (clinical) evaluation of their herbs, their possible medical applications, as well as the safety and efficacy of their usage is relatively rather scarce. Without well documented information on safety, efficacy and phytochemical characteristics, it is difficult to assess the virtual value of some new raw materials and extracts of African origin. In order to address these shortcomings, the Association of African Medicinal Plants Standards is developing an African Herbal Pharmacopeia with trading standards which provide information and technical data on some 50 important medicinal plants (Gurib-Fakim and Kasilo 2010).

The objective of developing these monographs is to ensure that these plants become visible on the world market. The monographs lay emphasis on the quality control issues, dosage, use, efficacy, pharmacology and safety of important African medicinal plants. Despite the increasing popularity of phytomedicines of African origin, the number of publications dealing with the results of controlled clinical trials is relatively modest. Consequently, the African pharmacopoeia initiative can be rightly regarded as the first African, interdisciplinary effort to compile information in the form of a pharmacopeia. Ultimately, a reliable, authoritative compendium of standards and analytical methods will be available for the increasingly interested herb community, from policy-makers to entrepreneurs. It is also expected to contribute to bringing economic benefit to African nations in the sustainable exploitation of their natural resources, and create an environment conductive to the growth of the plant-based medicine industry.

## 2 Conclusion

MAPs have been accompanying mankind from the early times of civilizations. Our forefathers, who had their origin in Africa, used natural substances of plant origin, in order to ease their pains, heal their wounds, etc. These healers can be regarded as forefathers of Traditional Medicine. A heritage of Africa, that is still servicing approx. 80% of the population of this huge continent that is home to 55 countries, 800 languages and an unmeasurable wealth of plant diversity.

The continent with a varied flora of an estimated 68,000 plant species is home to about 35,000 endemic (Cunningham 1993) species, growing in seven centers of endemism.

Due to the nearly permanent renewed interest in the utilization of MAPs, Africa and it's not yet fully explored medicinal plant resources, frequently come to the forefront of interest.

There is a tremendous potential for developing new crops and new products, in Africa. In view of the recently significantly increased requirements for quality, safety and efficacy of both botanicals and their products, there remains much to be done to provide convincing proofs through clinical studies on the varied pharmacological activities of African MAPs (e.g.: antimalarial and other Antiprotozoal products, anti-proliferative potential, anti-inflammatory and analgesic activities, anti-diabetes activity antioxidant activity and activity on the reproductive cardiovascular and central nervous systems, etc.). These trials seem to offer a nearly unlimited scope for exploration. This is witnessed by the fact that ca. 60% of all recent publications on African medicinal plants, appeared during the last decade.

Despite a marked recent increase in the number of publications on indigenous medicinal plants and dietary supplements, Africa is lagging behind Europe and Asia in terms of the number of products that have been commercialized and the percentage of the flora that is utilized for international trade.

The growing demand on fewer resources, has resulted, in some cases, in the local disappearance of favored and effective sources of botanicals and reduced species diversity. By elaborating international and national standards of sustainability, these dangers can be either eased or eradicated. Domestication of wild-crafted species can also contribute to preserve the genetic resources and to reduce the sometimes imminent threats of extinction. The recent initiative to develop national and ultimately an African Herbal Pharmacopoeia is likely to mark an important milestone in meeting quality, safety and efficacy standards both in African and international perspectives.

**Acknowledgements** Thanks are due to Ms. J.V. van Hal, ProFound – Advisers In Development, The Netherlands for her comments.

## References

Amoa Onguéné P et al (2013) The potential of anti-malarial compounds derived from African medicinal plants, Part I: A pharmacological evaluation of alkaloids and terpenoids. Malar J 12:449

Anonymous (2005) Conservation international 2005 annual report, Washington, DC

Atawodi SE (2005) Antioxidant potential of African medicinal plants. J Biotechnol 4(February):128–133

Bakewell-Stone P, Lieblein G, Francis C (2008) Potentials for organic agriculture to sustain livelihoods in Tanzania. Int J Agric Sustain 6(1):22–36

Barrett HR et al (2001) Smallholder farmers and organic certification: Accessing the EU market from the developing world. Biol Agric Hortic 19(2):183–199

Blumenthal M (2011) African herbal pharmacopoeia – book review. HerbalGram 90:71–72

Brummitt RK, Pando F (2001) World geographical scheme for recording plant distributions, 2nd edn. Hunt Institute for Botanical Documentation Carnegie Mellon University, Pittsburgh

Brummitt R et al (2001) World geographical scheme for recording plant distributions. International Working Group on Taxonomic Databases for Plant Sciences (TDWG), Pittsburgh

Canter PH, Thomas H, Ernst E (2005) Bringing medicinal plants into cultivation: opportunities and challenges for biotechnology. Trends Biotechnol 23(4):180–185

Cocks M, Dold A (2000) The role of 'African chemists' in the health care system of the Eastern Cape province of South Africa. Soc Sci Med 51(10):1505–1515

Cunningham AB (1989) Herbal medicine trade: a hidden economy. Indic South Africa 6(3):51–54

Cunningham A (1993) African medicinal plants. Setting priorities at the interface between conservation. UNESCO, Paris

Cunningham A (1997) An Africa-wide overview of medicinal plant harvesting, conservation and health care. In: Bodeke G, KKS B, Burly J, Vantomme P (eds) Medicinal plants for forest conservation and health care. FAO, Rome, pp 116–129

Dzoyem JP, Tshikalange E (2013) Medicinal plant research in Africa. Elsevier, Oxford

Dzoyem JP, Tshikalange E, Kuete V (2013) Medicinal plants market and industry in Africa. In: Medicinal plant research in Africa. Elsevier, Oxford, pp 859–890

Elujoba A, Odeleye O, Ogunyemi C (2005) Traditional medicine development for medicinal and dental primary health care delivery system in Africa. African J Tradit Complemen Altern Med 2(1):46–61

FAO (2012) FAO statistical yearbook 2012 – Africa food and agriculture. FAO, Rome

Fennell CW et al (2004) Assessing African medicinal plants for efficacy and safety: Pharmacological screening and toxicology. J Ethnopharmacol 94(2–3):205–217

Gurib-Fakim A, Kasilo M (2010) Promoting African medicinal plants through an African herbal pharmacopoeia. Afr Health Moni Spec Issue Decade Afr Tradit Med 14:63–66

IndexMundi, 2016. Endangered plant species in Africa. (Retrieved 06 May 2017. at: http://www.indexmundi.com/facts/indicators/EN.HPT.THRD.NO/map/africa)

Iwu M (2014) Handbook of African medicinal plants, 2nd edn. CRC Press, Boca Raton

Jimu L (2011) Threats and conservation strategies for the African cherry (Prunus africana) in its natural range – a review. J Ecol Nat Environ 3(4):118–130
Juliani R, Simon JE, Ho CT (2007) African natural plant products: new discoveries and challenges in chemistry and quality. American Chemical Society, Washington, DC
Klopper R, Smith GF, Van Rooy J (2002) The biodiversity of Africa. In: Baijnath H, Singh Y (eds) Rebirth of science in Africa: a shared vision for life and environmental sciences. Umdaus Press, Hatfield, pp 60–86
Kuete V (2013) Medicinal plant research in Africa pharmacology and chemistry. In: Kuete V (ed) Elsevier insights. Elsevier, London/Waltham, pp 251–260
Kuete V, Viertel K, Efferth T (2013) Antiproliferative potential of African medicinal plants. In: Medicinal plant research in Africa: pharmacology and chemistry. Elsevier, Oxford, pp 711–724
Makunga N, Philander L, Smith M (2008) Current perspectives on an emerging formal natural products sector in South Africa. J Ethnopharmacol 119(3):365–375
Mander M, Le Breton G (2005) Southern African trade directory of indigenous natural products. In: McKenzie MMM (ed) . Commercial Products from the Wild Group, Stellenbosch
Mander M, Steytler N, Diederichs N (2006) The economics of medicinal plant cultivation. In: Diederichs N (ed) Commercializing medicinal plants. A Southern African guide. Sun Press, Stellenbosch, p 215
Market Analysis and Research, I.T.C. (ITC) Trade map – list of exporters for the selected product group (spices and herbs). Title – Trade map trade statistics for international business development. Available at: http://www.trademap.org/cbi/Country_SelProduct_TS.aspx. Accessed 12 Feb 2016
Máthé Á (2015) Botanical aspects of medicinal and aromatic plants. In: Máthé Á (ed) Medicinal and aromatic plants of the world. Springer, Dordrecht, pp 13–33
McGaw L, Ee E, Eloff J (2014) Toxicological survey of African medicinal plants. In: Kuete V (ed) Toxicological survey of African medicinal plants. Elsevier, London/Waltham, pp 181–233
Myers N (1988) Threatened biotas: "hot spots" in tropical forests. Environmentalist 8(3):187–208
Myers N et al (2000) Biodiversity hotspots for conservation priorities. Nature 403(6772):853–858
Okigbo R, Mmeka E (2006) An appraisal of phytomedicine in Africa. KMITL Sci Technol J 6(2):83–94
Okigbo R, Eme U, Ogbogu S (2008) Biodiversity and conservation of medicinal and aromatic plants in Africa. Biotechnol Mol Biol Rev 3(6):127–134
Olson DM et al (2001) Terrestrial ecoregions of the world: a new map of life on earth. BioScience 51(11):933–938
Profound Advisers in Development (2014) Wild-collected botanicals and the EU market, BTC, Belgian Development Agency
Rukangira E (2001) The African herbal industry: constraints and challenges|Elton Mukama – Academia.edu. Erboristeria Domani. Available at: https://www.academia.edu/3443773/The_African_herbal_industry_constraints_and_challenges. Accessed 13 Feb 2016
Simon J et al (2007) Issues in new crops and uses. In: Janick J, Whipkey A (eds) Issues in new crops and uses. ASHS Press, Alexandria, pp 322–331
Stewart KM (2003) The African cherry (Prunus africana): can lessons be learned from an over-exploited medicinal tree? J Ethnopharmacol 89(1):3–13
Tanzanian Organic Agriculture Movement (TOAM). (Retrieved: 06 May 2017. at: http://www.kilimohai.org/?L=0)
Tekwu EM, Beng VP, Kuete V (2013) African medicinal plants acting on the reproductive, cardiovascular, and central nervous systems. In: Medicinal plant research in Africa: pharmacology and chemistry. Elsevier, Oxford, pp 805–841
Udvardy M (1975) A classification of the biogeographical provinces of the world. IUCN, Morges
van Seters A (1997) Forest based medicines in traditional and cosmopolitan health care. In: Bodeker G et al (eds) Medicinal plants for forest conservation and health care. FAO, Rome, pp 5–11

Van Wyk BE (2011) The potential of South African plants in the development of new medicinal products. S Afr J Bot 77(4):812–829
Van Wyk BE (2015) A review of commercially important African medicinal plants. J Ethnopharmacol 176:118–134
Van Wyk BE et al (2009) Medicinal plants of South Africa, 2nd edn. Briza Publications, Pretoria
White F (1983) The vegetation of Africa: a descriptive memoir to accompany the UNESCO/AETFAT/UNSO vegetation map of Africa. Nat Resour Res 20:1–356
Wiersum KF (1997) From natural forest to tree crops, co-domestication of forests and tree species, an overview. Neth J Agric Sci 45:425–438
Wiersum K, Shackleton C (2005) African forests between nature and livelihood resource: interdisciplinary studies in conservation and forest management. In: Ros-Tonen M, Dietz T (eds) African studies. Edwin Mellen Press, Lewiston, pp 67–91
Wiersum K et al (2006) Medicinal and aromatic plants. Agricultural, commercial, ecological, legal, pharmacological and social aspects. In: Bogers R, Craker L, Lange D (eds) Frontis. Springer, Dordrecht, pp 43–58

# A Review of African Medicinal and Aromatic Plants

**Ben-Erik Van Wyk**

**Abstract** The botanical and cultural diversity of the African continent provides numerous opportunities for the development of innovative new crops and new products. However, many of African medicinal and aromatic plants remain scientifically poorly known and in need of detailed investigation. To contribute to this process a broad review of commercially important African medicinal and aromatic plants is presented. The following data are given for a selection of 246 indigenous and 12 naturalised/commonly cultivated species: Scientific name, author citation, plant family, trade name and/or vernacular name(s), part(s) used, main uses, region of origin within Africa, African countries involved in local, regional or international trade and recent references (general publications since 2000). Important publications at global, continental and regional levels are briefly mentioned. The categories of use include traditional medicines, phytomedicines, dietary supplements and functional foods, sources of extracts or chemical entities and sources of essential oils. Examples of several African plants and their products in each of these categories are illustrated in colour.

**Keywords** Aromatic plants • Central Africa • Chemical entities • Dietary supplements • East Africa • Essential oils • Mascarene Islands • Medicinal plants • North Africa • Phytomedicines • Southern Africa • Traditional medicine • West Africa

## 1 Introduction

Africa is a continent of tremendous biological and cultural diversity that have remained poorly explored from a scientific and commercial perspective. In recent years however, there has been a surge of research interest in Africa as a source of natural products. A recent review of African medicinal plants (Van Wyk 2015) showed that 60% of the scientific papers on the most important plants appeared in

B.-E. Van Wyk (✉)
Department of Botany and Plant Biotechnology, University of Johannesburg,
P.O. Box 524, Auckland Park, 2006 Johannesburg, South Africa
e-mail: bevanwyk@uj.ac.za



M. Neffati et al. (eds.), *Medicinal and Aromatic Plants of the World - Africa Volume 3*, Medicinal and Aromatic Plants of the World,
DOI 10.1007/978-94-024-1120-1_2

the last decade. African medicinal plants and plant-derived chemical entities represent about 10% of the total diversity in international markets. *Coffea arabica* and *Aloe vera* are arguably the most valuable of all plant-based natural products in global trade. At least 10 African plant species feature prominently in international markets as functional foods and nutraceuticals (Van Wyk 2015). The African flora undoubtedly represents a rich source of inspiration for cultural, scientific and commercial innovations.

While the explicit aim of this volume is to demonstrate the wealth and diversity of medicinal and aromatic plants of the African continent, this chapter is aimed at providing the reader with a bird's eye view of the most prominent and popular African species.

## *Diversity of the African Flora*

The total number of plant species in Africa has not yet been accurately recorded and new species are continuously being described. However, a fairly accurate estimate is possible. In their checklist for sub-Saharan Africa, Klopper et al. (2006) listed 50,136 species, with 32,424 taxa (including subspecies and varieties) in tropical Africa and 22,755 taxa in southern Africa. The Cape Flora represents a special hotspot of diversity and endemism, with 9000 species (Manning and Goldblatt 2012). The North African flora comprises an estimated 10,000 species (Dobignard and Chatelain 2010–2013), while the figure for Madagascar is estimated at 11,350 species (Madagascar Catalogue 2013). The total number of African species is, therefore, close to 71,500. In a recent publication from the Royal Botanic Gardens Kew, it was estimated that there are 391,000 vascular plants known to science (Royal Botanic Gardens Kew 2016). The African flora represents roughly 18% of the total species diversity of the world.

## *Popular and Commercially Relevant African Plant Species*

Based mainly on the literature discussed below, and the results of recent review papers, an attempt was made to provide a summary of the most important and well-known African medicinal and aromatic plants (Table 1). The list focuses strongly on indigenous species (246 are listed) and includes only 12 very prominent exotic species that are naturalized and/or widely cultivated. Naturalized species are commonly used in traditional medicine in Africa and many non-indigenous species are often included in publications (e.g. Vasisht and Kumar 2004 and Iwu 2014). However, most of these are excluded here. The regional classification of African countries follows Van Wyk (2015) and author citations for scientific names follow The Plant List (2013). Examples of well-known species are illustrated in Figs. 1, 2, 3, 4, 5.

**Table 1** Commercially relevant African medicinal and aromatic plants

| Species, author citation and family (non-indigenous species)* | Trade name and/or vernacular name | Part(s) used | Main uses (see caption) | Origin (see caption) | African countries involved in local, regional or international trade | References (see caption) |
|---|---|---|---|---|---|---|
| *Abrus precatorius* L. (Fabaceae) | Crab's eye vine, coral pea | Leaves (seeds are extremely toxic) | TM (asthma, bronchitis, coughs, stomach cramps, malaria, physical and sexual asthenia; wounds) | PA, Ma | Many countries | 1, 2, 3, 11 |
| *Acacia drepanolobium* Sjöstedt (Fabaceae) | Whistling thorn | Gum | TM (root: diuretic; bark: pharyngitis); gum as for gum Arabic | EA | Ethiopia (gum) | 1, 3 |
| *Acacia karroo* Hayne (Fabaceae) | Sweet thorn | Gum | TM (bark: diarrhea; root: colic); gum as for gum Arabic | SA, EA | South Africa, Zimbabwe | 1, 2, 3 |
| *Acacia nilotica* (L.) Willd. ex Delile (Fabaceae) | Egyptian thorn | Gum | TM (all parts); gum as for gum arabic | EA | Algeria, Mozambique, South Sudan, Tanzania | 1, 2, 5 |
| *Acacia senegal* (L.) Willd. (Fabaceae) | Gum arabic | Gum (exudate) | TM (bark, root); gum: emollient, wound healing); pharmaceutical aid (emulsifier, stabilizer, additive) | WA, EA | Central African Republic, Chad, Eritrea, Ethiopia, Ghana, Madagascar, Mali, Nigeria, Rwanda, South Sudan, Tanzania | 1, 2, 3, 8, 10, 11 |
| *Acacia seyal* Delile (Fabaceae) | Red acacia | Gum | TM (bark, root, leaf); gum as for gum Arabic | EA | Ethiopia, South Sudan | 1, 3, 5 |

(continued)

**Table 1** (continued)

| Species, author citation and family (non-indigenous species)* | Trade name and/or vernacular name | Part(s) used | Main uses (see caption) | Origin (see caption) | African countries involved in local, regional or international trade | References (see caption) |
|---|---|---|---|---|---|---|
| *Acanthosicyos horridus* Welw. ex Hook.f. (Cucurbitaceae) | !Nara | Seeds | TM (root); seed oil (edible, cosmetics) | SA | Namibia | 1, 2 |
| *Achillea fragrantissima* (Forssk.) Sch. Bip. (Asteraceae) | Lavender cotton; yarrow | Herb | TM (general medicine) | NA | Egypt | 3, 5 |
| *Achyranthes aspera* L. (Amaranthaceae) | Chaff flower; devil's horsewhip | Leaves | TM (many ailments) | PA | Many countries | 1, 2, 10 |
| *Acokanthera oppositifolia* (Lam.) Codd (Apocynaceae) | Bushman poison | All parts | TM (snake bite, headache, many uses) | EA, SA | Kenya, South Africa | 1, 2, 5, 6 |
| *Adansonia digitata* L. (Malvaceae) | Baobab | Dry fruit pulp | TM (all parts); DS (fruit pulp: anti-oxidant) | PA | Algeria, Burkina Faso, Madagascar, Malawi, Nigeria, South Africa, South Sudan, Tanzania, Zimbabwe | 1, 3, 4, 8, 11 |
| *Aerva javanica* (Burm.f.) Juss. (Amaranthaceae) | Desert cotton | Leaves, roots | TM (skin ailments) | NA, EA | Egypt, Kenia | 1, 2, 3 |

| | | | | | | |
|---|---|---|---|---|---|---|
| *Aframomum melegueta* K.Schum. (Zingiberaceae) | Melegueta pepper | Seeds | TM (leaf, fruit, seed); DS (spice); ES | WA, CA | Cameroon, Democratic Republic of the Congo (hereafter D.R. Congo), Equatorial Guinea, Gabon, Ghana | 1, 3, 11 |
| *Agathosma betulina* (Berg.) Pillans (Rutaceae) | Round leaf buchu | Leaves | TM (urinary antiseptic); DS (bitter tonic); ES (food flavourant) | SA | South Africa | 1, 2, 3, 7, 8, 10, 11 |
| *Ageratum conyzoides* L. (Asteraceae)* | Goat weed, billy-goat weed | Leaves | TM (wound healing, diarrhea, dyspepsia, as febrifuge; many uses) | WA, CA, EA | Burundi, D.R. Congo, Ethiopia, Ghana, Nigeria | 1, 2, 4 |
| *Albizia adianthifolia* (Schumach.) W.Wight (Fabaceae) | Flat crown, *umgadankawu* | Bark | TM (diversity of ailments) | PA | Kenya, South Africa | 1, 2 |
| *Alchornea cordifolia* (Schumach. & Thonn.) Muell.-Arg.; *A. floribunda* Muell.-Arg.; (Euphorbiaceae) | Christmas bush, dovewood | Leaves, roots, fruit | TM (many uses: antimicrobial, antispasmodic; antidepressant) | PA | Burkina Faso, Ghana | 1, 6 |
| *Alepidea amatymbica* Eckl. & Zeyh., *A. cordifolia* B.-E. van Wyk and other spp. (Apiaceae) | *Ikhathazo* | Rhizome and roots | TM (respiratory ailments – colds and influenza) | SA, EA | South Africa, Zimbabwe | 1, 2, 3 |
| *Allanblackia floribunda* Oliv. (Clusiaceae) | Tallow tree | Seeds | TM (bark, fruit, leaves); seed oil (edible, cosmetics) | CA, WA | Cameroon, Côte d'Ivoire | 1, 3 |
| *Aloe arborescens* Mill. (Asphodelaceae) | Krantz aloe, kidachi aloe | Leaf | DS (bitter tonic) | SA | Japan | 1, 2, 6 |

(continued)

**Table 1** (continued)

| Species, author citation and family (non-indigenous species)* | Trade name and/or vernacular name | Part(s) used | Main uses (see caption) | Origin (see caption) | African countries involved in local, regional or international trade | References (see caption) |
|---|---|---|---|---|---|---|
| *Aloe ferox* Mill. and other species (Asphodelaceae) | Cape aloe | Exudate, leaf gel | TM (laxative); DS (gel drinks, bitter tonic, flavourant) | SA | South Africa | 1, 2, 6, 8, 10, 11 |
| *Aloe secundiflora* Engl.; *A. lateritia* Engl.; *A. turkanensis* Christian (Asphodelaceae) | Kenya aloes | Exudate | TM (laxative) | EA | Kenya | 1, 6 |
| *Aloe vera* L. (Asphodelaceae) | Aloe vera | Leaf gel | DS (tonic drinks); leaf gel (cosmetic, wounds) | Cu | South Africa, Nigeria | 1, 6, 10, 11 |
| *Ammi majus* L. (Apiaceae) | Bishop's weed | Fruits | TM (many uses); CH (xanthotoxin: psoriasis, leukoderma, vitiligo) | EA | South Sudan, Ethiopia, Somalia (India) | 2, 5, 11 |
| *Ammi visnaga – see Visnaga daucoides* | | | | | | |
| *Ammodaucus leucotrichus* Coss & Durr. (Apiaceae) | Cumin du Sahara | Leaves, fruits | TM (digestive system ailments) | NA | Morocco, Algeria | 5 |
| *Anabasis articulata* (Forssk.) Moq. (Amaranthaceae) | Jointed anabasis | Herb | TM (skin diseases, diabetes); soap industry | NA | | 3 |
| *Anacyclus pyrethrum* L. (Asteraceae) | Spanish pellitory | Root | TM (diversity of uses; kidney ailments; promotion of salivation) | NA | Morocco and Algeria | 1, 5 |
| *Anogeissus leiocarpa* (DC.) Guill. & Perr. (Combretaceae) | African birch, *ngalama* | All parts | TM (anthelmintic, chewing sticks, veterinary medicine) | WA | Senegal to Cameroon | 1, 3 |

| | | | | | | |
|---|---|---|---|---|---|---|
| *Antidesma madagascariensis* Lam. (Euphorbiaceae) | *Bois bigaignon* | Leaves, bark | TM (dysentery, fever, diabetes) | Ma | Madagascar | 4, 8 |
| *Aphloia theiformis* Benn. (Flacourtiaceae) | *Fandamane* | Leaves, roots | TM (leaves: malaria, fever; root: worms) | Ma | Madagascar | 1, 2, 4, 8 |
| *Argania spinosa* (L.) Skeels (Sapindaceae) | Argan (oil) | Seeds | TM: skin conditions, cosmetic; edible oil | NA | Morocco | 5 |
| *Artemisia afra* Jacq. ex Willd. (Asteraceae) | African wormwood | Herb | TM (digestive and respiratory ailments); ES (fragrance) | SA, EA | Ethiopia, South Africa, Tanzania, Zimbabwe | 1, 2, 8, 10, 11 |
| *Artemisia herba-alba* Asso (Asteraceae) | White wormwood | Herb | TM (antiseptic, antispasmodic, carminative, vermifuge); ES | NA | Tunisia, Egypt | 1, 5 |
| *Artemisia judaica* L. (Asteraceae) | Judean wormwood | Herb | TM (gastrointestinal ailments, anthelmintic); ES | NA | Egypt | 5 |
| *Artemisia rehan* Chiov. (=*A. absinthium* L.?) (Asteraceae) | *Ariti* | Herb | Ritual incense; ES (fragrance) | EA | Ethiopia | 4 |
| *Aspalathus linearis* (Burm.f.) R. Dahlgren (Fabaceae) | Rooibos tea | Herb | TM (antispasmodic); DS (health tea); cosmetics | SA | South Africa | 1, 2, 3, 7, 8, 10, 11 |
| *Athrixia phylicoides* DC. (Asteraceae) | Bush tea | Herb (tea) | DS (health tea) | SA | South Africa | 2 |
| *Aucoumea klaineana* Pierre (Burseraceae) | *Oukoumé* | Resin | TM (general health) | WA | Gabon | 1 |
| *Azadirachta indica* A.Juss.* (Meliaceae) | Neem | All parts | TM (malaria and many other ailments); seed oil (antifertility agent) | Asia | Chad, Kenya, Madagascar, Niger, South Sudan, Tanzania | 1, 3, 10, 11 |

(continued)

**Table 1** (continued)

| Species, author citation and family (non-indigenous species)* | Trade name and/or vernacular name | Part(s) used | Main uses (see caption) | Origin (see caption) | African countries involved in local, regional or international trade | References (see caption) |
|---|---|---|---|---|---|---|
| *Baillonella toxisperma* Pierre (Sapotaceae) | African pearwood tree, moabi tree | Bark, seeds | TM (bark); seed oil (cosmetics) | CA | Cameroon | 1, 3 |
| *Ballota africana* L. (Lamiaceae) | African horehound | Herb | TM (colds, fever, cough, insomnia); ES | SA | South Africa | 1, 2, 10 |
| *Ballota undulata* (Sieber ex Fresen.) Benth. (Lamiaceae) | Horehound | Herb | TM (skin allergy); ES | NA | Egypt | 3 |
| *Balanites aegyptiaca* (L.) Delile (Zygophyllaceae) | Desert date | All parts, fruits, seeds | DS (fruits); seed oil (food, cosmetics) | NA, CA, EA | Algeria, Chad, Ethiopia, Madagascar, South Sudan | 1, 5, 8 |
| *Bersama abyssinica* Fresen. (Melianthaceae) | Winged bersama, *umukaka* | Bark, leaves, roots | TM (anthelmintic, dysentery, roundworm, hemorrhoids, aphrodisiac) | WA, EA, SA | Ethiopia and many other countries | 1, 6 |
| *Boophone disticha* (L.f.) Herb. (Amaryllidaceae) | Poison bulb | Bulb | TM (analgesic, wound healing; hallucinogen) | SA | Lesotho, South Africa, Swaziland | 1, 2, 11 |
| *Boswellia frereana* Birdw. (Burseraceae) | Maydi frankincense | Resin | TM (topical); traditional incense; ES (fragrance) | EA | Somalia only | 3 |
| *Boswellia neglecta* S.Moore (Burseraceae) | Olibanum | Gum-resin | TM (topical); traditional incense | EA | Ethiopia, Kenya | 1 |
| *Boswellia papyrifera* (Del. ex Caill.) Hochst. (Burseraceae) () | Ethiopian frankincense | Gum-resin | TM (topical); traditional incense; ES (fragrance) | EA | Ethiopia, Eritrea, South Sudan | 1 |
| *Boswellia rivae* Engl. (Burseraceae) | Frankincense | Gum-resin | TM (topical); traditional incense; ES (fragrance) | EA | Ethiopia | |

| | | | | | | |
|---|---|---|---|---|---|---|
| *Boswellia sacra* Flueck. (=*B. carteri* Birdw.) (Burseraceae) | Arabian frankincense | Gum-resin | TM (topical); traditional incense; ES (fragrance) | EA | Somalia | 8, 10, 11 |
| *Bowiea volubilis* Harv. ex Hook.f. (Hyacinthaceae) | Climbing potato, *igibisila* | Bulb | TM (headache, constipation, oedema, infertility, cystitis) | SA | South Africa | 1, 2 |
| *Bridelia atroviridis* Muell.-Arg. (Phyllanthaceae) | Fever leaf, *mkarati* | Bark | TM (purgative and diuretic) | WA | Nigeria | 1, 6 |
| *Brucea antidysenterica* J.F.Mill. (Simaroubaceae) | *Abalo*, *waginos* | Leaves, roots | TM (diarrhea, indigestion, skin problems, leprosy) | EA | Ethiopia, Kenya | 1, 6 |
| *Bulbine frutescens* (L.) Willd. (Asphodelaceae) | Bulbine, burn jelly plant | Leaf gel | TM (topical); cosmetics | SA | South Africa | 1, 2, 8, 11 |
| *Cajanus cajan* (L.) Millsp.* (Fabaceae) | Pigeon pea | Leaves, seeds, roots | TM (leaves: sickle-cell anaemia, numerous ailments); DS (seeds) | India | Kenya, Malawi, Nigeria, Uganda, Tanzania | 1, 2, 8, 11 |
| *Calotropis procera* (Aiton) W.T. Aiton (Apocynaceae) | Giant milkweed, *arka* (Ayurveda) | Roots, bark, leaves | TM (root: diarrhea, dysentery, dyspepsia; leaves or latex: ulcers, leprosy) | NA, WA, EA | Egypt, Libya, Benin, Ethiopia, Ghana, Kenya, Somalia, Sierra Leone, South Sudan, Togo | 1, 5, 9 |
| *Cananga odorata* (Lam.) Hook.f.* (Annonaceae) | Ylang-ylang | Fresh flower | ES (fragrance) | Asia | Madagascar | 1, 3, 4 |
| *Cannabis sativa* L.* (Cannabaceae) | Marijuana | Herb | TM (analgesic, anti-emetic, asthma); intoxicant | Asia | Ethiopia, South Africa and other countries (usually illegal; major trade item) | 1, 2, 6, 7, 10, 11 |

(continued)

**Table 1** (continued)

| Species, author citation and family (non-indigenous species)* | Trade name and/or vernacular name | Part(s) used | Main uses (see caption) | Origin (see caption) | African countries involved in local, regional or international trade | References (see caption) |
|---|---|---|---|---|---|---|
| *Capparis spinosa* L. and other species (Capparaceae) | Caper bush | Herb, roots, fruits | TM (herb: rheumatism; roots: diuretic, astringent, tonic) | NA | North African countries | 3, 5 |
| *Caralluma adscendens* (Roxb.) R.Br. (Apocynaceae) | Caralluma | Stems | TM (tonic and other uses); DS (putative appetite suppressant) | WA, EA, NA | India | 9 |
| *Carapa procera* DC. (Meliaceae) | African crabwood | Seeds | seed oil: cosmetic uses; TM (bark, leaves, seed oil: fever, respiratory and digestive ailments) | CA, WA | Cameroon, Congo, Gabon, Ivory Coast, Mali, Senegal | 1 |
| *Carissa edulis* (Forssk.) Vahl. (Apocynaceae) | Simple spined carissa | Roots, leaves, fruits | TM (numerous ailments); DS (fruits: appetizer for invalids) | PA | Cameroon, Ethiopia, Kenya, Madagascar, Namibia, South Africa, Tanzania, Zimbabwe | 1, 2, 4, 8 |
| *Catha edulis* (Vahl.) Forssk. (Celastraceae) | Khat | Roots, bark, leaves | TM (roots, bark); DS (fresh leaves: stimulant) | SA, EA | Ethiopia, Somalia | 1, 2, 10, 11 |
| *Catharanthus roseus* (L.) G.Don (Apocynaceae) | Madagascar periwinkle | Herb | TM (diabetes and many other uses); CH (vinca alkaloids – vinblastine, vincristine and others: leukemia and Hodgkin's lymphoma) | Ma | Burundi, Madagascar, Rwanda and other countries | 1, 2, 4, 6, 8, 10, 11 |
| *Cedrelopsis grevei* Baill. (Rutaceae) | *Katrafay* | Bark | TM (fever, back pain) | Ma | Madagascar | 1, 4 |

| | | | | | | |
|---|---|---|---|---|---|---|
| *Cedrus atlantica* L. (Pinaceae) | Atlantic ceder | Wood | ES (Atlantic ceder wood oil) | NA | Morocco | |
| *Centella asiatica* (L.) Urb. (Apiaceae) | Indian pennywort | Herb | PH (topical: prevention of scar tissue formation); CH (asiaticoside: vulnerary) | Co | Madagascar, Tanzania | 1, 2, 4, 7, 8, 10, 11 |
| *Ceratonia siliqua* L. (Fabaceae) | Carob tree | Fruit pulp, seeds | TM (diarrhea, vomiting); seed gum: thickener, dietary supplement | NA | Arabia, Mediterranean countries; widely cultivated elsewhere | 2, 5, 11 |
| *Chenopodium ambrosioides* L.* (Amaranthaceae) | Wormseed | Seeds | TM (anthelmintic, fever); ES (anthelmintic) | SAm, CAm | Many countries (Egypt, Burundi, Rwanda, South Africa, Zimbabwe) | 1, 5, 10, 11 |
| *Chironia baccifera* L. (Gentianaceae) | Christmas berry | Herb | TM (hemorrhoids, ulcers, acne) | SA | South Africa | 1, 2 |
| *Cinchona ledgeriana* (How.) Bern.Moens. and other spp.* (Rubiaceae) | Quinine tree | Bark | TM (malaria, appetite stimulant) | SAm | Burundi, D.R. Congo, Guinea, Kenya, Madagascar, Rwanda, Tanzania | 1, 10, 11 |
| *Cissampelos pareira* L., *C. mucronata* A.Rich. (Menispermaceae) | Velvetleaf | All parts | TM (gout, syphilis and numerous other ailments); CH (cissampeline: skeletal muscle relaxant) | PA | Almost all countries | 1, 2, 3, 4, 6 |
| *Citrullus colocinthus* (L.) Schrad. (Cucurbitaceae) | Colocynth | Fruit | TM (purgative) | NA | Egypt and all other North African countries | 1, 5, 11 |
| *Citrullus lanatus* (Thunb.) Matsum. & Nakai (Cucurbitaceae) | Kalahari melon | Seeds | Seed oil (edible, cosmetics); also TM | SA | Botswana | 1, 2 |

(continued)

**Table 1** (continued)

| Species, author citation and family (non-indigenous species)* | Trade name and/or vernacular name | Part(s) used | Main uses (see caption) | Origin (see caption) | African countries involved in local, regional or international trade | References (see caption) |
|---|---|---|---|---|---|---|
| *Cladanthus mixtus* (L.) Chevall. (Asteraceae) | *Moroccan* chamomile | Herb | TM (topical); ES (cosmetics) | NA | Morocco, Algeria, Tunisia, Libya | 5 |
| *Clausena anisata* (Willd.) Hook.f. (Rutaceae) | Horsewood | Root, leaves | TM (numerous ailments; chewing sticks) | PA | Many countries | 1, 2, 4, 9 |
| *Cleome chrysantha* Decne (Brassicaceae) | Spider flower | Herb | TM (anthelmintic, infantile convulsions) | NA | Egypt | 3 |
| *Cleome droserifolia* (Forssk.) Delile (Brassicaceae) | Spider flower | Herb | TM (diabetes and numerous other ailments) | NA, EA | Egypt, Libya, Ethiopia | 3, 5 |
| *Clivia miniata* (Lindl.) Regel (Amaryllidaceae) | Bush lily | Whole plant | TM (uterotonic) | SA | South Africa | 1, 2 |
| *Coffea arabica* L. (Rubiaceae) | Coffee | Seeds | CH (caffeine: CNS stimulant) | EA | Ethiopia, Kenya, South Africa, Tanzania | 1, 3, 10, 11 |
| *Cola acuminata* (P.Beauv.) Schott. & Endl. (Malvaceae) | Cola nut tree, abata kola | Seeds | TM (masticatory, tonic, many uses); CH (caffeine, theobromine: stimulant) | WA, CA | Cameroon; also D.R. Congo, Gabon, Ghana, Northern Nigeria, Senegal, Sudan | 1, 3, 7, 10, 11 |
| *Cola nitida* (Vent.) Schott. & Endl., (Malvaceae) | Cola nut tree, gbanja kola | Seeds | TM (masticatory, tonic, many uses); CH (caffeine, theobromine: stimulant) | WA, CA | Ghana, Sierra Leone, Northern Nigeria, Burkina Faso, Côte d'Ivoire, Sudan | 1 |

| | | | | | | |
|---|---|---|---|---|---|---|
| *Colophospermum mopane* (J.Kirk ex Benth.) J. Léonard (Fabaceae) | Mopane | Seeds | ES (fragrance); TM (bark: diarrhoea; many uses) | SA, EA | Namibia | 1, 2 |
| *Combretum coccineum* (Sonn.) Lam. (Combretaceae) | Flame vine | Seeds | TM (seeds, "voan seeds", *voanta menaka*: anthelmintic) | Ma | Madagascar | 1, 4, 9 |
| *Combretum micranthum* G.Don (Combretaceae) | *Kinkeliba* | Leaves | DS (tonic tea); also TM (bark, leaves: cholagogue, antipyretic, tonic; many uses) | WA | Côte d'Ivoire, Mali, Senegal | 1, 4, 8, 9 |
| *Combretum nigricans* Lepr. ex Guill. & Perr. (Combretaceae) | Gum combretum | Gum | TM (root, leaves); edible gum, similar to gum Arabic | WA | Burkina Faso, Nigeria | 1, 3 |
| *Commiphora africana* (A.Rich) Engl. (Burseraceae) | African myrrh, bdellium | Gum-resin | Incense; TM (all parts: numerous ailments) | PA | Algeria, Chad | 1, 2 |
| *Commiphora erythraea* Engl. (Burseraceae) | Opopanax tree | Gum-resin | Incense, TM, ES (fragrance) | EA | Somalia | 7 |
| *Commiphora guidottii* Chiov. (Burseraceae) | Scented myrrh, bisabol | Gum-resin | Incense, TM, ES (fragrance) | EA | Ethiopia, Somalia | 1 |
| *Commiphora myrrha* (Nees) Engl. (Burseraceae) | Myrrh tree | Gum-resin | TM (antiseptic); incense; ES (fragrance) | EA | Ethiopia, Somalia | 1, 3, 8, 10, 11 |
| *Commiphora wildii* Merxm. (Burseraceae) | Namibian myrrh tree, *omumbiri* | Gum resin | ES (fragrance) | SA | Namibia | |
| *Cryptolepis sanguinolenta* (Lindl.) Schltr. (Apocynaceae) | Ghana quinine | Roots | TM (malaria) | WA, CA | Ghana, Madagascar, Mali | 1, 8, 9 |
| *Cullen plicatum* (Delile) C.H.Stirt. (Fabaceae) | Scurfy pea | Herb | TM (antibiotic, antiemetic) | NA | Egypt | 3 |

(continued)

**Table 1** (continued)

| Species, author citation and family (non-indigenous species)* | Trade name and/or vernacular name | Part(s) used | Main uses (see caption) | Origin (see caption) | African countries involved in local, regional or international trade | References (see caption) |
|---|---|---|---|---|---|---|
| *Curtisia dentata* (Burm.f.) C.A.Sm. (Curtisiaceae) | *Assegai*, *umlahleni* | Bark | TM (tonic, aphrodisiac, diarrhea) | SA | South Africa | 1 |
| *Cyclopia genistoides* (L.) R.Br. and other spp. (Fabaceae) | Honeybush | Herb | TM (health tea); CH (mangiferin: antioxidant) | SA | South Africa | 2, 8, 11 |
| *Cymbopogon schoenanthus* (L.) Spreng. (Poaceae) | Camel grass | Herb | TM (antispasmodic); ES: commercial skin care products | NA | North African countries | 1, 3, 5 |
| *Cyperus rotundus* L. (Cyperaceae) | Nutsedge | Tubers | TM (numerous ailments); ES | NA | North African countries | 1, 2, 5 |
| *Dacryodes edulis* (G.Don) H.J.Lam (Burseraceae) | *Safou*, *safu*, butterfruit | Seed | DS (seed oil); also TM (all parts: many ailments) | CA | Cameroon, Gabon | 1 |
| *Danais fragrans* C.F.Gaertn. (Rubiaceae) | *Liane jaune* | Roots, bark | TM (analgesic, skin infections) | Ma | Madagascar | 1, 4, 8 |
| *Detarium microcarpum* Guill. & Perr. (Fabaceae) | Sweet dattock | Root, bark, leaves | TM (diuretic, astringent; many uses) | WA, CA | Mali, Senegal and other countries | 1, 6 |
| *Dichrostachys cinerea* (L.) Wight & Arn. (Fabaceae) | Sickle bush | Roots | TM (numerous ailments) | SA, EA, WA | Many countries (South Africa, Zambia, Mozambique, Zimbabwe) | 1, 2, 6, 9 |
| *Dicoma anomala* Sond. (Asteraceae) | *Hloenya* | Root (leaves) | TM (numerous ailments) | SA, EA | Lesotho, South Africa, Zimbabwe | 1, 2 |

| | | | | | | |
|---|---|---|---|---|---|---|
| *Dicoma capensis* Less. (Asteraceae) | *Koorsbossie* | Leaves | TM (fever, diuretic) | SA | South Africa | 1, 2 |
| *Dodonaea viscosa* Jacq. (Sapindaceae) | Sand olive, *umusasa* | Leaves, roots | TM (many ailments) | EA, CA, SA | Many countries | 1, 2, 4 |
| *Drimia maritima* (L.) Stearn (Asparagaceae) | Squill, sea squill | Bulb | TM (heart tonic, diuretic, expectorant) | NA | All North African countries | 5, 10, 11 |
| *Drosera madagascariensis* DC. (Droseraceae) | Sundew | Herb | TM (respiratory ailments) | Ma | Madagascar | 2, 4 |
| *Echinops kebericho* Mesfin (Asteraceae) | *Kebericho* | Roots | TM (diarrhoea, stomach ache, fever, typhus) | EA | Ethiopia | |
| *Embelia schimperi* Vatke (Primulaceae) | *Inkoko* | Seeds | TM (anthelmintic) | EA | Ethiopia, Tanzania | 1 |
| *Eucomis autumnalis* (Mill.) Chitt. (Asparagaceae) | Pineapple flower, *amathunga* | Bulb | TM (bone fractures, numerous ailments) | SA | South Africa | 1 |
| *Euphorbia hirta* L. [=*Chamaesyce hirta* (L.) Millsp.] (Euphorbiaceae) | Asthma herb | Leaves | TM (amebic dysentery and numerous other ailments) | Co | Madagascar, Mauritius and many other African countries | 1, 2, 4, 6, 8 |
| *Euphorbia peplus* L. (Euphorbiaceae) | Petty spurge | Latex | CH (ingenol mebutate; skin ailments, solar keratosis) | NA, Asia, Europe | cultivated in Australia | 2, 6, 11 |
| *Fadogia agrestis* Schweinf. ex Hiern (= *Vangueria agrestis*) (Rubiaceae) | Baraboro | Bark | TM (aphrodisiac, erectile dysfunction) | WA | Nigeria | 1 |
| *Faidherbia albida* (Delile) A. Chev. (Fabaceae) | Ana tree | Bark | TM (toothache, cough, diarrhea and other ailments) | NA, WA, EA, SA | Algeria, Mali, South Sudan, Ethiopia | 1, 2 |
| *Ferula communis* L. (Apiaceae) | Giant fennel | Root, inflorescence | TM (many topical uses of the resin gum – called *fessoukh*) | NA, EA | Morocco and other North African countries | 1, 5 |

(continued)

**Table 1** (continued)

| Species, author citation and family (non-indigenous species)* | Trade name and/or vernacular name | Part(s) used | Main uses (see caption) | Origin (see caption) | African countries involved in local, regional or international trade | References (see caption) |
|---|---|---|---|---|---|---|
| *Garcinia afzelii* Engl., *G. mannii* Oliv. and other spp. (Clusiaceae) | Garcinia | Stems, roots | Oral hygiene (toothbrush sticks) | WA, CA | Cameroon, Côte d'Ivoire, Equatorial Guinea, Gabon, Ghana, Nigeria | 1 |
| *Garcinia kola* Heckel (Clusiaceae) | Bitter kola | Seeds | TM (numerous ailments); DS (antioxidant, aphrodisiac, immunostimulant) | WA, CA | Cameroon, D.R. Congo, Equatorial Guinea, Nigeria, Sierra Leone | 1, 8 |
| *Glinus lotoides* Loefl. (Molluginaceae) | Damascisa, lotus sweetjuice, *metera* | Fruits | TM (tapeworm, diarrhea) | NA, EA | Ethiopia, Egypt | 1 |
| *Gloriosa superba* L. (Colchicaceae) | Flame lily | Tuber, seeds | TM (tuber; many uses); CH (seeds; colchicine) | PA | Cameroon, Nigeria, Zimbabwe | 1, 2, 6, 11 |
| *Gnetum africanum* Welw. (Gnetaceae) | *Eru* | Leaves | DS (leaves); TM (numerous uses) | CA | Cameroon, D.R. Congo, Gabon | 1 |
| *Griffonia simplicifolia* (Vahl ex DC.) Baill. (Fabaceae) | Griffonia | Seeds | PH/CH (5-hydroxytryptophan: neurological and psychiatric disorders) | WA | Côte d'Ivoire, Ghana | 1, 6, 8, 11 |
| *Guiera senegalensis* J.F.Gmel. (Combretaceae) | *Guiera* | Leaves | TM (many uses; cough syrup) | WA, NA | Senegal and other countries | 1, 9 |
| *Gunnera perpensa* L. (Gunneraceae) | River pumpkin, *ugobho* | Rhizome | TM (impotence, infertility and many other uses) | SA | South Africa | 1 |

| | | | | | | |
|---|---|---|---|---|---|---|
| *Gymnema sylvestre* (Retz.) R.Br.ex Schult. (Apocynaceae) | Miracle fruit | Leaves | DS (diabetes, weight loss); CH: gymnemic acid (antisweet activity) | PA | India, USA | 1, 9 |
| *Hagenia abyssinica* (Bruce) J.F.Gmel. (Rosaceae) | *Kosso* | Female flowers | TM (anthelmintic) | EA | Ethiopia | 1 |
| *Harpagophytum procumbens* DC. ex Meisn., *H. zeyheri* Decne (Pedaliaceae) | Devil's claw | Secondary roots | TM (anti-inflammatory; bitter tonic); PH (low back pain) | SA | Botswana, Namibia, South Africa | 1, 2, 3, 7, 8, 10, 11 |
| *Harrisonia abyssinica* Oliv. (Simaroubaceae) | *Baingou* | Root | TM (many uses) | PA | Kenya | 1, 6 |
| *Harungana madagascariensis* Lam. ex Poir. (Hypericaceae) | Haronga | Bark, leaves | TM (digestive ailments, topical uses) | PA | Madagascar | 1, 4, 8, 10, 11 |
| *Helichrysum odoratissimum* (L.) Sweet (Asteraceae) | *Imphepho* | Leaves | TM; ritual incense | SA | South Africa | 1, 2 |
| *Heteromorpha arborescens* (Spreng.) Cham. & Schltdl. (Apiaceae) | Parsley tree, *mkatlala* | Stems | TM (aphrodisiac, hypertension, headache) | SA, EA | Lesotho, South Africa, Zimbabwe | 1, 2 |
| *Hibiscus sabdariffa* L. (Malvaceae) | Hibiscus, red sorrel | Calyx and epicalyx | DS (antioxidant); TM (numerous uses; also topical) | CA | Mali, Senegal, South Sudan | 1, 2, 3, 7, 8, 10, 11 |
| *Hoodia gordonii* (Masson) Sweet ex Decne (Apocynaceae) | Hoodia, *ghaap*, *ghôba* | Stem | DS (functional food, to suppress hunger and thirst) | SA | Botswana, Namibia, South Africa | 8, 9, 11 |
| *Hyoscyamus muticus* L. (Solanaceae) | Egyptian henbane | Stems and leaves | CH (tropane alkaloids – hyoscyamine and scopolamine); bronchial diseases, ophthalmology | NA | Egypt | 3, 5, 6 |

(continued)

**Table 1** (continued)

| Species, author citation and family (non-indigenous species)* | Trade name and/or vernacular name | Part(s) used | Main uses (see caption) | Origin (see caption) | African countries involved in local, regional or international trade | References (see caption) |
|---|---|---|---|---|---|---|
| *Hypoxis hemerocallidea* Fisch. & C.A.Mey. (Hypoxidaceae) | Hypoxis, African potato | Corm | DS (general tonic; prostate hyperplasia) | EA, SA | Mozambique, South Africa | 1, 2, 8, 10, 11 |
| *Ipomoea pes-caprae* L. (Convolvulaceae) | Goat's foot | Leaves, roots | TM (antihistaminic, antimicrobial, anti-inflammatory) | Co | All coastal countries | 1, 2, 4, 8, 9 |
| *Irvingia gabonensis* Baill. (Irvingiaceae) | Dika nut tree | Seed oil | DS (edible nut, oil); also TM (bark, leaves: numerous ailments, weight loss) | CA, WA | Cameroon, Equatorial Guinea, Gabon | 1 |
| *Jateorhiza palmata* (Lam.) Miers (Menispermaceae) | Calumba | Root (*radix calumbae*) | TM (diarrhoea, dysentery and other uses); DS (bitter tonic, anorexia nervosa) | EA | Malawi, Mozambique | 1,6 |
| *Jatropha curcas* L.* (Euphorbiaceae) | Purging nut | Seeds, leaves | TM (purgative; numerous other uses) | SAm | Many countries | 1, 2, 4, 6, 11 |
| *Juniperus oxycedrus* L. (Cupressaceae) | Prickly cedar, cade | Wood, fruits, oil | ES (cade oil, obtained from the wood): topical antiseptic, psoriasis, chronic eczema | NA | Morocco, Algeria, Tunisia | 5 |
| *Juniperus phoenicia* L. (Cupressaceae) | Phoenician juniper | Leaves, fruits | TM (leaves: digestive and urinary ailments); ES: antiseptic | NA | Algeria, Morocco, Tunisia | 5 |
| *Juniperus procera* Hochst. ex Endl. (Cupressaceae) | African cedar | Twigs and leaves | TM (antifertility, expectorant, rheumatism); ES (fragrance, medicinal) | EA | Kenya | 1 |

| | | | | | | |
|---|---|---|---|---|---|---|
| *Kigelia africana* (Lam.) Benth. (Bignoniaceae) | Sausage tree | Fruit (bark, roots) | TM (antimicrobial, antineoplastic; numerous traditional uses) | PA | Ethiopia, South Africa and many other countries | 1, 2, 8, 10, 11 |
| *Lavandula antineae* Maire (Lamiaceae) | Desert lavender | Herb | ES | NA | Algeria and Libya | 5 |
| *Lavandula multifida* L. (Lamiaceae) | Fernleaf lavender | Herb | TM (numerous ailments); ES | NA | Tunisia, Egypt | 5 |
| *Lavandula stoechas* L. (Lamiaceae) | Spanish lavender | Herb | TM (catarrh, colds, abdominal pain); ES (aromatherapy) | NA | Morocco and other countries | 5 |
| *Laurus nobilis* L. (Lauraceae) | Laurel, sweet bay | Leaves, fruits | Culinary herb; ES (topical use, flavourant) | NA | Tunisia and other countries (widely cultivated) | 5 |
| *Leonotis leonurus* (L.) R.Br. (Lamiaceae) | Wild dagga | Herb | TM (respiratory ailments, numerous other uses) | SA | South Africa | 1, 2 |
| *Lepidium sativum* L. (Brassicaceae) | Garden cress | Seeds | DS (colic, dysentery, headache, asthenia, measles, rheumatism) | EA | South Sudan, Ethiopia | 1, 11 |
| *Lessertia frutescens* (L.) P.Boldblatt & J.C.Manning [=*Sutherlandia frutescens* (L.) R.Br.] (Fabaceae) | Sutherlandia, cancer bush | Leaves | DS (bitter tonic; diabetes, numerous ailments) | SA | Lesotho, Namibia, South Africa, | 1, 2, 8, 10, 11 |
| *Lippia javanica* (Burm.f.) Spreng. (Verbenaceae) | Fever tea | Herb | TM (coughs, colds, headache, numerous other ailments); ES (mosquito repellent) | SA, EA | Mozambique, South Africa, Zimbabwe | 1, 2 |
| *Lippia scaberrima* Sond. (Verbenaceae) | *Mosukujane* | Herb | DS (stomachic, tonic, health tea) | SA | Botswana | 1, 2 |
| *Lophira lanceolata* Van Tiegh. (Ochnaceae) | Dwarf red ironwood | All parts | TM (diversity of ailments) | WA | Guinea | 1, 3 |

(continued)

**Table 1** (continued)

| Species, author citation and family (non-indigenous species)* | Trade name and/or vernacular name | Part(s) used | Main uses (see caption) | Origin (see caption) | African countries involved in local, regional or international trade | References (see caption) |
|---|---|---|---|---|---|---|
| *Malva sylvestris* L. (Malvaceae) | Common mallow | Leafy stems, flowers | TM (antitussive, topical antipruritic); flowers: natural colourant | NA | Tunisia and other countries | 2, 5, 7, 11 |
| *Massularia acuminata* (G.Don) Bullock ex Hoyle (Rubiaceae) | Massularia | Bark, roots | TM (chewing sticks, aphrodisiac) | CA, WA | Cameroon, Sierra Leone | 1 |
| *Mentha longifolia* (L.) Huds. (Lamiaceae) | Horse mint | Herb | TM (antispasmodic); ES | NA, SA | Egypt, South Africa | 1, 2, 3 |
| *Merwilla plumbea* (Lindl.) Speta (=*Scilla natalensis* Planch.) (Hyacinthaceae) | *Inguduza* | Bulb | TM (fertility, purgative, wound healing) | SA | South Africa | 1, 2 |
| *Momordica charantia* L. (Cucurbitaceae) | Bitter melon | Leaves | TM (numerous ailments) | PA | Many countries | 1, 2, 4, 7, 10 |
| *Mondia whitei* (Hook.f.) Skeels (Apocynaceae) | White's ginger, mondia, tonic root | Root | TM (aphrodisiac; numerous ailments; spice) | PA | Benin, Kenya, South Africa, Uganda, Zimbabwe | 1, 8, 9, 11 |
| *Monodora myristica* (Gaertn.) Dunal (Annonaceae) | Calabash nutmeg, African nutmeg | Fruit, seeds | Fruit: essential oil, spice; ES (flavourant); TM (stimulant, stomachic, many ailments) | PA | Benin, Cameroon, Equatorial Guinea, Gabon | 1 |
| *Morinda lucida* Benth. (Rubiaceae) | Brimstone tree | Leaves, roots, bark | TM (fever, diabetes and numerous other ailments) | CA, WA | Madagascar, Nigeria, D.R. Congo | 1 |

| | | | | | | |
|---|---|---|---|---|---|---|
| *Moringa oleifera* Lam.* (Moringaceae) | Drumstick tree, horseradish tree | all parts | DS (health food); TM (many and diverse ailments) | India | Madagascar, South Sudan and most tropical parts | 1, 4, 8, 11 |
| *Moringa peregrina* (Forssk.) Fiori (Moringaceae) | Wild drumstick tree | Leaves, seeds | Seed oil: edible, cosmetic uses | NA | Egypt, Sudan | 5 |
| *Moringa stenopetala* (Baker f.) Cufod. (Moringaceae) | Ethiopian horseradish tree, African moringa | Leaves | DS (health food) | EA | Ethiopia, Kenya | |
| *Myrianthus arboreus* P. Beauv. (Urticaceae) | Giant yellow mulberry, bush pineapple | Leaves, bark, roots | DS (fruits, leaves); TM (headache, pains, numerous ailments) | WA, EA | Cameroon, Congo, Ghana, Ivory Coast, Nigeria, Sierra Leone, Tanzania | 1 |
| *Myrothamnus flabellifolius* (Sond.) Welw. (Myrothamnaceae) | Resurrection plant | Herb | TM (respiratory ailments); ES (antimicrobial) | SA | Angola, Mozambique, Namibia, South Africa, Zimbabwe | 1, 2 |
| *Myrtus communis* L. (Myrtaceae) | Myrtle | Leaves, fruits | TM (expectorant, astringent, antiseptic); ES (oil of myrtle) | NA | Egypt, Tunisia | 5 |
| *Nauclea* – see *Sarcocephalus* | | | | | | |
| *Neobeguea mahafalensis* J.-F.Leroy (Meliaceae) | *Andy* | Bark | TM (rheumatism, asthenia, erectile dysfunction) | Ma | Madagascar | 1 |
| *Nerium oleander* L. (Apocynaceae) | Oleander | Root, leaves | TM (topical applications); cardiotonic (toxic!) | NA | Morocco, Algeria, Tunisia, Libya | 1, 2, 5, 10, 11 |
| *Ocimum gratissimum* L. (Lamiaceae) | African basil, *romba* | Herb | TM, ES (colds, respiratory ailments, headache, fever and diversity of other uses) | PA | Burundi, Madagascar, Nigeria | 1, 2, 4 |

(continued)

**Table 1** (continued)

| Species, author citation and family (non-indigenous species)* | Trade name and/or vernacular name | Part(s) used | Main uses (see caption) | Origin (see caption) | African countries involved in local, regional or international trade | References (see caption) |
|---|---|---|---|---|---|---|
| *Ocotea bullata* (Burch.) Baill. (Lauraceae) | Stinkwood, *unukani* | Bark | TM (headache, stomach complaints, nervous disorders, infantile diarrhoea) | SA | South Africa | 1, 2 |
| *Okoubaka aubrevillea* Pellegr. & Normand (Santalaceae) | Okoubaka | Bark | TM (many uses); DS (stomachic and tonic) | WA, CA | Ghana, Nigeria, Cameroon, D.R. Congo | 1, 6 |
| *Olea europaea* L. subsp. *africana* (Mill.) P.S.Greene; subsp. *europaea* (Oleaceae) | Wild olive, olive | Leaves, bark, roots | TM (urinary tract, colic, ophthalmia, tapeworm); TM (olive leaf, olive oil) | EA, CA, SA | Many countries | 1, 2, 4, 7, 10, 11 |
| *Origanum compactum* Benth. (Lamiaceae) | Moroccan wild oregano | Herb | TM; ES | NA | Morocco | |
| *Origanum syriacum* L. (Lamiaceae) | *Za'ater, zatar*, bible hyssop | Herb | TM (many uses), ES | NA | Egypt | 5 |
| *Parinari excelsa* Sabine (Chrysobalanaceae) | Guinea plum | Roots, bark | TM (stomach complaints, headache, wounds) | WA, CA | Central African Republic, Ghana, Ivory Coast, Mali, Senegal, | 1 |
| *Parkia biglobosa* (Jacq.) R. Br. ex Don (Fabaceae) | African locust bean | Seeds (all parts) | TM (wide diversity of uses) | WA | Chad, Nigeria | 1 |
| *Parquetina nigrescens* (Afzel.) Bullock (Apocynaceae) | African parquetina, *ogbo* | Leaves | TM (anaemia and many other uses) | PA | Nigeria | 1, 9 |

| | | | | | | |
|---|---|---|---|---|---|---|
| *Pauridiantha paucinervis* (Hiern) Bremek. (Rubiaceae) | *Tamirova* | Roots, leaves | TM (headache, purgative, wounds) | Ma | Madagascar | 1 |
| *Pausinystalia johimbe* (K.Schum.) Pierre (Rubiaceae) | Yohimbe tree | Bark | CH: yohimbine (aphrodisiac) | WA, CA | Cameroon | 1, 3, 10, 11 |
| *Peganum harmala* L. (Zygophyllaceae) | Harmal, African rue, wild rue | Roots, leaves, seeds | TM (high blood pressure, haemorrhoids, bronchitis, rheumatism); CH:beta-carboline alkaloids) | NA | Egypt, Morocco, Tunisia and other countries | 2, 5, 10, 11 |
| *Pelargonium* cv. Rosé ("*P. graveolens*" = *P. capitatum* (L.) L'Hérit. *x P. radens* H. Moore) (Geraniaceae) | Rose geranium | Leaves | ES (fragrance) | SA | Egypt, Tunisia, Kenya, Reunion, South Africa | 2, 5 |
| *Pelargonium sidoides* DC. (Geraniaceae) | *Rooirabas*, umckaloabo | Roots | PH (bronchitis, immune stimulant); TM (dysentery; general tonic) | SA | Lesotho, South Africa | 1, 2, 3, 7, 8, 10, 11 |
| *Pentadiplandra brazzeana* Baill. (Pentadiplandraceae) | *L'oubli* | Roots, fruits | TM (roots: pre- and postnatal care); CH (brazzein: protein sweetener) | WA, CA | Cameroon, D.R. Congo, Nigeria | 1, 6 |
| *Pentanisia prunelloides* (Klotzsch ex Eckl. & Zeyh.) Walp. (Rubiaceae) | *Icimamlilo* | Roots (leaves) | TM (headache, pain, purgative, wounds) | SA | South Africa | 1,2 |
| *Physostigma venenosum* Balf. (Fabaceae) | Calabar bean | Seeds | CH (physostigmine: cholinesterase inhibitor; Alzheimer's) | WA | Cameroon | 1, 10 |
| *Phytolacca dodecandra* L'Hérit. (Phytolaccaceae) | Endod | All parts | TM (numerous ailments; control of snails and bilharzia) | EA | Burundi, Ethiopia, Madagascar | 1, 2, 6 |

(continued)

**Table 1** (continued)

| Species, author citation and family (non-indigenous species)* | Trade name and/or vernacular name | Part(s) used | Main uses (see caption) | Origin (see caption) | African countries involved in local, regional or international trade | References (see caption) |
|---|---|---|---|---|---|---|
| *Piper guineense* Schumach. & Thonn. (Piperaceae) | West African pepper | Fruit; roots, leaves | Spice (fruit, essential oil); ES (flavourant); TM (numerous ailments) | WA, CA | Cameroon, Central African Republic, Congo, D.R. Congo, Equatorial Guinea | 1 |
| *Pistacia lentiscus* L. (Anacardiaceae) | Mastic tree, lentisc | Gum resin (roots, leaves, fruits) | TM (stomach ulcers); gum resin (mastic): dietary and cosmetic preparations | NA | Tunisia, Morocco (commercial production mostly on Chios, Greece) | 1, 5 |
| *Plantago afra* L. (Plantaginaceae) | Fleawort | Ripe seeds | TM (chronic constipation, demulcent) | NA | Egypt | 5, 10, 11 |
| *Prosopis africana* (Guill. & Perr.) Taub. (Fabaceae) | African mesquite | Seeds | DS (fermented seeds); TM (bark, roots, leaves: numerous uses) | WA | Mali, Nigeria | 1 |
| *Prunus africana* (Hook.f.) Kalkman (Rosaceae) | Red stinkwood, pygeum | Bark | PH (benign prostate hyperplasia); TM (pain, fever, inflammation and other uses) | PA | Cameroon, D.R. Congo, Equatorial Guinea (& Bioko), Ethiopia, Kenya, Madagascar, Tanzania, Uganda | 1, 2, 3, 4, 8, 10, 11 |
| *Rauvolfia media* Pichon (Apocynaceae) | Madagascar serpent-wood | Bark, roots | CH (indole alkaloids – reserpiline) | Ma | Madagascar | 6 |

| | | | | | | |
|---|---|---|---|---|---|---|
| *Rauvolfia vomitoria* Afzel. (Apocynaceae) | African serpent-wood | Roots, bark | TM (mild essential hypertension); CH: isolation of ajmaline | CA, WA, EA | Cameroon, Central African Republic, Guinea, Kenia, Madagascar, South Sudan | 1, 6, 8 |
| *Ravenala madagascariensis* Sonn. (Strelitziaceae) | Traveller's palm | Leaves | TM (antiseptic, fever) | Ma | Madagascar | 1, 4, 8 |
| *Rhamnus prinoides* L'Hérit. (Rhamnaceae) | *Geisho*, *umunanira* | Roots, bark, leaves | TM (colic, fever, pneumonia, headache and other uses) | EA | Ethiopia, Rwanda | 1, 2 |
| *Ricinodendron heudelotii* (Baill.) Pierre (Euphorbiaceae) | *Erimado*, ground nut tree, *musodo* | Seeds | DS (seeds: nuts); TM (roots, bark, leaves: many uses) | WA, CA, EA | Cameroon, Equatorial Guinea | 1 |
| *Ricinus communis* L. (Euphorbiaceae) | Castor oil plant | Cold-pressed seed oil | TM (seed oil: purgative; leaves, roots: many other uses) [biodiesel] | PA | Burundi, Ethiopia, Mozambique, Nigeria, Rwanda, South Africa | 1, 2, 5, 10, 11 |
| *Rosa canina* L. (Rosaceae) | Wild rose, dog rose | Flower buds, fruits (rosehips) | TM (numerous ailments; tonic); | NA | Algeria, Tunisia, Morocco | 5, 7, 10, 11 |
| *Rosa gallica* L. (Rosaceae)* | Common rose | Flower buds, petals | TM (numerous ailments; tonic); ES, source of rose water | NA | Egypt, Algeria, Tunisia, Morocco | 5 |
| *Rosmarinus officinalis L.* (Lamiaceae) | Rosemary | Herb | TM (antispasmodic, antiseptic, anti-inflammatory); ES | NA | Morocco, Algeria, Tunisia, Libya (also widely cultivated) | 5, 7, 10, 11 |
| *Ruta chalepensis* L. (Rutaceae) | Aleppo rue, Syrian rue | Herb, roots | TM (respiratory ailments, rheumatism and many other uses); ES | NA | Morocco, Algeria, Tunisia, Libya (also widely cultivated) | 1, 5, 9 |

(continued)

**Table 1** (continued)

| Species, author citation and family (non-indigenous species)* | Trade name and/or vernacular name | Part(s) used | Main uses (see caption) | Origin (see caption) | African countries involved in local, regional or international trade | References (see caption) |
|---|---|---|---|---|---|---|
| *Salvadora persica* L. (Salvadoraceae) | Mustard tree, toothbrush tree | Roots, stems | TM (toothbrush sticks, oral hygiene) | PA | Algeria, Chad, Egypt, Ethiopia, Tanzania | 1, 2, 3, 10, 11 |
| *Sarcocephalus latifolius* (Sm.) Bruce (= *Nauclea latifolia* Sm.) (Rubiaceae) | African peach, African quinine, *opepe* | Roots, bark and leaves | TM (fever, malaria and many other uses) | WA, CA | Benin, Burkina Faso, Cameroon, D.R. Congo, Gabon, Gambia, Ghana, Ivory Coast, Mali, Nigeria, Sierra Leone | 1, 8 |
| *Sceletium tortuosum* (L.) N.E.Br. (= *Mesembryanthemum tortuosum* L.) (Aizoaceae) | Sceletium, *kanna* | Whole plant | TM/PH (hypnotic, sedative; neurological and psychiatric disorders) | SA | South Africa | 2, 8, 10, 11 |
| *Sclerocarya birrea* (A.Rich) Hochst. (Anacardiaceae) | Marula, *umganu* | Seeds; bark, fruits | Seed oil (edible, cosmetic); TM (bark: diarrhoea, dysentery, malaria and many other uses) | SA, EA, NA | Algeria, Botswana, Mozambique, Namibia, South Africa, Swaziland, Zimbabwe, Tanzania | 1, 2 |
| *Sclerochiton ilicifolius* A.Meeuse (Acanthaceae) | *Molomo monate* | Root bark | CH (monatin/arruva: sweetener, now synthesized) | SA | South Africa | |

| | | | | | | |
|---|---|---|---|---|---|---|
| *Securidaca longepedunculata* Fresen. (Polygalaceae) | Violet tree | Root bark (leaves, stem bark) | TM (a wide diversity of ailments in most parts of Africa) | PA | Ethiopia, Madagascar, Malawi, Mali, Mozambique, Tanzania, Zimbabwe | 1, 2, 10, 11 |
| *Senna alata* (L.) Roxb.* (Fabaceae) | Ringworm cassia | Leaf | TM (purgative, antifungal) | Trop. Am. | Benin, Congo, Ghana, Ivory Coast, Nigeria, Sierra Leone, Uganda | 1, 6 |
| *Senna alexandrina* Mill. (Fabaceae) | Senna, Alexandria senna | Leaf, fruit | TM (stimulant laxative) | NA, EA | Sudan, Egypt, Libya, Algeria, South Sudan, Ethiopia, Somalia, Kenya | 1, 5, 6, 10, 11 |
| *Senna occidentalis* (L.) Link (Fabaceae) | Coffeeweed, *tsotsorinangatra* | Roots, leaves | TM (many ailments) | PA | Madagascar and many other countries | 1, 2 |
| *Senna italica* Mill. (Fabaceae) | Senegal senna | Leaves, fruits | TM (purgative); hair care | PA | Many countries; Egypt and Sudan (hair conditioner) | 1, 2, 6 |
| *Sesamum indicum* L. (= *S. orientale* L.) (Pedaliaceae) | Sesame | Seeds, seed oil | DS (functional food: reduction of cholesterol and blood sugar levels) | EA | Ethiopia, South Sudan, | 1, 2, 3, 11 |
| *Siphonochilus aethiopicus* (Schweinf.) B.L.Burtt (Zingiberaceae) | African ginger | Rhizome and roots | TM (anti-inflammatory, anti-malarial); ES (flavourant) | PA | South Africa, Mozambique, Swaziland | 1, 2, 8, 10 |

(continued)

**Table 1** (continued)

| Species, author citation and family (non-indigenous species)* | Trade name and/or vernacular name | Part(s) used | Main uses (see caption) | Origin (see caption) | African countries involved in local, regional or international trade | References (see caption) |
|---|---|---|---|---|---|---|
| *Solanum incanum* L. (Solanaceae) | Bitter apple | Fruit, leaves, roots | TM (wounds, pain, toothache, venereal diseases and numerous other ailments) | EA, CA, SA | Botswana, Djibouti, Ethiopia, Namibia, Rwanda, South Africa, Tanzania and other countries | 1, 2, 6 |
| *Solenostemma argel* (Delile) Hayne (Apocynaceae) | Arghel | Herb, root | TM (colds, indigestion, skin ailments, eye lotion) | NA | Egypt, Sudan, Algeria, Chad, Libya | 5, 9 |
| *Stachys aegyptiaca* Pers. (Lamiaceae) | Egyptian stachys | Herb | TM (antispasmodic, antiallergenic); ES | NA | Egypt | 3 |
| *Sterculia setigera* Delile (Malvaceae) | Karaya | Exudate; roots, bark | Exudate (gum, gum karaya); TM (roots, bark: diversity of uses) | EA | South Sudan (gum); Benin, Cameroon, Central African Republic, Senegal | 1 |
| *Strophanthus gratus* (Wall. & Hook.) Baill. (Apocynaceae) | Strophanthus, poison arrow vine | Seeds; root, stem leaf | CH (seeds: ouabain, a cardiotonic); TM (gonorrhea and numerous other uses) | WA, CA | Cameroon, Nigeria, Gabon | 1, 6, 8, 10, 11 |
| *Strophanthus hispidus* DC. (Apocynaceae) | Brown strophanthus | Seeds | CH: strophanthidine glycosides (chronic heart weakness) | CA, EA | Many countries | 1, 6 |
| *Strophanthus kombe* Oliv. (Apocynaceae) | Strophanthus | Seeds | CH (*k*-strophanthin: cardiotonic) | CA, SA, EA | Madagascar | 2, 6 |

| | | | | | | |
|---|---|---|---|---|---|---|
| *Sutherlandia* – see *Lessertia* | | | | | | |
| *Synsepalum dulcificum* (Schum. & Thonn.) Daniell (Sapotaceae) | Miracle fruit, miraculous berry | Fruit | CH (miraculin: sweetener; taste modifier) | WA | Congo, Ghana | 1, 11 |
| *Syzygium cordatum* Hochst. (Myrtaceae) | Water berry, *umdoni* | Bark | TM (diarrhea, stomach ailments and other uses) | SA, EA | Kenya, South Africa, Zimbabwe | 1, 2 |
| *Tabernaemontana crassa* Oliv. (Apocynaceae) | Adam's apple flower | Bark | TM (many uses) | WA, CA | Regional trade (many countries) | 1, 6 |
| *Tabernanthe iboga* Stapf (Apocynaceae) | Iboga | Root | CH (ibogaine, a stimulant used for the treatment of drug addiction); TM (hallucinogen, aphrodisiac, asthenia; conjunctivitis and other ailments) | CA, WA | Gabon, Cameroon, Central African Republic, Congo, D.R. Congo, Equatorial Guinea | 1, 6 |
| *Tamarindus indica* L. (Fabaceae) | Tamarind | Fruit pulp | DS (fruit pulp: mild laxative, general tonic); TM (roots, bark, leaves: diverse uses) | EA | Chad, Ethiopia, South Sudan | 1, 2, 10 |
| *Tamarix gallica* L. (Tamaricaceae) | French tamarisk, saltcedar | Leaves, bark | TM (astringent, tonic, liver stimulant, stomachic) | NA | Morocco, Algeria | 5 |
| *Tanacetum cinerariifolia* (Trevir.) Sch.Bip. (=*Chrysanthemum cinerariifolium* (Trevir.) Vis.* (Asteraceae) | Pyrethrum | Flower heads | CH (source of pyrethrins, natural insecticides) | Balkans | Cultivated: Kenya, Tanzania | 10 |
| *Taverniera abyssinica* Rich. (Fabaceae) | Taverniera, *dingetegna* | Root | TM (headache, stomach ache, fever – root chewed) | EA | Ethiopia | |
| *Terminalia sericea* Burch. ex DC. (Combretaceae) | Silver cluster-leaf, *mususu* | Roots, leaves, bark | TM (general tonic, diarrhea, many other ailments) | SA, EA | Botswana, Mozambique, South Africa, Zimbabwe | 1, 2, 8 |

(continued)

**Table 1** (continued)

| Species, author citation and family (non-indigenous species)* | Trade name and/or vernacular name | Part(s) used | Main uses (see caption) | Origin (see caption) | African countries involved in local, regional or international trade | References (see caption) |
|---|---|---|---|---|---|---|
| *Tetraclinis* articulata (Vahl.) Masters (Cupressaceae) | Arar tree | Leaves, fruits, wood, resin | TM (numerous ailments); ES: distilled from sandarac(h) or gum sandarac (the resin) | NA | Morocco, Algeria, Tunisia | 5 |
| *Tetradenia riparia* (Hochst.) Codd (Lamiaceae) | *Iboza*, *umuravumba* | Leaves | TM (coughs, colds, bronchitis, respiratory ailments, diarrhea, angina, venereal diseases) | CA, SA, EA | Burundi, D.R. Congo, Rwanda, South Africa, Zimbabwe | 1, 2, 4 |
| *Tetrapleura tetraptera* (Schumach. & Thonn.) Taub. (Fabaceae) | Aridan | Fruits, bark | DS (fruit: spice, tonic); TM (fruit, bark: numerous ailments) | CA, WA, EA | Cameroon, Congo, D.R. Congo, Equatorial Guinea, Gabon, Ivory Coast, Senegal, Togo, Uganda | 1 |
| *Teucrium polium* L. (Lamiaceae) | Felty germander | Herb | TM (dysmenorrhea, digestive ailments, antispasmodic, skin allergy) | NA | Egypt, Tunisia | 5 |
| *Thaumatococcus daniellii* (Bennett) Benth. (Marantaceae) | Miraculous fruit, *ewe eran* | Seed aril | CH (source of thaumatin, a protein sweetener) | WA, CA | Côte d'Ivoire, D.R. Congo, Nigeria | 1 |
| *Thymbra capitata* (L.) Cav. [=*Coridothymus capitatus* (L.) Reichenb.f.] (Lamiaceae) | Capitate thyme | Herb | TM (various ailments); ES (antiseptic, antispasmodic) | NA | Tunisia (widely cultivated) | 5 |

| | | | | | | |
|---|---|---|---|---|---|---|
| *Thymus decussatus* Benth. (Lamiaceae) | Sinai thyme | Herb | TM (antispasmodic, toothache); ES | NA | Egypt | 3 |
| *Thymus satureioides* Coss. & Ball. (Lamiaceae) | Moroccan wild thyme | Herb | TM (topical anti-inflammatory); ES (aromatherapy) | NA | Morocco | 5 |
| *Toddalia asiatica* (L.) Lam. (Rutaceae) | Orange climber, Lopez root | Leaves, roots | TM (antimalarial, anti-inflammatory, spasmolytic) | EA, CA, SA, Ma | Uganda, Gabon, Kenya, Madagascar, South Africa, Tanzania | 1, 2, 4, 8 |
| *Tribulus terrestris* L. (Zygophyllaceae) | Caltrops | Fruits | TM (urinary disorders and many other uses); nutritional supplement | PA | North African countries | 1, 2, 5 |
| *Trichilia emetica* Vahl (Meliaceae) | Cape/Natal mahogany | Bark; seeds | TM (bark: many uses); seeds: cosmetic oil (mafura butter, mafoureira tallow) | PA | Ethiopia, Mozambique, South Africa, Zimbabwe | 1, 2, 8 |
| *Trigonella foenum-graecum* L. (Fabaceae) | Fenugreek | Seeds | DS, PH (hypoglycaemic, cholesterol-lowering); TM (convulsions, stomach disorders; topical for ulcers) | EA | Ethiopia, South Sudan | 1, 2, 7, 10, 11 |
| *Tylosema esculentum* (Burch.) Schreib. (Fabaceae) | Marama bean | Seeds | DS (seeds); seed oil (edible, cosmetics) | SA | Namibia | |
| *Urera acuminata* Gaudichaux (Urticaceae) | *Sampivato* | Root | TM (gynecological use) | Ma | Madagascar | 4 |
| *Vepris madagascarica* (Baill.) H.Perrier (Rutaceae) | *Anzéty* | Bark, leaves | ES (anethole) | Ma | Madagascar | 9 |

(continued)

**Table 1** (continued)

| Species, author citation and family (non-indigenous species)* | Trade name and/or vernacular name | Part(s) used | Main uses (see caption) | Origin (see caption) | African countries involved in local, regional or international trade | References (see caption) |
|---|---|---|---|---|---|---|
| *Vernonia amygdalina* Delile, *V. colorata* (Willd.) Drake (Asteraceae) | Bitter leaf | Leaves, roots | TM (general tonic, numerous uses) | PA | D.R. Congo, Ethiopia, Kenya, Mozambique, Nigeria, Rwanda, Tanzania, Uganda, Zimbabwe | 1, 2, 8 |
| *Visnaga daucoides* Gaertn. [=*Ammi visnaga* (L.) Lam.] (Apiaceae) | Khella | Fruit | TM; ES | NA | Morroco, Egypt | 1, 7, 11 |
| *Vitellaria paradoxa* Gaertn.f. (=*Butyrospermum parkii* Kotschy) (Sapotaceae) | *Karate*, shea tree, shea butter tree | Seeds | DS (seed oil: edible shea butter, cosmetics); TM (leaf, bark: several uses) | WA, CA | Burkina Faso, Cameroon, Chad, Mali, Nigeria | 1 |
| *Voacanga africana* Stapf *and V. thouarsii* (Rubiaceae) | Voacanga | Seeds | CH (isolation of tabersonine; synthesis of vincamine, a cerebral stimulant); TM (bark, roots: many uses) | PA | Cameroon, Madagascar; also Côte d'Ivoire, Ghana, D.R. Congo, Sao Tome and Principe, Tanzania | 1, 6, 8 |
| *Warburgia salutaris* (G. Bertol) Chiov. (=*W. ugandensis* Sprague) (Canellaceae) | Pepperbark tree | Bark, leaves | TM (bark: antibiotic, many uses; leaves: tablets, spice) | EA, SA | South Africa, Kenya, Madagascar, Mozambique, Tanzania | 1, 2, 8, 10, 11 |
| *Withania somnifera* (L.) Dunal (Solanaceae) | Indian ginseng, *ashwagandha* | Roots | PH (physical and psychological stress); TM (roots, leaves: many uses, also topical for sores) | Co, PA | Almost all countries | 1, 2, 4, 6, 10 |

| | | | | | | |
|---|---|---|---|---|---|---|
| *Ximenia americana* L. (Olacaceae) | Small sourplum | Seeds | Seed oil: cosmetic; TM (roots, leaves: many uses) | PA | Ethiopia, Madagascar, Mozambique, Namibia | 1, 2 |
| *Xylopia aethiopica* (Dunal) A.Rich. (Annonaceae) | Guinea pepper | Fruits | Fruits: essential oil, spice; ES (flavourant) TM (fruits, leaves, bark: diverse uses) | CA, WA | Benin, Cameroon, Central African Republic, Congo, Equatorial Guinea, Sierra Leone | 1, 8 |
| *Xysmalobium undulatum* (L.) Aiton f. (Apocynaceae) | Uzara, *ishongwe* | Roots | TM/PH (diarrhea, dysentery, dysmenorrhea, colic; root powder: headache) | SA, EA | Namibia, South Africa, Zimbabwe | 1, 7, 8, 9, 10, 11 |
| *Zanthoxylum zanthoxyloides* (Lam.) Zepern. & Timler (Rutaceae) | Senegal prickly-ash | Roots, bark, leaves | TM (many ailments); DS (roots: sickle cell anaemia) | WA | Côte d'Ivoire, Mali, Burkina Faso, Ghana, Nigeria | 9 |
| *Ziziphus mucronata* Willd. (Rhamnaceae) | Buffalo thorn, *mokgalo* | Bark (roots, leaves, fruits) | TM (numerous ailments, including stomach and urinary tract; topically for ulcers) | WA, EA, SA | Many African countries | 1, 2 |

The list includes 258 species (246 indigenous and 12 naturalised and/or commonly cultivated)

Main uses: *PH* phytomedicine, *DS* dietary supplement and/or functional food, *CH* source of a useful chemical compound(s), *ES* source of essential oil, *TM* traditional medicine

Origin (following Van Wyk 2015): *NA* North Africa, *WA* West Africa, *EA* East Africa, *SA* Southern Africa, *PA* Pan-African, *Ma* Madagascar and Indian Ocean Islands, *Co* Cosmopolitan, *Cu* cultigen of unknown/uncertain origin

General references (1–11): 1 = Neuwinger (2000), 2 = Arnold et al. (2002), 3 = Vasisht and Kumar (2004), 4 = Gurib-Fakim and Brendler (2004), 5 = IUCN (2005), 6 = Schmelzer and Gurib-Fakim (2008) [PROTA 11(1)], 7 = Wichtl (2009), but see also Wichtl and Bisset (2000), 8 = African Herbal Pharmacopoeia (Brendler et al. 2010), 9 = Gurib-Fakim and Schmelzer (2013) [PROTA 11(2)], 10 = Van Wyk et al. (2015), 11 = Van Wyk and Wink (2015)

Exotic (non-African) taxa are indicated with an asterisk (*)

Table 1 provides a brief overview of the most important species of African origin, their main uses and region(s) of origin. The list includes products of international trade, such as phytomedicines (Fig. 1), traditional medicines (Fig. 2), dietary supplements (Fig. 3), and sources of pure chemical entities (Fig. 4) and essential oil (Fig. 5), but also some species of regional or local importance. The choice of species is partly based on the statistics provided by Van Wyk (2015), who used a combination of Google Scholar and Scopus to determine the relative popularity and scientific interest in 175 species from sub-Saharan Africa. It is also based on 11 general publications and reviews that have appeared since 2000, namely Neuwinger (2000), Arnold et al. (2002), Vasisht and Kumar (2004), Gurib-Fakim and Brendler (2004), IUCN (2005), Schmelzer and Gurib-Fakim (2008), Wichtl (2009), (Brendler et al. 2010), Gurib-Fakim and Schmelzer (2013), Van Wyk et al. (2015) and Van Wyk and Wink (2015). These publications can be consulted for detailed information on the botany, chemistry, traditional uses and biological activities of the plants.

**Fig. 1** Examples of African phytomedicines. *1*, *Centella asiatica* leaves; *2*, *Griffonia simplicifolia* seeds; *3*, *Harpagophytum procumbens* (*3a*, leaves and flowers; *3b*, fruit; *3c*, sliced and dried secondary roots); *4*, *Pelargonium sidoides* (*4a*, flowering plant; *4b*, flowers; *4c*, fresh roots); *5*, *Prunus africana* (*5a*, leaves and fruits; *5b*, bark); *6*, *Sceletium tortuosum* (*6a*, flowering plant; *6b*, traditional dried product); *7*, *Trigonella foenum-graecum* (*7a*, leaves, flower and fruit; *7b*, seeds); *8*, *Withania somnifera* (*8a*, leaves, flowers and fruits; *8b*, roots); *9*, *Xysmalobium undulatum* (*9a*, leaves and flowers; *9b*, dried roots) (Copyright: B.-E. Van Wyk)

**Fig. 2** Examples of African medicinal plants. *1*, *Aloe ferox* (*1a*, flowering plant; *1b*, from top to bottom: spray-dried leaf juice, powdered aloe lump, aloe lump); *2*, *Bulbine frutescens* (*2a*, flowering plant; *2b*, flowers; *2c*, leaves showing gel); *3*, *Catha edulis* (*3a*, flowers; *3b*, fresh leaves); *4*, *Harungana madagascariensis* leaves and fruits; *5*, *Kigelia africana* flowers and fruit; *6*, *Lessertia frutescens* (*6a*, flowers and fruits; *6b*, dried leaves); *7*, *Mondia whitei* (*7a*, flowers; *7b*, dried roots and powdered root); *8*, *Phytolacca dodecandra* flowers; *9*, *Securidaca longepedunculata* (*9a*, leaves and flowers; *9b*, root bark); *10*, *Senna alexandrina* (*10a*, leaves and flowers; *10b*, dried leaves and fruits); *11*, *Warburgia salutaris* leaves and flower (Copyright: B.-E. Van Wyk)

## *Sources of Information on African Medicinal and Aromatic Plants*

Medicinal plants have been used in Africa since ancient times but the documentation of the knowledge generally dates back to the last two centuries. Colonial floras are often viewed as inventories of plants with potential commercial value and these publications are still of considerable historical value. Major contributions made during the twentieth century were those of Burkill (1937, 1985–2004) for West Africa and Watt and Breyer-Brandwijk (1962) for southern and eastern Africa. General works include Sofowora (1982) and Iwu (1993, 2014), as well as the first African pharmacopoeia in two volumes (African Pharmacopoeia 1981, 1986). Other noteworthy publications are Boulos (1983) for North Africa; Ayensu (1978), Oliver-Bever (1986) and the Ghana Herbal Pharmacopoeia (1992) for West Africa; Kokwaro (1993) and Abebe and Ayehu (1993) for East Africa, and Gelfand et al. (1985), Hedberg and Staugård (1989), Hutchings et al. (1996) and Van Wyk et al. (1997) for southern Africa.

The year 2000 marked a turning point in the tempo of publications on African medicinal and aromatic plants. Neuwinger (2000) made a major contribution by compiling all available knowledge on sub-Saharan Africa, which incorporated many

**Fig. 3** Examples of African dietary supplements and functional foods. *1*, *Acanthosicyos horridus* (*1a*, female plant; *1b*, fruit and seeds); *2*, *Adansonia digitata* (*2a*, tree; *2b*, fruit, seeds and fruit pulp); *3*, *Aloe vera* (*3a*, plants; *3b*, flowers; *3c*, fresh gel and spray-dried gel); *4*, *Aspalathus linearis* (*4a*, plant; *4b*, leaves, flowers and fruits; *4c*, rooibos tea (fermented and green)); *5*, *Balanites aegyptiaca* (*5a*, tree; *5b*, leaves and thorns; *5c*, fruits); *6*, *Combretum micranthum* (whole and cut leaves); *7*, *Cyclopia genistoides* (*7a*, leaves and flowers; *7b*, cut stems and leaves); *8*, *Hibiscus sabdariffa* (*8a*, flower; *8b*, dried calyces); *9*, *Hoodia gordonii* (*9a*, flowering plant; *9b*, stems); *10*, *Hypoxis hemerocallidea* (*10a*, flowering plant; *10b*, corms); *11*, *Lepidium sativum* (*11a*, plants; *11b*, flowers and fruits; *11c*, seeds); *12*, *Moringa stenopetala* (*12a*, tree; *12b*, flowers; *12c*, leaves); *13*, *Sesamum indicum* (*13a*, leaves and flower; *13b*, seeds); *14*, *Tamarindus indica* (*14a*, tree; *14b*, leaves and fruits; *14c*, fruits; *14d*, fruit pulp) (Copyright: B.-E. Van Wyk)

**Fig. 4** Examples of African sources of chemical compounds and extracts. *1*, *Acacia senegal* (*1a*, leaves and fruits; *1b*, gum); *2*, *Catharathus roseus* flowers; *3*, *Coffea arabica* (*3a*, plant; *3b*, flower and fruits); *4*, *Gloriosa superba*; *5*, *Pausinystalia johimbe* (*5a*, leaves; *5b*, bark); *6*, *Physostigma venenosum* (*6a*, leaves; *6b*, seeds); *7*, *Rauvolfia vomitoria* leaves and fruits; *8*, *Sclerochiton ilicifolia*; *9*, *Strophanthus gratus*; *10*, *Vitellaria paradoxa* seed oil (shea butter); *11*, *Voacanga africana* seeds (Copyright: B.-E. Van Wyk)

poorly known publications of the twentieth century, including those from French-speaking regions of the continent. An ambitious project to document the useful plants of tropical Africa was initiated in 2000. This was the PROTA program (Plant Resources of Tropical Africa), aimed at treating 8681 useful plant species (PROTA 2010a) in separate commodity groups (see http://www.prota4u.org). Medicinal plants have been published in two volumes (out of a planned total of four), namely volumes 11(1) and 11(2) of the PROTA series. No less than 894 primary use species were included in Volume 1 (Schmelzer and Gurib-Fakim 2008), while 409 species were included in 146 reviews in Volume 2 (Gurib-Fakim and Schmelzer 2013). Another useful book dealing with the medicinal and aromatic plants of the various countries and regions of the African continent (Vasisht and Kumar 2004) was published by the United Nations Industrial Development Organization (UNIDO). Detailed information on the chemistry and biological activity of African plants can also be obtained from the database developed by Ermias Dagne at Addis Ababa University (Natural Database for Africa 2011). Of special interest is the African herbal pharmacopoeia (Brendler et al. 2010), representing a collective effort from several African scientists which comprises monographs of 51 carefully selected African medicinal plants.

**Fig. 5** Examples of African aromatic plants. *1*, *Aframomum melegueta* seeds; *2*, *Agathosma betulina* (*2a*, leaves and flower; *2b*, dried leaves); *3*, *Artemisia afra* leaves and flower heads; *4*, *Boswellia neglecta* (*4a*, leaves and fruits; *4b*, gum-resin); *5*, *Boswellia papyrifera* (*5a*, tree; *5b*, leaf and flowers; *5c*, gum-resin); *6*, *Boswellia rivae* gum-resin; *7*, *Boswellia sacra* (*7a*, leaves; *7b*, gum-resin); *8*, *Commiphora guidottii* gum-resin; *9*, *Commiphora myrrha* (*9a*, leaves and fruits; *9b*, gum-resin); *10*, *Commiphora wildii* gum-resin; *11*, *Lippia javanica* leaves and flower heads; *12*, *Myriothamnus flabellifolius* leaves and flowers; *13*, *Pelargonium* cv. Rosé; *14*, *Siphonochilus aethiopicus* (*14a*, flowering plant; *14b*, rhizomes); *15*, *Xylopia aethiopica* dried fruits (Copyright: B.-E. Van Wyk)

Books and publications with a global scope also included several African species of commercial relevance. Twelve of a total of 118 species (i.e. more than 10%) treated in the four volumes of the WHO Selected Monographs of Medicinal Plants (WHO 1999, 2003, 2007, 2009) are indigenous to Africa, namely *Aloe ferox, A. vera*, *Ammi majus*, *Centella asiatica*, *Commiphora myrrha* (as *C. molmol*), *Harpagophytum procumbens, Momordica charantia*, *Prunus africana, Ricinus communis*, *Senna alexandrina* (as *Cassia senna*), *Visnaga daucoides* (as *Ammi visnaga*) and *Withania somnifera. Aloe vera* is an ancient cultigen of North African and Arabian origin, while *Ricinus communis* is almost certainly of Ethiopian origin, given the extreme morphological variation in this region (Van Wyk 2015).

Another global review of medicinal plants was published by Van Wyk and Wink (2004). In the most recent revised edition of this book (Van Wyk et al. 2015; Van Wyk and Wink 2017), 36 of the 350 monographs (roughly 11%) represent African plants. The summary table of 911 species shows 111 species of African origin (roughly 12%).

Seven African species were listed amongst the 43 phytomedicines presented by Van Wyk and Wink (2015), namely *Centella asiatica*, *Griffonia simplicifolia*, *Harpagophytum procumbens*, *Pelargonium sidoides*, *Prunus africana*, *Trigonella foenum-graecum* and *Withania somnifera*. In addition, *Sceletium tortuosum* and *Xysmalobium undulatum* may also qualify for inclusion in the list of phytomedicine. Ten of the 56 species (nearly 18%) listed as functional foods and nutraceuticals are indigenous to Africa: *Adansonia digitata*, *Aloe ferox*, *Aspalathus linearis*, *Hibiscus sabdariffa*, *Hoodia gordonii*, *Hypoxis hemerocallidea*, *Lepidium sativum*, *Sesamum indicum*, *Synsepalum dulcificum* and *Tamarindus indica*. Thirteen of the 135 species (roughly 10%) listed as sources of commercially relevant chemical compounds are African: *Catharanthus roseus*, *Centella asiatica*, *Coffea arabica*, *Cissampelos pareira*, *Cyclopia genistoides*, *Pausinystalia johimbe*, *Physostigma venenosum*, *Sclerochiton ilicifolius*, *Strophanthus gratus*, *Synsepalum dulcificum*, *Thaumatococcus daniellii*, *Voacanga africana* and *V. thouarsii*.

Numerous other scientific publications and books provide additional insights into the most popular and important African medicinal and aromatic plants (Van Wyk 2015, and references cited therein). These include the treatment of 25 African plants with exceptional potential, published by PROTA (2010b), IUCN (2005) for North Africa, the Nigerian Natural Medicine Development Agency (2004, 2006) for West Africa.

Hedberg et al. (2009) and Asfaw and Demissew (2009) for East Africa, Latham and Konda ku Mbuta (2014) for Central Africa, Van Wyk and Gericke (2000), Von Koenen (2001), Arnold et al. (2002) and Van Wyk et al. (2009) for Southern African and Gurib-Fakim and Brendler (2004) for Madagascar and the Indian Ocean Islands. Surveys of the informal and unregulated markets that exist in African countries are a valuable source of information on the most popular medicinal plants and the large volumes of material that are traded. Examples include Cunningham (1988), Mander (1998), Williams et al. (2000) and Von Ahlefeldt et al. (2003) for South Africa, Van Andel et al. (2012) for Ghana, Jusu and Sanchez (2014) for Sierra Leone, Quiroz et al. (2014) for Benin, Towns et al. (2014) for Gabon, Otieno et al. (2015) for Tanzania and Randriamiharisoa et al. (2015) for Madagascar.

## *Conclusions and Future Prospects*

Although some African medicinal and aromatic plants have become important items of international trade, the full diversity and commercial potential of the entire flora have yet to be explored. The strong scientific interest in African plants, as reflected in the large numbers of recent publications, indicates an awareness of the potential for developing novel treatments and for making new discoveries that can be of benefit to mankind. Much remains to be discovered, especially in poorly known plants from remote regions of the African continent. The botanical literature, incomplete as it may be, already provides a glimpse of the exciting research opportunities that are presented by the rich diversity of useful African plants. Traditional uses are based on oral-traditional knowledge that has been passed down through many generations. This cultural context adds to the complexity of the patterns of plant use and is particularly relevant to understanding the origins of medicinal plant use and aromatherapy as practiced by our early ancestors in Africa. The available wisdom and experience should not only be studied and documented but also used to inspire new innovations. It seems reasonable to expect that African medicinal and aromatic plants will gradually rise to more international prominence and that they will eventually rival the large number of herbal products of European and Asian origin. The rapid development of more sophisticated technologies will open up new frontiers of scientific exploration. It can be anticipated that new tools and more holistic approaches will become available for gaining insights into the efficacy and pharmacological activities of many of the popular African plants that have not yet been subjected to rigorous scientific study.

**Acknowledgements** Thanks are due to the University of Johannesburg and the National Research Foundation of South Africa for providing logistic and financial support to the National Research Chair in Indigenous Plant Use.

## References

Abebe D, Ayehu A (1993) Medicinal plants and enigmatic health practices of Northern Ethiopia. B.S.P.E, Addis Ababa

African Pharmacopoeia (1981, 1986) (vols 1, 2). OAU/STRC Publications Division, Laos

Arnold TH, Prentice GA, Hawker LC, Snyman EE, Tomalin M, Crouch NR, Pottas-Bircher C (2002) Medicinal and magical plants of southern Africa: an annotated checklist. Strelitzia 13. National Botanical Institute, Pretoria

Asfaw N, Demissew S (2009) Aromatic plants of Ethiopia. Shama Books, Addis Ababa

Ayensu ES (1978) Medicinal plants of West Africa. Reference Publications Inc, Algonac

Boulos L (1983) Medicinal plants of North Africa. Algonac, Michigan

Brendler T, Eloff JN, Gurib-Fakim A, Phillips LD (eds) (2010) African herbal pharmacopoeia. Association of African Medicinal Plant Standards, Port Louis

Burkill HM (1985–2004) The useful plants of West Tropical Africa, 2nd edn, 6 vols. Royal Botanic Gardens, Kew

Cunningham AB (1988) An investigation of the herbal medicine trade in Natal/KwaZulu. Investigational Report No. 29. Institute for Natural Resources, University of KwaZulu-Natal, Pietermaritzburg

Dobignard A, Chatelain C (2010–2013) Index synonymique flore d'Afrique du Nord, vols 1–5. Conservatoire et jardin Botaniques, Genève

Gelfand M, Mavi S, Drummond RB, Ndemera B (1985) The traditional medicinal practitioner in Zimbabwe. Mambo Press, Gweru

Ghana Herbal Pharmacopoeia (1992) Policy Research and Strategic Planning Institute, Accra

Gurib-Fakim A, Brendler T (2004) Medicinal and aromatic plants of Indian Ocean islands: Madagascar, Comores, Seychelles and Mascarenes. Medpharm Scientific Publishers, Stuttgart

Gurib-Fakim A, Schmelzer GH (eds) (2013) The plant resources of Tropical Africa 11(2), Medicinal Plants 2. Wageningen, CTA PROTA

Hedberg I, Staugård F (1989) Traditional medicinal plants: traditional medicine in Botswana. Ipeleng, Gaborone
Hedberg I, Friis I, Persson E (2009) Flora of Ethiopia and Eritrea, vol 8. The National Herbarium, Addis Ababa University/Department of Systematic Botany, Uppsala University, Addis Ababa/ Uppsala
Hutchings A, Scott AH, Lewis G, Cunningham A (1996) Zulu medicinal plants. Natal University Press, Pietermaritzburg
IUCN (2005) A guide to medicinal plants in North Africa. International Union for Conservation of Nature and Natural Resources, Málaga
Iwu MM (1993) Handbook of African medicinal plants. CRC Press, Boca Raton
Iwu MM (2014) Handbook of African medicinal plants, 2nd edn. CRC Press, Boca Raton
Jusu A, Sanchez AC (2014) Medicinal plant trade in Sierra Leone: threats and opportunities for conservation. Econ Bot 68:16–29
Klopper RR, Chatelain C, Bänninger V, Habashi C, Steyn HM, De Wet BC, Arnold TH, Gautier L, Smith GF, Spichiger R (2006) Checklist of the flowering plants of Sub-Saharan Africa. An index of accepted names and synonyms. South African Botanical Diversity Network Report No. 42. SABONET, Pretoria
Kokwaro JO (1993) Medicinal plants of East Africa, 2nd edn. East Africa Literature Bureau, Nairobi
Latham P, Konda ku Mbuta A (2014) Useful plants of Bas-Congo Province, Democratic Republic of Congo, 2nd edn, 2 vols. Published by the Authors
Madagascar Catalogue (2013) Catalogue of the vascular plants of Madagascar. Missouri Botanical Garden, St. Louis, U.S.A. & Antananarivo, Madagascar. http://www.efloras.org/madagascar. Accessed May 2013
Mander M (1998) Marketing of indigenous medicinal plants in South Africa: a case study in KwaZulu-Natal. Food and Agricultural Organization of the United Nations, Rome
Manning J, Goldblatt P (2012) Plants of the Greater Cape Floristic Region 1: The Core Cape Flora. Strelitzia 29. South African National Biodiversity Institute, Pretoria
Natural Database for Africa (2011) Version 2.0. NDA is a species based versatile database developed by Ermias Dagne, Ethiopia (edagne@gmail.co)
Neuwinger HD (2000) African traditional medicine – a dictionary of plant use and applications. Medpharm Scientific Publishers, Stuttgart
Nigerian Natural Medicine Development Agency (2004) Medicinal plants of Nigeria: South-West Nigeria. Nigerian Natural Medicine Development Agency, Lagos
Nigerian Natural Medicine Development Agency (2006) Medicinal plants of Nigeria: North Central Zone. Nigerian Natural Medicine Development Agency, Lagos
Oliver-Bever B (ed) (1986) Medicinal plants of tropical West Africa. Cambridge University Press, Cambridge
Otieno J, Abihudi S, Veldman S, Nahashon M, Van Andel T (2015) De Boer HJ (2015) Vernacular dominance in folk taxonomy: a case study of ethnospecies in medicinal plant trade in Tanzania. J Ethnobiol Ethnomed 11:10. doi:10.1186/1746-4269-11-10
Prota (2010a) The plant resources of tropical Africa. Updated list of species and commodity groupings. Backhuys Publishers and CTA PROTA, Wageningen
Prota (2010b) Review of 25 African plants with special potential. Backhuys Publishers and CTA PROTA, Wageningen
Quiroz D, Towns A, Legba SI, Swier J, Brière S, Sosef M, Van Andel T (2014) Quantifying the domestic market in herbal medicine in Benin, West Africa. J Ethnopharmacol 151:1100–1108
Randriamiharisoa MN, Kuhlman AR, Jeannoda V, Harison Rabarison H, Rakotoarivelo N, Randrianarivony T, Raktoarivony F, Randrianasolo A, Bussmann RW (2015) Medicinal plants sold in the markets of Antananarivo, Madagascar. J Ethnobiol Ethnomed 11:60. doi:10.1186/ s13002-015-0046-y
Royal Botanic Gardens Kew (2016) The state of the world's plants report – 2016. Royal Botanic Gardens, Kew. ISBN:978-1-84246-628-5

Schmelzer GH, Gurib-Fakim A (eds) (2008) The plant resources of tropical Africa 11(1), Medicinal Plants 1. Backhuys Publishers and CTA PROTA, Wageningen
Sofowora A (1982) Medicinal plants and traditional medicine in Africa. Wiley, Chichester
The Plant List (2013) The plant list. A working list of all plant species, version 1.1. Royal Botanic Gardens, Kew and Missouri Botanical Garden. http://www.theplantlist.org/
Towns AM, Quiroz D, Guinee L, De Boer H, Van Andel T (2014) Volume, value and floristic diversity of Gabon's medicinal plant markets. J Ethnopharmacol 155:1184–1193
Van Andel T, Myren B, Van Onselen S (2012) Ghana's herbal market. J Ethnopharmacol 140:369–378
Van Wyk B-E (2015) A review of commercially important African medicinal plants. J Ethnopharmacol 176:118–134
Van Wyk B-E, Gericke N (2000) People's plants – a guide to useful plants of Southern Africa. Briza Publications, Pretoria
Van Wyk B-E, Wink M (2004) Medicinal plants of the world. Timber Press/Briza Publications, Portland/Pretoria
Van Wyk B-E, Wink M (2015) Phytomedicines, herbal drugs and poisons. Kew Publishers, Kew and University of Chicago Press, Chicago
Van Wyk B-E, Wink M (2017) Medicinal plants of the world, 2nd edn. CABI/Briza Publications, Wallingford/Pretoria
Van Wyk B-E, Van Oudtshoorn B, Gericke N (1997) Medicinal plants of South Africa. Briza Publications, Pretoria
Van Wyk B-E, Van Oudtshoorn B, Gericke N (2009) Medicinal plants of South Africa, 2nd edn. Briza Publications, Pretoria
Van Wyk B-E, Wink C, Wink M (2015) Handbuch der Arzneipflanzen. Wissenschaftliche Verlagsgesellschaft, Stuttgart
Vasisht K, Kumar V (2004) Compendium of medicinal and aromatic plants, vol 1. Trieste, ICS-UNIDO
Von Ahlefeldt D, Crouch NR, Nichols G, Symmonds R, McKean S, Sibiya H, Cele MP (2003) Medicinal plants traded on South Africa's eastern seaboard. Ethekwini Parks Department and University of Natal, Durban
Von Koenen E (2001) Medicinal, poisonous and edible plants in Namibia. Klaus Hess Publishers, Windhoek/Göttingen
Watt JM, Breyer-Brandwijk MG (1962) The medicinal and poisonous plants of Southern and Eastern Africa, 2nd edn. Livingstone, London
WHO Monographs on selected medicinal plants, (vols 1–4) (1999, 2003, 2007, 2009) World Health Organisation, Geneva
Wichtl M (2009) Teedrogen und Phytopharmaka. Wissenschaftliche Verlagsgesellschaft, Stuttgart
Wichtl M, Bisset NG (eds) (2000) Herbal drugs and phytopharmaceuticals. CRC Press, Boca Raton
Williams VL, Balkwill K, Witkowski ETF (2000) Unravelling the commercial market for medicinal plants and plant parts on the Witwatersrand, South Africa. Econ Bot 54:310–327

# Traditional Use of Medicinal and Aromatic Plants in Africa

**Olufunke O. Fajinmi, Olaoluwa O. Olarewaju, and J. Van Staden**

**Abstract** Traditional medicine and the use of medicinal and aromatic plants in Africa are linked to social life, culture and daily activities of the African people. The celebration of life events involve the presentation and sharing of several culturally significant items such as honey, salt, Cola and other medicinal plants and their products. In this chapter, the rich and unique culture of African countries is discussed in relation to the use of medicinal and aromatic plants. As a special feature, the major medicinal and aromatic plants endemic to Western, Northern and Southern Africa are reviewed with regards to their manifold uses. The disposition of African people to traditional medicine, formulation of policies to support the growth of traditional medicine and the progress made so far are highlighted. As a result of the acceptance of African traditional medicine by a vast majority of African populace and support gained from the African Union, World Health Organization and the United Nations; African traditional medicine will in the near future be highly valued just like the Chinese traditional medicine system.

**Keywords** African culture • Traditional medicine • Aloe species • Wonder nut (*Cola nitida*) • Henna • Marula • Acalypha species • Aromatic plants • Buchu

## 1 Africa, Its People, Culture and Traditional Medicine

Africa is one of the seven continents of the world, the second largest after Asia, with about a billion people of diverse tribes, language, tradition and culture. It comprises 54 countries, with a total land area of 30 Million Square Kilometers. Africa, as a

O.O. Fajinmi • J. Van Staden (✉)
Research Centre for Plant Growth and Development, School of Life Sciences, University of KwaZulu-Natal, Private Bag X01, Scottsville, Pietermaritzburg 3209, South Africa
e-mail: vanstadenj@ukzn.ac.za

O.O. Olarewaju
Department of Horticultural Science, School of Agricultural, Earth and Environmental Science, University of KwaZulu-Natal, Private Bag X01, Scottsville, Pietermaritzburg 3209, South Africa



M. Neffati et al. (eds.), *Medicinal and Aromatic Plants of the World - Africa Volume 3*, Medicinal and Aromatic Plants of the World,
DOI 10.1007/978-94-024-1120-1_3

continent, is rich in various and interesting cultures and as such is regarded as the most diverse amongst the seven continents of the world. These cultures and traditions have been passed from one generation to another and are still maintained to date, irrespective of colonization and civilization.

One of the most important aspects of African culture is its use of traditional medicine (TM) or medicinal plants (MP) which is an ancient practice. African TM is one of the oldest and unique systems across the world. It is diverse, as it is shaped by cultural beliefs of different countries in Africa. Some African countries have similarities in the use and practice of TM and therefore utilize similar plants for the cure of diseases and infections. This, however, is common in countries with linked histories. For example, some African countries were formed as a result of migration from another African country during wars and tribal conflicts arising from royal battles and land ownership. This migration to exile led to the birth of subtribes, tribes and countries. While in exile, diseases and infections were treated either with plants found in their place(s) of refuge or dried MP taken from their home country.

Furthermore, this similarity often arises from the common diseases that affected certain regions of Africa. These have been the cause of high mortality. For example, malaria in West Africa, tuberculosis in the Southern African Development Community (SADC) and North Africa, dengue fever in East Africa and hepatitis in Northern Africa. Although, treatment for these diseases have been developed in clinical practice, MP endemic to Africa offer varieties of bioactive compounds capable of addressing such health issues at an affordable price. In addition to affordability, TM is more accessible, as compared to orthodox medicine. Hence, more preferred by most African people, especially those living in the rural areas.

## 2 Africa, a Continent Blessed with Plants of High Medicinal Value

Africa is blessed with a vast amount of plant resources with an estimated number of species between 40,000–45,000, out of which some 5000 species are used medicinally (Mahomoodally 2013). Several patents have been developed from MP endemic to Africa and formulations (tinctures, teas and infusions) derived from African MP are utilized both across Africa and the developed world. Some of the highly utilized MPs constitute a crucial part of African pharmacopeias. *Acacia senegal*, *Artemisia herba-alba*, *Aspalathus linearis*, *Centella asiatica*, *Catharanthus roseus*, *Cyclopia genistoides*, *Momordica charantia*, and *Pelargonium sidoides* are some MP with commercial importance that form part of the African herbal pharmacopeia (Mahomoodally 2013). Apart from the above species, there are also several other plants that are highly valued by African TM and revered as export materials which add to the economy of several African nations. Some of these species are endemic (e.g. *Acalypha* sp.) or are commonly found in specific regions of Africa (e.g. *Aloe* ssp.). The Aloe species exist abundantly in SADC countries and are also medicinally utilized extensively in these countries.

## 3 Aloe Species, an Integral Part of the Southern African Traditional Medicine System

*Aloe ferox*, popularly known as Cape Aloe is revered in South African TM system for its high medicinal value. It is regarded as the most common *Aloe* species in South Africa and native to South Africa and Lesotho. It has been used since ancient times and is one of the few plants found in San rock paintings (Mahomoodally 2013). Globally, it is a valuable raw material used in the pharmaceutical, food, and cosmetic industries. Both aloe production and tapping form an important means of livelihood in some rural communities of Southern Africa. The finished product obtained from aloe tapping, aloe bitters, has remained an important South African export product since 1761 when it was first exported to Europe (Mahomoodally 2013).

Till date, it has remained one of the highly utilized MPs within SADC countries and beyond. *Aloe* is harvested extensively and used in various ways in Zambia and Zimbabwe (Shumba et al. 2009) and other Southern African countries. The fleshy leaf is the main part of the plant utilized for medicinal and cosmetic purposes. It is sliced open and the fresh gelatinous juice is applied topically for the treatment of wounds, rashes, skin irritation and burns or can be boiled and drunk as a cough remedy (Shumba et al. 2009). Apart from the medicinal value of Aloe, it is also used in preparing young girls for marriage ceremony (Shumba et al. 2009). Wedding is an important traditional ceremony among Africans.

## 4 Medicinal Plants Used During Wedding Ceremonies in the West African Culture

In West Africa, traditional wedding takes place before the court wedding as the couple are regarded as husband and wife once the traditional wedding has taken place. Traditional wedding is done in a unique way in West Africa. During the event, the elders of both families will be present together with the parents of the couples' or their representatives if they are deceased. The use of certain MPs during the ceremony is common in almost all West African countries, as they constitute an important part of the culture of these countries. Kola nuts (*Cola nitida*), nuts of bitter kola (*Garcinia kola*) and alligator pepper (*Aframomum meleguata*) are frequently used during traditional weddings across West Africa.

The genus *Cola* belongs to the family Sterculiaceae which is one of the most important genera of this family. It is endemic to tropical Africa and has its greatest diversity in West Africa (Asogwa et al. 2006). Kola nut is highly revered during traditional ceremonies in West African culture. It has a unique place amongst West Africans as a result of its specific importance in the social life and religious customs of the people in the tropics of West Africa (Adebayo and Oladele 2012). It is an important commodity presented before and during traditional wedding ceremonies in West African culture. When the family of the groom sends Kola nuts to the family of the bride and it is accepted, this act symbolizes their support and readiness for the union of the couple. During the wedding ceremony, Kola nut is broken to seal the friendship between the two families and blessings are pronounced on the couple. In some West African communities, it is believed that marriages would crash if Kola is not presented during the wedding ceremony.

Bitter kola (*Garcinia kola*) is an indigenous medicinal plant found in rain forest of Central and Western African countries such as Benin Republic, Cameroon, Democratic Republic of Congo (DRC), Cote d'Ivoire, Gabon, Ghana, Liberia, Nigeria, Senegal and Sierra Leone (Esiegwu et al. 2014). *Garcinia kola* is cultivated in West African countries for its edible fruit and seeds which serve as rejuvenating agent for masticatory purposes (Ibiblio 1983). It is popularly chewed in Nigeria for nervous alertness and induction of insomnia (Esiegwu et al. 2014). The nut of bitter kola is mostly eaten to stimulate the flow of saliva (Omode et al. 1995; Leakey 1999) and, hence, are widely traded in the market.

The dried nut of bitter kola is ground and mixed with honey to make a mixture used for cough treatment (Esiegwu et al. 2014). In West Africa, *Garcinia kola* dried peels are used to treat cuts and sore throats, roots soaked in gin for tooth decay, cough and gonorrhoea, tea (steamed leaves) for stomach ache, hepatitis, respiratory tract inflammation, diarrhea and as an aphrodisiac, seed (nuts) for liver cirrhosis, dried seed for asthma and fresh fruit as a broncho-dialator (Lewis 2001). The roots of the plant are commonly used as chewing stick which are mostly sold in local markets across West Africa (Iwu 2014).

*Aframomum melegueta* (grains of paradise, alligator pepper, melegueta pepper) is an aromatic plant with edible spicy fruit. It is a tropical, herbaceous, perennial plant belonging to the family Zingiberaceae (Doherty et al. 2010). The seeds have a pungent peppery taste as a result of its aromatic ketones (Galal 1996). *Aframomum melegueta* has both medicinal and nutritive values and is commonly found in rain forests (Doherty et al. 2010). The plant is extensively utilized in West and Central African TM systems (Iwu 2014). It is widely spread across tropical African countries such as Nigeria, Ghana, Cote D'Ivoire, Liberia, Cameroon, Sierra Leone, Ghana and Togo (Doherty et al. 2010). In the southern part of Nigeria, it is chewed with Kola nut as a masticatory agent. The fresh fruit is used as an aphrodisiac and leaves for measles and leprosy (Iwu 2014). Alligator pepper is used in the production of cosmetics and perfumes (Burkill 1985). During various traditional ceremonies in Africa, bitter kola and alligator pepper are presented together with Kola nut.

## 5 Wonder Nut (*Cola nitida*) and Its Importance in the Cultural, Social and Everyday Life of the African People

Kola nut is presented when welcoming important guests, elders or a chief in Nigeria. The significance of Kola nut among the Igbo people of Nigeria are: ushering in a new day, divination, oath taking and welcoming of guests. Breaking of Kola nut is believed to usher in peace and unity. During settlement of disputes between family members or communities involved in land disputes, kola is presented to the representatives of the families or communities before the meeting is commenced. The nut when chewed helps to stay awake, quenches thirst and hunger. For this reason, it is widely used in West African countries as an alternative to coffee drinking.

As a result of its ability to suppress sleep, kola nut is now popularly consumed by higher institution students in West Africa. In the olden days, Kola nut was used in exchange for money in some West African countries. The nut is also of great economic importance as it is a cash crop of commercial importance which is cultivated for local consumption and export purposes like cocoa. The produce is consumed locally and exported outside Africa.

There are other endemic plants used during ceremonies and festivals in Africa such as Henna which is known globally for its use in the cosmetic industry.

# 6 Legendary Plants from North Africa

## *Henna (*Lawsonia inermis*): Its Cultural, Traditional Importance and Its Cosmetic Use in Africa*

The Henna plant, which is used across Africa during wedding ceremonies, is also utilized in the cosmetic industry. Henna is known globally for its use as hair dye and stain for the skin and nails. *Lawsonia inermis Linn* (Henna) belongs to the family Lythreaceae with the genus *Lawsonia* as its sole member. It is a glabrous branched shrub with small, sub-sessile, greenish brown leaves and small, red or rose color (Chauhan and Pillai 2007), scented flowers (Simon et al. 1984). It tolerates extreme heat and long droughts and grows wild in semi-arid regions (Kidanemariam et al. 2013). It is a perennial shrub native to North Africa (Vasudevan and Laddha 2003) and it is naturalized and cultivated in other parts of Africa where it is utilized for cultural and medicinal purposes.

The species is sometimes classified as *Lawsonia reba* with height up to 6 m (Simon et al. 1984). The leaves of the plant contain a dye which is used in making patterns on the hand and nails. The chemical compound responsible for the colouring effect of the dye is contained by the leaves. It is known as Hennotannic acid or Lawsone (2-hydroxy-1, 4-napthoquinone) (Kidanemariam et al. 2013). According to Semwal et al. (2014), Henna can be applied to the hands and feet to prevent fungal infections and to the hair to combat lice and dandruff.

Lawsone levels of Henna is determined by the soil and its moisture content. Dry, hot, iron bearing soils produce Henna with high Lawsone levels (Kidanemariam et al. 2013) while moist, fertile soils produce Henna with a lower level of Lawsone (Yogisha et al. 2002). Bridal Henna nights remain a popular tradition in Northern African countries. In Algeria, the bride's mother-in-law paints the Henna on the bride's hands as a tradition. In Ethiopia (Kidanemariam et al. 2013), Sudan, Eritrea,

Nigeria (Northern Nigeria), Somalia and several other African countries, *L. inermis* is traditionally used to make patterns on the hands, feet, nails and colourings on the hair for occasions such as weddings and religious festivals. In the last fifteen years, Henna body art has, recently, become very popular, spreading across the globe from its areas of origin, and has changed from being just a traditional bridal and festival adornment to an exotic fashion accessory (Semwal et al. 2014).

Henna dye is prepared by mixing a mildly acidic liquid with powdered Henna leaves to prevent oxidation of Lawsone, and left to soak for several hours, (Pradhan et al. 2012). The product is then applied with tools which place the wet paste in contact with hair, nails or skin (Semwal et al. 2014). The colourant penetrates, stains, and binds to keratin in the skin, nails and hair, without heating and the wet paste is then scraped or rinsed off the skin (Semwal et al. 2014). A high percentage of the Lawsone remains in the outermost, dead skin layer, while about 1% moves into the blood-bearing, living layers and carried away by urine (Kraeling et al. 2007). Its flowers are scented and are used to extract perfume which is used as base for local scents.

*Lawsonia inermis* (Henna) is also a medicinal plant (Mitscher et al. 1972; Cajkovac et al. 1996; Singh and Singh 2001; Azaizeh et al. 2003) as its leaves are used in the treatment of wounds, anaemia, ulcers, lumbago, rheumatagia, inflammations, bronchitis, diarrhoea, leucoderma, scabies, boils, haemorrhages, cough, fever and dysentery (Shihata et al. 1978; Vaidyratnam 1995). It is used for alleviating jaundice, venereal diseases, skin diseases, smallpox and spermatorrhoea (Chaudhary et al. 2010). An infusion of the flowers, when applied to bruises can facilitate healing processes (Chaudhary et al. 2010). It is used in the treatment of liver and digestive disorders, diabetic foot disorders, reduction of tissue loss in leprosy, and ulcers (Semwal et al. 2014). Apart from Henna, there are several other valuable MP endemic to Northern Africa.

## *Highly Utilized Medicinal Plants Across the Northern African Countries*

Several plant species endemic to Northern African countries are utilized in these countries for similar ailments/diseases and form an integral part of their TM systems. *Ephedra alata* is used in Algeria, Libya, and Morocco to treat asthma, hypertension, headaches and pulmonary diseases (Boulos 1983). *Lepidium sativum* seeds crushed with honey is used as a treatment for cough and pulmonary affections, and skin problems (De Natale and Pollio 2012). The leaves are effective in the prevention of scurvy and are often used as a condiment (Boulos 1983; Bellakhdar et al. 1991). The milk infusion of the seeds is used to cure migraines (De Natale and Pollio 2012). Seeds boiled in oil are used to treat diarrhea while the powdered seeds are applied topically to treat skin ulcers and warts (De Natale and Pollio 2012). The seeds also form part of a preparation used to treat sexual impotence (Merzouki et al. 2000).

*Globularia alypum* branch and leaf decoctions are prepared for the treatment of intermittent fevers, arthritis and rheumatisms (Boulos 1983; Bellakhdar et al. 1991). The decoction of the aerial parts is used for constipation, fever, and mycosis (Hammiche and Maiza 2006). *Cyperus rotundus* tubercles are used to cure a wide range of infections (Boulos 1983). In Sudan, a tuber decoction is used to treat stomach problems (El-Kamali and El-Khalifa 1999). Its stem galls is mixed with leaves of *Lawsonia inermis*, powdered, kneaded with water and applied as a hair tonic (Merzouki et al. 2000). In Morocco, the plant is used as a cosmetic for hair care (Bellakhdar et al. 1991).

*Balanites aegyptica* leaf decoction is used for medicinal purposes in the central Sahara (Algeria) and in other Northern African countries, as an anthelminthic and for skin infections (De Natale and Pollio 2012). It is also used as a cure for stomach, liver, pulmonary and spleen infections (Boulos 1983; Hammiche and Maiza 2006). In Sudan, the fruit is used as a purgative and a bark decoction against jaundice (El-Kamali and El-Khalifa 1999). It is also used as a purgative and for bilharzias (Hussein and Svendsen 1981).

## *Marula, a Revered South African Native*

Marula tree (*Sclerocarya birrea*) is a member of Anacardiaceae family (Pretorius et al. 1985). It is distributed across Africa with its southern-most location in the lowlands of KwaZulu-Natal (South Africa) and extends northwards through tropical Africa into Ethiopia and Sudan (Van Wyk 1974). In the aspect of domestication and commercialization, Marula has received most attention of all the fruit trees endemic to South Africa (Shackleton 1996). It is among the most valued of Southern African's trees as a result of its multipurpose use and its edible fruit. There is currently an increasing demand for *S. birrea* fruit in various parts of Southern Africa and beyond. As a result of the high and increasing demand of Marula beer and nut oil the plant have attained high economic value and serve as source of livelihood in many rural communities within SADC. *Sclerocarya birrea* is highly valued for its nutritional and financial benefits to several poor communities within SADC.

The Tonga people celebrate feast of the First Fruits by pouring drink offering of fresh juice from the fruit on the tombs of dead chiefs (Palgrave 1983). The pulp of the fruit is considered delicious and the large nut is also edible (Eloff 2001). Some tribes such as the Pedi also make delicacies from the leaves (Fox and Young 1982). Traditional Zulu healers wash themselves in the bark decoctions of Marula before commencing treatment of patients with gangrenous rectitis and at the same time use the decoction to treat the patient (Bryant 1966). To treat dysentery and diarrhoea in some parts of Southern Africa, 300 ml doses of Marula bark decoction is taken (Watt and Breyer-Brandwijk 1962).

The bark of Marula has also been used for the treatment of proctitis (Eloff 2001) and the Vhavenda people use it to treat fevers and ulcers (Mobogo 1990). The roots of the plant are used for many purposes including sore eyes in Zimbabwe

(Gelfand et al. 1985). In East Africa, Marula is an ingredient of an alcoholic medicine used for treating an internal ailment known as kati while its bark is used for stomach disorders (Kokwaro 1976). The Hausa people in West Africa use a cold infusion of the bark together with native natron as a traditional remedy to treat dysentery (Oliver-Bever 1986).

Products such as beer, juice, jam and jelly have been developed from the mesocarp of Marula fruit and are successfully marketed (Burger et al. 1987) within South Africa and beyond. Ripe fruits of *Marula* can also be consumed by biting or cutting through the thick leathery skin and sucking the juice or chewing the mucilaginous flesh after the skin has been removed (Viljoen et al. 2008). The leaf and bark infusions are often used to treat diarrhoea, boils and malaria while the oil from the seeds is used to preserve meat and utilized as a cosmetic (Shone 1979).

## **Acalypha** *species Utilized in the African Traditional Medicine Systems*

The genus *Acalypha* belong to the family Euphobiaceae and is widespread across Africa. About 100 species of *Acalypha* are known from Africa (Levin et al. 2007). In the African TM, *Acalypha* species is of high relevance (Adesina et al. 2000; Akinyemi et al. 2005; Wiart et al. 2004; Motsei et al. 2003). *Acalypha manniana* is a slender climbing shrub that grows to about 6 m high in marshy places and found in scattered locations from Rwanda, Burundi, Uganda, across northern DRC, Cameroun, Nigeria, Benin, and Ghana (Noumedem et al. 2013) while *A. guineensis* is restricted to the mountains of Upper Guinea (Levin et al. 2007).

## *Medicinal Uses of* **Acalypha manniana,** *and* **A. wilkesiana** *in Africa*

Traditional healers found in the Western region of Cameroon, use decoction of the leaves for the treatment of mycosis and various skin diseases (Noumedem et al. 2013). A leafy stem decoction of *A. manniana* is taken in some African countries such as Ivory Coast, Ghana, Uganda, Rwanda, Burundi and Cameroon to treat diarrhoea (Schmelzer 2007; Schmelzer and Gurib-Fakim 2008). *Acalypha wilkesiana* is used in West Africa for the treatment of headache and cold while the cold leaf extract is popularly used in Nigeria to bath babies with skin infection (Adesina et al. 2000) such as eczema.

## 7 Aromatic Plants, an Integral Part of Human Day to Day Life

The appeal of scented flowers and leaves and their by-products are as irresistible today as it was in the first fragrant plot, the "Garden of Eden" (Allardice 1992). A garden should appeal to all senses especially the sense of smell and sight (La Croix 1984). Aromatic plants have been used for innumerable purposes among which are food and drink flavouring, scenting of clothes and to uplift the spirits of humans for many centuries (Allardice 1992). Aromatic plants possess volatile substances in the form of essential oils in one or more parts of the plant (Skaria et al. 2007). Many essential oils are used commercially for the production of perfumes (La Croix 1984).

The use of fragrances for pleasure is very old; an Egyptian papyrus from around the sixteenth century BC gives recipes for perfumes, cosmetics and deodorants (La Croix 1984). Perfume was integral to ancient Egypt civilizations and the perfumes were derived from aromatic plants (Allardice 1992). The Egyptians used perfumes as an offering to their Gods and the aromatic products were prepared and kept in a small room in their temples (Skaria et al. 2007). According to Skaria et al. (2007), incense and aromatic oils were used in religious ceremonies and to ward off evil spirits.

The oil of cedar wood collected from the dense forests of Mount Lebanon was in much demand by the Egyptians who used it for religious rites, fragrant cosmetics and ointments and they believed that cedarwood was imperishable and could preserve life (Allardice 1992). The Hebrews got the knowledge about aromatic perfumes from their countrymen who were held as prisoners by Egyptians (Skaria et al. 2007). In Elizabethan times and earlier, floors were strewed with aromatic herbs such as *Acorus calamus*, chamomile, thyme, hyssop and lavender (La Croix 1984). For several thousands of years, the "Incense Road" between Arabia and Egypt was travelled by the wealthy spice caravanserai and the rewards for a successful incense merchant were great and the entire economy of the kingdom of Sheba in the southern part of Africa depended on the sale of Frankincense and Myrrh (Allardice 1992). In the Middle East, Egypt is one of the most important producers of aromatic and medicinal crops (Elsabawy 2012). The Egyptians also had knowledge about the efficacy of aromatic plants in healing diseases and this pool of knowledge was applied in medical care (Skaria et al. 2007).

When plague was a fact of life, lice and fleas were man's constant companions and waste was disposed through the nearest open window, scented flowers were used for their antiseptic qualities and also to mask the odour (Verey 1981). In an attempt to stop the spread of plague in Athens, Hippocrates fumigated streets with aromatic plants (Skaria et al. 2007). Herbs were grown and utilized not only for their scents but also for their medicinal uses (La Croix 1984) and some of them have remarkable application as therapeutic agents in pharmaceutical and drug industries (Skaria et al. 2007). A rhythm exhibiting calmness is produced when essential oil with a sedative effect is inhaled as it stimulates the sense of smell and the emotional centre of the brain (Skaria et al. 2007). In light of this, aromatherapy has become a lucrative business and the production and export of essential oil from plants is on the increase in Africa.

## 8 Northern Africa at the Forefront of Aromatic Plant Production and Export

"The breeze of essential oils of the world has blown from yesterday's narrow definition as a symbol of luxury to indispensable necessities for both the rich and poor from the cradle of infancy to the silence of the grave" (Skaria et al. 2007). In ancient Egypt, women used cinnamon oil (from Nepal) combined with honey, myrrh and almonds to make a perfume for the feet and legs (Allardice 1992). At present, Egypt is one of the leading African exporters of aromatic plant material to the international market (Tsey 1997). In Morocco, sale of aromatic MP is a very lucrative business as a result of which 14,468 tons of essential oils and various plant material worth Dh 300,000 is exported annually (Kouhila et al. 2002). The highly utilized plants are *Rosmarinus officinalis*, from which 60 tons of essential oil are extracted and exported, Thyme and Lavender species, *Artemisia herba-alba*, *Mentha pulgemium*, *Origanum compactum* and *Coriander sativum* (Montanari 2004).

In Sudan, an incense mixture called bakour, composed of gum, exudates, balsam and dried plant material which is burnt in a traditional clay incense burner to give a fragrant smoke believed to repel insects, clean air and make the environment smell pleasant (Khalid et al. 2012). In Sudan, there is a unique tradition of preparing several different local perfumes of which the basic compounds are Khumra, which is prepared from the seeds of *Prunus mahleb* and *Santalum album* powder (Khalid et al. 2012). Khumra can be classified as Khumra zeit which is oil-based, Khumra dufra includes shells and Khumra misk contains extracts of crocodile sweat glands (Khalid et al. 2012).

In Algeria and Tunisia, *Thymus algeriensis* is used as a culinary herb (Chaieb and Boukhriss 1998). *Thymus algeriensis* which is endemic to Lybia, Tunisia, Morocco and Algeria (Houmani et al. 2002), is an herbaceous fragrant plant largely used, fresh or dried, as a culinary herb (Chaieb and Boukhriss 1998). This plant is widely used in TM against illnesses of the digestive tube (Le Floch 1983). Globally, Morocco is the ninth largest exporter of medicinal and aromatic plants (Ozhatay et al. 1997). Egypt and Morocco are among two major countries from Africa involved in the export of essential oil into the European Union (EU) (Montanari 2004). In SADC, South Africa is one of the leading countries involved in the growth, sale and export of indigenous aromatic plants and their products. The Rutaceae family which has a high percentage of species endemic to South Africa is among the family with species which their essential oil and products are marketed locally.

## 9 *Buchu*, a Genus of Dynamic Aromatic Plants of the South African Flora

The family Rutaceae to which citrus belong to are easily identified on the field as a result of dotted oil glands on the reverse side of their leaves. As a result of this and scents associated with species of the Rutaceae, several species are utilized in the production of perfumes, flavors and spices. In South Africa, members of the Rutaceae, most especially species of the genus *Agathosma* are referred to as *Buchu*. '*Buchu*' (*Agathosma* species) has been an indispensable part of the San and Khoi traditional healing culture in the Cape and is still being used for this purpose throughout South Africa (Moolla et al. 2007).

The San people used aromatic plants lubricated with fat to keep their skin soft and moist in the desert climate, and as an antifungal agent to promote good health through the uptake of aromatic substances by the skin (Simpson 1998). *Agathosma ovata* is still used for medicinal purposes amongst the Cape people (Van Wyk and Gericke 2000). The South African indigenous plant, *Agathosma betulina,* is among the most popularly used medicinal plant in the family Rutaceae in the South African TM system. This has resulted in the local cultivation of these plants by farmers, indigenous nurseries and National conservation bodies.

Essential oil of *Agathosma betulina* and other *Agathosm*a species are marketed by several local and international natural product companies. *Buchu*, *Agathosma betulina* oil is used for medicinal and cosmetic purposes (Van Wyk et al. 1997). Globally, Buchu is highly sought after for its essential oil which has a high diosphenol content (Moolla et al. 2007). Moolla and Viljoen (2008) reported that a kilogramme of *A. betulina* oil sells on the international market for about 700 Euro.

Irrespective of the vast amount of medicinal and aromatic plants Africa is blessed with and uniqueness of its TM system, the future of Africa's TM is dependent on its integration into the healthcare system of African countries and the formulations of policies which will encourage its growth. Moreover, its sustainability is also dependent on the incorporation of health policies which can support the growth and documentation of TM and MP used for specific health conditions.

## 10 A Transition in the Acceptance of African Traditional Medicine from the Days of the Colonial Masters to the Present Age

In Africa, during colonisation; colonial governments connected TM Practitioners (TMPs) to the use of witchcraft, and TM was discredited and placed on legal bans (Stangeland et al. 2008). TM was outlawed (Bodeker 1994; Obbo 1996; Svarstad and Dhillion 2000) and in several African countries, modern medicine was officially recognised while TM and its practitioners were not tolerated (Tabuti et al. 2003). When colonization ended, independence brought about a more tolerant disposition towards TM, which helped to regain African identity and cultural values (Stangeland et al. 2008). The passion of African leaders and world health organization (WHO) towards the development of TM cannot be underestimated (Kasilo et al. 2010). This is evident by the declarations made at African Summits of Head of States and Government such as the declaration of 2001–2010 as the Decade for African TM, and the institution of African TM Day for Advocacy by WHO (Kasilo et al. 2010).

Throughout the Third World, TM is being promoted by WHO and other international agencies (WHO 1978; Leslie 1980; Jingfeng 1987; Launs 1989; Chavunduka 1994). The Inter-African Experts Committee on African TM and MP in the Organization of African Unity (OAU) led to proclamation of the Decade of African Medicine from 2001 to 2010 (Mahunnah 2002), while the African TM Day was set to be on the 1st of September (Stangeland et al. 2008). The facilitation, recognition, acceptance, development and institutionalization of TM are part of the key priorities of the African Union action plan (Kasilo et al. 2010).

## 11 Progress Made so far by African Countries

As at 2010, few countries have developed national policies, legal frameworks, and codes of ethics regulating the practice of TM (WHO Regional Office for Africa 2010). Diverse forces are responsible for the integration of services of Traditional Health Practitioners (THPs), they include the cultural acceptance of TM, which majority of African population use to cater for their primary health care needs (Kasilo et al. 2010). Allopathic health care providers in some African countries such as Ghana, Senegal, Burkina Faso, Kenya, Uganda, South Africa, Tanzania and Zimbabwe have been collaborating with THPs in the use of TM and HIV/AIDS prevention activities. Researchers have observed a reduction in viral load, an improvement in the clinical condition of HIV/AIDS patients, and in some cases, weight increase (Kasilo et al. 2010).

A number of countries have made good progress in the area of TM policy formulation but some of the policies are yet to be translated into implementation (Kasilo et al. 2010). In 2010, 39 of the 46 African countries already had established national offices for TM (Kasilo et al. 2010). According to an AU report (2007), Benin Republic, Burkina Faso, Cameroon, Cote d Ivoire, Equatorial Guinea, Ethiopia, Kenya, Ghana, Guinea, Madagascar, Malawi, Mali, Mozambique, Nigeria, Senegal, South Africa, Tanzania, Togo, Uganda, Zambia and Zimbabwe have enacted some laws and regulations to govern TM while Angola, Central African Republic, Chad, DRC, Equatorial Guinea, Gambia, Rwanda, Sao Tome and Principe, and Seychelles are developing laws and regulations.

TM products form a crucial part of the healthcare system of SADC countries and there is a cooperation project to establish a regional databank of MP and develop policies and legal frameworks for its practice within the region (Lezotre 2014). The progress made so far coupled with support from African leaders could make African TM accepted globally just like for Chinese TM.

**Acknowledgement** O.O. Fajinmi gratefully acknowledges J. Van Staden for his mentorship over the years.

# References

Adebayo SA, Oladele OI (2012) Medicinal values of Kolanut in Nigeria: implication for extension service delivery. Life Sci J 9:887–891

Adesina SK, Idowu O, Ogundaini AO, Oladimeji H, Olugbade TA, Onawunmi GO, Pais M (2000) Antimicrobial constituents of the leaves of Acalypha wilkesiana and Acalypha hispida. Phytother Res 14:371–374

Akinyemi KO, Oladapo O, Okwara CE, Ibe CC, Fasure KA (2005) Screening of crude extracts of six medicinal plants used in South-West Nigerian unorthodox medicine for anti-methicillin resistant *Staphylococcus aureus* activity. BMC Complement Altern Med 5:6–10

Allardice P (1992) The scented garden. A complete guide to growing and using fragrant plants. Collins Angus and Robertson Publishers, Australia.

Asogwa EU, Anikwe JC, Mokwunye FC (2006) Kola production and utilization for economic development. Afr Sci 7:217–222

Azaizeh H, Fulder S, Khalil K, Said O (2003) Ethnobotanical knowledge of local Arab practitioners in the Middle Eastern region. Fitoterapia 74:98–108

Bellakhdar J, Claisse R, Fleurentin J, Younos C (1991) Repertory of standard herbal drugs in the Moroccan pharmacopoea. J Ethnopharmacol 35:123–143

Bodeker G (1994) Traditional health knowledge and public policy. Nat Res Forum 30:5–16

Boulos L (1983) Medicinal plants of North Africa. Algonac, Michigan

Bryant AT (1966) Zulu medicine and medicine men. Struik, South Africa

Burger AEC, De Villiers JBM, Du Plessis LM (1987) Composition of the kernel oil and protein of the marula seed. S Afr J Sci 83:733–735

Burkill HM (1985) The useful plants of West Tropical Africa, 2nd edn. Pergamon Press, Richmond

Cajkovac M, Oremovic L, Cajkovac V (1996) The effect of hair colours and perm products on the state of hair. Acta Pharma 46:39–49

Chaieb M, Boukhriss M (1998) Flore suscinte et illustrée des zones arides et sahariennes de Tunisie. Association de la Protection de la Nature et de l'Environnement, l'Or du Temps (ed), Sfax, Tunisia, 42

Chaudhary G, Goyal S, Poonia P (2010) *Lawsonia inermis* Linnaeus: a phytopharmacological review. Int J Pharm Sci Drug 2:91–98

Chauhan MG, Pillai AP (2007) Microscopic profile of powdered drug used in Indian system of medicine, vol II. Gujarat Ayurved University, Jamnagar, India

Chavunduka GL (1994) Traditional Medicine in modern Zimbabwe. University of Zimbabwe Publications, Zimbabwe.

De Natale A, Pollio A (2012) A forgotten collection: the Libyan ethnobotanical exhibits (1912–14) by A. Trotter at the Museum O. Comes at the University Federico II in Naples, Italy. J Ethnobiol Ethnomed 8:1

Doherty VF, Olaniran OO, Kanife UC (2010) Antimicrobial activities of Aframomum melegueta (Alligator pepper). Int J Biol 2:126

El-Kamali HH, El-Khalifa KF (1999) Folk medicinal plants of riverside forests of the Southern Blue Nile district, Sudan. Fitoterapia 70:493–497

Eloff JN (2001) Antibacterial activity of Marula (*Sclerocarya birrea* (A. rich.) Hochst. subsp. caffra (Sond.) Kokwaro) (Anacardiaceae) bark and leaves. J Ethnopharmacol 76:305–308

Elsabawy MNE (2012) Medicinal and aromatic crops in Egypt: a study in medical geography. ARPN J Sci Technol 2:207–221

Esiegwu AC, Okoli AC, Emenalom OO, Esonu BO, Udedibie ABI (2014) The emerging nutricitical benefits of the african wonder nut (*Garcinia Kola Heckel*): a review. Glob J Anim Sci Res 2:2
Fox FW, Young MEN (1982) Food from the veld. Edible wild plants of Southern Africa. Delta Books, Johannesburg
Galal AM (1996) Antimicrobial activity of 6-paradol and related compounds. Int J Pharmacogn 34:64–69
Gelfand M, Mavi S, Drummond RB, Ndemera B (1985) The traditional medical practitioner in Zimbabwe. Mambo Press, Gweru
Hammiche V, Maiza K (2006) Traditional medicine in Central Sahara: pharmacopoeia of Tassili N'ajjer. J Ethnopharmacol 105:358–367
Houmani Z, Azzoudj S, Naxakis G, Skoula M (2002) The Essential Oil Composition of Algerian Zaâtar: *Origanum* spp. and *Thymus* spp. J Herbs Spices Med Plants 9:275–280
Hussein SM, Svendsen AB (1981) Medicinal and aromatic plants in the Sudan: usage and exploration. Fitoterapia 52:243–246
Ibiblio JO (1983) Some medicinal plants of Nigeria. Ibadan publishers Nigeria Limited, Nigeria, pp 132–133
Iwu MM (2014) Handbook of African medicinal plants. CRC Press/Taylor and Francis Group, Boca Raton/London
Jingfeng C (1987) Towards a comprehensive evaluation of alternative medicine. Soc Sci Med 25:659–667
Kasilo MJO, Trapsida JM, Mwikisa CN, Lusamba-Dikassa PS (2010) An overview of the traditional medicine situation in the African region (Decade of African Traditional Medicine, 2001–2010). African Health Moni Spec Issue 14
Khalid H, Abdalla WE, Abdelgadir H, Opatz T, Efferth T (2012) Gems from traditional north-African medicine: medicinal and aromatic plants from Sudan. Nat Prod Bioprospect 2:92–103
Kidanemariam TK, Tesema TK, Asressu KH, Boru AD (2013) Chemical investigation of *Lawsonia inermis* L. leaves from Afar region, Ethiopia. Orient J Chem 29:1129–1134
Kokwaro JO (1976) Medicinal plants of East Africa. East African Literature Bureau, Nairobi
Kouhila M, Kechaou N, Otmani M, Fliyou M, Lahsasni S (2002) Experimental study of sorption isotherms and drying kinetics of Moroccan *Eucalyptus globulus*. Dry Technol 20:2027–2039
Kraeling ME, Bronaugh RL, Jung CT (2007) Absorption of lawsone through human skin. Cutan Ocul Toxicol 26:45–56
La Croix IF (1984) Scented plants in Southern Africa. Macmillan South African, Johannesburg
Launs L (1989) Integrated medicine-a challenge to the health-care system. Acta Sociologica 32:237–251
Le Floch E (1983) Contribution à une étude ethnobotanique de la flore tunisienne. Ministère de l'Enseignement Supérieur et de la Recherche Scientifique (ed), Imprimerie Officielle de la République Tunisienne, 402
Leakey RRB (1999) Potential for novel food products from agroforestry trees: a review. Food Chem 66:1–14
Leslie C (1980) Medical pluralism in world perspective (1). Soc Sci Med Part B Med Anthropol 14:191–195
Levin GA, Morton JK, Robbrecht E (2007) Two new species of Acalypha (Euphorbiaceae) from Tropical Africa, and a review of some Robyns names for cupricolous plants from the Democratic Republic of the Congo. Syst Bot 32:576–582
Lewis R (2001) Discovery: science as a window to the world. Black Science Property Inc, USA
Lezotre PL (2014) International cooperation, convergence and harmonization of pharmaceutical regulations: a global perspective. Academic, Oxford
Mahomoodally MF (2013) Traditional medicines in Africa: an appraisal of ten potent African medicinal plants. Evid Based Complement Alternat Med 2013:1–14
Mahunnah RLA (2002) Ethnobotany and conservation of medicinal plants in Africa: the way forward in the next decade. In: Proceedings of the 15th meeting of the Inter-African Experts Committee on African Traditional Medicine and Medicinal Plant, OAU Lagos, Nigeria

Merzouki A, Ed-Derfoufi F, Mesa JM (2000) Contribution to the knowledge of Rifian traditional medicine. II: Folk medicine in Ksar Lakbir District (NW Morocco). Fitoterapia 71:278–307

Mitscher LA, Leu RP, Bathala MS, Wu WN, Beal JL, White R (1972) Antimicrobial agents from higher plants. I: Introduction, rational, and methodology. Lloydia 35:157–166

Mobogo DEN (1990) Unpublished M.Sc., thesis, University of Pretoria

Montanari B (2004) Alternative paths to commercialising medicinal and aromatic plants in a community of the High Atlas Mountains, Morocco. University of Kent, Canterbury

Moolla A, Viljoen AM (2008) 'Buchu'-*Agathosma betulina* and *Agathosma crenulata* (Rutaceae): a review. J Ethnopharmacol 119:413–419

Moolla ASF, Van Vuuren RL, Van ZYL, Viljoen AM (2007) Biological activity and toxicity profile of 17 *Agathosma* (Rutaceae) species. S Afr J Bot 73:588–592

Motsei ML, Lindsey KL, Van Staden J, Jäger AK (2003) Screening of traditionally used South African plants for antifungal activity against *Candida albicans*. J Ethnopharmacol 86:235–241

Noumedem JA, de Dieu TJ, Teke GN, Momo RC, Kuete V, Kuiate JR (2013) Phytochemical analysis, antimicrobial and radical-scavenging properties of Acalypha manniana leaves. Springerplus 2:1–6

Obbo C (1996) Healing: cultural fundamentalism and syncreticism in Buganda. Africa 66(2):183–201

Oliver-Bever BEP (1986) Medicinal plants in tropical West Africa. Cambridge University Press, Cambridge

Omode AA, Fatoki OS, Olagun KA (1995) Physico-chemical properties of some underexploited and unconventional oilseeds. J Agric Food Chem 43:2850–2853

Ozhatay N, Koyuncu M, Atay S, Byfield AJ (1997) The wild medicinal plant trade in Turkey. DHKD/WWWF. TRAFFIC, Istanbul

Palgrave KC (1983) Trees of Southern Africa. C. Struik, Cape Town

Pradhan R, Dandawate P, Vyas A, Padhye S, Biersack B, Schobert R, Ahmad A, Sarkar F (2012) From body art to anticancer activities: perspectives on medicinal properties of henna. Curr Drug Target 13:1777–1798

Pretorius V, Rohmer E, Rapp A, Holtzhausen LC, Mandery H (1985) Volatile flavour components of marula juice. Zeitschrift für Lebensmitteluntersuchung and Forschung 181:458–461

Schmelzer GH (2007) *Acalypha ornata* Hochst. ex A.Rich. In*:* Schmelzer GH, Gurib-Fakim A (eds) Medicinal plants/Plantes médicinales. Prota, Wageningen

Schmelzer GH, Gurib-Fakim A (2008) Ressources végétales de l'Afrique tropicale 11. Plantes Médicinales 1:869

Semwal RB, Semwal DK, Combrinck S, Cartwright-Jones C, Viljoen A (2014) *Lawsonia inermis* L. (henna): Ethnobotanical, phytochemical and pharmacological aspects. J Ethnopharmacol 155:80–103

Shackleton CM (1996) Potential stimulation of local rural economies by harvesting secondary products: a case study of the central Transvaal Lowveld, South Africa. Ambio 25:33–38

Shihata IM, Hassan AG, Mayah GY (1978) Pharmacological effects of *Lawsonia inermis* leaves (el-henna). Egypt J Vet Sci 15:31

Shone AK (1979) Notes on the marula. Bulletin 58, Department of Forestry, Pretoria, South Africa

Shumba E, Carlson A, Kojwang H, Sibanda M, Masuka M, Moyo N (2009) Traditional medicinal plant practice in Southern Africa: a situation analysis in Zambia and Zimbabwe. WWF-World Wide Fund for Nature (Formerly World Wildlife Fund) Harare, Zimbabwe

Simon JE, Cahadwick AF, Craker LE (1984) Herb: an indexed bibliography, 1971–1980. The scientific literature on selected herbs, aromatic and medicinal plants of the temperature zone. Archon Books Press, Purdue University, Hamden

Simpson D (1998) Buchu – South Africa's amazing herbal remedy. Scott Med J 43:189–191

Singh A, Singh DK (2001) Molluscicidal activity of *Lawsonia inermis* and its binary and tertiary combinations with other plant derived molluscicides. Indian J Exp Biol 39:263–268

Skaria BP, Joy PP, Matthew S, Mathew G, Joseph A, Joseph R (2007) Aromatic plants. New India Publishing Agency, New Delhi

Stangeland T, Dhillion SS, Reksten H (2008) Recognition and development of traditional medicine in Tanzania. J Ethnopharmacol 117:290–299
Svarstad H, Dhillion SS (2000) Responding to bioprospecting: from biodiversity in the south to medicines in the north. Spartacus Forlag, Oslo
Tabuti JRS, Dhillion SS, Lye KA (2003) Traditional medicine in Bulamogi county, Uganda: its practitioners, users and viability. J Ethnopharmacol 85:119–129
Tsey K (1997) Traditional medicine in contemporary Ghana: a public policy analysis. Soc Sci Med 45:1065–1074
Vaidyratnam PS (1995) Indian medicinal plant; a compendium of 500 Species, vol III. Botanical Survey of India, Ministry of Environment and Forests, India, New Delhi, pp 303–304
Van Wyk P (1974) *Sclerocarya caffra* trees of the Kruger National Park. Purnell, Cape Town
Van Wyk BE, Gericke N (2000) People's plants: a guide to useful plants of Southern Africa. Briza Publications, Pretoria
Van Wyk BE, Van Quadshoom B, Gericke N (1997) Medicinal plants of South Africa. Briza Publication, Pretoria, p 236
Vasudevan TN, Laddha KS (2003) Herbal drug microscopy. Yucca Publishing House, New Delhi
Verey R (1981) The scented garden; choosing, growing and using the plants that brings fragrance to your life, home and table. Michael Joseph Limited, London
Viljoen AM, Kamatou GPP, Başer KHC (2008) Head-space volatiles of marula (*Sclerocarya birrea* subsp. *caffra*). S Afr J Bot 74:325–326
Watt JM, Breyer-Brandwijk MG (1962) The medicinal and poisonous plants of southern and eastern Africa. E.S. Livingstone, Edinburgh
WHO Regional Office for Africa (2010) Decade of African TM (2001-2010). African Health Monit Spec Issue 14:4–103
Wiart C, Hannah A, Yassim M, Hamimah H, Sulaiman M (2004) Antimicrobial activity of *Acalypha siamensis* Oliv. ex Gage. J Ethnopharmacol 95:285–286
World Health Organization (1978) The promotion and development of traditional medicine: report of a WHO meeting (held in Geneva from 28 November to 2 December 1977)
Yogisha S, Samiulla DS, Prashanth D, Padmaja R, Amit A (2002) Trypsin inhibitory activity of *Lawsonia inermis*. Fitoterapia 73:690–691

# Medicinal and Aromatic Plants of the World – Africa

**Ossy M.J. Kasilo, Mawuli Kofi-Tsekpo, and Francis Gachathi**

**Abstract** The medicinal and aromatic plants of Africa are found in the continent from Cape Town to Cairo and have been used in the African people's traditional culture of health for centuries. The declaration of Alma Ata by the International Conference on Primary Health Care in 1978 was a significant turning point in the development of health systems in the developing countries in Africa. This declaration provided for incorporation of traditional medicine in health systems. Medicinal plants and aromatic plants form the major component of African traditional medicine practice and these have become very important in the development of health systems in Africa.

Therefore, in 2001 the Regional Committee of the WHO African Region adopted a regional strategy on promoting the role of traditional medicine in health systems (WHO, Promoting the role of traditional medicine in health systems: a strategy for the African Region (Strategy documents, AFR/RC50/9 and Resolution AFR/RC65/R3). WHO Regional Office for Africa, Harare, Zimbabwe, 2001). In 2013 the Regional Committee further emphasized the enhancement of the role of traditional medicine in health systems in the African Region (WHO, Enhancing the role of traditional medicine in health systems: a strategy for the African Region (Strategy documents, AFR/RC63/6 and Resolution AFR/RC65/R3). WHO Regional Office for Africa, Brazzaville, Republic of the Congo, 2013). If medicinal and aromatic plants are to be incorporated in health systems then, their identity and quality must be assured in order to establish the safety and efficacy of the medicines they are used to manufacture.

O.M.J. Kasilo (✉)
Health Technologies and Innovation (HTI), Health Systems and Services Cluster, World Health Organization, Regional Office for Africa, P.O. Box 6, Brazzaville, Republic of the Congo
e-mail: kasiloo@who.int

M. Kofi-Tsekpo
Consolata Hospital, P.O. Box 2005, 60200 Nkubu, Meru, Kenya
e-mail: Mawule.mawule.kofitse18@gmail.com

F. Gachathi
Kenya Forest Research Institute, P.O. Box 200412, 00200 Nairobi, Kenya
e-mail: gachathif@yahoo.com



M. Neffati et al. (eds.), *Medicinal and Aromatic Plants of the World - Africa Volume 3*, Medicinal and Aromatic Plants of the World,
DOI 10.1007/978-94-024-1120-1_4

This chapter contains monographs on 15 medicinal plants which are of major health and commercial importance in the world today. Each monograph provides the following information: Botanical name, Synonyms, Geographical Distribution, Description, Parts used for medicine preparation, Chemical constituents identified, Pharmacology and therapeutics and References.

**Keywords** Africa • Traditional medicine • Monographs • Pharmacology

## 1 Introduction

Medicinal and aromatic plants have, for centuries, constituted an inseparable aspect of the African traditional culture of health. They are part of the nutritive, preventive, curative, and spiritual components of the African way of life and the African traditional health systems. The documentation of these concepts of the African health systems have been given through oral traditions for centuries. However, it was only in the last century that some written accounts began to appear and such notable publications include those of Watt and Breyer-Brandwijk (1962), Iwu (1993), Kokwaro (2009) and Gachathi (2007).

*Information on African Health Systems* The information on the African health systems particularly, the one involving medicinal and aromatic plants has remained in the secret domain through oral communication up to now. This has been partly due to the abolition of traditional medicine during the colonial period and also due to the secret taboos associated with the practice of African traditional medicine. The need for some openness arose when the World Health Organization proposed the incorporation of traditional medicine in health systems.

## 2 The Role of Medicinal and Aromatic Plants in Health Systems in Africa

The declaration of Alma Ata in 1978 jointly sponsored by the World Health Organization (WHO) and the United Nations International Children's Fund (UNICEF) as decided by the World Health Assembly (WHO resolutions) brought into focus the practice of African traditional medicine and, by default, the significance of medicinal and aromatic plants on the continent. The declaration provided for the use or incorporation of African traditional medicine in the western-style health systems in order to achieve a greater health care coverage in the communities on the continent. However, this brought out some logistical challenges. The WHO Regional Office for Africa took the challenge to rationalize the incorporation of traditional medicine in health systems. This would require the establishment of quality standards for: (a) the mode of practice of traditional medicine, (b) the authentication and quality of the practitioners, and (c) the medicinal and aromatic plants used for treating patients. In this respect the WHO has been assisting the member states in the African region to carry out these rationalizations. The materials from medicinal and aromatic plants are critical elements for the rationalization of the use of the culture-based practices in contemporary evidence-based health systems.

# 3 Monographs, Pharmacopoeias, and Formularies

## *Monographs and Pharmacopoeias*

Most of the African cultures have a verbal tradition and, therefore, written information on cultural features in the past are not so readily available from Africa as from many other parts of the world. The ethnobotanical information on the uses of these plants was sometimes documented on herbarium labels and, in this way, ethnobotanical information on a number of medicinal and aromatic plants began to accumulate. Systematic accounts in written form dealing with medicinal and aromatic plants in Africa are of a fairly recent date, while reports dealing with ethnopharmacological aspects are more recent.

## *The Need for the Creation of Formularies*

It is becoming evident that formularies are required for the quality control of the preparations in African traditional medicine. The herbal preparations are almost invariably formulated from more than two or more materials from different plant materials. This practice has pharmaceutical, pharmacological and pharmacodynamics implications. The creation of official formularies are desirable from the viewpoint of drug interactions, and also in order to reduce the confounding factors that bedevil the western-trained physician who has to reconcile his diagnosis and treatment with a herbal preparation with little available information.

# 4 Some Selected Medicinal Plant Monographs

The monographs of the following 15 medicinal and aromatic plants are described below as examples of the uniqueness and richness of the African flora.

## Catha edulis *Forssk,*

Family, CELASTRACEAE

**Synonyms** *Catha forsskalii, Catha inermis, Celastrus edulis, Celastrus tsaad, Dillonia abyssinica, Methyscophyllum glaucum.* Some common names include: Miraa (Kenya); qat, kat (Ethiopia) and Bushman's tea (South Africa); and the dried leaves are also known as Abyssinian tea and Arabian tea (WHO 1980, 2006).

**Geographical Distribution** *Catha edulis* is a flowering plant which is native to the horn of Africa and the Arabian peninsula, namely, Ethiopia, Somalia, Yemen, South Sudan, Djibouti, Eritrea, Uganda and Kenya. Khat is also scattered in KwaZulu-Natal and Eastern Cape, Western Cape and Mpumalanga in South Africa as well as Swaziland and Mozambique. It has now been cultivated in other countries in Africa (WHO 1980, 2006).

**Habitat and Ecology** *Catha edulis* is an evergreen hardy tree which normally grows on mountain slopes or hill sides. It has been cultivated in a variety of climates and soils and accordingly yielding wide varieties (WHO 1980, 2006). *Catha edulis* grows often as a shrub but it can grow to over ten meters in a fertile mountainous environment at altitudes of 4000–5000 m

**Chemical Constituents** Over 40 alkaloids have been identified in the leaves of which two are of biological importance, namely, cathinone and cathine. The leaves also contain other alkaloids that include kathedulins, flavonoids, glycosides, and terpenoids (WHO 1980, 2006; Wabel 2011).

**Uses in Traditional Medicine** That is a traditional ritual plant of which the fresh young shoots and leaves are chewed. The dry leaves are also smoked (WHO 1980, 2006; Wabel 2011). The leaves are also brewed as tea.

**Biological Activities** The chewing of khat in order to experience its stimulant effects occurs over a short period because cathinone and cathine are readily metabolized into nonstimulant derivaties. Cathinone is enolisable to the enol which is easily metabolized by the cytochrome P450 superfamily. The enolic form is then reduced to cathine which metabolized. This explains why the users of khat have to chew the material continuously in order to maintain the stimulant effect. The World Health Organization therefore does not consider *khat* as an addictive drug.

It would be of interest to note that synthetic cathinones (which should be called "cathinoids") cannot undergo enolization and they therefore produce persistent hallucinogenic effects.

## Catharanthus roseus *(L.) G. Don*

Family, APOCYNACEAE

**Synonyms** *Vinca rosea*; Madagascar periwinkle; Rose periwinkle.

**Geographical Distribution** This is an ornamental herb which grows everywhere in Africa. It is particularly endemic in Madagascar where is cultivated as a commercial crop. It grows in the tropical and sub-tropical belt of the world.

**Habitat and Ecology** An evergreen, herbaceous, ornamental plant growing up to 1.5 m. There are varieties of this plant with different colored flowers: white, rose-pink, and in different shades.

**Chemical Constituents** The leaves and flowers have been reported to contain indole alkaloids, often called the vinca alkaloids, which include vinblastine and vincristine (Van der Heijden et al. 2004; Sain and Sharma 2013).

**Uses in Traditional Medicine** All parts of the plant have been used in regional herbal medicine, including the dried root, leaves, flowers and stalks. Alkaloids used in modern medicine are extracted from the whole dried plant (Van der Heijden et al. 2004; Sain and Sharma 2013; Tiong et al. 2013; Gajalakshmi et al. 2013).

**Biological Activities** The vinca alkaloids vinblastine and vincristine are useful drugs in cancer chemotherapeutics, particularly in the treatment of leukemia (Van der Heijden et al. 2004; Sain and Sharma 2013). The traditional use of the hot water extract for the management of diabetes mellitus has also been documented (Van der Heijden et al. 2004; Sain and Sharma 2013).

## Corynanthe pachyceras *K. Schum.*

Family, RUBIACEAE

**Synonyms** *Pausinystalia johimbe, Corynanthe yohimbe; Corynanthe johimbe K.Schum; Pausinystalia trillesii Beille; Pausinystalia yohimbe Beille.*

**Geographical Distribution** The plant occurs in West and Central Africa: Cameroon, Congo, Democratic Republic of the Congo, Equatorial Guinea, Gabon, Ghana, Liberia and Nigeria

**Habitat and Ecology** The tree is an evergreen tree which grows from 9 to 30 m high.

**Chemical Constituents** The bark is a natural source of the alkaloid yohimbine (Putheti and Okigbo 2008; EFSA 2009; Burkill 2010)

**Uses in Traditional Medicine** The bark extract is traditionally used for the treatment of erectile dysfunction (Putheti and Okigbo 2008)

**Biological Activities** Yohimbine is an alkaloid used for treatment of male impotence (EFSA 2009).

## Erythrina abyssinica *Lam ex DC,*

Family, LEGUMINOSAE

**Synonyms** Red hot poker tree; Lucky bean tree, *Erythrina tomentosa* R.Br. ex A. Rich.

**Geographical Distribution** The plant is common tree on the African continent.

**Habitat and Ecology** It is a medium-sized tree of height between 5 and 15 m.

**Chemical Constituents** The bark extract contains flavonoids and iso-flavonoids.

**Uses in Traditional Medicine** Antioxidant and antimalarial properties against *Plasmodium falciparum* (crude extract of the root bark); antiplasmodial activity including activity against chloroquine-resistant strains of *Plasmodium falciparum* (stem bark extracts) and a curare-like poison (seeds) (Burkill 2010; Yenesew et al. 2009; Kebenei et al. 2011; PROTA 2015)

**Biological Activities** The bark is most commonly used in traditional medicine, to treat snakebites, malaria, sexually transmittable diseases such as syphilis and gonorrhoea, amoebiasis, cough, liver inflammation, stomach-ache, colic and measles. Roasted and powdered bark is applied to burns, ulcers and swellings. The liquid from crushed bark of green stems is used to cure conjunctivitis caused by Chlamydia trachomatis (trachoma), whereas bark sap is also drunk as an anthelmintic. The bark is also applied against vomiting. Pounded flowers serve to treat dysentery. A maceration of the flower is drunk as an abortifacient, and applied externally to treat earache. Roots are taken to treat peptic ulcers, epilepsy, malaria, blennorrhagia and schistosomiasis. Leaves are taken to treat peptic ulcers; they are also used for treatment of diarrhoea. A leaf decoction serves as an emetic. Leaves are applied externally to wounds and painful joints. Fruit extracts are taken to treat asthma and meningitis (Yenesew et al. 2009).

## Gymnema sylvestre *R. Br,*

Family, ASCLEPIADACEAE

**Synonyms** *Asclepias geminate*, *Gymnema melicida*, *Periploca sylvestris*.

**Geographical Distribution** A wide distribution on the African continent.

**Habitat and Ecology** A woody climbing plant.

**Chemical Constituents** The main constituents in the leaves are oleanane-type triterpene saponins called gymnemic acids. Other constituents include anthraquinones and flavonoids (Najafi and Deokule 2011; Anjum and Hasan 2013).

**Uses in Traditional Medicine** The extracts of G. sylvestre have been used for centuries in the management of diabetes mellitus, although scientific proof has not been adduced (Najafi and Deokule 2011; Anjum and Hasan 2013).

**Biological Activities** Antidiabetic activities have of the extracts have been demonstrated in humans and laboratory animals (Anjum and Hasan 2013).

## Hypoxis hemerocallidea, *Fisch, CA Mey & Ave-Lall.*

Family, HYPOXIDACEAE

**Synonyms** *Hypoxis rooperi*, African potato.

**Geographical Distribution** Southern Africa: Botswana, Lesotho, Mozambique, Namibia, South Africa, Swaziland and Zimbabwe.

**Habitat and Ecology** This is a perennial geophyte with a tuberous rootstock.

**Chemical Constituents** Beta-Sitosterol, Ergosterol, Stigmasterol, Hypoxoside (Boukes et al. 2008; Mkhize et al. 2013).

**Uses in Traditional Medicine** Antineoplastic properties, build up the immune system of patients suffering from cancer and HIV and is effective in the treatment of inflammations and the urinary system (Boukes et al. 2008; Mkhize et al. 2013).

**Biological Activities** The extract of the tuber has been reported to be used in the treatment of various forms of cancer and for treatment of AIDS.

## Kigelia africana *(Lam.) Benth,*

Family, BIGNONIACEAE

**Synonyms** *Kigelia pinnata* (Jacq.) DC, Sausage tree.

**Geographical Distribution** This plant is widespread across the African continent.

**Habitat and Ecology** Kigelia Africana is a flowering evergreen tree growing over 20 m tall. Its fruits are cylindrical, sausage-like, about 30–60 cm long, hence its common name.

**Chemical Constituents** The following groups of chemical compounds have so far been identified: naphthaquinones, iridoids, monoterpenoids, isocoumarins, lignans, sterols, and flavonoids (Gabriel and Olubunmi 2009; Agyare et al. 2013).

**Uses in Traditional Medicine** The leaves, the fruits, the stem bark, and the roots are all used in African traditional medicine for a variety of conditions

**Biological Activities** Many therapeutic uses have been reported in African traditional medicine; these include: anticancer properties, anti-psoriasis conditions, antimicrobial properties, and anti-inflammatory properties (Saini et al. 2009; Oyedeji and Bankole-Ojo 2012; Oyelami et al. 2012).

## Maytenus buchananii *(Loes.) Wikzek.*

Family, CELASTRACEAE

**Synonyms** None.

**Geographical Distribution** The genus is endemic in the Eastern, Central and Southern Africa.

**Habitat and Ecology** An evergreen perennial shrub usually found in the forest habitat.

**Chemical Constituents** The chemical constituents include the following: Pristimerin; The ANSA MACROLIDES which include: Maytansine, Maytanprine, Maytanbutine (USA Dept of Agric 1978).

**Uses in Traditional Medicine** The roots are employed for the treatment of rheumatoid arthritis and cancer.

**Biological Activities** Anticancer properties (USA Dept of Agric 1978; Handbook of African Meduicinal plants).

## Mondia whitei *(Hook.f.) Skeels*

Family, APOCYNACEAE

**Synonyms** Common names: White's ginger; Tonic root.

**Geographical Distribution** The plant is native to all parts of Africa.

**Habitat and Ecology** The plant is a perennial, woody climber with a tuberous rootstock. The roots are aromatic.

**Chemical Constituents** The chemical compounds isolated so far include 2-hydroxy-4-methoxybenzaldehyde; methyl salicylate; beta-sitosterol; alpha-beta-amyrin acetate (Watcho et al. 2007; Malviya et al. 2011; Watcho et al. 2012).

**Uses in Traditional Medicine** Enhancement of sexual function in the male; Treatment of erectile dysfunction;

**Biological Activities** The aphrodisiac. properties have not been clinically proved.

## Prunus Africana *(Hook.f.) Kalkman,*

Family, ROSACEAE

**Synonyms** *Pygeum africanum*, Stinkwood.

**Geographical Distribution** *Prunus Africana* can be found in the forests of countries of Africa south of the Sahara. It is most common in the congo basin, eastern and southern Africa (Cunningham and Mbenkum 1993; Thorne Research Inc 2002; Kadu et al. 2012).

**Habitat and Ecology** It is an evergreen tree that is native to montane regions of and the neighboring African islands. It grows at altitudes between 900 and 3400 m, reaching 25 m high.

**Chemical Constituents** Phytosterols; Pentacyclic triterpenoids; long chain aliphatic alcohols; long fatty acids; Trans-ferulic acid esters (Thorne Research Inc 2002; Kadu et al. 2012).

**Uses in Traditional Medicine** The stem bark is used in combination with other plant materials in traditional medicine.

**Biological Activities** The bark extracts have been used for the treatment of benign prostate hyperplasia.

## Sclerocarya birrea *(A. Rich.) Hochst.*

Family, ANACARDIACEAE

**Synonyms** *Poupartia birrea* (A. Rich.) Aubrev.; *Spondias birrea* (A. Rich.), Common name, Marula.

**Geographical Distribution** The plant is native to the following countries: Botswana, Democratic Republic of Congo, Eritrea, Ethiopia, Gambia, Kenya, Malawi, Mozambique, Namibia, Niger, Senegal, Somalia, South Africa, Sudan, Swaziland, Tanzania, Uganda, Zambia, Zimbabwe.

**Habitat and Ecology** It is a medium-sized tree with a wide spreading crown and grows to 18 m high. It grows well in dry land habitat.

**Chemical Constituents** Oils from seeds.

**Uses in Traditional Medicine** The bark and fruits are used.

**Biological Activities** Antiparastic and antimicrobial properties; Gastrointestinal and antihypertensive activities (Hall et al. 2002; Wynberg et al. 2002; Mariod and Abdelwahab 2012).

## Sutherlandia frutescens *(L.) R.Br.*

Family, FABACEAE

**Synonyms** *Colutea frutescens* L.; *Lessertia frutescens* (L.) Goldblatt & JC Manning.; Cancer bush.

**Geographical Distribution** South-west of the Western Cape of South Africa.

**Habitat and Ecology** This is a small plant that grows wild as bush to about 1 m high.

**Chemical Constituents** A triterpene called SU1, Canavanine, a polyhydroxy cyclitol, pinitol; and gamma-Aminobutyric acid, GABA (Chinkwo 2005; Van Wyk and Albrecht 2008).

**Uses in Traditional Medicine** The leaves are used to prepare medicines for cancer.

**Biological Activities** Treatment of the following conditions: HIV/AIDS, cancer, diabetes, stress, inflammatory diseases (Chinkwo 2005; Van Wyk and Albrecht 2008).

## Tabernanthe iboga *(L.) Nutt.*

Family, APOCYNACEAE

**Synonyms** Iboga.

**Geographical Distribution** West-Central Africa: Gabon, Cameroon, Republic the Congo.

**Habitat and Ecology** A perennial rainforest shrub growing to 2 m, although it may under favourable conditions, grow to 10 m high.

**Chemical Constituents** Indole alkaloids, Ibogaine and derivatives (Alper 2001; Daniel 2006; Koenig and Hilber 2015).

**Uses in Traditional Medicine** Roots and stem bark are used in Bwiti cultural traditions.

**Biological Activities** Effects on the central nervous system; treatment of effects due to heroin withdrawal and similar effects of other drugs (Koenig and Hilber 2015).

## **Toddalia asiatica** *(L.) Lam.*

Family, RUTACEAE

**Synonyms** *Paullinia asiatica.*

**Geographical Distribution** This plant is native to many countries in Africa.

**Habitat and Ecology** It is woody and thorny climber reaching up to 10 m in length.

**Chemical Constituents** Benzophenanthridine Alkaloids. Nitidine, Chelerythrine, Norchelerythrine, Chelerythrine_cyanide, Avicine (Timberlake and Kativu 1999; Rajkumar et al. 2008; Nyahanga et al. 2013).

**Uses in Traditional Medicine** The roots are used in traditional treatment of malaria and sickle cell anemia.

**Biological Activities** No confirmed clinical studies have been done.

## **Voacanga africana** *Stapf,*

Family, APOCYNACEAE

**Synonyms** Voacanga.

**Geographical Distribution** The plant is native to most countries in Africa.

**Habitat and Ecology** This is a small tropical tree that grows up to 6 m in height in the forest vegetation.

**Chemical Constituents** The plant contains indole alkaloids which include voacangine and Ibogaine (Thomas and Biemann 1968; De Ruijter 2008).

**Uses in Traditional Medicine** The roots, stem bark and leaves are used in traditional medicine.

**Biological Activities** The constituents have been demonstrated to exhibit anti-inflammatory, and antioxidant effects.

# 5 Conclusions

The above account has summarized the situation analysis on medicinal and aromatic plants in Africa with particular reference to their usefulness in addressing their application in providing health on the African continent. In the short term, the account has identified some areas which would need to be addressed whilst adopting the herbal medicine for the incorporation strategy. The strategy has raised some fundamental issues on implementation. On the other hand, there is abundant evidence of the therapeutic usefulness of medicinal and aromatic plants in therapy.

## References

Agyare C, Dwobeng AS, Agyepong N, Boakye YD, Mensah KB, Patrick George Ayande P, Adarkwa-Yiadom M (2013) Antimicrobial, antioxidant, and wound healing properties of Kigelia africana (Lam.) Beneth. and Strophanthus hispidus DC. Adv Pharmacol Sci 2013: Article ID 692613, 10 pages. Available at: http://www.hindawi.com/journals/aps/2013/692613/. Accessed on 13 May 2015

Alper KR (2001) Chapter 1: Ibogaine: a review. In: The alkaloids, vol 56. Academic, San Diego, pp 1–38

Anjum T, Hasan Z (2013) Gymnema sylvestre. Plant used by peoples of Viidisha District for the treatment of diabetes. Int J Eng Invent 2:98–102

Boukes GJ, Van de Venter M, Oosthuizen V (2008) Quantitative and qualitative analysis of sterols/sterolins and hypoxoside contents of three *Hypoxis* (African potato) spp. Afr J Biotechnol 7(11):1624–1629

Burkill HM (2010) The useful plants of West Tropical Africa, complete set of volumes 1–6, 2nd edn. Royal Botanic Gardens, Kew. ISBN-13:9781842461327

Chinkwo KA (2005) *Sutherlandia frutescens* extracts can induce apoptosis in cultured carcinoma cells. J Ethnopharmacol 98:163–170

Cunningham AB, Mbenkum FT (1993) People and plants working paper 2 May, Geographical distribution and values of Prunus Africana, pp 4–11

Daniel M (2006) Medicinal plants. Chemistry and Properties. Science Publishers, – Science – 250 pages

De Ruijter A (2008) VOACANGA AFRICAN A Stapf. In: Medicinal plants – vol 1 – Page 623. Protologue J Linn Soc Bot 30:87 (1894). Available at: https://books.google.cg/books?isbn=9057822040. Accessed on 14 May 2015

European Food Safety Authority (2009) Guidance on safety assessment of botanicals and botanical preparations intended for use as ingredients in food supplements

Gabriel OA, Olubunmi A (2009) Comprehensive scientific demystification of Kigelia Africana: a review. Afr J Pure Appl Chem 3(9):158–164

Gachathi FM (2007) Kikuyu botanical dictionary, Revised 2nd edn. Tropical Botany, Gituamba

Gajalakshmi S, Vijayalakshmi S, Devi Rajeswari V (2013) Pharmacological activities of Catharanthus roseus: a perspective review. Int J Pharm Bio Sci 4(2):431–439

Hall JB, O'Brien EM, Sinclair FL (2002) Sclerocarya birrea: a monograph. School of Agricultural and Forest Sciences Publication Number 19, University of Wales, Bangor. 157 pp (Edited by John B. Hall, E. M. O'Brien and Fergus L. Sinclair)

Iwu M (1993) Handbook of African Medicinal plants. CRC Press, Boca Raton, Feb 18, – Science – 464 pages.

Kadu CAC, Parich A, Schueler S et al (2012) Bioactive constituents in *Prunus africana:* Geographical variation throughout Africa and associations with environmental and genetic parameters. Phytochemistry 83:70–78

Kebenei JS, Ndalut PK, Sabah AO (2011) Synergism of artemisinin with – V from *Erythrina abyssinica* (Lam ex) against *Plasmodium falciparum* parasites: a potential anti-malarial combination therapy. J Med Plants Res 5(8):1355–1360

Koenig X, Hilber K (2015) The anti-addiction drug Ibogaine and the heart: a delicate relation. Molecules *20*:2208–2228

Kokwaro JO (2009) Medicinal plants of East Africa, 3rd edn. University of Nairobi Press, Nairobi

Malviya N, Jain SS, Gupta VB, Vyas S (2011) Recent studies on aphrodisiac herbs for the management of male sexual dysfunction – a review. Acta Poloniae Pharmaceutica ñ Drug Research 68(1):3–8

Mariod AA, Abdelwahab SI (2012) *Sclerocarya birrea* (Marula), an African tree of nutritional and medicinal uses: a review. Food Rev Int 28:375–388

Mkhize N, Mohantalall V, Odhav B (2013) Isolation and Quantification of β-sitosterol, ergosterol and stigmasterol from *Hypoxis rigidula* Baker var. *rigidula* and *Hypoxis hemerocallidea* Fisch., C.A.Mey. & Avé-Lall (Hypoxidaceae). Int J Sci 2:118–134

Najafi S, Deokule SS (2011) Studies on gymnema sylvestre – a medicinally important plant of the family asclepiadaceae. Trakia J Sci 9(2):26–32. Available online at: http://www.uni-sz.bg. Accessed on 14 May 2015

Nyahanga T, Jondikoa JI, Manguroa LOJ, Orwa JA (2013) Antiplasmodial and larvicidal compounds of *Toddalia asiatica* root bark. J Sci 125(5):1115–1121

Oyedeji FO, Bankole-Ojo OS (2012) Quantitative evaluation of the antipsoriatic activity of sausage tree (Kigelia Africana). Afr J Pure Appl Chem 6(13):214–218

Oyelami OA, Yusuf KO, Oyelami AO (2012) The use of Kigelia Africana in the management of polycystic ovary syndrome (PCOS). Chin Med 3:1–3

PORTA: Erythrina abyssinica (PROTA). Available at: http://uses.plantnet-project.org/en/Erythrina_abyssinica_%28PROTA%29. Accessed on 14 May 2015

Putheti R, Okigbo RN (2008) Effects of plants and medicinal plant combinations as anti-infectives. Afr J Pharm Pharmacol 2(7):130–135

Rajkumar M, Chandra RH, Kaleab Asres K, Veeresham C (2008) Toddalia asiatica (Linn.) Lam. – a comprehensive review. Pharmacogn Rev 2(4):386–397

Sain M, Sharma V (2013) Catharanthus roseus (an anti-cancerous drug yielding plant) – a review of potential therapeutic properties. Int J Pure Appl Biosci 1(6):139–142

Saini S, Kaur H, Verma B, Ripudaman, Singh SK (2009) Kigelia Africana (Lam.) Benth. – an overview. Nat Prod Radiance 8(2):190–197

Thomas DW, Biemann K (1968) The hydroxyindolenine derivative of Voacangine, a new indole alkaloid from Voacanga africana. Tetrahedron 24:4223–4231

Thorne Research Inc (2002) Monograph. Pygeum africanum. Altern Med Rev 7(1):71–74

Timberlake J, Kativu S (1999) African plants: biodiversity, taxonomy and uses: proceedings of the 1997 AETFAT congress, Harare, Zimbabwe, 581 pages. Royal Botanic Gardens, Kew

Tiong SH, Looi CY, Hazni H et al (2013) Antidiabetic and antioxidant properties of alkaloids from Catharanthus roseus (L.) G. Don. Molecules 18:9770–9784

USA DEPT OF AGRIC (1978) Review of taxonomy, distribution and ecology of Maytenus buchananii. Maurice Iwu M (1993)

Van der Heijden R, Jacobs DI, Snoenijer W, Hallard D, Verpoorte R (2004) The Catharanthus alkaloids: pharmacognosy and biotechnology. Curr Med Chem 11:607–628

Van Wyk B-E, Albrecht C (2008) A review of the taxonomy, ethnobotany, chemistry and pharmacology of *Sutherlandia frutescens* (Fabaceae). J Ethnopharmacol 119:620–629

Wabel NT (2011) Psychopharmacological aspects of *Catha edulis* (Khat) and consequences of long term use: a review. J Mood Disord 1(4):187–194

Watcho P, Djoukeng C, Zelefack F et al (2007) Relaxant effect of *Mondia whitei* extracts on isolated Guinea Pig Corpus Cavernosum. Pharmacol Online 2:44–52

Watcho P, Zelefack F, Ngouela S et al (2012) Enhancement of erectile function of sexually naïve rats by beta-sitosterol and alpha-beta-amyrin acetate isolated from the hexane extract of *Mondia whitei*. Asian Pac J Trop Biomed. S1266–S1269

Watt JM, Breyer-Brandwijk MG (1962) The medicinal and poisonous plants of Southern and Eastern Africa, 2nd edn. E & S Livingstone Ltd, Edinburgh/London

WHO (1980) KHAT: pharmmacological aspects. UN Bulletin on Narcotics, Issue 3

WHO Expert Committee on Drug Dependence (2006) Assessment of khat (Catha edulis. Forsk). 34th report. WHO Technical Report Series No. 942, pp 1–25

World Health Organization (2001) Promoting the role of traditional medicine in health systems: a strategy for the African Region (Strategy documents, AFR/RC50/9 and Resolution AFR/RC65/R3). WHO Regional Office for Africa, Harare, Zimbabwe

World Health Organization (2013) Enhancing the role of traditional medicine in health systems: a strategy for the African Region (Strategy documents, AFR/RC63/6 and Resolution AFR/RC65/R3). WHO Regional Office for Africa, Brazzaville, Republic of the Congo

Wynberg R, Cribbins J, Leakey R et al (2002) Knowledge on Sclerocarya birrea subsp. caffra with emphasis on its importance as a non-timber forest product in South and southern Africa: a summary. South Afr For J 196:67–70

Yenesew A, Twinomuwhezi H, Kiremire BT et al (2009) 8-Methoxyneorautenol and radical scavenging flavonoids from Erythrina abyssinica. Bull Chem Soc Ethiop 23(2):205–210

# Some Important Aromatic and Medicinal Plants of Morocco

**Saadia Zrira**

**Abstract** The most important characteristic of the Moroccan Aromatic and Medicinal Plants sector is its diversity and its richness. Because of the ecological heterogeneity and climatic variations, 800 of the more than 4200 plant species growing in this country are of aromatic and/or medicinal interest. This diversity is more marked when compared with other Mediterranean countries: Algeria (3150), Tunisia (2200), Libya (1800), Egypt (2100), Mauritania (1100), Lebanon (2100), Jordan (2500), Syria (2100) and Portugal (3100). In the European countries France and Spain, the number of native species is 4500 and 4900, respectively. Turkey is the country in this region that shelters the greatest number of species (9000). The Moroccan production of herbs and their extracts comes from both wild crafted and farmed species. It is estimated that only 280 species are currently evaluated. Almost 100 species are exported in the form of dried herbs for food herb trade. More than 20 species are used for the production of essential oils or other aromatic extracts intended primarily for the perfumery and cosmetic industry and for the preparation of hygienic products and the formulation of flavours.

**Keywords** Morocco • Biodiversity • Aromatic and medicinal plants • *Argania spinosa* • *Artemisia herba-alba* • *Cedrus atlantica* • *Laurus nobilis* • *Laurus azorica* • *Myrtus communis* • *Rosa damascena* • *Rosmarinus officinalis*

## 1 Introduction

Morocco is a traditional supplier of the world market in aromatic and medicinal plants. This activity concerns the exploitation of both wild-crafted and cultivated species. Several products (more than 70) are exported in the form of dried plants for the food herbs trade. More than twenty species are used for the production of essential oils or other aromatic extracts intended primarily for the perfumery and

S. Zrira (✉)
Département des Sciences Alimentaires et Nutritionnelles, Institut Agronomique et Vétérinaire Hassan II, BP 6202, Rabat Instituts, Rabat, Morocco
e-mail: Saadia.zrira@gmail.com

 
M. Neffati et al. (eds.), *Medicinal and Aromatic Plants of the World - Africa Volume 3*, Medicinal and Aromatic Plants of the World,
DOI 10.1007/978-94-024-1120-1_5

cosmetic industry and also for the preparation of hygienic products and flavour formulations (Zrira 2011a, b).

The harvesting of wild growing species represents more than 90% of national production. The number of wild-crafted species is large. Their most important representatives are rosemary, thyme, pennyroyal, Artemisia, oregano, myrtle and carob seeds.

The wild-crafted species can be classified into the following three categories:

- Plants grown in the forest, such as rosemary and myrtle. In order to exploit these resources, it is necessary to have authorisation from the Moroccan forestry administration, either after a deal or by direct negotiation. In both cases the recipient pays a royalty to rent the right to exploit an area for 3 years.
- Plants grown on local communities' land. The exploitation of this resource is subject to the same requirements and same steps. The only difference lies in the ministry responsible for the supervision; in this case it is the Moroccan Interior Ministry. Artemisia is typically exploited in this way. Plant harvesting is undertaken by the local population.
- Plants grown on private land include pennyroyal, oregano, the annual species *Tanacetum annuum*, *Ormenis*, etc. In general, harvesting is carried out by the local population without any particular administrative procedures.

The most important cultivated species are verbena, coriander, sage, mint, lavender, geranium, bergamot, citrus, basil and saffron (Zrira 2010a, 2011a, b).

In this chapter, we will give a brief description of some important species in Morocco such as: *Argania spinosa*, *Artemisia herba-alba, Cedrus atlantica, Laurus nobilis, Rosa damascena, R. centifolia and Rosmarinus officinalis.*

## 2 Argan (*Argania spinosa* L.)

The Argan tree (*Argania spinosa* (L.) Skeels; Sapotaceae is an 80-million-year-old relic tree species that has been known since the time of the Phoenicians, who established trade centers throughout the Mediterranean Sea and along the coast of Morocco, in 600 BC. It is generally accepted that all Argan trees disappeared from Northern Africa during the Quaternary glaciations, except in the Souss valley where optimum conditions for the trees' survival remained (Kenny and De Zborowski 2007). The Argan tree is the only species of the family Sapotaceae that grows in the subtropical zone. The Argan forest covers about 8280 km$^2$, mostly in the dry lowlands of the Souss valley and on the sunny mountain spurs of the Anti-Atlas. The Argan tree grows very slowly. It takes 15 years to mature. But it is extremely resistant. Argan trees can live for 150 years, and sometimes more than 200 years, and because of its deep root system, it can survive long periods of drought (Kenny and De Zborowski 2007; Morton and Voss 1987). Therefore, the Argan tree is often the ultimate warrior when the desert is encroaching on the Souss valley.

**Picture 1** Argan tree

From an ecologic and economic standpoint, the Argan tree (1) stabilizes the soil, reduces erosion, maintains moisture in the air, shelters a large variety of small and wild animals, and (2) shades domestic cultures and furnishes via its fruit the major part of the Argan forest dwellers lipid diet. The quasi-symbiotic development between the Souss valley, the Argan forest and its dwellers has resulted in an equilibrated rural micro-society that has been lasting for centuries but that is totally dependent on the Argan tree (Achhal et al. 1980).

## *Taxonomy and Distribution*

The Argan tree, *Argania spinosa* (L.) Skeels belongs to a tropical family, *Sapotaceae*, which includes about 10 genera and 600 species. Specifically endemic, Moroccan Argan tree is a fruit, forest tree whose size hardly exceeds 8–10 m whose height, dense and rounded, has spiny branches (Picture 1) (M'Hirit et al. 1998).

Leaves alternate, lanceolate, dark green on the upper surface and clear in below, are subpersistant. The hermaphrodite flowers are pretty and complete (Picture 2).

The fruit, called "Argan nut" is a green bay of variable size and shape containing a seed (Picture 2). Though raised by several authors (Boudy 1951; Boulos 1983; Dupin 1949; Emberger 1938; Rieuf 1962), taxonomy of the Argan tree has not been biosystematically studied to better define the degree of polymorphism of this species.

The flowering is in May–June. The species is drought resistant, shedding foliage and remaining in a state of dormancy for several years during prolonged drought. Production of fruits is at its maximum when the tree is 60 years old. The average yield of fruit per tree is estimated to be 8 kg per annum.

The current geographical area of Argan covers more than 800,000 ha in the southwest Morocco, extending from Safi at north to Saharan coast. This area is one of the most remarkable part of Moroccan territory, both by its flora and vegetation.

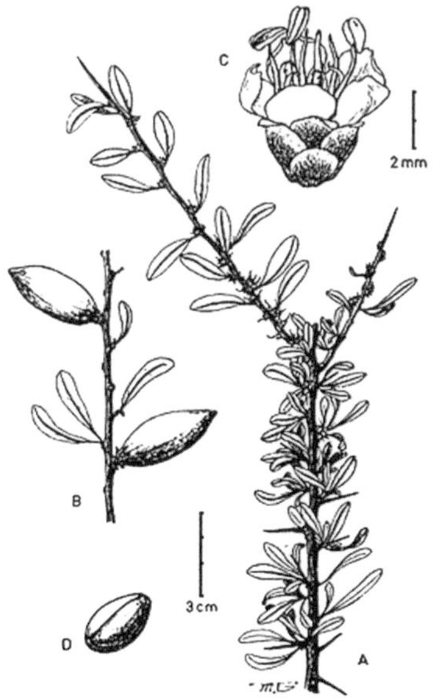


**Picture 2** Botanical characteristics of Argan tree (**a**) branch with inflorescences; (**b**) branch with fruit; (**c**) flower; (**d**) seed

In northern, two stations quite special: that of Oued Grou near Rabat, and that of the bulk of Beni Snassen near Oujda, reflect the extension of the Argan in tertiary and early Quaternary.

## Ecology

The trees cling to the slopes of rough hills and seem to thrive between the rocks on poor soil. Growing up to an elevation of 1500 m, they only need annually from 100 to 200 mm of rain. Unable to withstand cold, the Argan tree is resistant to extreme heat and drought, making it ideal for the climate of the arid southeastern Moroccan coast. The associated species in the Argania woodlands includes *Periploca laevigata, Senecio anthephorbium, Launaea arborescens, Warionia saharae, Acacia gummifera, Balanites aegyptiaca, Maerua crassifolia, Rhus trpartitum, Withania frutescens, Euphorbia officinarum, Cytisus albidus, Ephedra altissima* and *Tetraclinis articulata.*

The mean annual temperature is 18–20 °C and the mean annual rainfall is 100–500 mm. The altitude is 0–1500 m.

Soil type: It is well suited to calcareous soils, sandy deposits and relatively poor semi desert soils conditions but not drifting sands and water-logged soils (Debouzie and Mazih 1999).

## *The Argan-Oil*

The Argan tree (*Argania Spinosa* (L.)) Skeels provides a nut sized fruit containing kernels that are used to produce a range of cosmetic and edible Argan-oils available for national and export markets. Argan-oil has a high percentage of unsaturated fatty acids, which makes it of high nutritional quality (Yaghmur et al. 1999; Khallouki et al. 2003). It is relatively stable in storage (Khallouki et al. 2003; Farines et al. 1981; Chimi et al. 1994) and frying (Yaghmur et al. 2001) because of its natural antioxidants, mainly tocopherols.

In Argan-oil production, the ripe fruit pulp and peel are discarded and the nuts are carefully broken manually with stones to separate the kernels (or "almonds"). Argan-oil extracted from raw kernels by press extraction is referred as cosmetic grade while edible Argan-oil is extracted from roasted kernels. Kernel roasting is used to give the oil its desirable organoleptic characteristics and to improve oil extraction yield. Edible Argan-oil, can be produced by a traditional (manual), or a semi mechanical process. In the traditional technology, women use a multi-step process that includes kernels roasting in clay containers followed by grinding the roasted kernels using a millstone to give a brownish viscous mix. The mix is hand-worked with the concomitant addition of water to give dough, which is, and then hand pressed to give an emulsion and a cake. Oil is separated from the emulsion by decantation. This hand-press oil extraction technology is time consuming which gives low oil yields and a product with poor shelf-life (Charrouf and Guillaume 1999). According to the mechanical technology, kernels are roasted in a rotating oven using gas burners at the base of the cabinet for heat followed by the extraction of the oil by mechanical pressing. The introduction of the mechanical technology by several women cooperatives and industrial units has improved oil quality and extraction yield, and significantly reduced the time required for oil extraction (Charrouf and Guillaume 1999; Fellat et al. 1987).

### Chemical Composition of Argan-Oil

The edible Argan-oil density at 20 °C, relative to the density of water at the same temperature, ranges from 0.906 to 0.919. At the same temperature, its refractive index is 1.463–1.472, and its acid value between 0.8 and 2.5 [Service de normalisation industrielle (SNIMA) 2003]. The acid value makes it easy to distinguish extra-virgin Argan-oil, fine-virgin Argan-oil, virgin Argan-oil, and lampante Argan-oil; the acid value of extra-virgin Argan-oil must be lower than 0.8 (Service de normalisation industrielle (SNIMA 2003). The variability of the extra-virgin Argan-oil acid value has been studied as a function of its various parameters. Together with the seed origin, the extraction technology is also a parameter possibly modifying the Argan-oil acid value (Hilali et al. 2005). Comparing Argan-oil samples prepared from roasted versus non-roasted kernels, mechanically peeled versus goat-digested fruits, and fruits originating from different geographical regions, the acid values ranged between 0.15 and 0.9 (Hilali et al. 2005).

Acylglycerols, including 95% of triacylglycerols, constitute 99% of the Argan-oil. The remaining 4% are composed of monoacylglycerols (0.27–0.65%), diacylglycerols (0.68–1.53), and free fatty acids (1.1–2%) (Maurin 1992). The two main fatty acids found in Argan-oil acylglycerols are oleic acid (O) (46–48%) and linoleic acid (L) (31–35%), α mono- and α-diunsaturated fatty acid, respectively. The third and fourth main fatty acids are palmitic acid (P) (11–14%) and stearic acid (S) (4–7%) (SNIMA 2003; Hilali et al. 2005). These two latter acids are saturated fatty acids.

The major triacylglycerols found in Argan-oil include two or more oleic acid residues. Other frequently encountered triacylglycerols include two linoleic acid residues and one oleic acid moiety or two oleic acid residues and one palmitic acid moiety (Rahmani 2005).

Unsaponifiable matter constitutes 1% of Argan-oil. It is made of carotenes (37%), tocopherols (8%), triterpene alcohols (20%), sterols (29%), and xanthophylls (5%). In extra virgin Argan-oil, the levels of tocopherols are between 600 and 900 mg/kg. The main tocopherol found in Argan-oil is γ-tocopherol (between 81 and 92%); it is a strong antioxidative agent. α-, β-, and δ-tocopherols represent 2.4–6.5%, 0.1–0.3%, and 6.2–12.8%, respectively, of the total tocopherol fraction. Other phenols have also been identified as traces using mass spectroscopy; they might also act as antioxidants. These phenol derivatives include caffeic acid, oleuropein, vanillic acid, tyrosol, ferulic acid, syringic acid, catechol, resorcinol, (−)-epicatechin, and (1)-catechin (Rojas et al. 2005; Charrouf and Guillaume 2008). Five ubiquitous triterpene alcohols have also been identified: tirucallol (27.9%), β-amyrine (27.3%), butyrospermol (18.1%), lupeol (7.1%), and 24-methylene cycloartanol (4.5%) (Charrouf and Guillaume 2008).

Edible Argan-oil has a rich aroma and flavor that confer it a high culinary value as a seasoning and cooking oil. Pyrazines, aldehydes, ketones, hydrocarbons, alcohols, pyrroles and furans were the main aroma compounds of commercial edible Argan-oils (Zahar et al. 2007). In addition, qualitative and quantitative differences in aroma profile were observed between commercial oils and these differences were attributed to differences in the roasting step and/or extraction techniques used in their preparation. In a study of the effect of oil extraction on the quality of edible Argan-oil during storage, Matthaus et al. (2010) found no change in the sensory characteristics of Argan-oil obtained by mechanical extraction after 20 weeks at 20 °C; however, the oil obtained by traditional extraction, developed a Roquefort cheese taste after 12 weeks of storage.

These authors found also that sensory attributes of oil from unroasted seeds changed very fast and within 4 weeks attributes like moldy/musty and Roquefort cheese taste appeared. Hilali et al. (2005) found that the processing method had no, or little, influence on the overall oil composition (fatty acid profile, antioxidants, pigments, etc.) and that both press-extracted and traditionally prepared oil had similar compositions. However, oils obtained from roasted seeds had lower acidity values and α-tocopherol contents than oils extracted from non-roasted seeds.

Zahar et al. (2011) found that aroma compounds identified in Argan-oil prepared from unroasted kernels (cosmetic grade) were limited in number and dominated by alcohols and few aldehydes and ketones. Volatile compounds identified in edible Argan-oils prepared from roasted kernels were derived primarily from Maillard reaction, lipid oxidation, sugar degradation and/or their interactions. Edible oils prepared by the traditional manual process or the semi-mechanical process had similar aroma profiles but they showed differences in aroma compounds levels. The oil obtained by the semi-mechanical process contained significantly ($p < 0.05$) higher levels of pyrazines, aldehydes and ketones than the edible traditional oil. No significant differences were observed for furans, furanones, pyrroles, sulfuric compounds and terpenes between the two edible oils.

The olfactometric profile of cosmetic grade oil revealed mainly fruity/green/herbal notes. In edible Argan-oils, in addition to odours detected in the cosmetic grade oil, roasty/toasty/nutty and burnt sugar/caramel notes were also present. Traditional oil presented more fruity and burnt sugar/caramel notes than the oils obtained by the semi-mechanical process.

### Medicinal Properties of Argan-Oil

The Argan-oil, extracted from Argan-tree fruits, has been known for its various pharmacological properties and used in traditional medicine like a natural remedy since several centuries. Argan-oil is traditionally used for skin, nail, and hair care, cooking, massage and healing. Its chemical composition highlights the interest of many laboratories to use it in their best selling products. The remarkable properties of Argan-oil evaluated by numerous laboratories are: restoration of the skin water-lipid layer and an increase of nutrients in the skin cell, stimulation of intracellular oxygen, neutralization of free radicals, and protection of the conjunctive tissue. Recently, various studies were realised *in vitro* or on human and animal models suggesting that Argan-oil could play a beneficial role in cardiovascular prevention and its consumption could protect against atherosclerosis *via* a variety of biological mechanisms. Virgin Argan-oil is characterized by high contents of antioxidants and mono and polyunsaturated fatty acids. These data, in addition to its cosmetic benefits give to Argan-oil a unique value as natural source with high economic value (Adlouni et al. 2009).

## 3 *Artemisia herba-alba* Asso.

## *Botanical Description*

*Artemisia herba-alba* is an herbaceous plant with woody and branched stems, 30–50 cm, very leafy with thick stock. The leaves are small, sessile, pubescent and silvery appearance. The flowers are grouped in clusters, with very small heads (3/1.5 mm) and ovoid. The involucre bracts are overlapping the orb and pubescent

**Picture 3** *Artemisia herba-alba*

outside. The floral receptacle is naked with 2–5 yellowish flowers per head all hermaphrodites (Picture 3).

## *Distribution and Ecology*

The largest populations of this species are found in the arid and semi-arid bioclimatic zones. They grow on various types of soils that are generally rich in limestone. *Artemisia herba-alba* leak sites where rainwater accumulates and stagnates (Benjilali and Richard 1980).

The sagebrush stands cover an estimated area of 10 million hectares in North Africa. This is the most common aromatic plant in Morocco. It occupies about 2.5 million hectares. The sagebrush stretches on the high plateau of the East, the High Moulouya, the foothills south of the Atlas and the valley of Oued Souss, with representatives in the arid western region of Haouz and arid valleys High Atlas and Souss continental. It grows, both on the plates in the plains and troughs.

*Artemisia herba-alba* is a Xerophil chamephyte; it forms more or less open steppes. Its range, extends from Spain to the Middle East, via North Africa and Egypt; it is also present in Iran, Iraq and Afghanistan.

In Morocco, it prefers hot, dry climates, where it is widespread in many areas and especially in the blast eastern plateaus, the region of Boulmane, Midelt, High Atlas and Anti-Atlas (Bruneton 1993; Fennane 1987).

## *Chemical Composition of the Essential Oil*

The qualitative and quantitative analysis of *Artemisia herba-alba* essential oils shows an extraordinary polymorphism. Sixteen "chemical varieties" (chemotypes) were distinguished in Moroccan *A. herba-alba*. The most important chemotypes are as follows:

- α-thujone chemotype: Essential oils of that chemotype are characterized by a high content of α-thujone, usually greater than 50% and may reach 80%. This chemotype characterized *A. herba-alba* of the Dades Valley and the plain of Rachidia.
- β-thujone chemotype: This chemotype is characterized by a high content of β-thujone (50%–80%). It is especially abundant in the region of Boulmane and Haute Moulouya.
- Chemotype with camphor: In this chemotype, camphor alone accounts for over 50% of the total essential oil. It's found in the Anti-Atlas and Souss region.
- Davanon chemotype: The essential oil of this chemotype is rich in davanon (40–70%). This chemotype is found in the region of Saghro and High Atlas.
- Chrysanthenone chemotype: The major constituent is chrysanthenone. This chemotype is scarce. It is found mainly in the Upper Moulouya and the North-East of Morocco.
- Chrysanthényle acetate chemotype: In this chemotype chrysanthenyle acetate can reach 60% of the essential oil. It's a relatively rare chemotype, which is found in the northeast of the country.
- α-thujone-camphor chemotype: This chemotype is characterized by an average content of about 30% in α-thujone, camphor 36% and 6% β-thujone. *A. herba-alba* of Marrakech is characteristic of this chemotype (Benjilali et al. 1982).

It's important to notice that the commercial *A. herba-alba* essential oil is the α-thujone/camphor chemotype (32 and 37% respectively) (Benjilali et al. 1984, 1985, 1986; Bellakhdar 1978; Bellakhdar et al. 1987).

## *Use of* A. herba-alba *and Its Essential Oil*

In Morocco, *Artemisia herba-alba* is used extensively both in traditional medicine to treat helminthiasis, diabetes mellitus and other conditions such as jaundice. Also, the antihyperglycaemic, antimicrobial, antioxidant, antispasmodic, anti-venom, nematicidal, anthelmintic, anti-leishmanial, neurological, pesticidal and antibiotic resistant inhibitor activities of this plant have previously been reported (Imelouane et al. 2010; Mohamed et al. 2010). Moreover, the aqueous extract significantly increased gastrointestinal transit time and the reaction time to thermal stimuli (Husnia Marrif et al. 1995). Furthermore, the species of this genus are widely used in the pharmaceutics, cosmetics and food industry. They are also used to treat hair and the preventive effect of diseases of the cold season, mainly colds and rheumatism. They are also fortifying species, antiseptic, healing and warming character. A decoction or infusion based on *Artemisia* species can relieve abdominal pain, vomiting and digestive disorders. This recipe is also recommended in the case of fever, diarrhea, upset stomach, and liver. The same preparation is also used as an anthelmintic.

*Artemisia herba-alba* Asso, known also as desert wormwood (known in Arabic as shih, Armoise blanche (Fr.)), has been used in folk medicine by many cultures since ancient times. In Moroccan folk medicine it has been used to treat arterial hypertension and/or diabetes. Herbal tea from this species has been used as analgesic, antibacterial, antispasmodic, and hemostatic agents. This plant is also suggested to be important as a fodder for sheep and for livestock in the plateau regions of Morocco where it grows abundantly.

## 4 Atlas Cedar (*Cedrus atlantica* Mannerti)

### *Taxonomy*

The genus *Cedrus*, family Pinaceae, subfamily Abieteas, is an old genus, known since the tertiary. Widespread, it includes several species whose taxonomic value has long been controversial. First integrated into a single species (*Cedrus libanotica* Link), cedars evolved into the following four species, depending on biogeographical conditions:

- The Atlas cedar (*Cedrus atlantica* Manetti)
- The Lebanon cedar (*Cedrus libani* Loudon)
- Cyprus Cedar (*Cedrus brevifolia* Henry);
- Himalayan cedar (*Cedrus deodora* Loudon) (M'Hirit 2006).

### *Morphological Description*

It's a majestic allure tree that can easily exceed 40 m in height (Picture 4). Its longevity is important; the branches are very long and horizontal. The needles of 25 mm, bouquet on short shoots, green or glaucous, apex acute. The cones of 5–8 cm are cylindrical flat-topped, green and brown (Pictures 4, 5 and 6).

Atlas cedar lives in mountainous areas and cedar forests grow between an altitude of 1500 and 2500, with a preference for the northern and western slopes that are watered; it has several local varieties, differing in size, appearance, and especially the color.

### *Geographical Distribution*

Cedar spontaneously constitutes three distinct biogeographic blocks (M'Hirit 1982, 1994):

- North Africa, with *Cedrus Atlantica* (180,000 ha).

**Picture 4** Atlas cedar tree

**Picture 5** The cones containing the seeds

**Picture 6** The leaves of Atlas cedar

- Minor Asia, where live two species *Cedrus libani* in Lebanon (1700 ha), Syria some hundreds of hectares and more than 160,000 ha in Turkey and about ten hectares *of Cedrus brevifolia* in Cyprus.
- The third non-Mediterranean bloc in Himalayas with *Cedrus deodora* representing large areas of India and Afghanistan (500,000 ha).

The genus *Cedrus* is subdivided into four distinct species: *Cedrus atlantica* Manetti, *Cedrus libanotica* Barr, *Cedrus deodora* Loudon, *Cedrus brevifolia* Henry (Moufaddal 1983).

In Morocco the cedar exists in the Middle Atlas, the Rif, and the High Atlas (M'Hirit 1982). In the Rif, cedar occurs sporadically in the Limestone Mountains of Jbala but its density increases rapidly eastward. In the high Atlas and the Middle Atlas, stands form two blocks, one occupies the high mountains, south of Taza, the other is between the Middle central Atlas and the eastern High Atlas.

## *Ecological Requirements*

Atlas cedar is a large evergreen tree gymnosperm, bole squat, whose bark, initially smooth, greyish brown cracks with age in small brown or dark gray scales. Very long branches spread horizontally or slightly raised, form a whole by a large conical or ovoid crown with a wingspan can exceed 100 m. The arrow takes in older tree, a horizontal direction. The cones are smaller (5–6 cm), often with a small hollow in the center, yellowish green, they become purple before maturity. The leaves (needles) evergreen, leathery, acute, green or glaucous, isolated on elongated branches gathered in dense rosette on numerous short branches that fill the entire length of the branches. Seeds, triangular, are 10–15 mm long, dull colors are tending toward brown and provided with long wing and are rich in resins.

Height growth of Atlas cedar is continuing into old age, at least 200–300 years. The growth rate is weak the early years.

## *Chemical Composition*

Morocco is the main supplier of the international market by the Atlas cedar essential oil extracted from sawdust. It is a yellowish orange yellow liquid, viscous and sometimes a little trouble. Its smell is very special, slightly camphor with a sweet and woody notes reminiscent Cassie and mimosa.

The essential oil extracted from the Moroccan *Cedrus atlantica* sawdust is rich in α-himachalene (17.19%), γ-himachalene (9.53%) and β-himachalene (41.69%) and its also contains γ-Atlantone (2.92%) and Cedryl methylique cetone (3.52%) (Zrira 2010b).

Chalchat et al. (1994) analyzed the essential oil extracted from the Moroccan *Cedrus atlantica* sawdust, and identified 53 compounds. The special particularity of this essential oil is its himachalène content in the form of three isomers: α, β, and γ, which constitute the majority of the essential oil, compounds (70%). Other characteristic compounds of this essential oil are atlantone (E and Z) are also present but in much lower concentrations.

Based on the work done in the GEF-Rif project, essential oil extracted from the leaves of Atlas cedar located in Ghomara region has a chemical composition that is completely different from that extracted from sawdust. The essential oil extracted from the thick boughs was rich in monoterpenic hydrocarbons (57%): limonene (29.2%) ocimene + α-pinene (18.6%) and myrcene (16%).

### *Use of Atlas Cedar and Its Essential Oil*

Cedar makes multiple services and provides useful products: lumber, firewood and sawdust as energy, essential oils (wood, leaves and branches) and wood tar and is home to a large pasture.

It provides a fairly brittle wood, which limits its use for framing. Due to its imputrescibility it is popular in shipbuilding. Cedar poles have good elasticity, prevent and withstand before breaking sufficient evidence to be used as mine timbers. Traditionally destined for uses structure, cedar wood has been used in woodworking, fine woodworking and running both inside and outside (El Abid 2006).

The Atlas cedar essential oils are used in perfumery for his role as fixative and its unique scent that blends well with labdanum products (rock rose) and perfume materials woody woody-floral type. It is used in the manufacture of scented soaps, deodorants and other flavors (Arcander 1960).

The aromatic essential oil of Atlas cedar has antifungal, antiseptic, healing, astringent and decongestant to respiratory system. It is reported to have also relaxing, draining, lymphotonic and diuretic, lipolytic (fat burning) effects. It is said to improve the quality of the arterial network and scalp (dandruff as cade oil). It repels mosquitoes and moths.

Among the ancient Egyptians, cedar oil and resin were added to the preparations used to embalm mummies. Even today this natural attribute is exploited in cedar hangers and balls hanging in wardrobes. The aromatic substances released from the oil have excellent repellent properties against mites (Duquenois 1968).

## 5 Laurel (Laurus nobilis and Laurus azorica)

### *Taxonomy*

The *Laurus* genus belongs to the family *Lauraceae* that are usually green and wild growing trees, represented by 2000–2500 species. Popular representative of the genus are, e.g.: cinnamon (from China, Ceylon), bay leaves, rosewood, the ravensara, sassafras.

The genus *Laurus* contains three species that are evergreen trees with very close key (Mabberley 1997):

**Picture 7** Leaves, flowers and fruits of *Laurus nobilis*

- *Laurus azorica* (Seub) Franco, also called laurel Azores;
- *Laurus nobilis* L., commonly known as the "Noble Bay Laurel", "Sauce Laurel", "franc laurel" or Apollon laurel is known in arabic as "الرند" or "ورق سيدنا موسى";
- *Laurus novocanariensis*, formerly included in *Laurus azorica*, is a native of Madeira and the Canary (Costa et al. 2004).

## *Morphological Description*

### *Laurus nobilis*

The laurel is a shrub native to Minor Asia and up to 10 m high. The plant has a smooth black bark, evergreen quite variable form, alternate, coriaceous, entire, lanceolate or oblong, 4–10 cm/2 to 3 cm (Picture 7). It is a dioecious plant that blooms in early spring (April, May) in the Mediterranean countries.

The female plants produce fruits in abundance: these are black drupes, ovoid, very bright.

### Laurus azorica

A native of the Azores laurel forests, *Laurus azorica* also called the bay of the Azores is a massive tree with a height of 15 m, evergreen, alternate and elliptical 4–15 cm long, smooth and dark green. Its flowers are greenish yellow. The fruits of a length of 12 mm are black when ripe (Picture 8). The *Laurus azorica* leaves are less aromatic than the laurel noble and are still used in cooking.

**Picture 8** Leaves and flowers of *Laurus azorica*

## *Geographical Distribution*

### *Laurus nobilis*

*Laurus nobilis* is widespread in many parts of the world of warm and temperate climate. In particular, it is abundant in the Mediterranean countries, including Morocco, Algeria, Spain, southern France, Italy, Palestine and Cyprus.

### *Laurus azorica*

*Laurus azorica* is currently stationed tree in the Atlantic islands (Canaries; Madeira, Azores). This species also existed in Europe, where it was known as fossils in many localities (Roiron 1979).

In Morocco, the species is found in the forest communities of Jebel Tazerkount (Atlas Ksiba) at the forest with deciduous oaks, around 1650 m (Barbero et al. 1981), so it takes a very different ecological situation that of *Laurus nobilis*. It is indeed full floor above Mediterranean deciduous at high altitudes and at minor limestone cliffs (Barbero et al. 1981).

Many representatives of this tree were also observed in the Atlas near Beni Mellal in Jebel Ighnayene. It is therefore in the same habitat, at a much higher altitude, between 1900 and 2000 m. In this zone, *Laurus azorica* always colonizes small limestone cliffs facing north. Deciduous oaks no longer exist at this level, probably because of too low winter temperatures, 3 or 4 observed localities lie substantially at the upper limit of the upper Mediterranean sclerophyllous oak or mountain Mediterranean.

## Ecological Requirements

### Laurus nobilis

Suitable for light sandy soils, loamy soil medium and heavy clay and prefers well-drained soil. It prefers soils with acidic, neutral and alkaline pH. It can grow in the shade or no shade half. It prefers dry or wet soils. The plant can tolerate strong winds.

### Laurus azorica

*Laurus azorica* occupies a very different ecological situation from *Laurus nobilis*. It's indeed, open floor above Mediterranean deciduous at high altitudes and at minor limestone cliffs (Barbero et al. 1981).

## Chemical Composition

### Chemical Composition of *Laurus* Leaves

The leaves of *Laurus nobilis* contain polar flavonoids (glycosylated derivatives of quercetin, kaempferol and catechin) and nonpolar (four acyl derivatives of kaempferol) (Fiorini et al. 1997), sesquiterpene lactones, alkaloids of isoquinoline (Simic et al. 2003), in addition many researchers showed the richness of its leaves in vitamin E.

By steam distillation, the leaves provide about 10–30 ml/kg (1–3%) of essential oil (Bruneton 1999), whose major constituents include 1,8-cineole, α et β –pinene, sabinene, linalool, eugenol, terpineol, plus other esters and terpenoids, but the proportions vary by geographic origin (Sayyah et al. 2001).

The chemical composition of the essential oil of *Laurus nobilis*, harvested in the region of Tétouan (Benkarrich and Oued Laou) and *Laurus azorica*, collected in the region of Beni Mellal is rich in 1,8-cineole (39.0–40.2%) and (39.48–42.2) respectively, linalool (3.71–8.03) and (8.1–12.25) respectively and α-terpenyle acetate (4.67–11.86) and (2.04–6.26) respectively. They have relatively low content of eugenol and methyl eugenol (GEF-Rif 3 1998).

### Chemical Composition of *Laurus* Berries

- **The essential oil of *L. nobilis* berries**

100 parts of peeled berries contain 0.8% of a volatile oil obtained by steam distillation of berries, 0.5 % of a particular crystalline substance, called laurine, 6.4% of a green fatty oil, 3.5% of crystalline fat having more consistency, 0.8% a soft resin,

half fluid (containing volatile oil), 12.5% of starch, 8.6% of gum, 3.2% of plant mucilage, 0.2% of noncrystallizable sugar, traces of vegetable albumin, 9.4% of vegetable fiber, 0.72% of saline ash and 3.2% of water. To obtain the crystalline substance, the berries must be exhausted with boiling alcohol and the latter is distilled. The remaining liquor allowed settling, on cooling, crystalline needles, elongated and yellowish. The liquor was composed of two layers, an upper, consisted of fatty oil, while the other was an alcoholic solution.

The fatty oil contains, like that of the leaves, 1,8-cineole in smaller amounts (12.8%), alcohol (10.6%), esters (17.9%), cinnamic acid (1.5%), phenol (2%), hydrocarbon terpene (15.4%) and carbonyl compounds and sesquiterpenes.

- ***Laurus* butter**

Laurel berries have the same culinary uses than the leaves; they are dried and ground such as nutmeg. Harvested in the fall, they are rich in fats.

The berries are also used to obtain, by hot expression, fatty oil, called "laurel butter" and whose yield is 8–9% of the berries weight. It's a creamy green mass, solid at room temperature, with a pleasant smell. It consists of a fat blend in which lauric acid (54.2%), oleic acid (15.1%) and linoleic acid (17.2%) dominate.

The unsaponifiable fraction contains undecanone, α-terpineol, α-terpényle acetate, β-elemene and β-sitosterol (Hafizoglu and Reunanen 1993).

## *Use of Laurel and Its Extracts*

The leaves of *Laurus nobilis* are among the most popular seasoning in all countries, they are generally used as a culinary spice (in soups, stews, sauces, *etc.*) and flavoring agent in food industry. Laurel is mainly used orally, for the symptomatic treatment of disorders of the upper digestive tract such as epigastric bloating, slow digestion, belching and flatulence.

The aqueous extract is used in traditional Turkish medicine as an anti-hemorrhoidal, antirheumatic, diuretic and as an antidote to snake bites and for the treatment of stomach ache. In Iranian traditional medicine, the leaves of this plant have been used to treat epilepsy and Parkinsonism.

Everywhere in Morocco, the infusion of the leaves is used in indigestion and concentrated decoction is used to gargle against diseases of the oral pharyngeal sphere (tonsillitis, mouth ulcers, bad mouth odor, gingivitis). The bay is used in chewing or gargle decoction, against the evils of the mouth, teeth, gums or tonsils. The leaves and berries are used as an aromatic condiment for the preparation of fish and meat. They are also used to flavor the traditional vinegar and canned olives and peppers.

The essential oil obtained from the leaves has been used for the relief of hemorrhoids and rheumatic pain (Sayyah et al. 2001). In addition, the essential oil is used by the cosmetics industry in perfumery and in the manufacture of soaps. It is one of the best ways to keep away annoying insects.

**Picture 9** Plant of myrtle (*Myrtus communis* L.)

The bay is mainly used to treat disorders of the upper digestive system and arthritis pain. Moreover, it stimulates the appetite and the secretion of gastric juices. Used as a condiment, leaves aid digestion and assimilation of food. They have the same beneficial effect as mint (*Mentha spicata*) and rosemary (*Rosmarinus officinalis*).

Laurel promotes the onset of menstruation. Diluted in neutral oil, the essential oil is mainly used as an ointment to rub the muscles and painful joints. Added to bath water decoction of the leaves soothes aching limbs.

## 6 Myrtle (*Myrtus communis* L.)

### *Taxonomy*

The *Myrtus* word is derived from the Greek word "myron", which is the fragrance: aromatic plant (Coste 1937). Also it can be derived from the Arabic word "Mayroune." The *Myrtus* genus belongs to the Myrtaceae family. A native of the Mediterranean basin, it includes about fifty species, some of which are also present in Western Asia, South America and Australia (Zrira 2002).

### *Description*

Evergreen shrub, 2–3 m high, Smooth, aromatic, evergreen; leaves opposite, very close together, ovate-lanceolate acute, entire, leathery, petiole reduced, smooth and shiny, not stipulated stipules. Flowers white, solitary, axillary, on long stalks, fragrant; calyx tube welded to the ovary, 5 spreading lobes; 5 petals; numerous stamens; 1 style, single stigma; inferior ovary. Fruit fleshy barely Bay, ovoid, bluish black, crowned by the calyx, few-seeded (Picture 9). The full maturity of the fruit is reached in November. Flowering occurs in summer from June to August (Fennane et al. 1987; Valdes et al. 1987).

## Ecology and Distribution

Myrtle is the only representative of the family Myrtaceae in North Africa. It grows in semi-arid bioclimatic, sub-humid and humid hot tempered variant on non-calcareous soils.

This species is very common in the Rif centro-occidental, the area between Larache, Tangier and Bab Taza in Mamora and Arbaa Sehoul. It is also present in the region of Amezmiz (High Atlas of Marrakech) (Barbero et al. 1981).

## Chemical Composition of Myrtle Essential Oil

The myrtle leaves contain tannin, resinous and bitter substances, and about 0.3% essential oil. The extraction of EO occurs by stripping with steam. Performance will vary depending on the scrub exposure level and the cutting period, technology and mining equipment. It is generally of the order of 0.7%.

Myrtle essential oil has a fairly simple chemical composition with generally α-pinene (8.0–31.4%, limonene, 1,8-cineole (20.5–31.0%), limonene (9.0–18.0%) and myrtenyl acetate (10.0–20.0%) as major constituents (Assaf 1995; Chalchat et al. 1998; Zrira et al. 2006).

The chemical composition of essential oils of myrtle is qualitatively stable. However, the percentage of a compound may vary widely (Chalchat et al. 1998).

## Uses of Myrtle

Myrtle is known since antiquities, as a symbol of love and peace. The uses of myrtle in traditional medicine and modern pharmacopoeia are numerous. It is still widely used to fight against diseases of the scalp, upper respiratory tract infections and gastrointestinal diseases against hemorrhoids. The fruits of the myrtle green or dried were used against bleeding; the juice of the berries is used as stomachic and diuretic (Fournier 1948). Their powder was used to prepare a macerate against felons and nail diseases and was also administered against seminal losses and sweat heart (Bellakhdar et al. 1994). Myrtle flowers are used against heart disease and liver. Their decoction is used to stop acute diarrhea, treatment of cough and rhinitis.

The myrtle essential oils pale yellow or amber colour with a camphor notes are very spicy. These are used in perfumery. They mix well with bergamot, lavender or lavender, rosemary, sage, hyssop and mugwort essential oil.

**Picture 10** *Rosa spp.*

This species is also used in formulations for flavouring sauces and meats mixed with spicy essential oils such as bay leaf, cloves and cardamom. It can identify the effect of ginger and positively changes the nutty flavour of nutmeg (Bellakhdar et al. 1994).

# 7 Rose (*Rosa damascena & Rosa centifolia*)

## *Taxonomy*

The rose belongs to the genus Rosa. The plant is in the form of a shrub climbing stems, erect or creeping, usually furnished with hairs or spines. The leaves are alternate, deciduous or evergreen, flowers terminal, solitary or in corymbs. The numerous carpels are inserted on a fleshy receptacle shaped urn. The genus Rosa has a large number of species, which are, derived the countless farmed varieties or hybrids. If more than five thousand roses are counted today, few of them have a strong fragrance. This fragrance is also very variable according to the variety. Only three roses are used for the isolation of their perfume: *Rosa damascena* Mill. cv. trigintipetala Dieck, *Rosa centifolia* L and *Rosa alba* L (Zrira 2006).

(a) ***Rosa damascena* Mill. variety *Trigintipetala* Dieck**

This flower, called the '*Damascus* Red Rose' and of unknown origin is cultivated as a hybrid of *R. gallica* L. and *R. canina* L. Nowadays, it can be found growing in the wild in regions of the Caucasus, Syria, Morocco and Andalusia. *Rosa Damascena*, very fragrant, contains an amount of essential oil that can be extracted by steam distillation. Being probably the most used in perfumery, it is grown extensively in Bulgaria and, on a smaller scale, in Turkey. It is a flower with regular petals (about thirty), long stamens with a piriform ovary, the elongate calyx. Yellow green stem is almost thornless (Picture 10). Damascus rose must have a suitable climate, temperature and a moderate humidity, on airy low calcium ground. Flowering begins in late April in the most exposed places and continues throughout the month of May (Zrira 2006).

(b) ***Rosa centifolia* L.**

Called the 'Rose of May', it's cultivated extensively in the Grasse region (south of France) and in North Africa. It contains a relatively large amount of essential oil that cannot however be economically isolated by steam distillation. Small amounts are used for the production of rose water, popular in the Mediterranean and Latin regions. Most of the production of *Rosa centifolia* L. is used to extract roses concrete and absolute, by volatile solvents. From a botanical point of view, the rose of May seems very close to *Rosa damascena Mill* (Zrira 2006).

(c) ***Rosa alba* L. ou *Rosa damascena* Mill. var. *Alba***

The white peasant rose 'contains much less essential oil than *Rosa damascena*'. This oil is, in addition, of lesser quality. Stronger than the red rose of Damascus, the peasant rose is cultivated in Bulgaria as a hedge around red roses and, at higher altitudes where the damask rose blooms more (Zrira 2006).

## *Cultivation of Rose*

*Rosa damascena* is cultivated in south of the High Atlas between the M'gouna chain and the Saghro in the commonly known M'gouna and Dadès valleys. Outside this area, the rose is found upstream of the two valleys and around Ouarzazate, Skoura and Agdz. Dadès belongs to the pre-Saharan bioclimatic costs. The crop soils consist of deep silt. Their alkaline pH varies between 7.4 and 7.8. The organic matter content is low. The concentration of nitrogen varies from medium to low.

Planting is done from November to February in 40 cm trenches deep rooted that receive plants, taken from tuft on old hedges.

The yield is about 0.8 kg of fresh roses per linear meter of hedge; it may reach 1.4 kg in good year for a production of 3500–5000 tons of fresh roses/year (Zrira 2006).

## *Extraction of Rose Essential Oil*

Rose essential oil is obtained by hydro-distillation, which gives rose water and essential oil. The essential oil is the most expensive and most popular material.

The method of distillation roses is different from those used for other plants.

Double distillation is practiced. The still must perform:

- Hydro-distillation;
- Reinjection of rose waters;
- Double distillation

25% of the essential oils is obtained in the first distillation, whereas 75% during the rest of the second distillation. The essential oil content in the rose flower is quite small, only about 2%.

## *Extraction of Rose Concrete and Absolute*

(a) **Extraction of concrete**

The extraction of concrete is processed by hexane as solvent. The raw material is loaded into the extractor and is then exhausted by successive washings with a suitable solvent. After passing through a settler and a hub, a partial distillation is performed. We obtain odorant molecules, waxes and pigments and the solvent to be reused.

Depending on the nature of the plant we could obtain:

- The resinoid obtained from dried matters (balms, gums and resins). It is used in their state;
- The concrete (treatment outcome of all fresh organs of the plant).

The solvent that dissolves the fragrant wax or miscella must be evaporated to obtain rose concrete. Obtaining the concrete is done in two phases:

1. A concentration under atmospheric pressure in evaporators. This operation stops at 60 °C in order not to alter the volatiles.
2. Final concentration under reduced pressure at low capacity condensers (50–100 l). This operation is difficult and must be carried out at the lowest temperature possible, all traces of solvent must be removed. Each industry has its own method for this phase.

(b) **Concrete conversion to absolute**

Although easy to carry and maintain thanks to the presence of waxy mass concrete are not directly used in perfume compositions. Thus, they must be transformed into absolute (real cleared of waxes). The dewaxing is carried out by successive washings of the concrete in the alcohol inactive polar solvent on waxes. Usually five to six washes are made by kneading in threshing and precipitation of wax is obtained at −20/−25 °C. Alcoholic or absolute solution is purified by filtration.

(c) **Concrete and absolute yield**

The concrete rose volatile solvent is extracted from *Rosa centifolia* and *Rosa damascena*. The yield varies between 0.17% and 0.27% in concrete that gives 55%–65% absolute.

## *Chemical Composition of Rose Extracts*

*Rosa damascena* Mill. contains carboxylic acid, terpene, myrcene, and vitamin C.

- Flavonoids such as kaempferol and quercetin glycosides were detected. (quercetin 3-O galactosid, quercetin 3-O-xyloside kaempferol glycosides, The kaempferol glycosides, along with the kaempferol aglycone, accounted for 80% of the

total compounds that were quantified, with kaempferol 3-O-glucoside being the predominant component. The high flavonol content of approximately 16 g/kg on a dry weight basis, rutin, quercitrin, myricetin, quercetin, apigenin, and kaempferol.
- Catechin, epicatechin, rutin
- Antocyanins (like cyanidin-3-O-β-glucoside)
- Proanthocyanidins kind of tannins
- Phenolic acids (gallic acid etc., m-coumaric acid)
- Major carotenoids as the β-carotene, lycopene, rubixanthin, zeaxanthin, lutein
- Vitamin C (European Medicines Agency 2013).

The basic character of rose oil, mostly dependent upon citronellol and geraniol, is further modified by nerol (5–11%) and farnesol (0.2–1.4%). Their contents are slightly higher in village oils. Higher farnesol content leads to the establishment of strong floral character and an overall improvement of body-note volume. Nerol not only adds to the rosaceous character but also to its freshness. In those cases where the geraniol content is low, however, the freshness of nerol manifests itself as slightly citrusy. When geraniol content is high, the combination of citronellol, geraniol, farnesol, and nerol results in a strong, sweet, floral, fresh rosaceous character. Other typical constituents of rose oil are geranyl acetate, nonanal, citronellyl formate, citronellyl acetate, eugenol, methyl eugenol, cis-rose oxide, α-terpineol, phenylethyl alcohol, and linalool. Damascenones and some sulfur compounds are among the minor components. Stearoptenes (paraffins) are natural constituents of rose oil (the major one being nonadecane) and due to their presence, rose oil solidifies at room temperature and when refrigerated. The chemical composition and physical properties of the rose absolute produced from rose concretes and the extract, showed that rose absolute consists mainly of β-phenylethyl alcohol, citronellol, geraniol, nerol, eugenol, methyl eugenol, geranyl acetate, benzyl alcohol, nonadecane, nonadecene and farnesol. It was observed that the solid residue contains mainly straight-chain saturated hydrocarbons with high molecular weight (C-15-C-31) and the esters of carboxylic acids and a homologous series between them exists. It was also concluded that the extract obtained by solvent extraction from the residue of rose flower (*Rosa damascena* Mill.), which was subjected to steam distillation is also, a valuable raw material (Aycı et al. 2005).

## *Properties and Uses*

For centuries, the Damascus rose (*Rosa damascena*) has been considered a symbol of beauty and love. The fragrance of the rose has been captured and preserved in the form of rose water by an ancient method that can be traced back to biblical times in the Middle East, and later to the Indian subcontinent. An Iranian doctor, Avicenna, is credited with the discovery of the process for extracting rose water from rose petals in the early eleventh century. Damascus roses were introduced into England during the reign of Henry VIII and were frequently displayed and scattered at weddings and festivals. Nowadays, they are popular in craft projects and as potpourri ingredients. They are used in wedding favors, gathered together in organza bags or favor boxes, and they replace the traditional Avola sugared almonds to make perfumed keepsakes. They are also used to decorate festive tables and as hair decorations when attached to hairpins.

The uses of the dried Damascus rose in beauty products are numerous. Soaking Damascus rosebuds in water for 3 or 4 days releases a rose essence which can be added to bath water or may be used to rinse hair after shampooing to leave the skin and hair soft with the fragrance of roses. As the gentlest of all astringents, rose water is often used as toner for fair and dry skin or as an anti-aging product in facial creams. Damascus rose oil also has therapeutic properties that sooth the mind and helps with depression, nervous tension and stress.

The flower buds and rose water are also used in herbal medicine. Indeed, everywhere in Morocco decoction of flower buds is used against stomach, and the infusion of rose petals henna mixture is administered as a laxative. To remedy the abdominal pain, rural women are preparing a dried flower decoction. Externally, the pale pink petals are chewed in toothache. They are also mixed with the myrtle, sweet clover, lavender and other plants to wet the Ghassul in the hair care. Rose water is used, in the treatment of fevers, nausea states, migraines and sunstroke (compresses on the head and forehead), otitis (ear drops) nervousness and anxiety (a spoon to drink twice a day). Rose water (hydrosol) is also used in cosmetics in the toilet water, cosmetics).

The rose odorous extracts (essential oil, concrete and absolute) have taken an important place in perfumery. The Grasse region is the main production area. These extracts are incopropated in the formulation of high quality perfumes (Channel, Yves Saint Laurent, Guerlain). Rose essence is used in many compositions, extracts, tinctures, beauty milk, and ointments, cosmetic oils. Dried rose petals musk is used for rose powder or the manufacture of scented sachets (Zrira 2006).

## *The Rose in Morocco*

The perfumed rose (*Rosa damascena*) is a specific crop to Ouarzazate region. Its cultivation is located in the Dadès Valley. Currently 4200 km linear of rose are under the form of hedges or fences around agricultural plots, representing about 1000 ha or 10% of the cultivated agricultural land. Roses' farming is often compromised by frost and cold that affect the flowering period and industrial quality of the rose. It varies greatly from year to year with an average of 4000 tons of fresh roses per year. Eleven hectares of Rose (*Rosa centifolia*) were cultivated in 1998 in the Tidass region for the production of concrete and rose absolute.

The rose is usually harvested in the first half of May, in the morning at sunrise. To produce 1 kg of dry rose, it takes 4–5 kg of fresh roses and we need 5000 kg of fresh roses to obtain 1 kg of rose essence. Pale rose petals contain 0.03–0.04% of an essential oil consisting of geraniol, nerol, citronellol, phenylethyl and aldehyde (soluble in water from which the fragrance release in rosewater).

The rose is used for the production of several products:

- Dried flower blossoms,
- Rose water,
- Rose essence and
- Concrete and absolute

Different rose aromatic extracts are produced by BIOLAND and "Arômes du Maroc" these are companies equipped with sophisticated modern processing units at Kelaat M'gouna in the Dades Valley, in the province of Ouarzazate. The companies also have processing units in the city of Khemisset (BIOLAND) and Tiddas "Les Arômes du Maroc".

BIOLAND produces rose essence and concrete r from *Rosa damascena* while "Les Arômes du Maroc" develop:

- Rose water, rose essence and concrete from *Rosa damascena*
- Concrete and absolute from *Rosa centifolia*

## 8 Rosemary (*Rosmarinus officinalis* L.)

### *Botanical Description*

Rosemary is an ornamental, aromatic and medicinal plant known in France under the name or marine rose (Fournier 1948) "Yazir" in the Moroccan vernacular or "Iklil al JBAL" in classical Arabic. *Rosmarinus* genus belongs to the Lamiaceae family and includes two species *Rosmarinus officinalis* L. and *Rosmarinus eriocalyx.*

Rosemary is a very aromatic shrub, from 0.50 to over 1 m, evergreen, very branching, very leafy. The leaves are evergreen, leathery, sessile, linear, rolled by the edges, green and aggrieved above, white-tomentose beneath. The flowers are pale blue or whitish, with peduncle very small. Small inflorescences axillary and terminal racemes, calyx bell bi-labiate, powder, bare throat, full oval upper lip, the lower 2 lanceolate lobes lipped corolla, with protruding tube to a larger helmet lip, cleft, the lower 3 lobes, the wide and concave means 2 stamens salient fillets, inserted in the groove of the corolla, fitted to the base of a small tooth. Fruits: achenes trines, smooth (Picture 11). Flowering occurs between April and July (Elamrani et al. 1994a, b).

**Picture 11** *Rormarinus officinalis* L.

## *Geographical Distribution*

Rosemary (*Rosmarinus officinalis* L) is found throughout the Mediterranean, in the scrub up to 1500 m altitude and gardens where it is used as an ornamental plant. It's mainly found in the Mediterranean (southern Europe, Asia Minor, Cyprus, North Africa, the Canary Islands, Madeira).

In Morocco, it's widespread in the Mediterranean part and especially in Eastern. It's found in the western Rif at the eastern part of the limestone ridge, and in the eastern Morocco at the wooded steppes of Debdou El Atef and Jerada. It's also reported in the regions of Bouarfa and Midelt.

## *Ecological Requirements*

Rosemary is a species, typical of the Mediterranean region, requiring light and heat, drought resistant. It acclimates particularly in the upper arid or areas under the hot sub-humid to fresh. It prefers calcareous soils and well drained. This species lives in a very dry sub-humid areas or regions.

## *Main Chemical Constituents of Rosemary*

The main constituents of rosemary are:

- Essential oil (1.0–2.5%)

- Methylated flavones: genkwanine, luteolin, diosmetin, apigenin, isoscutellaréine-7-O-glucoside
- Phenols acids: rosmarinic acid also called "tannin of Labiatae"
- Tricyclic phenolic Diterpenes: rosmaridiphenol, carnosolique acid (= carnosic acid) rosmadial (carnosol = picrosalvine, rosmanol)
- Polysaccharides acid (6%)
- Triterpenes and steroids: oleanolic acid, ursolic acid derivatives (Hoffler et al. 1986)

## The Essential Oil

Rosemary gives an essential oil (EO) pale yellow or light colored. It is a mobile liquid with a strong odor, fresh, herbaceous, woody, slightly minty.

The chemical composition of rosemary EO is relatively constant. However, the proportions of each component may vary significantly (Granger et al. 1973).

Chemical heterogeneity of rosemary populations was demonstrated. Thus chemical varieties (or chemotype) were differentiated on the basis of their EO. Three rosemary chemotypes were identified:

- 1.8-cineole chemotype, with 1,8-cineole as major constituent (Rosemary from Italy, Tunisia and Morocco).
- Camphre- borneol chemotype to present mainly in Spain
- α-pinene and verbenone chemotype to (Corsica, France and Algeria)

In Morocco, three rosemary chemotypes were identified:

- 1.8 cineole chemotype;
- Camphor chemotype;
- α-pinene chemotype (Elamrani et al. 1997a, b).

The commercial rosemary essential oil (1,8-cineole chemotype) of Morocco is characterized by the following components: 1,8- cineole: 54.4%, Camphor: 6.7%, α-pinene: 5.5% and β-pinene: 4.6% (Zrira et al. 1997).

## The Antioxidants

### (a) **Flavonoids**

Many flavonoid molecules have been isolated and identified: diosmetin, luteolin genkwanine the hispidulin and apigenin. Some of these flavonoids were isolated under hétérosidique form (diosmoside, methoxy-6-lutéoloside and apigénol-7-glucoside) and under glycosiliques form (Table 1) (Elamrani 1999).

### (b) **Phenolic acids:**

Three phenolic acids were identified in rosemary: rosmarinic acid, chlorogenic acid and caffeic acid (Table 2).

**Table 1** Chemical structures of flavonoids identified in *R. officinalis* L.

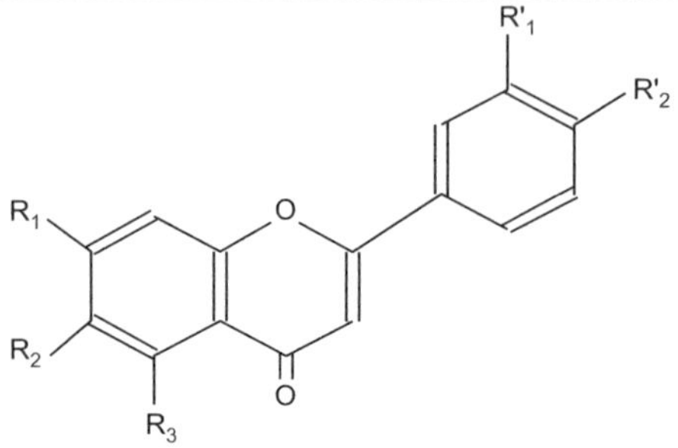


| | $R_1$ | $R_2$ | $R_3$ | R'1 | R'2 |
|---|---|---|---|---|---|
| Luteoline | OH | OH | OH | OH | OH |
| 6-Metoxy-luteoline | OH | $OCH_3$ | OH | OH | OH |
| 6-Metoxy-luteoline-7-methyl éther | $OCH_3$ | $OCH_3$ | OH | OH | OH |
| Apigenine | OH | H | OH | OH | H |
| Diosmetine | OH | H | OH | OH | $OCH_3$ |
| Hispiduline | OH | $OCH_3$ | OH | OH | H |
| Salvigenine | $OCH_3$ | $OCH_3$ | OH | H | $OCH_3$ |
| Genkwanine | $OCH_3$ | H | OH | OH | H |
| 6-Methoxy-Genkwanine | $OCH_3$ | $OCH_3$ | OH | OH | H |
| Genkwanine-4'-methyl éther | $OCH_3$ | H | OH | OH | $OCH_3$ |
| 5-Hydroxy-4'-7-dimethyl flavone | $OCH_3$ | H | OH | H | $OCH_3$ |

## Diterpenes, Diterpene Lactones and Quinones:

Diterpene lactones were isolated from the leaves of *R. officinalis*. Some authors have shown the presence of carnosolique acid and carnosol. Others have isolated the ether of carnosolique acid and rosmanol and rosmadial and have reported the presence of royléanone and its derivatives in the roots (Tables 3 and 4) (Elamrani 1999; Elamrani et al. 2000).

## Triterpenes

The ursolic acid was isolated from the leaves of *R. officinalis*. Then other terpenic acids were isolated: 19-α-acid hydroxyursolique, 2β-hydroxyoléanolique acid. Triterpene hydrocarbons adjacent sturctures were also isolated: α-amyrin, β-amyrin (Table 5).

**Table 2** Phenolic acids in *R. officinalis*: rosmarinic acid, chlorogenic acid and caffeic acid

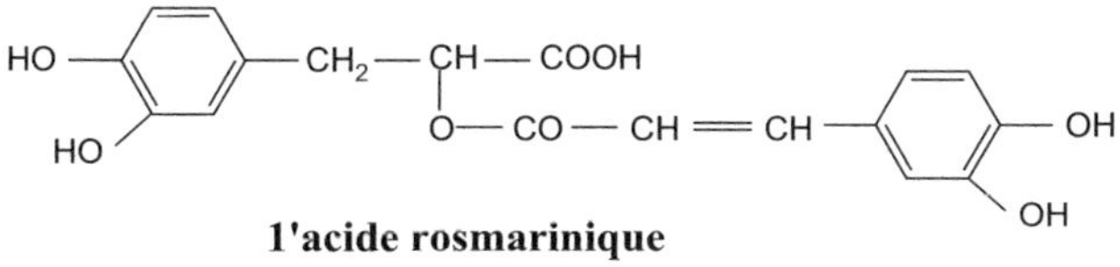


**l'acide rosmarinique**

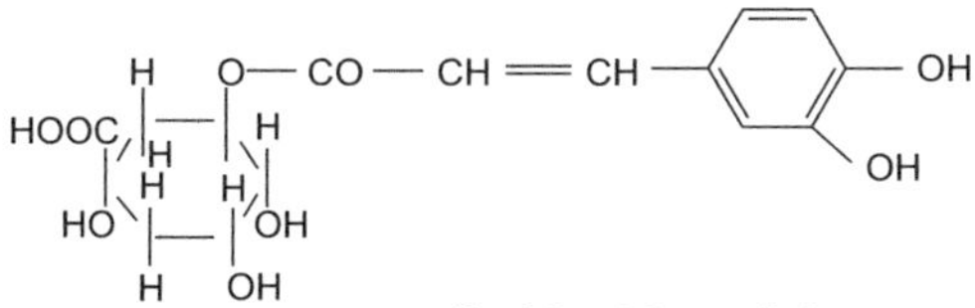


**l'acide chlorogénique**

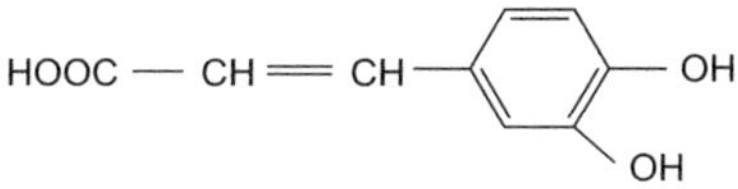


**l'acide caféïque**

**Table 3** Chemical structure of Diterpenes and diterpene lactones in *R. officinalis*

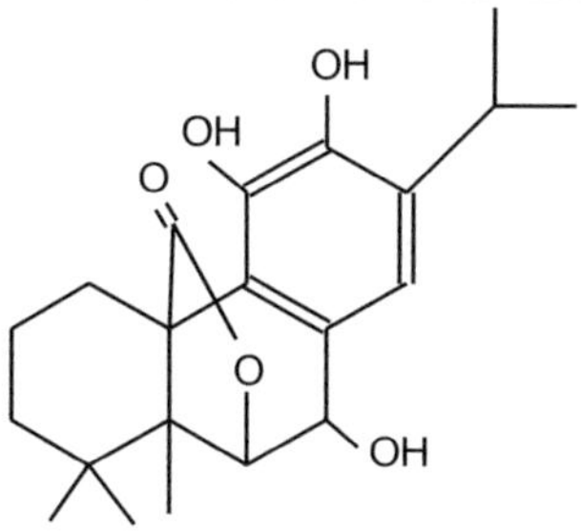


Rosmanol

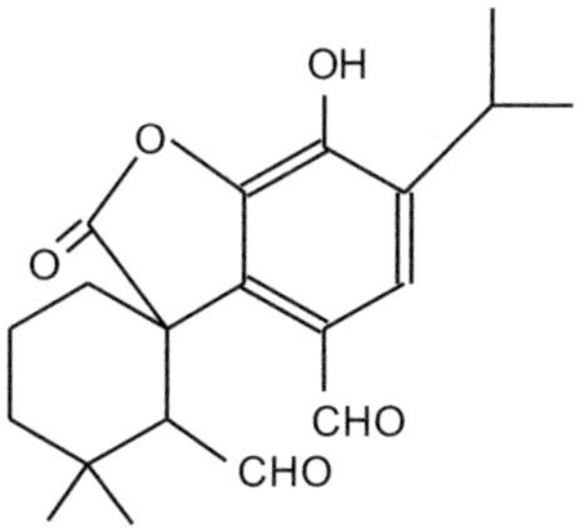


Rosmadial

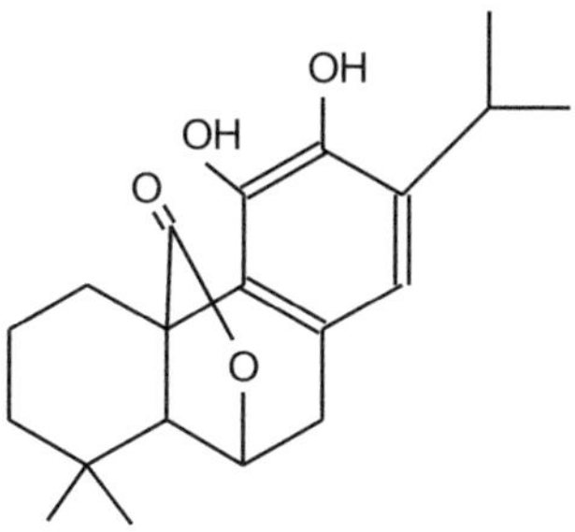


Carnosol

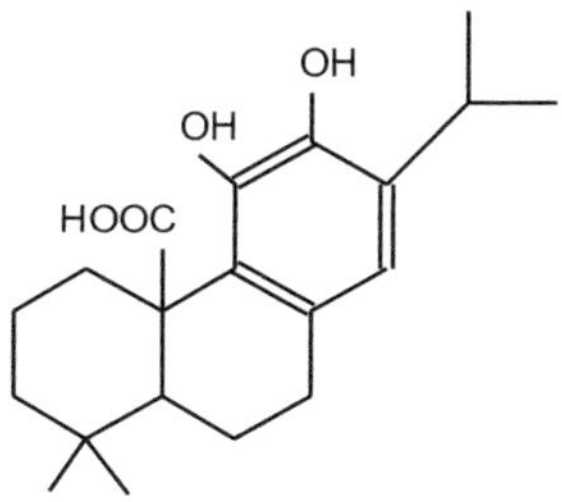


Acide carnosolique

**Table 4** Chemical structure of diterpenique quinones in *R. officinalis*

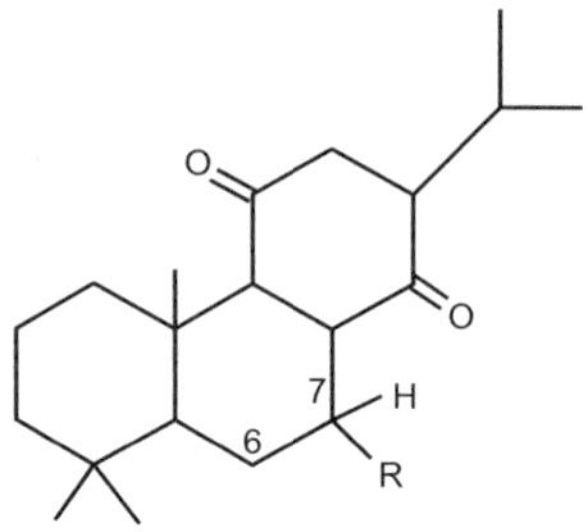


R = H: Royleanone
R = OH: 7-Hydroxy-royleanone (Horminone)
R = -O-CO-$CH_3$: 7-Acetoxy-royleanone

**Table 5** Chemical structure of some triterpene in *R.officinalis*

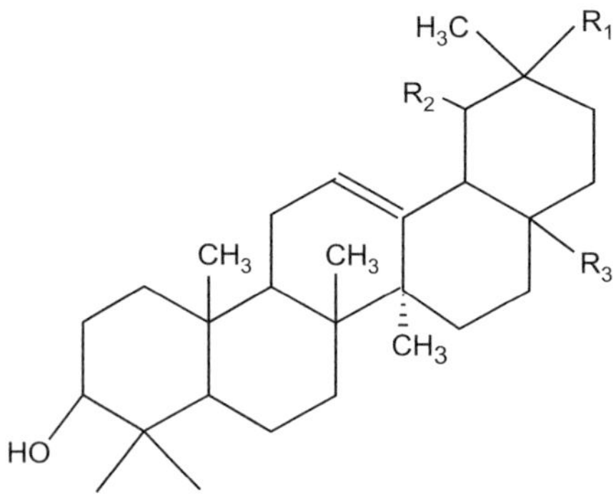


| | $R_1$ | $R_2$ | $R_3$ |
|---|---|---|---|
| α-Amyrine | H | $CH_3$ | $CH_3$ |
| β-Amyrine | $CH_3$ | H | $CH_3$ |
| Acide ursolique | H | $CH_3$ | COOH |
| Acide oléanolique | $CH_3$ | H | COOH |

## *Uses of Rosemary and Its Essential Oil*

Because of their aromatic flavor, rosemary leaves are used for culinary purposes. In the food industry, the dry leaves are used as food flavorings. Note that the Council of Europe in his collection of natural and artificial flavorings, consider adding rosemary to foods as a spice, is permitted without limitations.

Rosemary is a tonic, general stimulant including adrenocortical glands. It is also a pulmonary antiseptic. Rosemary is hypertensive, stomachic, antidiarrheal, due to its diuretic and sudorific qualities. It is cholagogue and choleretic to the point that the intravenous injection of the lyophilized extract prepared from the leaves in animals, double the bile secretion volume of the latter. Rosemary acts against the gallstones and stomach pains. In the intestine, it can be used in infections, colitis, diarrhea and flatulence. Rosemary leaves after powdering can be used to relieve stomach crises and strengthen this body. An infusion of the leaves has emmenagogue properties.

On the other hand external use of rosemary should not be underestimated. Thanks to its healing and fungicidal properties, it can beneficially intervene to heal wounds, burns, joint swelling and scabies.

Rosemary is widely used in folk traditions for its ability to keep the food against degradation by microorganisms and especially against oxidation of animal and vegetable fats. Its antiseptic properties are widely used in some traditional techniques of food preservation. Furthermore, the latest research showed that rosemary extracts exhibit anti-mutagenic activity and a marked antioxidant power due to flavonoids, polyphenols and phenolic acids it contains in significant amounts. The antioxidant properties of rosemary are linked essentially to the presence of rosmarinic acid (anti-inflammatory phenolic acid), rosmanol, rosmadial, carnasol, carnosolique acid and rofficerone, separated from dry leaves by extraction with hexane.

Essential oils of rosemary are currently used in the perfumery industry, cosmetics, soap and in the pharmaceutical industry. In perfumery, they are incorporated in the composition of most colognes. Combined with other essential oils, especially basil and patchouli, they provide fashionable perfumes. In the seventeenth century, rosemary essential oil was used to prepare the famous Queen of Hungary water.

This essential oil is used in deterpenated state in the preparation of soaps, shampoos and deodorants. It finds its application in agriculture as insecticides and disinfectants.

In pharmaceutical and therapeutic industry, rosemary essential oils are used as raw material in the formulation of drugs. The rosemary essential oils rich in cineole and pinene are used for medicinal purposes. Indeed, thanks to its anesthetic, antiseptic and stimulant properties, pure cineole is commonly used in the treatment of inflammation of the nose and throat as well as chronic bronchitis. They are used as stimulating the scalp and as pest control.

## References

Achhal A, Akabali A, Barbero M, Bernabid A, M'hirit A, Peyre C, Quezel P, Rivas-Martinez S (1980) A propos de la valeur bioclimatique et dynamique de quelques essences forestières au Maroc. Ecol Medit Marseilte 3:211–249

Adlouni A, Bennani H, El Messal M (2009) Argan-oil: From nutrition to Health. Congrès International Plantes Aromatiques et Médicinales, Maroc 2009, Ifrane du 24 au 26 septembre 2009

Arctander (1960) Perfume and flavor materials of natural origin. Edition Elisabeth New Jersey, pp 154–158, po 138–141

Assaf N (1995) Valorisation du *Myrtus communis* var. *italica* dans la zone de Beni Idder, province de Tétouan. Mémoire de troisième cycle pour l'obtention du diplôme d'Ingénieur des Eaux et Forêts, option "Ecologie et gestion des ressources naturelles". ENFI, Salé
Aycı F, Aydınlı M, Bozdemir A, Tuta M (2005) Gas chromatographic investigation of rose concrete, absolute and solid residue. Flav Fragr J 20(5):481–486
Barbero M, Fresneau M, Peyre C, Quezel P (1981) Sur la présence au maroc de Laurus azorica (Seub.) Franco. Acte III Congr. OPTIMA. Anales jard. Bot Madrid 37 (2):467–472
Bellakhdar J (1978) Médecine traditionnelle et toxicologie ouest saharienne. Contribution à l'étude de la pharmacopée marocaine. Edition: technique Nord-Africaine, Rabat. pp 55–71
Bellakhdar J, Baayaoui A, Kazdari A, Marechal J (1987) Herboristes et médecine traditionnelle à Tissint, oasis présaharien du sud marocain (Province deTata) Al Birunya. Rev Mar Pharm 3(1):7–50
Bellakhdar J, Miki W, Honda G (1994) Herboristes et pharmacopée à Marrakech et à Salé (Maroc). Simples et prescriptions. Al Biruniya, Rev. Mar. Pharm. Tome n°1. pp 26–27
Benjilali B, Richard H (1980) Étude de quelques peuplements d'armoise blanche du Maroc (*Artemisia herba-alba*). Rivista Italiana EPPOS 62:69–74
Benjilali B, Saris J, Richard H (1982) Nouveaux chemotypes d'armoise blanche du Maroc (*Artemisia herba-alba* Asso). Rivista Feraderno Agricolo Supplemento 2, atti delle giornata internationali Saint Vincent, 26–27 April 1982
Benjilali B, Richard H, Liddle P (1984) Chemotypes d'armoise blanche du Maroc '*Artemisia herba-alba*'. Rivista Feraderno Agricolo Supplemento 2, atti delle Giornata Internationali Saint-Vincent 26–27 April 1984
Benjilali B, Hammoumi M, M'Hamedi A, Richard H (1985) Analyse en composantes principales (ACP) et chimiotaxonomie de l'armoise blanche du Maroc (*Artemisia herba-alba*). Les cahiers du C.E.A.M.S 2:12–21
Benjilali B, Richard H, Baritaux O (1986) Étude des huiles essentielles de deux espèces d'origan du Maroc : Origanum compactum Benth et Origanum elongatum Emb. Et Maire Lebensem Wiss U Technol 19:22–26
Boudy P (1951) Economie forestière nord-africaine, II: Monographie et traitement des essences forestières, Larose, Paris, Fasc I: 382–414; III: Description forestière du Maroc, Stat. de rech. forest. Maroc, Rabat, pp 190–219
Boulos L (1983) Médicinal plants of North-Africa. Référence Publications Inc. 218 st. Clair River Drive, Box 344 Algonac, Michigan 48001
Bruneton J (1993) Pharmacognosie, Phytochimie, Plantes Médicinales. Tec. & Doc. Lavoisier, Paris, p 915
Bruneton J (1999) Pharmacognosie, Phytochimie, Plantes Médicinales, Tec. & Doc. Lavoisier, Paris, 2ème édition, pp 632–643
Chalchat JC, Garry RP, Michet A (1998) Essential Oils of Myrtle (*Myrtus communis* L.) of the Mediterranean Littoral. J Essent Oil Res 10(6):1998
Chalchat JC et al (1994) Essential oil components in sawdust of cedrus atlantica from Morocco. J Essen Oil Res 6(3):323–325
Charrouf Z, Guillaume D (1999) Ethnoeconomical, ethnomedical, and phytochemical study of Argania spinosa (L.) Skeels. J. Ethnopharmacol 67:7–14
Charrouf Z, Guillaume D (2007) Huile d'argan une production devenue adulte. Les Technologies de Laboratoire, n° 6 septembre-octobre 2007
Charrouf Z, Guillaume D (2008) Argan-oil: Occurrence, composition and impact on human health. Eur J Lipid Sci Technol 110:632–636
Chimi H, Cillard J, Cillard P (1994) Autooxydation de l'huile d'Argan. Argania Spinosa L du Maroc Sci Aliment 14:117–124
Costa J C, Capelo J, Jardim R, Sequeira M (2004) Catálogo Florístico do Arquipélago da Madeira. Quercetea 6:187–200
Coste H (1937) Flore descriptive et illustrée de la France de la Corse et des contrées limitrophes Tome 2, taxon n°1275. Paris: Librairies des Sciences et des Arts, p 75

Debouzie D, Mazih A (1999) Argan (Sapotaceae) trees as reservoirs for Mediterranean fruit fly (Diptera: Tephritidae) in Morocco. Environ Entomol 28(1):53–60

Dupin A (1949) L'Arganier survivant de la flore tertiaire, providence du sud marocain, Elevage et cultures

Duquenois P (1968) L'utilisation des huiles essentielles en pharmacie, leur normalisation et l'europe du medicament Parf. Cosmet Sav 11:132–137

Emberger L (1938) Les arbres du Maroc et comment les reconnaître. La rose, éditeurs, París, 318 p

El Abid A (2006) Le cèdre de l'Atlas mémoire du temps, Le bois de cèdre, structure, propriétés et utilisations, p 158

Elamrani A (1999) Les Huiles Essentielles des Romarins du Maroc (*Rosmarinus officinalis* et Rosmarinus eriocalix): Rendements, Chimie, Chimiotaxonomie, Facteurs influençant le rendement et la composition chimique des huiles essentielles. Thèse de Doctorat, Faculté des Sciences Ben M'sik Casablanca

Elamrani A, Zrira S, Ismaili Alaoui M, Berrada M, Benjilali B (1994a) L'huile essentielle de romarin du Maroc: effet de la période de récolte et du niveau de coupe sur les rendements et la composition chimique. Rivisita Italiana EPPOS N° spéciale, 1994, pp 238–245

Elamrani A, Zrira S, Ismaili Alaoui M, Berrada M, B (1994b) Le romarin du Maroc (*Rosmarinus officinalis* L.): Variabilité intraspécifique. Rivisita Italiana EPPOS N° spéciale, 1994, pp 226–236

Elamrani A, Zrira S, Ismaili Alaoui M, Berrada M, Benjilali B (1997a) Effet de la fréquence de coupe et de la période de récolte sur la production et la qualité de l'huile essentielle de romarin (*Rosmarinus officinalis* L.) du Maroc. Rivisita Italiana EPPOS N° spéciale, 1997, pp 655–667

Elamrani A, Zrira S, Ismaili Alaoui M, Berrada M, Benjilali B (1997b) Caractérisation chimique de l'huile essentielle de romarin (*Rosmarinus officinalis* L) du Maroc. Rivisita Italiana EPPOS N° spéciale, 1997, pp 646–654

Elamrani A, Zrira S, Berrada M, Benjilali B (2000) A study of Moroccan Rosemary oils. J Essent Oil Res 12:487–495

European Medicines Agency (2013) Assessment report on *Rosa gallica* L., *Rosa centifolia* L., Rosa damascena Mill., flos. Committee on Herbal Medicinal Products (HMPC), 15 December 2013 EMA/HMPC/137298/2013

Farines M, Charrouf Z, Soulier M (1981) The sterols of Argania spinosa seed oil. Phytochemistry 20:2038–2039

Fellat Zarrouk K, Smoughhen S, Maurin R (1987) A study of the pulp of the fruit of the Argan tree (Argania spinosa) of Morocco. Fats and latex. Actes de l'Institut Agronomique et Veterinaire, II. 7:3–4, 17–22

Fennane M (1987) La grande encyclopédie du Maroc: flore. Présentation du monde végétal GPE, Cremona, pp 7–13

Fennane M, Ibn Tattou M, Banabid A, Bellakhdar J, Lawalle (1987) La grande encyclopédie du Maroc. Flore et végétation – G.E.P. Crémona Italie

Fiorini C, Fouraste I, David B, Bessières JM (1997) Laboratoire de pharmacognosie, Faculté de pharmacie, Toulouse, Fr. Flav Fragr J

Fournier P (1948) Le livre des plantes médicinales et vénéneuses de France. Edition Lechevalier. Paris, 3. pp 335–336

Gef-Rif 3 (1998) Description des échantillons des produits de la distillation et développement des opérations d'analyse de ces échantillons, présentation des résultats d'analyse. Projet de production et de gestion participative des écosystèmes forestiers du Rif N° B 75040/I/94/02, Chefchaouen

Hafizoglu H, Reunanen M (1993) Fett ISSN 0931–5985 CODEN FWTEEG, Vol. 95, no 8, pp. 304–308 (49 ref.)

Imelouane B, El Bachiri A, Ankit M, Khedid K, Wathelet JP, Amhamdi H (2010) Essential oil composition and antimicrobial activity of *Artemisia herba-alba* Asso grown in Morocco. Banats J Biotechnol 1:48–55

Hilali M, Charrouf Z, El Aziz SA, Hachimi L, Guillaume D (2005) Influence of origin and extraction method on Argan-oil physico-chemical characteristics and composition. J Agric Food Chem 53:2081–2087

Hoffler C, Mortier F, Fleurintin J, Pelt JM, Guillemain J (1986) Composition chimique de *Rosmarinus officinalis* L. (Lamiacées). Al Biruniya. Rev Mar Pharm 2(2):79–89

Husnia Marrif I, Ali BH, Hassan KM (1995) Some pharmacological studies on *Artemisia herba-alba* (Asso.) in rabbits and mice. J Ethnopharmacol 49:51–55

Kenny L, De Zborowski I (2007) In Atlas de l'Arganier et de l'Arganeraie; IAV Hassan II: Rabat, Morocco; p 190

Khallouki F, Younos C, Soulimani R, Oster T, Charrouf Z, Speigelhader B, Bartsch H, Owen RW (2003) Consumption of Argan-oil (Morocco) with its unique profile of fatty acids, tocopherols, squalene, sterols, and phenolic compounds should confer valuable cancer chemopreventive effects. Eur J Cancer Prev 12(1):67–75

Mabberley DJ (1997) The plant book, a portable dictionary of the vascular plants, 2nd edn. Cambridge University Press, Cambridge

Maurin R (1992) L'huile d'Argan *Argania spinosa* (L.) Skeels sapotaceae. Rev Fr Corps Gras 39:139–146

Matthaus B, Guillaume D, Gharby S, Haddad A, Harhar H, Charrouf Z (2010) Effect of processing on the quality of edible Argan-oil. Food Chem 120:426–423

M'hirit O (1982) Etude écologique et forestière des cédraies du Rif marocain. Essai sur une approche multidimensionnelle de la phytoécologie et de la productivité du cèdre (Cedrus atlantica Manetti) Thèse es-Sciences Université d'Aix Marseille III

M'Hirit O (1994) Annales de la recherche forestière au Maroc, 27, vol 1. pp 5–15

M'Hirit O (2006) Le cèdre de l'atlas mémoire du temps, Taxonomie et répartition historique, Répartition actuelle du cèdre de l'atlas p 13, 39–42

M'Hirit O, Benzyane M, Benchekroun F, El Yousfi SM, Bendaanoun M (1998) L'Arganier. Une espèce fruitière-forestière à usages multiples; Mardaga: Sprimont, Belgium, p 150

Mohamed AE-HH, El-Sayed MA, Hegazy ME, Helaly SE, Esmail AM, Mohamed NS (2010) Chemical constituents and biological activities of *Artemisia herba-alba*. Rec Nat Prod 4:1–25

Morton JF, Voss GL (1987) The Argan tree (Argania sideroxylon, Sapotaceae). Desert Sour Edible Oil Econ Bot 41:221–233

Moufaddal (1983) Contribution à l'étude des possibilités d'extension du cèdre de l'Atlas (Cedrus Atlatica Manetti) en dehors de son aire actuelle. Thèse pour l'obtention de Doctorat en Ecologie générale et appliquée. IAV Hassan II, Maroc

Rahmani M (2005) Composition chimique de l'huile d'Argane "vierge". Cahiers Agric 14:461–465

Rieuf P (1962) Les champignons de l'Arganier. Cah. Rech. Agron., n° 15

Roiron P (1979) Recherches sur les flores Plio-Quaternaires méditerranéennes: la macroflore Pliocéne de Pichegru pres de Saint-Gilíes (Gard). Thése Doctorat 3éme cycle, U.S.T.L. Montpellier

Rojas LB, Quideau S, Pardon P, Charrouf Z (2005) Colorimetric evaluation of phenolic content and GC-MS characterization of phenolic composition of alimentary and cosmetic argan oil and press cake. J Agric Food Chem 53:9122

Sayyah M, Saroukhani G, Peirovi A, Kamalinejad M (2001) Analgesic and anti-inflammatory activity of the leaf essential oil of *Laurus nobilis* Linn

Service de Normalisation Industrielle (SNIMA) (2003) Huiles d'Arganes. Spécifications. Norme Marocaine NM 08.5.090. SNIMA, Rabat, Morocco

Simic M, Kundakovic T, Kovacevic N (2003) Department of Pharmacognosy, Faculty of Pharmacy, University of Belgrade, Vojvode Stepe 450, Belgrade 11221, Serbia and Montenegro

Valdes P, Tavavera S, Galiano EF (1987) Flora vascular de Andalucia occidental. Tome II. Ketres Editoria S.A

Yaghmur A, Aserin A, Mizrahi Y, Nerd A, Garti N (1999) Argan-oil -in-water emulsions: preparation and stabilization. J Am Oil Chem Soc 76(1):15–18

Yaghmur A, Aserin A, Mizrahi Y, Nerd A, Garti N (2001) Evaluation of Argan-oil for deep fat frying. Lebensm-Wiss Technol 34:124–130

Zahar M, Reineccius G, Schirle-Keller J-P (2007) Identification of aroma compounds in food grade Argan-oil. The 234th American Chemical Society meeting, August 19–23, Boston, MA, USA

Zahar M, Reineccius G, Schirle-Keller J-P (2011) Argan-oil aroma: effect of kernels roasting and oil extraction process. Actes du Premier Congrès International de l'Arganier, Agadir 15–17 Décembre 2011

Zrira S (2002) Etude des Plantes Aromatiques et Médicinales de la Région d'Aguelmame Sidi Ali-, Rapport AMO, Projet PMF/FEM, PNUD

Zrira S (2006) La rose du Dadès: Culture et technologies d'extraction des essences de rose. Bulletin Mensuel de Liaison et d'information Du PNTTA. Transfert De Technologie En Agriculture, MADER/DERD N°146 Novembre 2006 PNTTA

Zrira S (2010a) Moroccan essential oils: Some important issues. IFEAT International Conference in Marrakech, 26–30 Sept. 2010 'North African and Mediterranean Essential Oils and Aromas: 2010 Tales and Realities of our Industry – a new decade of challenges and opportunities'

Zrira S (2010b) Le cèdre de l'Atlas. 29èmes Journées Internationales Huiles essentielles et Extraits Digne les Bains 24 & 25 Juin 2010

Zrira S (2011a) Les plantes aromatiques et médicinales du Maghreb. 30èmes Journées Internationales Huiles essentielles et Extraits, DIGNE Les BAINS 14 & 15 Septembre 2011

Zrira S (2011b) The Value Chain of Aromatic & Medicinal Plants in the Maghreb. SIPAM, Tunisie, 2012

Zrira S, Elamrani A, Mejjati A, Belanger A, Ismaili Alaoui M, Benjlali B (1997) Rationalisation de l'exploitation industrielle des plantes aromatiques et médicinales: cas du romarin: Plantes Aromatiques et Médicinales et leurs Huiles Essentielles, Actes Edition pp 177–183 I.A.V Hassan II, Maroc

Zrira S, Menut C, Bessiere JM, Benjilali B (2006) Propriétés biologiques de quelques huiles essentielles marocaines. 25èmes Journées Internationales Huiles essentielles et Extraits, Digne les Bains 14 & 15 septembre 2006

# *Prunus africana* (Hook. f.) Kalkman (the African Cherry)

**Mélissa Rousseau, Claire Delvaux, Ezekiel Edward Mwakalukwa, Lawrence Mbwambo, Marie Caroline Momo Solefack, Harisoa Bako Ravaomanalina, Coert J. Geldenhuys, Ntamwira Niranda Seintsheng, Nils Bourland, and Hans Beeckman**

**Abstract** *Prunus africana* (Hook. f.) Kalkman, also known as *Pygeum africanum*, is the only species of the *Prunus* genus found in Africa. The reputation of this tree is due to the use of its bark extract in the treatment of benign prostatic hyperplasia (enlarged prostate). The present review compiles the knowledge of the medicinal uses, the extract composition and the threats to this overexploited tree but also possible solutions to sustainably produce its active compounds.

---

M. Rousseau (✉) • C. Delvaux (✉) • H. Beeckman
Wood Biology Service, Royal Museum for Central Africa,
Leuvensesteenweg 13, 3080 Tervuren, Belgium
e-mail: melissa.rousseau@africamuseum.be; claire.delvaux@africamuseum.be

E.E. Mwakalukwa
Department of Forest Biology, Faculty of Forestry and Nature Conservation, Sokoine University of Agriculture, P. O. Box 3010, Chuo Kikuu, Morogoro, Tanzania

L. Mbwambo
Tanzania Forestry Research Institute, P. O. Box 1854, Morogoro, Tanzania

M.C.M. Solefack
Department of Plant Biology, Faculty of Sciences, Dschang University,
B.P. 96, Dschang, Cameroon

H.B. Ravaomanalina
Department of Plant Biology and Ecology, Faculty of Sciences, Antananarivo University,
B.P 906 – 101, Antananarivo, Madagascar

C.J. Geldenhuys
Department of Forest and Wood Science, Stellenbosch University,
c/o Forestwood cc, P.O. Box 228, La Montagne 0184, South Africa

N.N. Seintsheng
Faculty of Sciences and Applied Sciences, Official University of Bukavu,
Bukavu, Democratic Republic of the Congo

N. Bourland
Wood Biology Service, Royal Museum for Central Africa,
Leuvensesteenweg 13, 3080 Tervuren, Belgium

Center for International Forestry Research, Situ Gede, Sindang Barang,
Bogor (Barat), 16115, Indonesia



M. Neffati et al. (eds.), *Medicinal and Aromatic Plants of the World - Africa Volume 3*, Medicinal and Aromatic Plants of the World,
DOI 10.1007/978-94-024-1120-1_6

**Keywords** *Prunus africana* • African cherry • Bark • Benign prostatic hyperplasia • Phytochemicals

## 1 Introduction

*P. africana* (Hook. f.) Kalkman is a multi-use species mainly known for its bark whose extracts are used in the treatment of benign prostatic hyperplasia, a disorder common in men over 50 years (Simons et al. 1998). In 1997, the annual trade of *P. africana* bark and its derivatives was estimated at more than $220 million (Cunningham and Mbenkum 1993).

Because the bark is harvested from the natural environment, the importance of its international trade leads to the overexploitation of the natural populations (Hall et al. 2000). This results in significant disturbances in the population dynamics (Cunningham and Mbenkum 1993; Sunderland and Tako 1999; Ndam and Ewusi 2000; Stewart 2003).

Concerns about the durability of the bark harvest have led to the inclusion of *P. africana* in Appendix-II of the Convention on International Trade in Endangered Species (CITES) in 1995 (Cunningham and Mbenkum 1993). The aim was to regulate the trade and to reduce the pressure on natural populations. However, in addition to the impacts of its trade, habitat loss due to agriculture and climate change is also threatening this species (Geldenhuys 1981; Mayaux et al. 2015). Moreover, the increase of the population concerned by prostatic hyperplasia and the renewed interest in herbal medicine make the bark of *P. africana* a product whose demand is likely to increase while the supply is remaining stable or even decreasing (Cunningham and Mbenkum 1993; Hall et al. 2000; Cunningham et al. 2014).

## 2 Geographical Distribution and Ecology

*P. africana* is a tree of the Rosaceae family, also known as *Pygeum africanum* or African cherry. It is the only species of the genus that is indigenous to Africa (Simons et al. 1998). This tree is restricted to high conservation value Afromontane forest islands (White 1983) in the African continent (Angola, Cameroun, Democratic Republic of the Congo, Ethiopia, Kenya, Malawi, Nigeria, Somalia, South Africa, Sudan, Swaziland, Tanzania, Uganda, Zimbabwe) and outlying islands (Bioko, Grand Comore, Madagascar, Sao Tome e Principe) (Kalkman 1965) (Fig. 1). Cameroon and Madagascar are the countries that contain the largest populations of the species because of their relatively large areas of montane forests (Stewart 2003).

*P. africana* is a light-demanding species growing within the secondary forests (Stewart 2003; Kadu et al. 2013). This tree is most abundant in open areas along

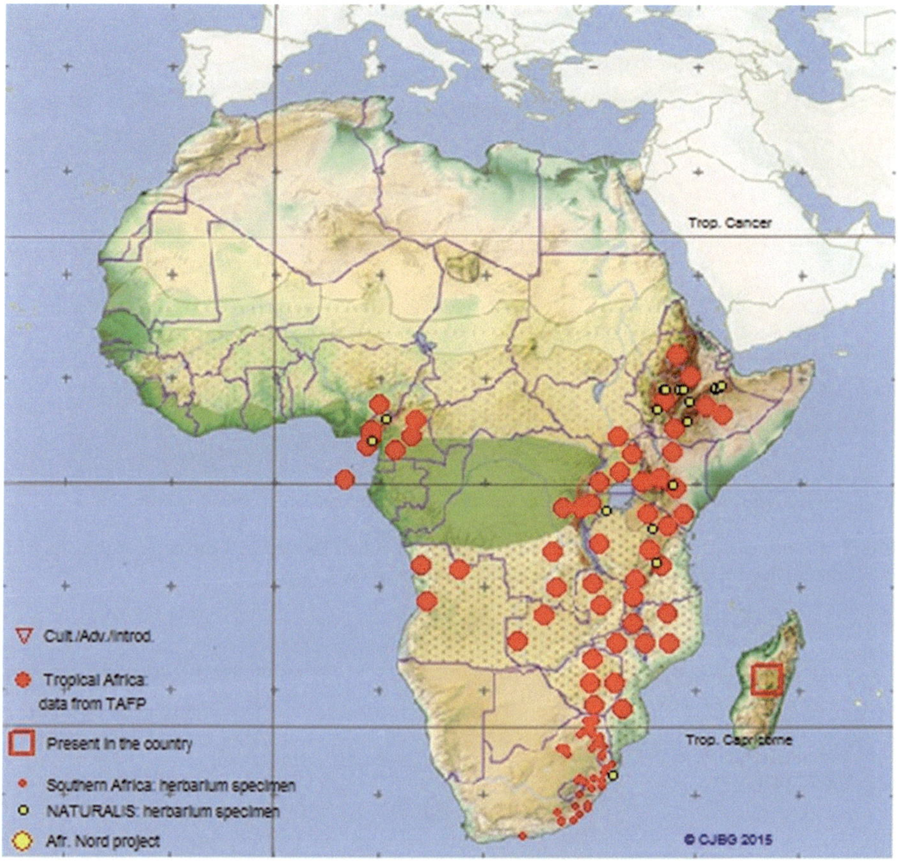


**Fig. 1** Geographical distribution of *Prunus africana* (Hook. f.) Kalkman in the African continent and outlying islands (Source: http://www.ville-ge.ch/musinfo/bd/cjb/africa)

forest margins and in disturbed areas (Geldenhuys 1981; Ndam 1996) and occurs at altitudes between 1000 and 2500 m (Cunningham and Mbenkum 1993; Sunderland and Tako 1999). It can reach 30–40 m in height and a diameter of 1 m. Its bark (Fig. 2) is light to dark reddish fissured (Stewart 2003).

Its evergreen canopy is medium to large with simple and alternate leaves (Stewart 2003) and small white flowers pollinated by insects (Farwig et al. 2008). The fruit is a drupe that becomes purple colored at maturity (Stewart 2003) and it is dispersed by birds and mammals such as red colobus and black and white colobus (Chapman and Chapman 1999; Chapman et al. 2003; Fashing 2004). Birds can disperse the relatively small seeds over distances of more than 300 km (White 1983). The different organs (Fig. 3) (Sim 1907) and the bark emit a characteristic "cherry" odor of the genus (Stewart 2003).

Three different zones can be distinguished based on flowering periodicity: the equatorial zone where some individuals are flowering almost continuously and the southern and northern zones where the flowering period seems to correspond with

**Fig. 2** *Prunus africana* (Hook. f.) Kalkman bark from a tree located in Tshivanga at the entry of the Kahuzi Biega National Park (Democratic Republic of the Congo)

**Fig. 3** Flowering branch (*1*) and fruit (*2*) of *Prunus africana* (Hook. f.) Kalkman (Adapted from Sim 1907 [14])

cool and dry conditions (Hall et al. 2000). *P. africana* starts to produce seeds at 15–20 years, a relatively late age (Simons et al. 1998). The fruiting period, generally associated with rainfall, seems to appear the 2–3 months after the flowering period (Hall et al. 2000).

## 3 Medicinal Uses

*P. africana* is multi-use species known mainly for the beneficial effects of its bark as treatment for prostate enlargement. Nevertheless, this species is also known for its timber and its leaves, roots and bark as traditional medicines in Africa (Cunningham and Mbenkum 1993; Kadu et al. 2012).

Before the 1960s, *P. africana* was only used for traditional purposes. In 1966, the effectiveness of its bark to treat prostate hypertrophy (an enlargement of the prostate) and Benign Prostatic Hyperplasia (BPH) was discovered (Debat 1966). Benign prostatic hyperplasia is a non-cancerous enlargement of the prostate that commonly affects men over 50 (Bombardelli and Morazzoni 1997; Simons et al. 1998). More than 50% of men over the age of 60 have urinary symptoms that can be attributed to BPH (Wilt and Ishani 1998). The number of side effects of other treatments of BPH (drugs and surgery) make phytotherapy an increasingly popular option (Stewart 2001). Moreover, *P. africana* is well tolerated and costs less than most prescripted medications (Wilt and Ishani 1998).

Though the mechanism of action of *P. africana* bark remains unclear, its efficacy to treat BPH symptoms is believed to be due to at least three groups of compounds: (a) phytosterols (β-sitosterol, β-sitostenone) having anti-inflammatory properties (Carbin et al. 1990), (b) pentacyclic triterpenoids (oleanolic and ursolic acids) providing anti-edematous activity (Bombardelli and Morazzoni 1997) and (c) ferulic acid esters (*n*-docosanol and *n*-tetracosanol) that lower blood levels of cholesterol, from which testosterone is produced (Bombardelli and Morazzoni 1997). These phytochemicals (Fig. 4) are believed to work synergistically to counteract the physiological changes associated with the disease (Bassi et al. 1987; Bombardelli and Morazzoni 1997; Simons et al. 1998).

Although the bark is the plant material commonly extracted, it has been found that wood, flowers and fruits have similar therapeutic values (Debat 1974).

The extract from the plant material is used in a variety of pharmaceutical forms such as tablets, capsules or cachets (Hall et al. 2000). To produce hundred 50-mg capsules, 5 g of extract are enough (Simons et al. 1998). Usually, *P. africana* extract is blended with other extracts known to have similar properties such as saw palmetto (*Serenoa repens*) (Lowe and Ku 1996).

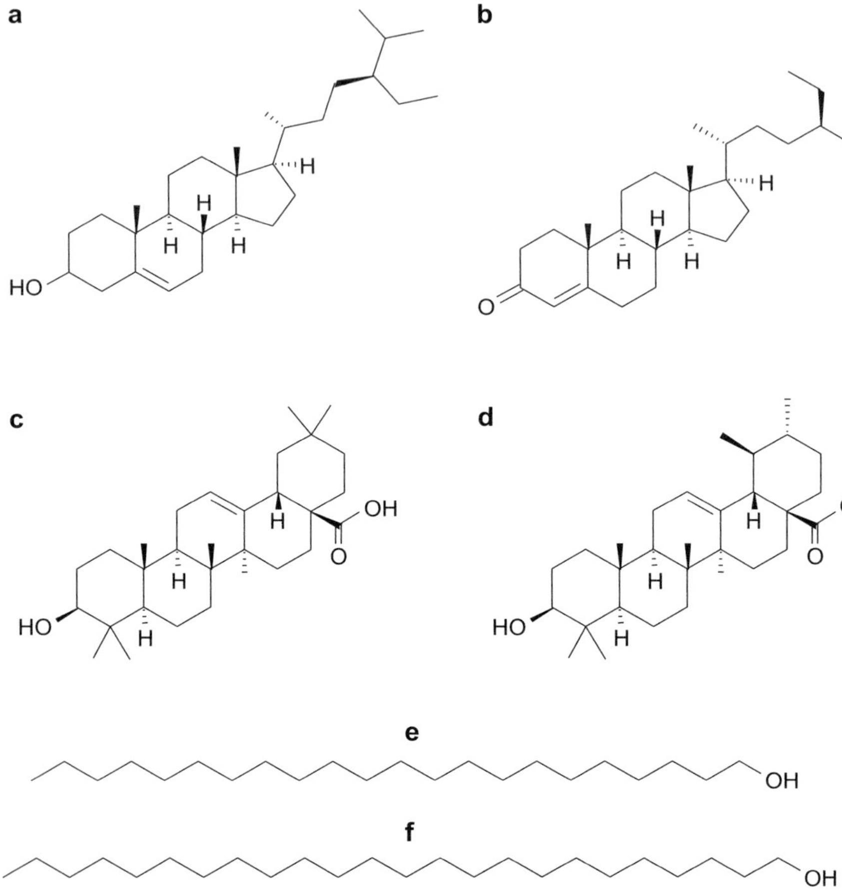


**Fig. 4** Some of the phytochemicals present in *Prunus africana* extract involved in the treatment of Benign Prostatic Hyperplasia (BPH). (**a**) β-sitosterol, (**b**) β-sitostenone, (**c**) oleanolic acid and (**d**) ursolic acid, (**e**) *n*-docosanol and (**f**) *n*-tetracosanol

## 4 International Trade

Following the discovery of the use of *P. africana* in the treatment of BPH, pharmaceutical companies began hiring people to harvest bark for exportation to European countries (Cunningham and Mbenkum 1993). The first countries to export from the early 1970s were Cameroon, Madagascar, Democratic Republic of the Congo (DRC), Kenya and Uganda (Cunningham et al. 1997). Cameroon, where the harvest began in 1972 around Mount Cameroon, has the longest history in *P. africana* bark trade (Stewart 2003).

In the 1990s, the annual quantity of dried bark or extract from *P. africana* exploited to export to Europe ranged between 3200 and 4900 tons, making it the

**Table 1** CITES national export quotas for *Prunus africana* (Source: https://cites.org/eng/resources/quotas/export_quotas)

| Country | Quotas | Unit | Specimens |
|---|---|---|---|
| Burundi | 0 | – | |
| Cameroon | 1,082,879 | kg | Dry bark |
| Democratic Republic of the Congo | 232 | kg | Dry bark |
| Kenya | 0 | – | |
| Madagascar | 0 | – | |
| Uganda | 176,179 | kg | Dry bark |

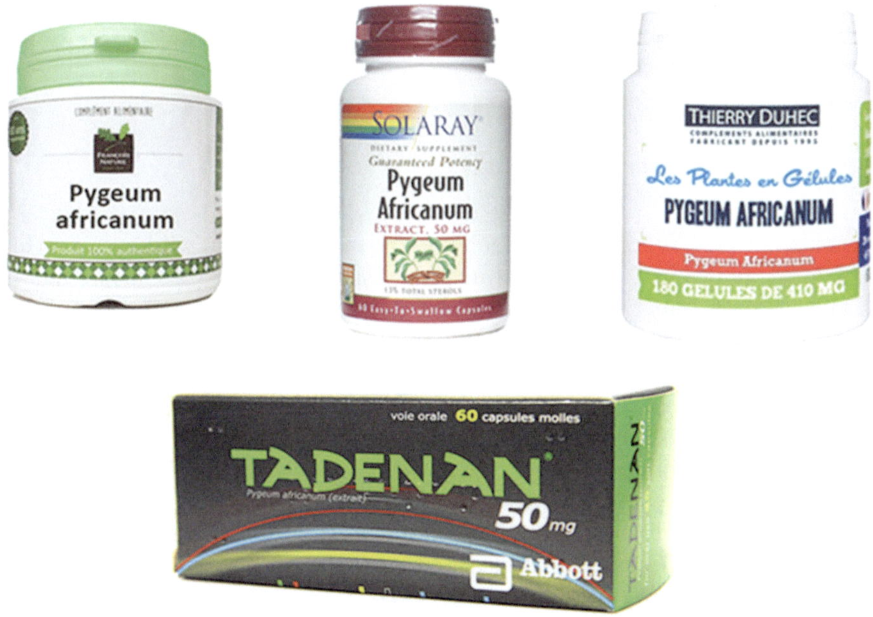


**Fig. 5** Products derived from *Prunus africana* extracts

first of any African medicinal plant in international trade (Cunningham et al. 2002). The global value of this international trade of products was estimated to be more than USD 220 million per year with Cameroon providing approximately 60% of the volume annually traded (Cunningham and Mbenkum 1993; Cunningham et al. 1997). However in the country, the price of the bark was less than USD 0.10 per kg compared to the international value of USD 2 per kg (Cunningham et al. 1997). The majority of the overall product value was taken by the pharmaceutical companies (Clark and Sunderland 2004).

In 2012, due to the establishment of quotas per exporting country, the volume traded was less than 1000 tons with Cameroon providing almost 75% of *P. africana* bark at the market price of USD 6 per kg (Cunningham et al. 2014). Over the last 40 years, the changing demand and markets have led to strong fluctuations in the values and volumes of bark traded (Ingram and Schure 2010).

Madagascar, Kenya and Burundi have a zero quota granted by CITES in 2016 so the global supply is supplied by Cameroon, DRC and Uganda (Table 1). Most of the bark harvested in Cameroon is exported to Italy and France while bark from DRC is exported to France and Belgium (Cunningham et al. 2002). The herbal products derived from *P. africana* (Fig. 5) are sold under its synonym, *Pygeum africanum* (Stewart 2003). Originally only two brand-name products were produced with *P. africana* extracts but now more than 40 brand-name products can be found on the market of ten countries (Wei et al. 2005). Capsules containing bark extracts have been marketed in Europe (mainly Austria, France, Italy and Switzerland) for over 40 years (Cunningham and Mbenkum 1993).

## 5 Threats

### *Unsustainable Harvest*

Currently, all the *P. africana* bark found on the market is from wild harvest (Hall et al. 2000). In theory, the harvest of its bark is supposed to be sustainable. Indeed the tree has the faculty to regenerate its bark as long as the vascular cambium is not damaged (Stewart 2003). Unfortunately, it is not the case in practice. The wild populations of *P. africana* are highly vulnerable due to bark overexploitation (Hall et al. 2000; Kadu et al. 2012).

Studies showed that most trees are even debarked from upper branches or felled and left to die to facilitate the access to the bark (Cunningham and Mbenkum 1993; Stewart 2003). Even partially strip barking living trees expose them to ring-barking and to insect attacks that can cause 50–90 % of post-harvest tree mortality (Ndam and Ewusi 2000). In Equatorial Guinea, 68% of exploited *P. africana* trees for local trade were experiencing death or canopy dieback (Sunderland and Tako 1999). Early studies trying to evaluate the impacts of bark harvest showed that it affected negatively the population structure, increased the mortality and decreased the fecundity (Parrot and Parrot 1989; Ewusi et al. 1992; Cunningham and Mbenkum 1993; Walter and Rakotonirina 1995). Death rates in commercially harvested wild populations of *P. africana* trees larger than 10 cm of diameter at breast height (dbh) can be 50–100 times higher than the natural mortality rate (Stewart 2003; Forboseh et al. 2011). Moreover, with the over-exploitation of large trees, there is a shift to trees with increasingly smaller diameters at breast height (Cunningham and Mbenkum 1993). The impacts on the population are ever higher since the density of mature trees is generally low in forests but also threaten ecosystem integrity (Ewusi et al. 1992; Hall et al. 2000).

The harvest of bark is geographically limited because of *P. africana* distribution in a specific ecosystem. It also increases the pressure on small areas (Cunningham et al. 2002). This harvest is still occurring even in our days, in Cameroon in areas with a protected status (Mount Oku, Mount Cameroon and Mount Kupe) endangering also other species (Cunningham and Mbenkum 1993). Moreover, in Madagascar, the prescription that two seed trees per hectare should be left in the exploitation area is not applied (Walter and Rakotonirina 1995; Hall et al. 2000). It is in Madagascar and in Cameroon that *P. africana* bark is under the greatest threat (Cunningham and Mbenkum 1993; Walter and Rakotonirina 1995). At an average *P. africana* density of 5.5 trees $ha^{-1}$ (Eben Ebai et al. 1992) and a bark yield of 55 kg per tree, the commercial harvesting of 1923 tons of bark each year would affect over 6300 ha of Afromontane forest annually (Cunningham and Mbenkum 1993).

With the world population aging, it is known that 88% of men in western countries will have the chance of developing histologic evidence of BPH (Cunningham and Mbenkum 1993). The demand of *P. africana* bark is then likely to be increasing which will strengthen the pressure on wild populations (Dawson et al. 2000; Hall et al. 2000).

## Loss of Habitat

*P. africana* is not only threatened by the unsustainable harvest of its bark but also by the loss of its habitat. For example, the forest cover of *P. africana* habitat in Ethiopia has been reduced from 35% in the beginning of the nineteenth century to less than 2.8% now (Mayaux et al. 2015). In Cameroon, the discovery of the potential of *P. africana* have resulted in a densification of the human populations in the concerned areas, causing a clearing of the forest for farming purposes (Cunningham and Mbenkum 1993).

There exist different major threats to montane ecosystems that are *P. africana* habitat: clearing for agriculture, timber extraction, browsing and trampling by livestock, and fire (Cunningham et al. 2008). Forest disturbances have impacted on species populations that can show an episodic natural regeneration as a result (Cunningham et al. 1997).

Considering the range of the species overall, the destruction of montane forests is probably the major factor affecting its conservation status (Hall et al. 2000).

## Climate Change

Climate change is another threat to *P. africana* populations that has to be considered. Sensitivity of Afromontane forest plants to small changes in temperature and rainfall can have high impacts on plants viability (Conway 2009). In South Africa, a study showed that 47% of *P. africana* standing stems of more than 10 cm diameter at breast height were dead due to increasing aridity in the Southern Cape region (Geldenhuys 1981). Moreover, climate changes can influence regeneration at the edges of the species' natural range (Geldenhuys 1981).

The countries that are more likely to be negatively affected by changes in mountain ecosystem climates are Tanzania, Madagascar, Cameroon, Democratic Republic of the Congo and Uganda (Ingram et al. 2015).

## 6 Protection Status

In 1995, growing concerns about the sustainability of *P. africana* harvest led to the species to be listed in CITES Appendix-II (Cunningham and Mbenkum 1993).

In 2005, Cunningham summarized in the CITES Significant Trade Review evidences of the destructive and highly unsustainable harvest of the bark (Ewusi et al. 1992; Sunderland and Tako 1999; Stewart 2001; Nkeng et al. 2009; Ingram 2014). This report was discussed at the CITES meeting in Lima in 2006. In 2007, the European Union banned the importation of *P. africana*. But because of the pressure from the private sector and from the Cameroon government, the ban was lifted in 2011 and quotas have been established for the main producing countries (Cunningham et al. 2014). This lift was mainly based on the "Management Plan for Prunus", a report published by different research organizations including the Center for International Forestry Research (CIFOR) (Ingram et al. 2009; Ingram 2014). However, there were many reservations about the poor quality of the report even among the authors (Cunningham et al. 2014).

Besides its status for CITES, *P. africana* is also listed in the Red List of International Union for Conservation of Nature (IUCN, www.iucnredlist.org) as a vulnerable species. Despite those designations, *P. africana* exported volumes are still making it Africa's most exported medicinal plant species (Cunningham et al. 2002).

## 7 Bark Harvesting and Composition

As mentioned previously, *P. africana* is among the few African tree species that exhibit complete bark regrowth and is even able to withstand complete bark removal (Stewart 2003). This ability offered the potential for sustainable harvesting of *P. africana* bark (Cunningham and Mbenkum 1993). The method currently presented as sustainable for bark removal is the two-quarter method (Fig. 6), with the procedure described by Ndam and Yogo (1999). Only the trees with a dbh of more than 30 cm can be debarked. The trees with a dbh below 50 cm should be debarked with two strips in opposite sides, each no wider than 1/4 of the tree circumference, from 1 m above the ground up to the first branch. From those with a dbh of 50 cm or more, four strips should be removed, each no wider than 1/8 of the tree circumference. Moreover each portion debarked on the tree should be separated from the next one by an untouched section of the bark of the same width. Lateral roots with a minimum diameter of 20 cm on trees with a dbh above 50 cm can also be debarked.

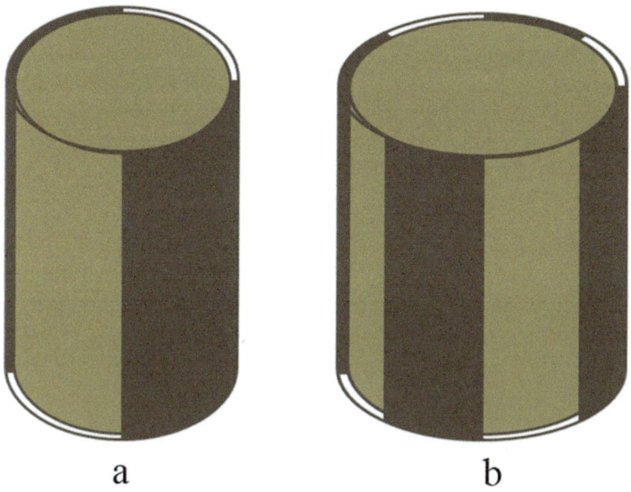


**Fig. 6** Two-quarter method for *Prunus africana* (Hook. f.) Kalman bark harvest (bark in *dark brown*) for trees with a diameter at breast height (dbh) of less than 50 cm (**a**) and for trees with a dbh of more than 50 cm (**b**) from 1.30 m height to the first branch

The trees should completely recover from the first debarking before subsequent debarking (Ndibi and Kay 1997).

The recommendations for a sustainable harvest of *P. africana* bark also include the time between the harvest of the first 50% of the bark and the harvest of the last part. The minimum time of rotation has been found to be 7–8 years (Ingram 2014). The problem for the harvesters is that the bark yield decreases with an increase of the rotation time. From a 5-year rotation to a 6-year rotation, the bark yield can be reduced by more than 20% (Cunningham et al. 2014).

In practice, the use of this harvest technique shows that the tree responds differently depending on the site. In moist sites, bark regrowth occurs but in dry sites, bark does not recover (Cunningham et al. 2014). This leads to wood-borer and fungal attack on the debarked trees (Cunningham et al. 2002). Insects' attacks also occur in lower altitude sites, even on healthy *P. africana* trees (Cunningham et al. 2014).

There is a tendency to harvest *P. africana* bark during dry or low rainfall periods (Cunningham and Mbenkum 1993; Walter and Rakotonirina 1995). Using the method described previously, between 15 and 100 kg (depending on the size of the tree) can be obtained from a tree at each harvest (Tsobeng et al. 2008). To produce 5 kg of extract 1 ton of dry bark (that corresponds approximatively to the double of fresh bark) is needed to be harvested from mature trees (Cunningham et al. 1997).

Differences in genetic and chemical constituents of the bark have been found across the distribution range of *P.africana* that may reflect environmental differences within the range of a species (Kadu et al. 2012). A study in Cameroun, Democratic Republic of the Congo and Madagascar has already shown that the chemical composition of the bark depends on its origin (Martinelli et al. 1986).

Madagascar, a region genetically distinct from the African continent, seems to have populations with very high chemicals constituents studied by Kadu et al. (2012).

## 8 Perspectives for a Sustainable Use

For a sustainable use of *P. africana*, it is important to focus on reducing the impacts of the current techniques of harvest on the populations' survival but also on looking for alternatives to wild harvest. A lot of research is absolutely necessary to underpin a sound harvest aiming at a sustained production of bark.

### *Bark, Cambium and Wood Anatomy*

The macroscopic phenomenon of *P. africana* bark recovery has been studied and it is known that after debarking, *P. africana* rapidly produces new bark on the wound surface (Cunningham and Mbenkum 1993; Vermeulen and Geldenhuys 2004; Geldenhuys et al. 2007). No information exists of this phenomenon at the microscopic level. However, to understand how *P. africana* can regenerate its bark, it is essential to study the tissues and the anatomical feature modification due to the injury. Delvaux et al. (2013) proposed the size of the conducting phloem as a key-factor for bark regeneration. Research should make it clear how, when and where new cambium is generated and how fast new tissues are being formed.

### *Inventory and Monitoring*

To assess the state of the *P. africana* populations and the impacts of the current techniques on their survival, a standardized and accepted method of sampling needs to be adopted. The National Management Plan for *P. africana* in Cameroon (Ingram et al. 2009) included inventories that used different methods, with very different results, even for the same locations (Cunningham et al. 2014). In their study, Morrison et al. (2008) found that grid-based systematic designs were the more efficient and practical method for sampling.

It is also important to assure that bark is not harvested from protected areas. A recent project carried out by ITTO-CITES has developed a DNA tracking technology for *P. africana* bark (Darren 2016). With this technique, it is possible to identify the origin of the harvested bark and if it was from authorized harvest zones.

## 9 Cultivation

Unlike *P. africana*, other tree species from which the bark is commercially harvested (cinnamon, cork oak, chestnut, *etc.*) have made the transition from wild harvest to production in agroforestry or plantation systems (Lubbe and Verpoort 2011).

In Cameroon, several attempts have started to cultivate *P. africana* more than 20 years ago (Cunningham and Mbenkum 1993). Small-scale farmers began producing the species in agroforestry systems (Cunningham and Mbenkum 1993; Tchoundjeu et al. 2002). In terms of industrial cultivation, only 9 ha of pure plantation have been established by the Cameroon Development Corporation (Ndam and Tonye 2004). The annual *P. africana* plantation targets of 2 ha set in 1986 and 5 ha set in 1992 (Ndibi and Kay 1997) was far too small to have a meaningful impact (Hall et al. 2000). Attempts at cultivation are also underway in Kenya (Dawson et al. 2000; Dawson and Powell 1999).

There are different cultivation practices to consider that can be found in Ingram et al. (2015). For example, Tchoundjeu et al. (2002) found that vegetative propagation is suitable for *P. africana*, giving other options to domestication. The harvest of leaves is another alternative to bark harvest, since both plant material have the same pharmaceutical properties (Debat 1974).

For *P. africana* conservation, it is important to have regional approaches because of the genetic variation across the range but even at the country level. Indeed, differences have been found in populations in Madagascar and in Cameroon (Dawson and Powell 1999). It would be especially interesting to do research on the potential of the species in agroforestry systems.

## 10 Conclusion

The growing interest for *P. africana* is strengthening the threats the species is facing. It is therefore necessary to protect wild populations and find other ways to produce *P. africana* bark in order to sustainably supply the pharmacological industry with its valuable extract.

## References

Bassi P, Artibani W, De Luca V, Al E (1987) Estratto standardizzato di Pygeum africanum nel trattamento dell'ipertrofia prostatica benigna. Studio clinico controllato versus placebo. Minerva Urol Nefrol 39:45–50

Bombardelli E, Morazzoni P (1997) *Prunus africana* (Hook.f.) Kalkm. Fitoterapia 68:205–218

Carbin B-E, Larsson B, Lindahl O (1990) Treatment of benign prostatic hyperplasia with phytosterols. Brit J Urol 66:639–641

Chapman CA, Chapman LJ (1999) Implications of Small-scale variation in ecological conditions for the diet and density of red colobus monkeys. Primates 40(1):215–231

Chapman CA, Chapman LJ, Rode KD, Hauckand EM, McDowell LR (2003) Variation in the nutritional value of primate foods: Among trees, time periods, and areas. Int J Primatol 24:317–333

CIFOR : Bogor, Indonesia. http://www.cifor.org/publications/pdf_files/WPapers/WP153Sunderland.pdf. Accessed 29 June 2015

Clark LE, Sunderland TC (2004) The key non-timber forest products of Central Africa: state of the knowledge. SD Publication Series Technical Paper No. 122. L. E. Clark and T. C. Sunderland, U.S. Agency for International Development, Office of Sustainable Development, Bureau for Africa: 1999

Conway G (2009) The science of climate change in Africa: impacts and adaptation. Discussion paper No 1. Grantham Institute for Climate Change. London Imperial College, UK

Cunningham AB, Mbenkum FT (1993) Sustainability of harvesting *Prunus africana* bark in Cameroon: a medicinal plant in international trade. People and plants Working Paper, vol 2, pp 1–28

Cunningham M, Cunningham AB, Schippmann U (1997) Trade in *Prunus africana* and the implementation of CITES. German Federal Agency for Nature Conservation, Bonn. Available via https://www.researchgate.net/publication/276028274_Trade_in_Prunus_africana_and_the_Implementation_of_CITES. Accessed 24 July 2015

Cunningham AB, Ayuk E, Franzel S, Duguma B, Asanga C (2002) An economic evaluation of medicinal tree cultivation: *Prunus africana* in Cameroon. In: People and plants Working Paper No. 10. UNESCO, Paris. Available via http://unesdoc.unesco.org/images/0013/001330/133098e.pdf. Accessed 29 June 2015

Cunningham AB et al (2008) Sustainable trade and management of forest products and services in the COMESA region. Cifor. Available via http://www.cifor.org/publications/pdf_files/Books/BCunningham0801.pdf. Accessed 27 July 2015

Cunningham A, Avana Tientcheu M-L, Anoncho V, Nkuinkeu R, Sunderland T (2014) Power, profits and policy: a reality check on the *Prunus africana* bark trade. In Working Paper 153. 10

Darren T (2016) A pilot study of *Prunus africana* shows that DNA tracking technology can help safeguard supply chains from illegality. ITTO Trop For Update 25(1):16–19

Dawson IK, Powell W (1999) Genetic variation in the Afromontane tree *Prunus africana*, an endangered medicinal species. Mol Ecol 8:151–156

Dawson I, Were J, Lengkeek A (2000) Conservation of *Prunus africana*, an over-exploited African medicinal tree. For Genet Resour FAO 28:27–33

Debat J (1966) Br. App. 15.893/66, 10 June 1966

Debat J (1974) *Prunus africana* extract. US Patent 3,856,946, 24 Dec 1974

Delvaux C, Sinsin B, Van Damme P, Beeckman H (2013) Size of conducting phloem: The « key » factor for bark recovery of 12 tropical medicinal tree species. Flora 208:111–117

Eben Ebai S, Ewusi BN, Asanga CA, Nkongo JBN (1992) An evaluation of the quantity and distribution of Pygeum africanum on the slopes of Mount Cameroon. Divisional Service of Forestry, Limbe

Ewusi BN, Eben-Ebai S, Asanga CA, Nkongo JBN (1992) An evaluation of the quantity and distribution of *Pygeum africanum* on the slopes of Mount Cameroon. Unpublished Report prepared for Plantecam-Medicam

Farwig N, Carsten B, Bohning-Gaese K (2008) Human disturbance reduces genetic diversity of an endangered tropical tree, *Prunus africana* (Rosaceae). Conserv Genet 9:317–326

Fashing PJ (2004) Mortality trends in the African cherry (*Prunus africana*) and the implications for colobus monkeys (*Colobus guereza*) in Kakamega Forest, Kenya. Biol Conserv 120:449–459

Forboseh PF, Sunderland TCH, Comiskey JA, Balinga M (2011) Tree population dynamics of three altitudinal vegetation communities on Mount Cameroon (1989–2004). J Mt Sci 8:495–504

Geldenhuys CJ (1981) *Prunus africana* in the Bloukrans River Gorge, Southern Cape. S A Fores J 118:61–66

Geldenhuys CJ, Syampungani S, Meke GS, Vermeulen WJ (2007) Response of different species to bark harvesting for traditional medicine in Southern Africa. In: Multiple use management of natural forests and woodlands: policy refinement and scientific progress. Department of Water Affairs and Forestry, Indigenous Forest Management, Pretoria, pp 55–62

Hall JB, O'Brien EM, Sinclair LF (2000) *Prunus africana*: a monograph. School of Agricultural and Forest Sciences. Publication No 18 University of Wales, Bango

Ingram V (2014) Profitability and value chain analysis of *Prunus africana* commercialisation from Mount Cameroon, Cameroon, P.-. GIZ (ed). Coopération Germano – Camerounaise. Ministère des Forêts et de la Faune (MINFOF). Ministère de l'Environnement et de la Protection de la Nature et du Développement durable (MINEP). Deutsche Gesellschaft für Internationale Zusammenarbeit (GIZ) GmbH: The Netherlands and Cameroon, p 100

Ingram V, Schure J (2010) Review of non timber forest products (NTFPs) in Central Africa: Cameroon. CIFOR, Yaoundé

Ingram V, Awono A, Schure J, Ndam N (2009) Guidance for a national *Prunus africana* management plan. Cameroon. Center for International Forestry Research, Food and Agriculture Organization, Cameroon, p 158

Ingram V, Van Loo J, Vinceti B et al (2015) Ensuring the future of the pygeum tree (*Prunus africana*). Factsheet (p 8). LEI Wageningen UR

Kadu CAC, Parich A, Schueler S, Konrad H, Muluvi GM, Eyog-Matig O, Muchugi A, Williams VL, Ramamonjisoa L, Kapinga C et al (2012) Bioactive constituents in *Prunus africana*: geographical variation throughout Africa and associations with environmental and genetic parameters. Phytochemistry 83:70–78

Kadu CAC, Konrad H, Schueler S, Muluvi GM, Eyog-Matig O, Muchugi A, Williams VL, Ramamonjisoa L, Kapinga C, Foahom B, Katsvanga C, Hafashimana D, Obama C, Geburek T (2013) Divergent pattern of nuclear genetic diversity across the range of the Afromontane *Prunus africana* mirrors variable climate of African highlands. Ann Bot 111:47–60

Kalkman C (1965) The Old World species of *Prunus* sub-genus *Laurocerasus* including those formerly referred to as *Pygeum*. Blumea 13:1–115

Lowe FC, Ku JC (1996) Phytotherapy in treatment of benign prostatic hyperplasia: a critical review. Urology 48:12–20

Lubbe A, Verpoort R (2011) Cultivation of medicinal and aromatic plants for specialty industrial materials. J Ind Crops Prod 34:785–801

Martinelli EM, Seraglia R, Pifferi G (1986) Characterization of Pygeum africanum bark extracts by HRGC with computer assistance. J High Resolut Chromatogr Chromatogr Comm 9:106–110

Mayaux P, Pekel J-F, Desclée B, Donnay F et al (2015) State and evolution of the African rainforests between 1990 and 2010. Philos Trans R Soc Lond Ser B Biol Sci 368(1625)

Morrison LW, Smith DR, Young CC, Nichols DW (2008) Evaluating sampling designs by computer simulation: a case study with the Missouri bladderpod. Popul Ecol 50(4):417–425

Ndam N. (1996) Recruitment patterns of *Prunus africana* (Hook f.) Kalkman on Mount Cameroon: a case study at Mapanja. In: A strategy for the conservation of *Prunus africana* on Mount Cameroon. Technical Papers and Workshop Proceedings

Ndam N, Ewusi B (2000) Case study: income from *Prunus africana*. In: Price MF, Butt N (eds) Forests in sustainable mountain development: a state of knowledge report for 2000. CAB International, Wallingford, pp 306–309

Ndam N, Tonye MM (2004) *Prunus africana* on Mount Cameroon: a case study of the production-to consumption systems. In: Sunderland T, Ndoye O (eds) Forest products, livelihoods and conservation. Case studies of non-timber forest product systems, vol 2 – Africa, pp 37–52

Ndam N, Yogo C (1999) A review of NFORKEMBA exploitation of *Prunus africana* in Cameroun

Ndibi BP, Kay EJ (1997) The regulatory framework for the exploitation of medicinal plants in Cameroon: the case of *Prunus africana* on Mount Cameroon. Biodivers Conserv 6:1409–1412

Nkeng PF, Ingram V, Awono A, Tientcheu MLA (2009) Assessment of *Prunus africana* bark exploitation methods and sustainable exploitation in the South west, North-West and Adamaoua regions of Cameroon, in Project GCP/RAF/408/EC « Mobilisation et Renforcement des Capacités des Petites et Moyennes Entreprises impliquées dans les Filières des Produits Forestiers Non Ligneux en Afrique Centrale », CIFOR, Editor. FAO-CIFORSNV- World Agroforestry Center-COMIFAC, Yaounde

Parrott J, Parrott H (1989) Report on the conservation of Prunus (Pygeum africanum) in Cameroon. Unpublished Report prepared for the International Council for Bird Preservation (Now Birdlife International)

Sim TR (1907) Forests and forest flora of the colony of the Cape of Good Hope. Taylor & Henderson, Aberdeen

Simons AJ, Dawson IK, Duguma B, Tchoundjeu Z (1998) Passing problems: prostate and *Prunus*. HerbalGram 43:49–53

Stewart KM (2001) The commercial bark harvest of the African cherry (*Prunus africana*) on Mount Oku, Cameroun: effects of traditional uses and population's dynamics. Dissertation, Florida international University

Stewart KM (2003) The African cherry (*Prunus africana*): can lessons be learned from an over-exploited medicinal tree? J Ethnopharmacol 89:3–13

Sunderland T, Tako CT (1999) The exploitation of *Prunus africana* on the island of Bioko, Equatorial Guinea. Report for the People and Plants Initiative, WWF-Germany and the IUCN/SSC Medicinal Plant Specialist Group, Bonn, p 13

Tchoundjeu Z, Avana ML, Leakey RRB, Simons AJ et al (2002) Vegetative propagation of *Prunus africana*: effects of rooting medium, auxin concentrations and leaf area. Agrofor Syst 54:183–192

Tsobeng A, Degrande A, Tientcheu M-LA, Tchoundjeu Z, Tabuna H (2008) Cultivation and sustainable bark collection of *Prunus africana*. In: Technical Note. ICRAF, Yaounde, p 15

Vermeulen WJ, Geldenhuys CJ (2004) Experimental protocols and lessons learnt from strip harvesting of bark for medicinal use in the southern Cape forests. In: FRP-DFIP Project R8305 Report, Wild Resources Limited, UK (unpublished)

Walter S, Rakotonirina J-CR (1995) L'exploitation de *Prunus africana* à Madagascar. PCDI Zahamena et la Direction des Eaux et Forêts, Antanarivo/Fénérive-Est, p 23

Wei JT, Clahoun E, Jacobsen SJ (2005) Urologic diseases in America project: benign prostatic hyperplasia. J Urol 173(4):1256–1261

White F (1983) Long-distance dispersal and the origins of the Afromontane flora. Sonderbände Naturwissenschaftliche vereins Hamburg 7:87–116

Wilt TJ, Ishani A (1998) *Pygeum africanum* for benign prostatic hyperplasia. Cochrane database of systematic reviews 1

# Bird Plum; *Berchemia discolor* (Klotzsch) Hemsl.: A Review of Its Ethnobotany, Phytochemistry and Pharmacology

**Ahmad Cheikhyoussef and Alfred Maroyi**

**Abstract** Bird plum (*Berchemia discolor*) is a shrub or tree distributed in the Arabian Peninsula and Africa characterized by edible fruits and its wood makes charcoal, building material, beehives, furniture and crafts. *Berchemia discolor* is also used as a dye, fodder, ornamental tree and as herbal medicine to treat several human and animal diseases and ailments. The objective of this study was to collate the fragmented information on the ethnobotanical uses, phytochemical and pharmacological properties of the Bird plum throughout its geographical range in the African continent so as to highlight research gaps and provide a foundation for further investigations on the plant species. The utilization of *Berchemia discolor* as a food plant, for various other uses including its medicinal applications forms the basis of the current demand for the plant species in Africa. Research on *Berchemia discolor* over the past decade on its health promoting properties have greatly contributed to the increased consumption of the species as a food plant and herbal medicine. Therefore, further studies are required to improve our knowledge about its phytochemistry, mechanisms of action, efficacy, toxicity and clinical relevance of the plant species.

**Keywords** Bird plum • *Berchemia discolor* • Africa • Food • Herbal medicine • Livelihood needs

---

A. Cheikhyoussef (✉)
Science and Technology Division, Multidisciplinary Research Centre (MRC), University of Namibia, Private Bag 13301, Windhoek, Namibia
e-mail: acheikhyoussef@unam.na

A. Maroyi
Medicinal Plants and Economic Development (MPED) Research Centre, Department of Botany, University of Fort Hare, Private Bag X1314, Alice 5700, South Africa
e-mail: amaroyi@ufh.ac.za



M. Neffati et al. (eds.), *Medicinal and Aromatic Plants of the World - Africa Volume 3*, Medicinal and Aromatic Plants of the World,
DOI 10.1007/978-94-024-1120-1_7

## 1 Taxonomy and Geographic Distribution

The genus *Berchemia* Neck. ex DC. is named after M. Berchem, a seventeenth century French botanist. The specific epithet *discolor* means the species has two different colours (Palmer and Pitman 1972) referring to the fact that the upper and the lower leaf surfaces are of different colours, that is, green and silver grey respectively. *Berchemia* is a member of the Rhamnaceae, characterized by 51 genera and over 600 species, distributed throughout the warm countries of the world (Palmer and Pitman 1972). It is native to Angola, Botswana, Eritrea, Ethiopia, Kenya, Madagascar, Malawi, Mozambique, Namibia, Somalia, South Africa, Swaziland, Tanzania, Uganda, Yemen, Zambia and Zimbabwe. *Berchemia discolor* is commonly known as *bird plum*; other vernacular names are given in Table 1.

*Berchemia discolor* is a semi-deciduous shrub or tree (Fig. 1a), usually between 3 and 20 m in height (Palmer and Pitman 1972). The trunk is a straight bole with rough, dark grey bark (Fig. 1b) that flakes longitudinally, with a dense and rounded crown. The branches are conspicuously covered by lenticelles. Branchlets are glabrous to densely pubescent with short, spreading whitish hairs. The leaves (Fig. 1c) are alternate or sub-opposite, entirely or obscurely crenate, shiny above, dull and dark green below with raised veins on the lower side. Flowers are small, solitary and thick, while fruits are datelike, yellow to pale orange with a single stone in sweet, edible flesh.

## 2 Traditional Uses

*Berchemia discolor* is one of the most valuable non-timber forest product in sub-Saharan Africa. The bark, fruit juice, leaves and roots are used to treat numerous health complaints including abdominal pains, abscesses, flu and cold, general body pains, infertility, malaria, menorrghagia, nose bleeding, skin itching, toothache and livestock diseases (Table 2).

The bark of *Berchemia discolor* is used as an alluring love charm in South Africa (Hutchings et al. 1996) and as poultices (Hutchings et al. 1996; van Wyk and Gericke 2007). *Berchemia discolor* is also used to produce alcoholic drink, animal food, gum or resin, human food, live fence, ornaments, tannin or dye, timber, tobacco pipes and windbreak (Table 2). The fruits are marketed mainly by women and children to generate income in Ethiopia, Namibia and Zimbabwe (Mukamuri and Kozanayi 1999; Feyssa et al. 2012a; Cheikhyoussef and Embashu 2013). Bille et al. (2013) reported on three value added products from *Berchemia discolor* fruits in Namibia, these products are: juice, jam and jelly which were scored through a sensory evaluation model with 3.2 ± 0.28 for juice, 2.95 ± 0.29 for jam and 2.31 ± 0.27 for jelly and they were significantly different (<0.05) from the other two fruits used in their study.

**Table 1** Vernacular names of *Berchemia discolor*

| Vernacular name(s) | Country | Reference(s) |
|---|---|---|
| Bird plum | English, South Africa | Palmer and Pitman (1972), Mabogo (1990), Hutchings et al. (1996), and van Wyk and Gericke (2007) |
| | Swaziland | Long (2005) |
| | English, Zimbabwe | Ganduri et al. (2014) |
| Brown ivory | English, South Africa | Hutchings et al. (1996) and van Wyk and Gericke (2007) |
| Bruinivoor | Afrikaans, South Africa | Hutchings et al. (1996) and van Wyk and Gericke (2007) |
| Deen | Somali, Kenya | Gachathi et al. (1994) |
| Eembe | Namibia | Barrion et al. (2001) and Bille et al.(2013) |
| Jajab | Orma, Kenya | Gachathi et al. (1994) |
| Jajabho | Malakote, Kenya | Gachathi et al. (1994) |
| Jejeba | Amharic, Ethiopia | Dejen and Merene (2013) |
| Matatya | Shona, Zimbabwe | Wild et al. (1972) |
| Mgandu | Gogo, Tanzania | Hines and Eckman (1993) |
| Mkuni | Nyamwezi, Tanzania | Hines and Eckman (1993) |
| Mmupudu | Botswana | Ohiokpehai (2003) |
| Mnago | Tanzania | Hines and Eckman (1993) |
| Monyee | Botswana | Kayombo et al. (2014) |
| Motsentsila | Tswana, South Africa | van Wyk and Gericke (2007) |
| Mountain date | English, South Africa | Palmer and Pitman (1972) and Hutchings et al. (1996) |
| Mozinzila | South Africa | Palmer and Pitman (1972) |
| Muchukwo | Kenya | Kipkorir and Kareithi (2012) |
| Mugaramhanga | Shona, Zimbabwe | Wild et al. (1972) |
| Muhukhuma | Venda, South Africa | Palmer and Pitman (1972) |
| Mukhukhuma | Venda, South Africa | Mabogo (1990) |
| Mujajabho | Malakote, Kenya | Gachathi et al. (1994) |
| Munhacha | Shona, Zimbabwe | Wild et al. (1972) |
| Munie | Venda, South Africa | Mabogo (1990) |
| Munyi | Shona, Zimbabwe | Wild et al. (1972) |
| Munyii | Tonga, Zimbabwe | Wild et al. (1972) |
| Munyiyi | Shona, Zimbabwe | Wild et al. (1972) |
| Mutsintsila | Tswana, South Africa | Palmer and Pitman (1972) |
| Muzinzila | Lozi, Zambia | Sorensen (1993) |

(continued)

**Table 1** (continued)

| Vernacular name(s) | Country | Reference(s) |
|---|---|---|
| Muwe | Tsonga, South Africa | Liengme (1981) |
| Mwii | Tonga, Zimbabwe | Mukamuri and Kozanayi (1999) |
| Nyahumbu | Pogoro, Tanzania | Hines and Eckman (1993) |
| Nyii | Hlengwe, Zimbabwe | Wild et al. (1972) |
| | Shona, Zimbabwe | Mukamuri and Kozanayi (1999) |
| Nyiri | Tsonga, South Africa | Liengme (1981) |
| Nyiyi | Tsonga, South Africa | Palmer and Pitman (1972) and Liengme (1981) |
| Okoo | Sandawi, Tanzania | Hines and Eckman (1993) |
| Omuve | Herero, Namibia | Palmer and Pitman (1972) |
| Omuye | Oshiwambo, Namibia | Cheikhyoussef and Embashu (2013) |
| Pink ivory | English, Zimbabwe | Ganduri et al. (2014) |
| Ubalatsheni likulu | Zulu, South Africa | Palmer and Pitman (1972) |
| Ubalatsheni omkulu | Zulu, South Africa | Hutchings et al. (1996) |
| Umadlozane | Zulu, South Africa | Hutchings et al. (1996) |
| Umhlungulo | Zulu, South Africa | Hutchings et al. (1996) |
| UmMumu | Zulu, South Africa | Palmer and Pitman (1972) |
| Umncaga | Ndebele, Zimbabwe | Wild et al. (1972) |
| Umnyi | Zimbabwe | Ganduri et al. (2014) |
| Umnyiyi | Ndebele, Zimbabwe | Wild et al. (1972) |
| Umumu | Zulu, South Africa | Hutchings et al. (1996) |
| Uvuku | Zulu, South Africa | Hutchings et al. (1996) |
| | Swaziland | Long (2005) |
| Voëlpruim | Afrikaans, South Africa | Carruthers (2000) |
| Wild almond | English, South Africa | Palmer and Pitman (1972) and Hutchings et al. (1996) |
| Wild almond | English, Tanzania | Hines and Eckman (1993) |
| Wilde dadel | Afrikaans, South Africa | Palmer and Pitman (1972) and Hutchings et al. (1996) |

## 3 Phytochemistry

The seed oil content from *Berchemia discolor* give a yield of 11% as reported by Sunder and Sino (1992). The unsaturated fatty acids (70%) (UNSAT), saturated fatty acids (30%) (SAT) with ω-6/ω−3 ratio of 17 (Bombarda et al. 2010). Polygalacturonase was extracted from ripe *Berchemia discolor* fruits in Zimbabwe

**Fig. 1** *Berchemia discolor* in Namibia at the main campus of the University of Namibia, Windhoek (**a**), barks on trunk (**b**) leaves and fruit (**c**) (Taken by author)

with optimum pH ranging from 4.5 to 5 and optimum temperature from 25 to 37 °C (Muchuweti et al. 2005).

The qualitative phytochemical analysis revealed the presence of tannins, flavonoids, saponins, total phenols, steroids, phlobatanins, terpenoids, cardiac glycosides and anthraquinones (Cheikhyoussef et al. 2010). The total phenolic content (TPC) in fruit extracts ranged between 33.98 ± 0.05 and 40.03 ± 0.04 mg gallic acid equivalent (GAE)/g. Total flavonoid content (TC) in bark extracts ranged from 3.829 ± 0.04 to 5.473 ± 0.08 mg equivalents to quercetin (QE)/g. Fruit extract had the highest content of saponins and the lowest content of tannins, meanwhile, bark extracts showed the highest content of tannins and the lowest content of saponins (Cheikhyoussef 2014). The presence of these phytochemicals in *Berchemia discolor* extracts is probably responsible for the medicinal efficacy of this plant in Namibian traditional healing system (Cheikhyoussef et al. 2011). Bark extracts exhibited the highest reducing power (508.36 %) followed by leaves (388.83%) then fruit (377%). The DPPH radical scavenging activity in bark extracts was very high (30.27%) compared to leaves and fruits extracts at 23.65% and 13.09% respectively (Cheikhyoussef 2014).

**Table 2** Ethnobotanical uses of *Berchemia discolor*

| Use | Plant part(s) used and preparation | Country practised | Reference(s) |
|---|---|---|---|
| **Medicinal uses** | | | |
| Abdominal pains | Roots extract used against abdominal pains medicine | Zimbabwe | Maroyi (2011) |
| Abscesses | Leaves used against abscesses | Ethiopia | Desta (1993) |
| Bleeding gums | Fruit juice used to treat bleeding gums | South Africa | van Wyk and Gericke (2007) |
| Flu and cold | Leaves used against flu and cold | Namibia | Cheikhyoussef and Embashu (2013) |
| General body pains | Body washed with root infusion as remedy for general body pains | Zimbabwe | Gelfand et al. (1985) |
| Infertility | Bark and fruits used for infertility | South Africa | Mabogo (1990) |
| Love charm | Bark used as an alluring | South Africa | Hutchings et al. (1996) |
| Malaria | Stem bark boiled with roots of *Cordia crenata* Delile and *Tamarindus indica* L. as remedy for malaria | Tanzania | Chin et al. (2006) |
| Menorrghagia | Smoke from burning roots used to treat menorrghagia | South Africa | Arnold and Gulumian (1984) |
| Nose bleeding | Bark used against nose bleeding | Namibia | Cheikhyoussef and Embashu (2013) |
| Poultices | Boiled leaves used as poultices | South Africa | Hutchings et al. (1996) and van Wyk and Gericke (2007) |
| Skin itching | Bark used against skin itching | Namibia | Cheikhyoussef and Embashu (2013) |
| Toothache | Bark used for toothache | Madagascar | Jenkins (1987) |
| Veterinary herbal medicine | Bark used as veterinary herbal medicine | Namibia | Cheikhyoussef and Embashu (2013) |
| **Other ethnobotanical uses** | | | |
| Alcohol | Alcoholic drink distilled from fruits | Namibia, South Africa | Barrion et al. (2001), van Wyk and Gericke (2007) and Cheikhyoussef and Embashu (2013) |
| Animal food | Fruit and leaves used as fodder for baboons, birds, camels, cattle, giraffe, goats and monkeys | Ethiopia; Kenya; South Africa; Tanzania | Palmer and Pitman (1972), de Groot and Hall (1989), Hines and Eckman (1993) and Feyssa et al. (2012a) |
| Apiculture | Bees attracted to the flowers; people hang beehives in the trees and tree trunks used for making beehives | Kenya, Tanzania | Hines and Eckman (1993) and de Groot and Hall 1989 |

(continued)

**Table 2** (continued)

| Use | Plant part(s) used and preparation | Country practised | Reference(s) |
|---|---|---|---|
| Gum or resin | Heartwood produces gum and resin | Tanzania | Hines and Eckman (1993) |
| Human food | Dried fruit flesh pounded to make a meal which, when mixed with seeds of Eleusine coracana Asch. & Gr. and Pennisitum typhoides Stapf. & C.E. Hubb. is kneeded into a biscuit dough and baked or steamed. Beverage similar to tea made from leaves, and fruits added to porridge | Ethiopia; Kenya, Nambia, South Africa, Swaziland; Tanzania, Zimbabwe | Palmer and Pitman (1972), Liengme (1981), de Groot and Hall (1989), Mabogo (1990), Hines and Eckman (1993), Mukamuri and Kozanayi (1999), Long (2005), Feyssa et al. (2012b) and Cheikhyoussef and Embashu (2013) |
| Income generation | Fruits marketed mainly by women and children to generate income | Ethiopia, Namibia, Zimbabwe | Mukamuri and Kozanayi (1999), Feyssa et al. (2012a) and Cheikhyoussef and Embashu (2013) |
| Live fence | Species conserved at farm boarders, live fences and enclosed pasture (kalo) areas | Ethiopia | Feyssa et al. (2012a) |
| Ornaments | Seeds used by Kwanyamas of Namibia to decorate hair of women | Namibia | Loeb et al. (1956) and Palmer and Pitman (1972) |
| Tannin or dye | Brown and purple dyes come from the wood and bark, these provide the colouring for basket fibres. Mauvish-brown colour given to fishing nets by boiling them in water mixed with bark of *Lannea stuhlmannii* | Botswana; Namibia; South Africa; Swaziland; Tanzania; Zimbabwe | Palmer and Pitman (1972), Mabogo (1990), Hines and Eckman (1993), Mukamuri and Kozanayi (1999), Long (2005), van Wyk and Gericke (2007), Shumba et al. (2010) and Ganduri et al. (2014) |
| Timber | Wood excellent for axe-handles, building material, charcoal, crafts, furniture, firewood, knife handles, knobkerries, mortar, poles, pestles and hair combs | Kenya; South Africa; Swaziland; Tanzania; Zimbabwe | Palmer and Pitman (1972), Liengme (1981), de Groot and Hall (1989), Hines and Eckman (1993), Mabogo (1990), Mukamuri and Kozanayi (2001), Long (2005), van Wyk and Gericke (2007), Kipkorir and Kareithi (2012) and Ganduri et al. (2014) |
| Tobacco pipes | Fashioned from wood | Botswana | Palmer and Pitman (1972) |
| Windbreaks | Spreading branches and heavy rounded crown makes *Berchemia discolor* an effective shade tree and act as windbreak | Tanzania | Hines and Eckman (1993) |

The fruit pulp has high sugar content (30%), and contains minerals (K, Ca, Mg, P) and vitamin C (65 mg/100 g) and seeds taste like walnuts (Venter and Venter 1996). The concentration of proteins and carbohydrates in *Berchemia discolor* were 18.21–19.59%, 16.72–16.89% respectively. There was a significant interactive effect ($p < 0.05$) between different land uses and nutrient contents for all the parameters, dry matter (DM), moisture, crude fiber (CF) and crude protein (CP) (Feyssa et al. 2012a). They further concluded that land use had significant effect on the nutritional content of *Berchemia discolor* indicating that land use should be considered as one factor of production and domestication of the species in Ethiopia (Feyssa et al. 2012a). The nutrients contents of fruits significantly varied for all variables analysed except for copper (Cu) and condensed tannin (CT). Phosphorus, calcium, zinc, magnesium, manganese are relatively higher in fruits collected from transhumance land use than settled farmers land use, while iron is relatively higher from fruits collected from settled farmers land use (Feyssa et al. 2012a). Vitamin-A content of *Berchemia discolor* significantly varied across land uses ($p < 0.05$) with higher mean value from samples collected from transhumance land use systems. There was no significant variation in vitamin-C content across land uses ($p > 0.05$). Therefore, land use is a factor to be considered in domestication of the species. Moreover, the high vitamin C indicates the potential that consumption of the fruits can enhance metallic nutrients absorption such as iron. These two vitamins are also among the critical vitamins focused by current human nutrition security (Feyssa et al. 2012a).

## 4 Pharmacology

Chin et al. (2006) isolated five new prenylated flavonoids (Fig. 2) from the root bark of *Berchemia discolor*, collected in Tanzania, along with ten known compounds, by bioactivity-guided fractionation. The isolated compound exhibited cytotoxic activity when evaluated against three human cancer cell lines (Lu1, LNCaP, and MCF-7) (Chin et al. 2006).

The structures of the known compounds were identified by physical and spectroscopic data measurement ([$\alpha$] D, CD, $^{1}$H NMR, $^{13}$C NMR, 2D NMR, and MS) and by comparing these with the data obtained from published studies, such as nitidulin, heminitidulan, leiocarpin, leiocin, leiocinol, and nitidulan (Fig. 3) (van Heerden et al. 1981), amorphigenin and dabinol (Abe et al. 1985), 3-hydroxy-4′-O-methylglabridin (Kinoshita et al. 1996) and 4′-hydroxycabenegrin A-I (Fig. 4) (Silva et al. 1997).

Four compounds namely, (3,S)-5,2′,3′-trihydroxy-4′-methoxy-3″-methyl-3″-(4-methylpent-3-enyl)-pyrano[7,8] isoflavanone, nitidulin, amorphigenin and dabinol exhibited cytotoxic activity ($ED_{50} \leq 5$ μg/mL) for one or more cancer cell lines. Among them, nitidulin inhibited the growth only of the LNCaP cell line (49–52%), propagated within the intraperitoneal site, at all doses tested. The compound was not active against any of the lines propagated subcutaneously (Chin et al. 2006).

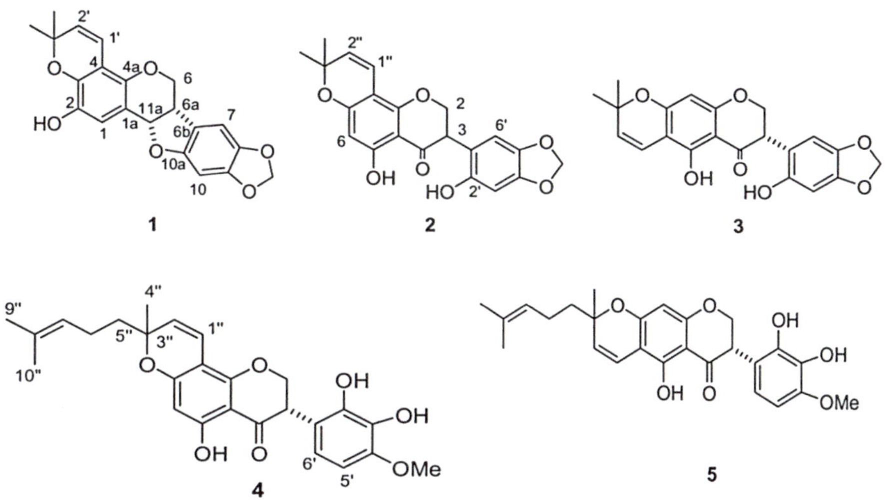


**Fig. 2** Five newly reported prenylated flavonoids from the root bark of *Berchemia discolor* collected in Tanzania (Adapted from Chin et al. 2006)

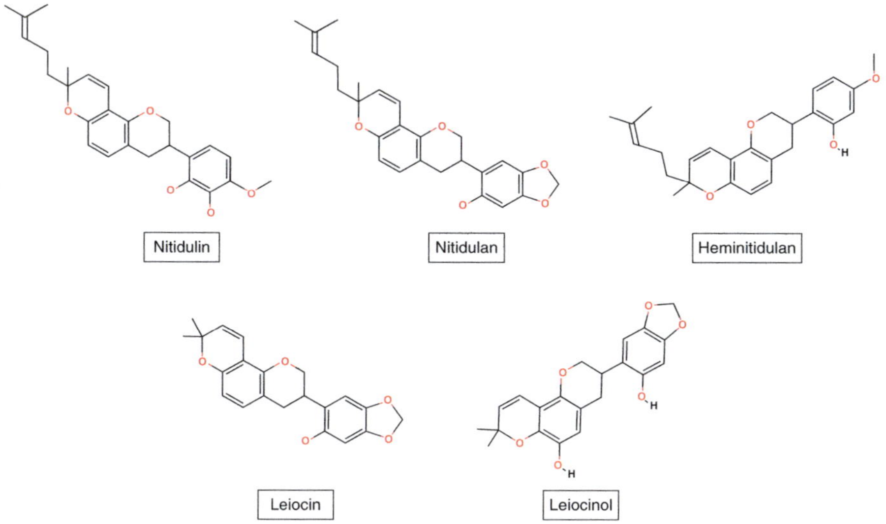


**Fig. 3** Prenylated flavonoids from *Berchemia discolor* (Group I)

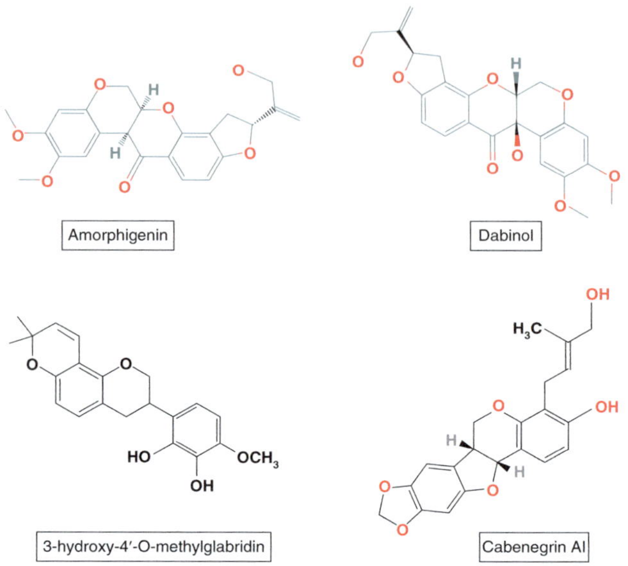


**Fig. 4** Prenylated flavonoids from *Berchemia discolor* (Group II)

Prenylated flavonoids have been identified in 37 of plant genera only and prenylation usually renders flavonoids with improved bioactivities (Yang et al. 2015). The mechanism of action is prenylation increases the lipophilicity of flavonoids, which results in a higher affinity to biological membranes and a better interaction with target proteins (Xu et al. 2012).

The antimicrobial activities of *Berchemia discolor* plant parts extracts were evaluated with well diffusion method (Perez et al. 1990). Ethanolic, methanolic and aqueous extracts of *Berchemia discolor* showed the strongest effects towards the indicator strains such as *Staphylococcus aureus* ATCC 25923 (inhibition diameter 9 ± 0.86 mm), *Bacillus subtilis* ATCC 11774 (13.5 ± 2.12 mm), *Alcaligenes faecalis* ATCC 8750 (11 ± 0.57 mm) *Clostridium difficile* ATCC 9689 (4.25 ± 1.06 mm), *Escherichia coli* ATCC 25922 (13 ± 1.52 mm) (Cheikhyoussef 2014) (Fig. 5).

*Lactobacillus plantarum* ATCC 8014 has not been affected by any extracts of *Berchemia discolor*. This could be explained by the fact that *Lactobacillus* ssp. have the ability to metabolize the phenolic compounds (Gyawali and Ibrahim 2012) and these compounds also serve as oxygen scavengers, which could have increased the growth of certain probiotic bacteria (Alberto et al. 2001).

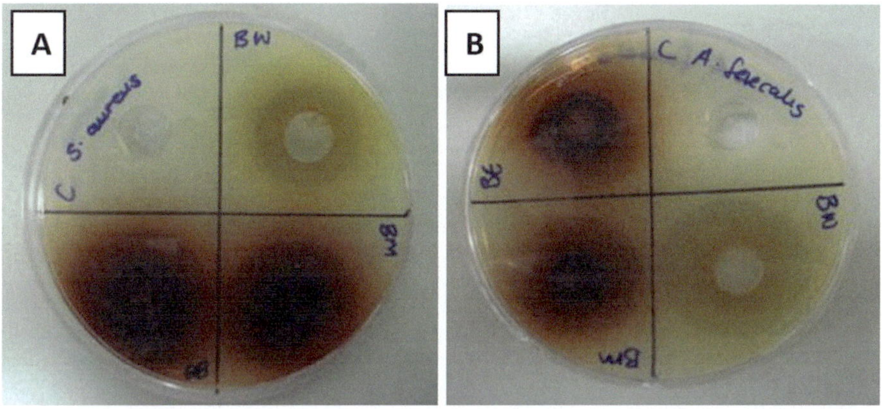


**Fig. 5** Antimicrobial activities of three extracts of *Berchemia discolor* using well diffusion method toward *Staphylococcus aureus* ATCC 25923 (**a**) and *Alcaligenes faecalis* ATCC 8750 (**b**) (Cheikhyoussef 2014)

## 5 Conclusion

*Berchemia discolor* is an important tree species characterized by edible fruits which are of social and economic value throughout its distributional range in African countries. *Berchemia discolor* is of high nutritional value but its consumption and use is not well promoted by the cultural production systems. *Berchemia discolor* is nutritionally rich in terms of major food nutrients such as carbohydrates, proteins, minerals, crude lipids, macro and micronutrients. Therefore, it can significantly contribute to human nutrition and ecosystem services to enhance human wellbeing. It provides livelihood needs in terms of nutrition, income generation, fuel wood, timber, fodder and medicine. *Berchemia discolor* extracts showed a broad inhibitory spectrum against food-borne bacteria and yeast. Scientifically based evidence on the health benefits of different parts of *Berchemia discolor* extracts may provide a farther value addition to this important fruit tree in the African continent.

## References

Abe F, Donnelly DMX, Moretti C, Polonsky J (1985) Isoflavonoid constituents from *Dalbergia monetaria*. Phytochemistry 24:1071–1176

Alberto MR, Farías ME, de Nadra MCM (2001) Effect of gallic acid and catechin on *Lactobacillus hilgardii* 5w growth and metabolism of organic compounds. J Agric Food Chem 49:4359–4363

Arnold HJ, Gulumian M (1984) Pharmacopoeia of traditional medicine in Venda. J Ethnopharmacol 12:35–74

Barrion S, Keya EL, Ngwira TN (2001) Country-wine making from Eembe fruit (*Berchemia discolour*) of Namibia. J Food Technol 6

Bille PG, Shikongo-Nambabi M, Cheikhyoussef A (2013) Value addition and processed products of three indigenous fruits in Namibia. Afr J Food Agric Nutr Dev 13(1):7192–7212

Bombarda I, Zongo C, McGill CR, Doumenq P, Fogliani B (2010) Fatty acids profile of *Alphitonia neocaledonica* and *Grevillea exul* var. *rubiginosa* seed oils, occurrence of an ω5 series. J Am Oil Chem Soc 87:981–986

Carruthers V (2000) Die natuurlewe van Suider-Afrika. Struik Uitgewes, Kaapstad

Cheikhyoussef A (2014) Ethnobotanical, phytochemical, antioxidant and antimicrobial activities of Namibian indigenous fruits and vegetables. Paper presented at the INSPIRE 2014; Sustainable Future for Green Medicine Symposium, 26–27 March 2014, Nelson Mandela Metropolitan University, George, South Africa

Cheikhyoussef A, Embashu W (2013) Ethnobotanical knowledge on indigenous fruits in Ohangwena and Oshikoto regions in Northern Namibia. J Ethnobiol Ethnomed 9:34

Cheikhyoussef A, Shapi M, Matengu K, Mu Ashekele H (2011) Ethnobotanical study of indigenous knowledge on medicinal plant use by traditional healers in Oshikoto region, Namibia. J Ethnobiol Ethnomed 7:10

Cheikhyoussef A, Naomab E, Potgieter S, Kahaka G, Raidron C, Mu Ashekele H (2010) Phytochemical properties of a Namibian indigenous plant; Eembe (Berchemia discolor). In: Proceedings of the first national research symposium, Windhoek, Namibia, pp 313–321

Chin Y-W, Mdee LK, Mbwambo ZH, Mi Q, Chai H-B, Cragg GM, Swanson SM, Kinghorn AD (2006) Prenylated flavonoids from the root bark of *Berchemia discolor,* a Tanzanian medicinal plant. Nat Prod 69:1649–1652

De Groot P, Hall D (1989) Back to grass roots. New Sci 121:44–47

Dejen A, Merene Y (2013) Ecology and field biology of the sorghum chafer, *Pachnoda interrupta* (Olivier) (Coleoptera: Scarabaeidae) in Ethiopia. J Entomol Nematol 5:64–69

Desta B (1993) Ethiopian traditional herbal drugs. Part 2: antimicrobial activity of 63 medicinal plants. J Ethnopharmacol 39:129–139

Feyssa DH, Njoka JT, Asfaw Z, Nyangito MM (2012a) Nutritional value of Berchemia discolour: a potential to food and nutrition security of households. J Biol Sci 12:263–271

Feyssa DH, Njoka JT, Asfaw Z, Nyangito MM (2012b) Comparative analysis of indigenous knowledge on use and management of wild edible plants: the case of Central East Shewa of Ethiopia. Ethnobot Res Appl 10:287–304

Gachathi FN, Johansson SG, Alakoski-Johansson GM (1994) A check-list of indigenous trees and shrubs of Bura, Tana river district, Kenya with Malakote, Orma and Somali names. J East Afr Nat Hist 83:117–141

Ganduri L, Nyathi JA, Nkiwane L, Nyoni AB (2014) Potential of natural dye processing in Matabeleland South Province: case study of Masendu Ward, Bulilima-Mangwe District Zimbabwe. Zimb J Sci Technol Spec Issue MS 11:1–7

Gelfand M, Drummond RB, Mavi S, Ndemera B (1985) The traditional medical practitioner in Zimbabwe: his principles of practice and pharmacopoeia. Mambo Press, Gweru

Gyawali R, Ibrahim SA (2012) Impact of plant derivatives on the growth of foodborne pathogens and the functionality of probiotics. Appl Microbiol Biotechnol 95:29–45

Hines DA, Eckman K (1993) Indigenous multipurpose trees of Tanzania: uses and economic benefits for people. Cultural Survival Canada and Development Services Foundation of Tanzania

Hutchings A, Scott AH, Lewis G, Cunningham AB (1996) Zulu medicinal plants: an inventory. University of Natal Press, Pietermaritzburg

Jenkins MD (1987) Madagascar: an environmental profile. IUCN, Gland

Kayombo B, Tsheko R, Semetsa S, Malepa D (2014) Documentation of indigenous knowledge and best-bet practices on the use of animals and plants for sustainable natural resources management in Botswana. Botswana J Agric Appl Sci 10(1):3–10

Kinoshita T, Kajiyama K, Hiraga Y, Takahashi K, Tamura Y, Mizutani K (1996) Isoflavan derivatives from Glycyrrhiza glabra (licorice). Heterocycles 43:581–588

Kipkorir D, Kareithi J (2012) Human and natural factors in the deterioration of indigenous irrigation Furrows in Marakwet, Kenya. Int J Humanit Soc Sci 2:113–128

Liengme CA (1981) Plants used by the Tsonga people of Gazankulu. Bothalia 13:501–518

Loeb EM, Koch C, Loeb EMK (1956) Kuanyama Ambo magic. Part 6. Medicinal, cosmetical and charm flora and fauna. J Am Folk 69:147–174

Long C Swaziland's flora (2005) siSwati names and uses. http://www.sntc.org.sz/flora/clbotalpha.asp?l=C&pg=10

Mabogo DEN (1990) The ethnobotany of the Vhavenda. MSc. Thesis. University of Pretoria. Pretoria, South Africa.

Maroyi A (2011) Ethnobotanical study of medicinal plants used by people in Nhema communal area, Zimbabwe. J Ethnopharmacol 136:347–354

Muchuweti M, Moyo E, Mushipe S (2005) Some properties of the polygalacturonase from four Zimbabwean wild fruits (*Uapaca kirkiana*, *Zizyphus mauritiana*, *Tamarindus indica* and *Berchemia discolor*) fruits. Food Chem 90:655–661

Mukamuri B, Kozanayi W (1999) Institutions surrounding the use of marketed bark products: the case of *Berchemia discolour*, *Warburgia salutaris* and *Adansonia digitata*. IES Working Paper 17. University of Zimbabwe, Harare.

Mukamuri BB, Kozanayi W (2001) Commercialization and institutional arrangements involving tree species harvested for bark by small-holder farmers in Zimbabwe. IES Working Paper 18. University of Zimbabwe, Harare.

Ohiokpehai O (2003) Promoting the nutritional goodness of traditional food products. Pak J Nutr 2:267–270

Palmer E, Pitman N (1972) Trees of southern Africa, covering all known indigenous species in the republic of South Africa, South-West Africa, Botswana, Lesotho and Swaziland. Balkema, Cape Town.

Perez C, Pauli M, Bazerque P (1990) An antibiotic assay by agar well diffusion method. Acta Biologica Medica Exp 15:113–115

Shumba E, Chidumayo E, Gumbo D, Kambole C, Chishaleshale M (2010) Biodiversity of plants. In: Chidumayo G (ed) The dry forests and woodlands of Africa: managing for products and services. Earthscan, London

Silva GL, de Abreu Matos FJ, Silveira ER (1997) 4′-dehydroxycabenegrin A-I from roots of *Harpalyce brasiliana*. Phytochemistry 46:1059–1062

Sorensen C (1993) Controls and sanctions over the use of forest products in the Kafue river basin of Zambia. Rural Development Forestry Network, ODI Forest Policy & Engineering Group, London

Sundar Rao K, Sino D (1992) Fatty acid composition of 20 lesser known Western Australian seed oils. J Sci Food Agr 58:585–587

van Heerden FR, Brandt EV, Ferreira D et al (1981) Metabolites from purple heartwoods of the *Mimosoideae*. Part 4. *Acacia fasciculifera* F. Muell ex. Benth: Fasciculiferin, Fasciculiferol, and the Synthesis of 7-Aryl and 7-Flavanyl-Peltogynoids. J Chem Soc Perkin Trans 1:2483–2490

Van Wyk BE, Gericke N (2007) People's plants: a guide to useful plants of southern Africa. Briza Publications, Pretoria

Venter F, Venter JA (1996) Making the most of Indigenous trees. Briza Publications, Cape Town/Pretoria

Wild H, Biegel HM, Mavi S (1972) A Rhodesian dictionary of African and English plant names. Government Printers, Salisbury

Xu MJ, Wu B, Ding T, Chu JH, Li CY, Zhang J, Wu T, Wu J, Liu SJ, Liu SL, Ju WZ, Li P (2012) Simultaneous characterization of prenylated flavonoids and isoflavonoids in *Psoralea corylifolia* L. by liquid chromatography with diode-array detection and quadrupole time-of-flight mass spectrometry. Rapid Commun Mass Spectrom 26:2343–2358

Yang X, Jiang Y, Yang J, He J, Sun J, Chen F, Zhang M, Yang B (2015) Prenylated flavonoids, promising nutraceuticals with impressive biological activities. Trends Food Sci Technol 44(1):93–104

# Aromatic and Medicinal Plants of Tunisian Arid and Desert Zone Used in Traditional Medicine, for Drug Discovery and Biotechnological Applications

**Hanen Najjaa, Abdelkarim Ben Arfa, Ákos Máthé, and Mohamed Neffati**

**Abstract** Tunisian has a high diversity of plants as a result of the variation in topography climate and edaphic conditions of the country. Among these plants several aromatic and medicinal species are traditionally used in folk medicines to treat and/or manage common ailments. Some others species are used in foods as spices and/or in aromatherapy. The present chapter endeavors to highlight the botanical description, geographical distribution, ethno pharmacological uses of 45 species spread over 22 botanical families among those the most used in traditional medicine. Information on the chemical composition and biological activities of these species are also included in this chapter.

**Keywords** MAPs • Tunisian • Arid land • Ethnopharmacology

## 1 Introduction

The global importance of Medicinal and Aromatic Plants (MAPs) is evident from the fact that in 2006, the world wide trade of MAPs touched upon 60 billions of US dollars. With the increase in demand, trade figures are expected to further grow to 5 trillion by the year 2050 (Lang 2008). It is also estimated that Europe alone annually imports about an average market value of 1 billion US$ from Africa and Asia (Larsen and Smith 2004; Sher and Hussain 2009). It is also worth mentioning that the use of wild edible herbs/culinary plants has been increasing in the global food market for diverse items ranging from salads to desserts. In recent years, there has

H. Najjaa (✉) • A.B. Arfa • M. Neffati
Laboratoire d'Ecologie Pastorale, Institut des Régions Arides, 4119 Médenine, Tunisia
e-mail: hanen.najjaa@yahoo.fr

Á. Máthé
Faculty of Agriculture and Food Science, University of West Hungary,
Vár 2, H-9200 Mosonmagyaróvár, Hungary

 
M. Neffati et al. (eds.), *Medicinal and Aromatic Plants of the World - Africa Volume 3*, Medicinal and Aromatic Plants of the World,
DOI 10.1007/978-94-024-1120-1_8

been an increasing interest also in the utilization of natural substances. Some questions concerning the safety of synthetic compounds have also encouraged the study of plant resources. Essential oils, i.e. volatile secondary substances of plant origin, can have wide ranging applications in both traditional folk medicine, and food, flavoring, cosmetic and fragrance industries. Recently, many essential oils and their constituents have been investigated for their multifunctional properties (Hajji et al. 2010). Authorities and consumers are equally concerned about food safety and the potential harmful effects of synthetic additives on health (Reische et al. 1998).

In addition to the increasing demand for safe and natural food in recent years, the great economic losses caused by the deterioration and poisoning of food products by food pathogens have motivated many researchers to explore new alternatives to traditional food preservation practices. Among these natural products, essential oils (EOs) from aromatic and medicinal plants have received particular attention as potential natural agents for food preservation (Burt 2004; Bakkali et al. 2008). In fact, EOs and their main components have been reported to possess a wide spectrum of antimicrobial, antiviral, insecticidal and antioxidant activities (Janssen et al. 1987; Rota et al. 2004; Gachkar et al. 2007).

The use of plants for medicines is by far the biggest use of plants in terms of the number of species specifically targeted. Plants provide the predominant ingredients of medicines in most traditional systems of healing and have been the source of inspiration for several major pharmaceutical drugs (Chevallier 1996; Sher et al. 2010a). According to Plant life International, the scale of trade in medicinal plants ranges from local to international. Much of this trade is unrecorded or poorly documented in official statistics. Due to the poor documentation, decision makers usually have little awareness of the significance of trade and consumption of medicinal plants, or of the problems of un-sustainability and sometimes deleterious impacts on natural habitats of wild collection (Hamilton and Hamilton 2006).

Numerous publications have yielded detailed knowledge of aromatic wild plants in certain locations in Africa. They have indicated that wild plants are essential constituents of many African drugs. The ethnobotanical survey and documentation of traditional knowledge, mainly on the medicinal uses of plants, have provided important clues for the discovery of new drugs nowadays. For many years, traditional plant remedies were empirically used in Africa for treating various diseases. According to the World Health Organization up to 80% of the population in Africa uses traditional medicine for their primary healthcare (Larsen and Smith 2004; Al-Quran 2008).

Tunisia has a high diversity of plans with several aromatic plant species traditionally used in folk medicines, in foods as spices, in massage and in aromatherapy. Among the 2250 species that compose Tunisia's vascular flora (Marrif et al. 1995), 1630 species are native to the arid and desert part of the country, which is characterized by low rainfall, high temperature and drying winds (Le Houérou 1959). Remarkably, a wide range of plant species thrive under these conditions, which is of high economic and ecological significance.

Despite the considerable medicinal plant potential of Tunisian flora, studies on the traditional uses of many species are relatively scarce. The present chapter aims at reviewing the traditional uses, chemical composition and pharmacological properties of Tunisian arid plants in order to provide information on their possible uses as preservatives in food products, in biotechnology or raw materials for pharmaceuticals and cosmetics. Our choice of investigated species is based on two criteria: the disponibility of ethnopharmacological data indicating their utilization in folk medicine (a) and the recent research results on their chemical composition and biological activities of their extracts (b).

## 2 Ethnopharmacological and Culinary Use of Arid Plants

In Tunisia, thanks to the diversity of its climate, the traditional pharmacopoeia consists of a wide arsenal of medicinal plants. Although, many of these species are nearly facing extinction (Ziyyat et al. 1997), only a small proportion of them has been scientifically studied.

In this review, the species are listed alphabetically by plant families. For each species the traditional medicinal and culinary uses are described. The surveyed species are listed in Fig. 1, where plant families and species within each family are cited in alphabetic order. Scientific names and local names are included, and information concerning the parts used and the methods of use are given.

Information on the therapeutic properties of plants mentioned in this review should not be considered as a practical manual for self-medication.

### *Amaryllidaceae*

This family comprises 700 species, widely distributed in the northern hemisphere (Jones et al. 2004; Tepe et al. 2005). *Allium* is the largest and the most important representative genus of the Amaryllidaceae. For many centuries, several species of Amaryllidaceae family have been used as vegetables and spices, and as folk medicines for curing various diseases (Hacıseferoğullari et al. 2005). Many *Allium* species (*Allium roseum* L., *Allium cepa* L., *Allium sativum* L.) were well studied in the world as well as in Tunisia (Najjaa et al. 2007, 2010; Leporatti and Ghedira 2009; Guetat et al. 2010).

*Allium* species have been found to shown various activities. For example, a wide array of therapeutic effects of garlic has attracted particular attention in modern medicines because of its widespread use as antiatherosclerotic, antidiabetic, antihypertensive, antimicrobial, anticancerous, antioxidant, antifungal and antiviral (Keusgen et al. 2003; Auger et al. 2004; Arnault et al. 2005; Bakri and Douglas 2005; El-Demerdash et al. 2005). According to Durenkamp and Dekok (2004) *Allium* species, namely, onion, garlic, leek and chive contain a variety of secondary sulphur compounds. Sulphur carrying flavour compounds are responsible for the characteristic smell and taste and are the source of major active compounds which provide the best known properties in *Allium* plants (Keusgen et al. 2003; Auger et al. 2004; Jones et al. 2004; Hacıseferoğullari et al. 2005).

**Fig. 1** Aromatic and Medicinal species: [**1**] *Allium roseum,* [**2**] *Allium ampeloprasum,* [**3**] *Rhus tripartitum,* [**4**] *Periploca laevigata,* [**5**] *Capparis spinosa,* [**6**] *Hernaria fontanesii,* [**7**] *Hamada scoparia = Arthrophytum scoparium,* [**8**] *Artemisia herba-alba,* [**9**] *Artemisia campestris,* [**10**] *Oudneya africana,* [**11**] *Diplotaxis harra,* [**12**] *Colocynthis vulgaris = Cistrullus colocynthis,* [**13**] *Juniperus phoenicea,* [**14**] *Ephedra alata,* [**15**] *Globularia alypum,* [**16**] *Hyparrhenia hirta,* [**17**] *Rosmarinus officinalis,* [**18**] *Thymus algeriensis,* [**19**] *Thymus vulgaris,* [**20**] *Thymus capitatus,* [**21**] *Ajuga iva,* [**22**] *Marribium deserti,* [**23**] *Marribum alysson,* [**24**] *Teucrium polium,* [**25**]

**Fig. 1** (continued) *Astragalus gombiformis,* [**26**] *Calicotome villosa,* [**27**] *Genista saharae,* [**28**] *Retama raetam,* [**29**] *Ceratonia siliqua,* [**30**] *Genista microcephala,* [**31**] *Cymbopogon schoenanthus,* [**32**] *Calligonum comesum,* [**33**] *Calligonum azel,* [**34**] *Calligonum arich,* [**35**] *Polygonum equisetiforme,* [**36**] *Ziziphus lotus,* [**37**] *Ruta graveolens,* [**38**] *Thymelaea hirsuta,* [**39**] *Crithmum maritimum,* [**40**] *Thapsia garganica,* [**41**] *Pituranthos chloranthus,* [**42**] *Pituranthos tortuosus,* [**43**] *Nitraria retusa,* [**44**] *Peganum harmala,* [**45**] *Zygophyllum album*

**Fig. 1** (continued)

***Allium roseum* L.,** a polymorphic species, is represented in North Africa by 12 different taxa: 4 varieties, 4 subvarieties and 4 forms (Cuénod 1954; Le Floc'h 1983). In Tunisia, the same authors mentioned the presence of only three varieties: var. *grandiflorum* Briq., var. *perrotii* Maire. and var. *odoratissimum* (Desf.) Coss. The last variety is considered as an endemic taxon in North Africa (Cuénod 1954). *A. roseum* var. *odoratissimum* (Desf.) Coss. prefers poor and sandy soils. It is also found in grassy and bushy places, cultivated fields and fallows, and roadsides. This taxon is used since ancient times as a vegetable, spice and herbal remedy (Le Floc'h 1983). Today its edible aerial parts are widely harvested and sold commercially. The fresh young leaves and bulbs of *A. roseum* or "lazoul", as it is called in Southern Tunisia, are consumed in salads and used as spice to prepare traditional recipes. Besides its culinary use, "lazoul" is also used in folk medicine. Le Floc'h (1983) reported its use for the treatment of headaches and rheumatism. It is also used for the treatment of respiratory tract infections, as an inhalation for colds, to diminish fever and enhance appetite.

Phytochemical screening of *A. roseum* using TLC showed the presence of antimicrobially active agents, such as saponins, tannins, flavonoids, coumarins, steroids, cardiac glycosides, free quinone and iridoids (Najjaa et al. 2011a, b). The dichloromethane, petroleum ether and ethanolic extracts of *A. roseum* exhibited a wide spectrum of antimicrobial activities against *Staphylococcus aureus*, *Staphylococcus epidermidis*, *Micrococcus luteus*, *Escherichia coli* and *Pseudomonas aeruginosa*. Such results may provide a good support for the use of this plant in herbal medicine. They may also serve as a base for the development of new drugs and phytomedicines and may be considered as functional food and/or served as a nutraceutical (Najjaa et al. 2007).

Recent studies revealed that *A. roseum* var. odoratissimum growing in Tunisia had high soluble carbohydrate, crude protein and dietary fibre contents, as compared to other *Alliums*. Its mineral content was high in potassium, and calcium. The mineral composition of 'rosy garlic' is sufficient in Ca, P, K, Cu, Fe, Zn and Mg so that it can meet many macronutrient and micronutrient requirements of the human diets (Najjaa et al. 2010)

***Allium ampeloprasum* L.,** locally known as "kurrat" in Tunisia, a wild leek species, originates from the Mediterranean region and southwest Asia (Figliuolo and Mang 2010).

Wild *A. ampeloprasum* in Tunisia is popularly appreciated as a medicinal plant or as a vegetable used to prepare local dishes, and it is widespread across the whole country landscape (Guenaoui et al. 2012). It is rich in flavonoids such as myricetin, kaemferol and quercetin (Horbowicz and Kotlińska 2000), responsible for anticarcinogenic and antithrombotic activities (Havey 1999).

## *Anacardiaceae*

Due to their high contents in phenolics, flavonoids and other phytochemicals, *Rhus* species are widely used in both modern and traditional medicine. The extracts showed antimalarial (Ahmed et al. 2001), antimicrobial (McCutcheon et al. 1992), antitumorigenic (Lee et al. 2004; Choi et al. 2012), antioxidant (Lee et al. 2002; Rima et al. 2011; Olchowik et al. 2012), antiviral (Lin et al. 1999), hypoglycaemic (Giancarlo et al. 2006), leukopenic (Yang and Du 2003), atheriosclerosis (Zargham and Zargham 2008) and anticonvulsant (Ojewole 2008) properties. In Tunisia, the genus *Rhus* is represented by two species: *Rhus tripartita* (Ucria) Grande [=*R. tripartitum* (Ucria) D.C. = *R. oxyacanthoides* Dum. Cours. = *R. oxyacantha* Shousb. Ex. Cav.] and *R. pentaphylla* Desf. (Pottier-Alapetite 1979; Le Floch and Boulos 2008). *R. pentaphylla* Desf. spreads in the north and the center of the country, and is relieved by *R. tripartita* from the center to the far southern region. The two species may be sympatric in several regions. They grow mainly on eroded substrates, under a rainfall ranging from 100 to 600 mm/year and at altitudes ranging from 10 to 500 m. The populations of both species are widely scattered, presumably as remnant sites of previously more continuously distributed species. The overharvesting of roots and other plant parts for traditional medicine, dyeing and tanning (Le Floc'h 1983) has led to a high death rate of trees and a decrease in the size of populations (Le Houerou 1969). *R. pentaphylla* Desf. is being lost at a higher rate than *R. tripartita*. Its habitat has been mainly cleared for agricultural and coal mining operations.

***Rhus tripartita (Rh. tripartitum)* (Ucria) Grande,** is a native presaharan Tunisian plant, with a great economic and ecological interest. *R. tripartitum* is classified in Tunisia among the rare plants that need preservation. This species is adapted to the sub-deserted climate (Quezel and Santa 1962). *R. tripartitum* is widespread in North Africa (Pottier-Alapetite 1981). In Tunisia, *R.tripartitum* has many traditional uses. In his ethnobotanical study, Le Floc'h (1983) cited many examples on the consumption of *R. tripartitum* species. Fruits of this plant are consumed fresh, stored (sweetened), or soaked in sour milk. They are added to drinking water to offer an acceptable taste (Le Floc'H 1983). The bark is used for tanning and the wood for the collery and to treat eye troubles. The infusion of the fruits, the leaves and the roots is recommended for gastric ulcer and intestinal ailments. The pulverized leaves or roots or barks are used on wounds to facilitate healing. Earlier investigations show the occurrence of vitamin E and flavonoid substances in *R. tripartitum* extracts (Mahjoub et al. 2004) and testify the importance of this plant as a source of polyphenolics as anti-inflammatory and antioxidant products. They alsodemonstrate the medicinal and pharmacological interest in *R. tripartitum* species. Recent work by chromatographic separation of Tunisian *R. tripartitum* revealed the presence of four main natural substances: a new biflavonoid masazinoflavanone (S1), myricetin (S2), (−)-lyoniresinol 3α-O-β-D-glucopyranoside (S3), and (−)-methyl shikimate (S4) (Mahjoub et al. 2005). A biflavonoid masazinoflavanone and the isobiflavonoid calodenone have been isolated and characterized in the methanolic aerial part extract of the plant *R. tripartitum* (Mahjoub et al. 2005). Compared with the standard drug dexamethasone, the biflavonoid substance masazinoflavanone (S1) inhibited, in a dose related manner, carrageenan-induced paw edema in rats. Mahjoub et al. (2005), using free radical forms of ABTS and DPPH, tested the antiradical scavenging capacities of extracts and natural substances from *R. tripartitum*. The Trolox Equivalent Antioxidant Capacity (TEAC) assay shows values of 1.92, 1.39, 1.25,

2.40 and 1.58 mM Trolox/mg sample respectively for the chloroformic extract, the ethyl acetate extract, (S1), (S2) and (S3). Values revealed from the total phenolic content assay were 299.3, 372.3, 99.1 and 235.7 mg Catechin/g extract respectively for the chloroformic, ethyl acetate, ethyl acetate/methanol and methanolic extracts (Mahjoub et al. 2010).

## *Asclepiadaceae*

***Periploca laevigata* Ait.** is native to the Mediterranean region and widely distributed in the Sahara area. *P. laevigata* is an important pastoral species in the dry season. This shrub grows mainly in the semiarid and sub-humid zones, between 400 and 1800 m. Having disappeared from large sites of its former distribution area but still being quite frequent on some islands, it is classified as endangered (EN). In Tunisia, it is predominantly found in the south of the country, especially in the mountains. It is used as an herbal preparation (tea) because of its reputed medicinal properties, e.g. for the treatment of headaches and diabetes (Askri et al. 1989). Its utilization for animal skin tanning mentioned by Chevalier (1935), has been confirmed by inhabitants of Fogo, but nowadays, it is rarely practiced. On Santiago it is used as a medicinal plant to treat fever and cough (Gomes et al. 1999). The root powder of *P. laevigata* is used for preparing soft drinks and as an aromatic, in Tunisia. The plant is used also to treat digestive disorders and hypertensive effects, as well as other health problems (Hajji et al. 2009). The most studied *Periploca* species, were reported to have various biological activities, such as anti-proliferative (Spera et al. 2007), antitumor (Itokawa et al. 1988) and hypotensive effects (Askri et al. 1989).

Many chemical compounds have been isolated and identified from species of the *Periploca* genus, such as α- and β-amyrin, lupeol β-sitosterol and elemane-type sesquiterpenes from *P. laevigata* (Askri et al. 1989). Gas chromatography/mass spectrometry was used to determine the composition of the P. laevigata root bark oil. Forty-three components were identified in the essential oil and the main compounds were benzaldehyde (56%), methyl 4-methoxysalicylate (6.55%) and carvacrol (4.75%). The *P. laevigata* root bark essential oil exhibited a dose-dependent manner of inhibitory activity toward ACE. The highest ACE inhibitory activity (54%) was observed at a concentration of 30 μg/mL. The *P. laevigata* root bark oil was also found to possess antioxidant activities, as evaluated by the 1,1-diphenyl-2-picrylhydrazyl (DPPH) radical method, β-carotene bleaching and reducing power assays. The antimicrobial activity of the essential oil was also investigated on several microorganisms. The inhibition zones and minimal inhibitory concentration (MIC) values of bacterial strains were in the range of 12–46 mm and 50–300 μg/ml, respectively. The inhibitory activity of the *P. laevigata* root bark essential oil against Gram-positive bacteria was significantly higher than against Gram-negative. It also exhibited remarkable activity against several fungal strains (Hajji et al. 2010). Our findings demonstrate that *P. laevigata* essential oil might be a good candidate for further investigations of new bioactive substances (Hajji et al. 2010).

## *Capparidaceae*

***Capparis spinosa* L.** is a perennial shrub and is the common name of the genus *Capparis,* family Capparidaceae. This genus is represented by several species (about 250) of which the most important economical species is *C. spinosa*. It is known by various names, e.g. Caper (English), Kabbar (Arab), Alcaparro (Spain), and Gollaro (Pakistan). Capers grow wildly in various regions of the world, especially in the Mediterranean basin, showing strong resistance to hard environmental conditions. Caper is adapted to poor soils, and is widespread on rocky areas, mountains and grows on numerous soil types, including alfisols, regosols and lithosols. It shows a good response to volcanic or alkaline soils. Soil pH from 6.1 to 8.5 is tolerated (Janick and Paull 2006). This perennial shrub with a height of 30–100 cm, has thick and deep roots, and can be thorny or inerm (Inocenio et al. 2006 and Saadaoui et al. 2007).

As a spontaneous plant, caper has a wide natural distribution. Despite adverse conditions, plants of *Capparis,* do not seem to show any water stresses or any symptoms of photo-inhibition. The plants efficiently utilize the high irradiance throughout the growth season (Levizon et al. 2004). In Tunisia the average annual caper production is estimated to be around 150 tonnes. The USA is one of the most important consumers (Janick and Paull 2006). The cultivation of this plant is limited. Most of the supplies for traditional and commercial uses of *Capparis* depend exclusively on the collection of wild plants. Capers are cultivated also in some other countries, e.g.: Spain, Italy, Turkey, Morocco, Tunisia, Algeria, Iran and the coastal areas of the Black Sea. Caper cultivation started around 1970 in Spain and Italy, and later in Morocco, with a maximum of about 4000, 1000, and 2500 ha in cultivation, respectively (Infantino et al. 2007). It is worth mentioning that various studies conducted on the pharmacological potential of *C. spinosa* have verified its local, folk medicinal uses.

The aromatic part of the caper plant is the floral bud, which is gathered just before it blossoms. Before commercial packaging, the buds are pickled in vinegar or preserved in granular salt (Germano et al. 2002). They have long been used in recipes of salads, pasta, meat, sauces and garnishes to add a pungent spicy flavor and aroma to food. Fici (2001) stated that the salted and pickled caper bud is often used as a seasoning and garnish. Furthermore, he also reported that Capers are common ingredients in the Mediterranean cuisine, especially Italian and it is also important in the commercial preparation of frozen food (Germano et al. 2002). Capers are graded and sold by their size, defined as follows, with the smallest sizes being the most valuable: Non-pareil (up to 7 mm), surfines (7–8 mm), capucines (8–9 mm), capotes (9–11 mm), fines (11–13 mm) and grusas (14+ mm). Capers are picked daily, since the youngest flower buds (about the size of peas) have the highest quality. The young fruits and tender branch tips can also be pickled and used as a condiment. More rarely, mature and semi-mature fruits are eaten as a cooked vegetable. The ripened fruits are rich from protein, lipid, carbohydrates, and vitamins and mineral (Fragiska 2005; Ozcan 2005) reported that outside of the Mediterranean and the Caucasus mountains, capers are not much known, although the pickled fruits of some Central Asian species (e.g., *Capparis aphylla*) are used as a vegetable in Afghanistan, Pakistan and North Western India (Sher and Alyemeni 2010).

Roots, leaves, buds, fruit, bark and seeds of caper were used by ancient peoples for medicinal purposes, to treat some diseases such as rheumatism, stomach problems, headache and toothache. The first use of *C. spinosa* was for medicinal purposes, in 2000 BC by the Sumerians and afterwards by Greeks and Romans (Romeo et al. 2007). Different parts of the caper plant can be used as a drug or a cosmetic. In ancient times, roots were consumed by the ancient Egyptians and Arabs to treat kidney disease, liver disease, stomach problems and scorpion's stings. Leaves were used by the ancient Arabs against skin diseases, to treat earaches and to kill worms in the ear. Buds were used by the ancient Arabs against spleen disease. Ancient Romans used this part of the caper plant to treat paralysis. Flowers could serve as a stimulant to increase erection and soothe pains. Fruits were used by ancient Greek to treat convulsions. Seeds served as a medication for gum problems by ancient Arab and to accelerate menstruation by ancient Egypt and ancient Greek (Rivera et al. 2003).

Actually, in Morocco the fruit of *C. spinosa* is used to cure diabetic (Eddouks et al. 2002) Iranian people use the roots, fruit and plant bark of *C. spinosa* as diuretics, tonics, against malaria and joint disease (Hooper 1937; Afsharypuor et al. 1998). In Pakistan, people use all parts of *C. decidua* as anti-rheumatic, as diuretic, as kidney disinfectant, tonics, to reduce flatulence, to improve liver functions, to cure arteriosclerosis and migraine (Jagtap et al. 2006; Pirzada et al. 2007); and they especially used leaves of *C. spinosa* as analgesic, antihemorrhoid, antirheumatic (Khakberdyev et al. 1968), aperients, deobstruent, depurative and diuretic (Chopra et al. 1986). In India, buds and roots of *C. spinosa* are useful in the treatment of boils; leaves are used as counter-irritant and as a cataplasm in swellings; roots are used to treat fever, rheumatism, paralysis, toothache and kill worm in the ear; the bark is used in the treatment of coughs, asthma and inflammation (Wealth of India 1992; Kiritikar and Basu 1987, 1991). These findings are in line with the study of Fragiska (2005), who reported that *C. spinosa* has the potential to support the formations of different agro-based industrial products. Moreover, Sushila et al. (2010) also reported that in ayurvedic medicine capers are recorded as hepatic stimulants and protectors, improving liver function and also act as a appetizer agent.

The presence of both α- and γ-tocopherol was found in buds. Moreover, *C. spinosa* contained an appreciable level of vitamin-C. Several studies have linked plants rich in these compounds with a reduced risk of several chronic and degenerative diseases including cancer, cardiovascular disorders and atherosclerosis (Brigelius-Flohé and Traber 1999; Sommer and Davidson 2002).

Tlili et al. (2010), show that leaves and flower buds of *C. spinosa* from different locations in Tunisia were very rich in total phenolic compounds (with an average of 3643.29 and 2621.14 mg/100 g FW, respectively), including rutin (with an average of 1352.71 and 693.14 mg/100 g FW, respectively). The highest value of phenolic compound and tocopherol was detected in samples from Tataouine, arid zone (3.96 ± 0.40 mg/100 g FW) while the lowest value was detected in samples from Ghar el Melh (2.19 ± 0.12 mg/100 g FW). This difference may be due to the effect of the location, and confirms our hypothesis that plants collected from stressed zones synthesize more bioactive substances. HPLC was used to determine carotenoids (β-carotene, lutein, neoxanthin, and violaxanthin) and tocopherols in leaves, buds, and flowers of Tunisian *C. spinosa* (Tlili et al. 2010). The means of total carotenoids were 3452.5; 518.5, and 342.7 μg/g fresh weights (FW) in leaves, buds, and flowers, respectively. The principal form of tocopherol detected in leaves was α-tocopherol (20.19 mg/100 g FW). In buds and flowers, they are also present at 49.12 and 28.68 mg/100 g FW, respectively, whileγ-tocopherol content amounted to 48.13 and 27.8 mg/100 g FW, respectively. The combined content of pro-vitamin-A and vitamin-E in capers encourages researchers to carry on with research and to explore this plant (Tlili et al. 2009). The glucosinolatecontent and composition of young shoots and raw flower buds of *C. spinosa*were investigated by HPLC with UV detection. Samples were harvested in August 2001 in Turkey. Twelve different glucosinolates were identified in the young shoots and buds of both species. Total content of glucosinolates ranged from 6.55 μmol/g of *C. spinosa* to 45.56 μmol/g (young shoots of *C. ovata*). The main glucosinolate was glucocapperin, which amounted to 90% of the total glucosinolates (Germano et al. 2002). Some publications refer to specific aspects of the qualitative composition of flavonoids, the occurrence of elemental sulfur, and physical and chemical properties (Rodrigo et al. 1992; Ozcan and Akgul 1998; Brevard et al. 1992; Inocencio et al. 2000), but up to now there is little available information about the glucosinolate content and composition of members of this family. The presence of different glucosinolates such as neoglucobrassicin, 4-methoxyglucobrassicinglucocapperin, glucoiberin, sinigrin, 1-methoxy-3-indolylymethyl, and glucobrassicin were described (Daxenbichler et al. 1991; Kjaer and Thomsen 1963; Schraudolf 1989), mostly in the roots or leaves.

Besides bitter flavonoid glycosides, rutin, quercetin, quercetin 3-O-glucoside and quercetin 3-O-glucoside-7-O-rhamnoside, quercetin 3-O-(6‴-α-L-rhamnosyl-6″-β-D-glucosyl)-β-D-glucoside (Sushila et al. 2010), and kaempferol glycosides, *C. spinosa* also contains lipids, glucocapparin (methyl glucosinolate), methyl isothiocynate, isopropyl isothiocyanate, *sec*-butyl isothiocyanate, benzyl-isothiocyanate, β-sitosterylglucoside-6′-octadecanoate, 3-methyl-2-butanyl-β-glucoside, stachydrine (a pyridine alkaloid), and cadabicine (a 24-membered polyamine lactam alkaloid). Furthermore, homologous polypreols namely; cappaprenol-12, cappaprenol-13, and cappaprenol-14 with 12, 13, and 14 isoprene units were also isolated from the alcoholic extract of *C. spinosa* (Al-Said et al. 1988; Inocencio et al. 2002). Later, new (6S)-hydroxy-3-oxo-α-ionol glucosides together with corchoinoside C (6S, 9S)-roseoside, and prenyl glucosides were isolated from the mature *C. spinosa* fruits (Calis et al. 2002). *C. spinosa* fruits also contained cappariloside-A, stachydrine, an adenosine nucleoside, hypoxanthine, β-sitosterol, vanillic acid, p-hydroxybenzoic acid, protocatechuric acid, daucosterrol, uracil, butanedioc acid, and uridine (Calis et al. 2002). In the volatile oils of *C. Spinosa* 145 compounds were identified. The major constituents were: aldehydes 22%, esters 21%, and sulfur containing compounds 8.42%. The presence of sesquiterpenes, monoterpenes, and capric acid was also confirmed (Arena et al. 2008). Chemical studies on caper have reported the richness of different partsin several beneficial chemical compounds. The significant level of all theses antioxidants highlight the important nutritional and medicinal value of this species used in the Mediterranean basin (Tlili et al. 2010).

Recent works reported that the *C. spinosa* extract has the ability to maintain skin homeostasis and can be used as a possible tanning agent or as a treatment for hair de-pigmentation (Matsuyama et al. 2009). Methanolic extract of *C. spinosa*buds may contribute to improve the immune surveillance of human peripheral blood mononuclear cells toward virus infection by up-regulating expression of peculiar pro-inflammatory cytokines; it suppressed the replication of herpes simplex virus type 2 and increased the expression of pro-inflammatory cytokines including interleukin-12, interferon-γ and tumor necrosis factor-α (Arena et al. 2008). More recently, a protein with potent antiproliferative activity toward tumor cells and inhibitory activity toward HIV-1 reverse transcriptase and some antifungal activity have been isolated from seeds of *C. spinosa* (Lam and Ng 2009). These important biological activities were probably due to the important quantity of the antioxidants which have an excellent activity against these diseases. It was reported that aqueous extracts of *C. spinosa*exhibit a potent anti-hyperglycemic activity in diabetic rats without affecting basal plasma insulin concentrations (Eddouks et al. 2004, 2005). These facts explain the use of caper by old people to treat diabetic disease (Jouad et al. 2001). Yue-lan et al. (2010) call attention to the notable protective activity of *C. spinosa* against oxidative stress and interrupting of ROS in systemic sclerosis dermal fibroblasts. Ethanolic extracts from *C. spinosa*can effectively inhibit fibroblast proliferation and type I collagen production in progressive systemic sclerosis (Cao et al. 2008). Aqueous extracts of *C. spinosa*can also act by decreasing the cholesterol biosynthesis (Eddouks et al. 2005). Gadgoli and Mishra (1999) suggested that *C. spinosa*extract possesses significant antihepatotoxic activity against carbontetrachloride and paracetamol which induced hepatotoxicity *in vivo* and thioacetamide and galactosamine which induced hepatotoxicity in isolated rat hepatocytes, using *in vitro* technique. This result supports the use of caper by the ancient people to treat liver disease and improve liver functions (Jagtap et al. 2006).

The anti-oxidant activity of cappariside, 4-hydroxy-5-methylfuran-3-carboxylic acid and several other organic acids from *C. spinosa* was confirmed (Tlili et al. 2009, 2010; Sher et al. 2010a). New anti-inflammatory and antioxidant compounds, lignin glucosides, 1H-indole-alkaloid glucosides, and phenolic glucosides were isolated from roots of another caper species *C. tenera*. Nonetheless, cappaprenol-13 isolated from *C. spinosa* showed significant anti-inflammatory activity and antioxidative potential (Al-Said et al. 1988). *C. spinosa* extract was shown to possess prominent activity against *Plasmodium falciparum* (Arena et al. 2008). The chonodroprotective effect of *C. spinosa* suggested that it might be used in the management of cartilage damage during the inflammatory process (Tlili et al. 2009). In diabetic rats, *C. spinosa* treatment, significantly reduced plasma glucose, triglyceride levels and plasma cholesterol levels (Sher et al. 2010a). On the other side, the extracts of *C. spinosa* flowering buds, showed the ability to protect against histamine-induced bronchospasm (Arena et al. 2008).

A popular compound finished herbal drug formulation Liv-52 (containing *C. spinosa* as its essential ingredient) was used in patients with liver cirrhosis following a hospital based clinical protocol. Liv-52 showed significant hepato-protective effect in cirrhotic patients. The protective effect of Liv-52 was attributed to the diuretic, anti-inflammatory, anti-oxidative, and immunomodulating properties of the component herbs (Arena et al. 2008). However, antihepatotoxic activity of p-methoxy benzoic acid isolated from *C. spinosa* was confirmed by earlier researchers (Gadgoli and Mishra 1999). It is worth mentioning that the ethanol extract of *C. spinosa* fruits exhibited a notable protective activity against oxidative stress and interrupting of ROS-ERK1/2- Ha-Ras signal loop in systemic sclerosis, signifying its potential protective effects against skin sclerosis (Sher et al. 2010a). The same researchers also reported that the ethanol extract from *C. spinosa* was found to effectively inhibit the fibroblast proliferation and type I collagen production in progressive systemic sclerosis (Zhang and Tan 2009). *Herpes simplex* virus type 1 (HSV-1) and type 2 (HSV-2) are common human pathogens that in particular cases, can also cause severe problems especially in immunodeficient patients. The methanol extract of *C. spinosa* contributed well to improving immune surveillance of PBMCs (human peripheral blood mononuclear cells) toward virus infection by up-regulating expression of peculiar pro-inflammatory cytokines; indicating the possibility of successful use of *C. spinosa* for treatment of HSV-2 infections in immunocompromised hosts (Arena et al. 2008). In a recent study, it was shown that the male flowers of *C. spinosa* save resources for the female function and that they primarily serve to attract pollinators as pollen donors (Zhang and Tan 2009). A monomeric protein with a molecular mass of 38 kDa was also purified from *C. spinosa* seeds. The protein exhibited an N-terminal amino acid sequence with some similarity to imidazo-leglycerol phosphate synthase. It inhibited HIV-1 reverse transcriptase and mycelia growth in fungus without having hemogglutinating, ribonuclease, mitogenic or protease inhibitor properties. Furthermore, a novel dimeric 62-kDa lectin was purified from caper (*C. spinosa*) seeds which inhibited HIV-1 reverse trans-criptase and proliferation of both hepatoma HepG2 and breast cancer MCF-7 cells (Lam et al. 2009). Based on the current detailed ethnopharmaceutical surveys and literature reviews, it is concluded that *C. spinosa.* Besides medicinal significance, various parts of *C. spinosa* were found to have food value, potential culinary value, and possessed also cosmetic ingredients. It is worthy to conclude that *C. spinosaa*s a key to resources for raw materials in the pharmaceutical, aroma and food industries, hence it is called "Plant of the Millennium" (Sher and Alyemeni 2010). Based on the present literature review *C. spinosa* is a safe plant and there are no reports in the scientific literature on its toxic manifestations after acute, sub-acute, or chronic treatment. However, allergic contact dermatitis caused by *C. spinosa* was reported (Angelini et al. 1991).

## *Caryophyllaceae*

***Herniaria fontanesii* J. Gay,** which have a widespread distribution in the Mediterranean area (Borkowski and Pasich 1958), have been until now little studied. In Tunisia, this plant has been used traditionally for Lithia's problems, as diuretic and for healing wounds and scars. In the traditional pharmacopoeia of Morocco, the aerial parts of this plant are used for the treatment of lithiasis (Mbark et al. 1999).

In a previous investigation into the secondary metabolites of *H. fontanesii*, a triterpenoid saponin (herniaria saponin A), flavonol derivative, isorhamnetin 3-(3′″-feruloylrhamnosyl(1-->6)galactoside), isorhamnetin 3-robinobioside and (−) catechin was also isolated from the aerial parts of *H. fontanesii* (Mbark et al. 1995, Mbark et al. 1996). The structures of the two saponins isolated from the aerial parts of this specieswere established as 28-O-{a-L-rhamnopyranosyl-(1-->2)-((α-L rhamnopyranosyl)-(1--->3))-(β-D-4-acetoxyfucopyranosyl}ester of 3-O-(α -L-rhamnopyranosyl-(1 --> 2)- β -D-glucopyranosyl uronic*)-2* β*-3* β*-16*-α -trihydroxyolean-12-ene-23, 28-dioic acid and 28-O-{α-L-rhamnopyranosyl-(1--> 2)-((α -L-rhamnopyranosyl (1--> 3))-(β -Dfucopyranosyl} ester of 3-O-(α -L-rhamnopyranosyl-(l--->2)- β -D-glucopyranosyl uronic)-2 β-3 β-16- α trihydroxyolean-12-ene-23, 28-dioic acid (Mbark et al. 1995) and oleanane saponins (Mbark et al. 1999).

## *Chenopodiaceae*

*Hammada scoparia* **(Pomel) Iljin** (=*Arthrophytum scoparia* **(Pomel) Iljin** = *Haloxylon articulatum* **subsp.** *scoparium* **(Pomel) Batt.**= *Haloxylon scoparium* **(Pomel)** is a small, highly-branched halophytic shrub distributed in south-east Spain, North Africa and parts of Iran, Turkey, Iraq and Syria (Irano-Turanian region) (Maire 1962). *A. scoparium* grows in North Africanregions, especially in Tunisia, Algeria and Morocco, with an extreme environment ofsalinity, dryness and sunlight. Given this type of environment, *A. scoparium* iswell-adapted to water shortages and exhibits adaptations for storing water; in addition, surviving harsh environmental conditions might cause it to produce bioactiveanti-stress factors. Conventionally, *A. scoparium* is usually used as traditional medicine to treat eye disorders, a poultice for mould, and as a snuff powder. The ash of *H. scoparia* or "Remeth" as called in Tunisia is blended with the powder of tobacco leaves for the preparation of the tobacco to appraise "Neffa". This tobacco has numerous therapeutic virtues of which the most important remains her action in the treatment of the scabies of the sheep. As an infusion, the plant is also used to struggle against the diarrhea and for treatment in case of bites of snakes and against rheumatisms. By blending dry leaves with *Pistacia atlantica* and *Cleome arabica*, this plant is used against headaches.

Recently, Aquoues extract of *A. scoparium* has been reported to show antioxidative activity, hepatoprotective effects (Bourogaa et al. 2012), antibacterial activity (Alghazeer et al. 2012) anti-cancer, anti-plasmodial and larvicidal activity (Bellei et al. 2010; Bourogaa et al. 2011). Bourogaa et al. (2012) tested the antioxidative activity and hepatoprotective effects of methanolic extract (ME) of *H. scoparia* leaves against ethanol-induced liver injury in male rats treated daily with 35% ethanol solution (4 g $kg^{-1}$ $day^{-1}$) during 4 weeks. This treatment led to an increase in the lipid peroxidation, a decrease in antioxidative enzymes (catalase, superoxide dismutase, and glutathione peroxidase) in liver, and a considerable increase in the serum levels of aspartate and alanine aminotransferase and alkaline phospahatase. However, this treatment protects efficiently the hepatic function of alcoholic rats by the considerable decrease in aminotransferase contents in serum of ethanol-treated rats. The glycogen synthase kinase-3 β was inhibited after ME administration, which leads to an enhancement of glutathione peroxidase activity in the liver and a decrease in lipid peroxidation rate by 76%.

*H. scoparia* has been described as a plant rich in alkaloids (Carling and Sandberg 1970; Benkrief et al. 1990; Jarraya et al. 1993, 2008; Jarraya and Damak 2001). The molluscicidal activity of *H. scoparia* leaf extracts and the principal alkaloids isolated from them (carnegine and *N*-methylisosalsoline) were tested against the mollusc gastropod, *Galba truncatula*, the intermediate host of *Fasciola hepatica* in Tunisia. The results indicated that the molluscicidal activity was correlated with the presence of alkaloids. A significant molluscicidal value, according to the World Health Organization, was found with the methanol extract ($LC_{50}$ = 28.93 ppm).

The effect of *A. scoparium* ethanol extract (ASEE) on melanogenesis regulation in B16 murine melanoma cells was investigated by Chao et al. (2013). Cells treated with 0.017% (w/v) ASEE showed a significant inhibition of melanin biosynthesis in a time-dependent manner without cytotoxicity. To clarify the mechanism behind the ASEE-treated melanogenesis regulation, the expressions of tyrosinase enzyme and melanogenesis-related genes were determined. Results showed that the expression of tyrosinase enzyme was significantly decreased and Tyr, Trp-1, Mitf and Mc1R mRNA expressions were significantly down-regulated. LC-ESI-TOF-MS analysis of the extract identified the presence of six phenolic compounds: coumaric acid, cinnamic acid, chrysoeriol, cyanidin, catechol and caffeoylquinic acid. The melanogenesis inhibitory effect of ASEE may therefore be attributed to its catechol and tetrahydroisoquinoline derivative content. Chao et al. (2013) report that the first report on the antimelanogenesis effect of *A. scoparium* and on its potential as a whitening agent. Flavonol triglycoside, isorhamnetin: 3-O-beta-D-xylopyranosyl-(1–3)-alpha-L rhamnopyranosyl-(1‴-6″)-beta-D-galactopyranoside, Isorhamnetin 3-O-beta-D-apiofuranosyl-(1–2) (alpha-L-rhamnopyranosyl-(1–6))-beta-D-galactopyranoside, Isorhamnetin3-O-alpha-L-rhamnopyranosyl-(1′-2″) (alpha-L-rhamnopyranosyl-(1′-6″))-beta-D-galactopyranoside have been isolated from leaves of *H. scoparia* collected from Tunisia (Ben Salah et al. 2002).

## *Compositeae/Asteraceae*

***Artemisia herba-alba* Asso.** is a greenish-silverperennial dwarf shrub growing in arid and semi-aridclimates. This species is known as "desert wormwood" in English, 'armoise blanche' in French, and 'chih' in Arab, and it is characteristic of the steppes and deserts of the Middle East, North Africa (Tunisia, Algeria and Morocco) and Southern Europe (Spain and Italy), extending into the Northwestern Himalaya (Vernin et al. 1995). In Tunisia, *A. herba-alba* is found from the mountains around Jebel Oust (Fahs) until the South of the country (Nabli 1989; Marrif et al. 1995). *A. herba-alba* is used as aromatisant for tea and in folk medicine for treatment of colds, coughing, and intestinal disturbances and as antidiabetic agent (Le Floc'h 1983).

This plant is widely used in the traditional medicine to treat diabetes, bronchitis, diarrhea, hypertension and neuralgias (Said et al. 2002; Tahraoui et al. 2007). The essential oil of this species was known for its therapeutic disinfectant, anthelminthic, antibacterial and antispasmodic virtues (Hatimi et al. 2001). Moreover, the oil exhibited antileishmanial (Hatimi et al. 2001), spasmolytic (Perfumi et al. 1999), antioxidant (Mighri et al. 2010b) and antimutagenic activity against the carcinogen benzopyren (Neffati et al. 2008). Numerous studies in the literature were focused on its essential oils and have reported the composition of *A. herba-alba* essential oil from different parts of the world and confirmed the many oil-dependent chemotypes assigned to the plant. According to the reviews of Lawrence (2003) and Salido et al. (2004), two types of oils could be distinguished: those whose composition is dominated by a major compound (camphor, β- or α-thujone, chrysanthenone, chrysanthenyl acetate, or davanone) and those characterized by the codominance of two or more of these compounds. According to Al-Mustafa and Al-Thunibat (2008), various compositions were observed for oils isolated from the aerial parts of *A. herba-alba* plants growing wild in Tunisia. In Tunisia (semi-arid and arid land), various compositions were observed, dominated either by a single component (α-thujone, β-thujone, 1,8-cineole, camphor, chrysanthenone or trans-sabinyl acetate) or characterized by the occurrence, at appreciable content, of two or more of these compounds (Akrout 2004; Boukrich et al. 2010; Mighri et al. 2010a). In Northern Tunisia, the major component was either α-thujone (48.7–79.9%), chrysanthenone (44.2–64.8%), or camphor (35.5 – 47.9%) for 13 oil samples out of 18, and in the five other samples, the main component did not exceed 34%. In Southern Tunisia, four oils out of eight contained a major component present at very high content, i.e., a-thujone (50.6–79.9%) or trans-sabinyl acetate (43.8%). The four other compositions were characterized by the occurrence, at appreciable contents, of two or more compounds (Al-Mustafa and Al-Thunibat 2008). Oxygenated monoterpenes were found to be the major components of *A. herba-alba* oil extracted from aerial parts of plants originated from arid regions of Tunisia (Akrout 2004; Neffati et al. 2008).

Studies from Spain (Salido et al. 2001, 2004) showed that monoterpene hydrocarbons and oxygenated monoterpenes are the most abundant skeletons in *A. herba-alba* oil, but large amounts of sesquiterpenes were found for some populations. Two oil types were found for plants grown in Sinai (Feuerstein et al. 1986) those of cineole-thujane-borane type and the pinane type with monoterpene skeletons. In Jordan regular monoterpenes were predominant and the principal components were α- and β-thujones, classifyng the plant as being a thujone chemotype (Hudaib and Aburjai 2006). In Morocco, the market leader in *A.herba-alba* essential oil exports, 16 chemotypes were found (Lamiri et al. 1997a, b), with 12 having monoterpenes as major components and for four, sesquiterpene skeletons represent the major fraction of the oil.

***Artemisia campestris*** **L.** "Field sagewort" is a perennial scarcely aromatic herb or small shrub. *A. campestris* "T'gouft" is medicinal plants commonly used by local population in the southern of Tunisia for several purposes.

The aerial part of this plant is used in popular medicine as anthelmintic, antiseptic, cholagogue, deobstruent, emmenagogue, stomachic, tonic, hypotensive and antivenin. This plant was used by some native North American Indian tribes as an abortifacient to terminate difficult pregnancies. This plant has been crushed and applied externally to rheumatic joints, eczema, bruises and sores. A poultice of the crushed leaves has been applied to sore eyes. An infusion of the roots has been used, especially on children, as a hair tonic and to treat scalp infections. It has been taken internally to promote urination and bowel movements (Le Floc'h 1983). The essential oils of *A. campestris* widespread in the south of Tunisia have been studied by several authors and were found to contain different compounds such as alpha and beta-pinenes, p-cymène, caryophyllene oxide, spathulenol, limonene, dehydro-1,8-cineole, cadin-4-en-7-ol, gamma-terpinene, (Z)-beta-ocimene, aromadendrene, germacrene D, bicyclogermacrene, myrtenol, p-cymen-8-ol, gamma-cadinene, ar-curcumene, delta-cadinene, calamenene, alpha-muurolene, gamma-muurolene, gamma-cadinene, bisabolene and endoperoxide, (Z,E)-farnesol, cedrol and verbenone (Neffati et al. 2008; Akrout et al. 2001, 2003). Solvent extracts and essential oils from *Artemisia campestris* have been shown to exhibit antioxidant, hepatoprotective, antibacterial, antiviral, insecticide and allelochemical activities (Hatimi et al. 2000; Djeridane et al. 2006; Aniya et al. 2000; Tharib et al. 1983).

## *Crucifereae*

***Oudneya africana*** **R. Br.,** a wild-growing plant in the arid region of Tunisia, is used in ethnomedicinal treatment of microbial infections. Later, Hammami et al. (2009) shown that compounds isolated from *O. africana* have broad-spectrum activity against both Gram-positive and Gram-negative bacteria.

Djeridane et al. (2006) considered this species as medicnal plant in Algeria to treat the digestive problem and they reported that the aqueous ethanol of *O. africana* characterised by 7.75 mg GAE/g dw ± 0.22 total phenolic content and 7.66 mg rutin equivalent/g dw ± 0.23 flavonoids content and has an antioxidant activity expressed as Trolox equivalent antioxidant capacity (TEAC) (16.30 (mmol TEAC/g dw).

***Diplotaxis harra* (Forsk.) Boiss.** an annual is widely distributed in many sandy and gypseous areas in southern Tunisia (Tlig et al. 2008). It is a ramified stem herb which occurs in both sandy and gypseous soils reaching a height of 50–60 cm (Chaieb and Boukhris 1998; Pottier-Alapetite 1979), and grows best in soils where NaCl levels are around 580 ppm.

When dried, this herb is browsed and appreciated by animals but its abuse seems to have certain toxicity; however, the fresh plants are generally refused by the animals (Le Floc'h 1983). It is also a medicinal plant used against the scabies of the animals. The decoction of the *D. harra* leaves was used against constipation (Boulos 1983) and showed higher activity against yeasts than bacteria (Hashem and Saleh 1999).

## *Cucurbitaceae*

***Colocynthis vulgaris* (L.) Schard.** = *Cistrullus colocynthis* **Schrad.** a plant indigenous to India, Pakistan and Sri Lanka have a variety of ethnomedical claims and finds uses as a purgative (Banarjee and Dandiya 1967), powerful cathartic (Adam 1978), abortifacient (Saha et al. 1961) and for amenorrhea (Al-Rawi and Chakravarty 1964). Extracts derived from fruits of C. *vulgaris* have been shown to possess cardiac depressant, smooth muscle relaxant, cytotoxic, and antitumour activities (Faust et al. 1958). Previous chemical investigations of the fruits of C. *vulgaris* by chromatography technic has revealed the presence of cucurbitacins B, D, E, I, L and the glucosides of cucurbitacins B, E and L (Darwish-Sayed et al. 1974).

C. *vulgaris* is used for rhemnatism, the Bedouins tie a slice of fresh, green gourd onto the heel before retiring. The gourds are eaten by domestic animals when green, according to an Abadi Bedouin of the Eastern Desert. Mason (1936) learned, during his travels in the Libyan Desert that they were eaten (when green) by gazelles, Barbary sheep and donkeys, probably, he postulated, for the water they contained. He tasted a juicy piece of green gourd and hod diarrhoea for 3 days. Tregenza (1955) guides told him that gazelles broke the dry gourds with their horns to get at the seeds. According to this author, the Tibbu people harvested the seeds of handal for food. The seeds were eaten after prolonged soaking in water. According to Hassanein Bey (1925) abra, the staple dish of the Tibbu people, is made from *handal*. The seeds are boiled thoroughly to get rid of the bitterness and then crushed with dates or locusts in a wooden mortar. Hassanein Bey recorded a conversation with a Bedouin who had once survived for 12 days on *handal* seeds which, the man said, had upset his digestion. The cathartic effect of the untreated seeds is known to all Bedouins.

In Bahariya Oasis learned that a purgative for intestinal parasites is milk that has stood overnight in a hollowed, dry ground gourd. Since the time of the Pharaohs Natives of Aswan dried gourds with a small opening in them have been placed among clothes to keep away insects.

## *Cupressaceae*

### *Juniperus phoenicea* L.

The genus *Juniperus* consists of approximately 60 species growing in the Northern Hemisphere and divided into three sections: Caryocedrus *Juniperus* (=Oxycedrus) and Sabina, the third being the most important one which comprises *J. phoenicea*. Recently, Adams (1999) showed on the basis of essential oil composition and RAPD DNA fingerprinting that *J. phoenicea* is distinguishable from other species of section Sabina. *J. phoenicea* is a shrub or a small tree from the Mediterranean region (Bonnier and Douin 1990). In Mediterranean countries the *Juniperus* flora is expanding to a great variety of natural habitats (Milos and Radonic 2000; Adams 2000; Munoz-Reinoso 2004) and has attracted special interest as constructional timber, because its wood is free of insults from pests and diseases. Another interesting use of Juniper wood was described by Herodotus, who connected the beneficial effects of plant resin with the preservation of dead human tissues in ancient Egypt. This activity was further confirmed by Dioscorides, who named Juniper as the "life of the dead". Later, the insect-repellent properties of Juniper were widely recognized through biological evaluation experiments that revealed its strong antibacterial and fungicide properties (Karaman et al. 2003). In addition, the EO of Juniper is known to possess significant insecticidal and repellence properties (Kanat and Alma 2004; Gao et al. 2004). Against mosquitoes the Eos of *Juniperus* sp. is reported to display potent repellent, insecticidal, oviposition-deterrent or larvicidal properties (Prajapati et al. 2005; Amer and Mehlhorn 2006a, b). The mixture of the leaves and cones of *J. phoenicea* is used as an oral hypoglycemic, whereas the leaves are used against bronchopulmonary diseases and as a diuretic (Bellakhder 1997).

There are few reports on the composition of *J. phoenicea* leaf oil, even though they concerned plants from various origins and a limited number of samples were analysed every time. The first studies date back to 1956 and 1973 and reported monoterpene-rich oils (subspecies not reported) (Gil de Meister and Hoffman 1956; Banthorpe et al. 1973). More recently, Vidrich and Michelozzi (1993) reported 1,8-cineole, α-pinene and borneol as major components for an oil from Italy (subspecies not reported). Adams et al. (1996) studied samples from different origins and different subspecies. They reported the following compositions: α-pinene for *J. phoenicea* (sensu stricto) from inland Greece and Spain and α-pinene/β--phellandrene/α-terpinyl acetate for *J. phoenicea* var. *turbinata* from coastal Spain and *J. phoenicea* subsp. *eu-mediterranea* from coastal Portugal. The composition of 50 samples of essential oil of individual plants of *J. phoenicea* subsp. *turbinate* from Corsica was investigated by GC, GC–MS and $^{13}C$ NMR. α-Pinene, β-phellandrene, α-terpinyl acetate, Δ-3-carene, myrcene and α-phellandrene were found to be the main constituents (Rezzi et al. 2001). Since species of *J. phoenicea* have already been studied how the geographical situation affects the production of secondary metabolites (Rezzi et al. 2001).

The essential oils of the leaves and cones have proven to possess antimicrobial activity (Stassi et al. 1996). These results indicate that the EOs of Juniper plants display potentials for use as natural alternative to synthetic agrochemicals, and research is needed for the most potent EO identification and the thorough study of its components (Vourlioti-Arapi et al. 2012). Hammami et al. (2009) mentioned that *J. phoenicea* extract have the most active antibacterial activity in comparison with two other plants from the arid regions of Tunisia; The essential oils, which were utilised centuries ago in cosmetics usually, show interesting biological features. The oils also help increase the flow of digestive fluids, improve digestion and eliminate gas and stomach cramping (Uphof 1968). *J. phoenicea* has been used for centuries as a steam inhalant for bronchitis and to control arthritis (Watt et al. 1962; Stassi et al. 1996). According to the results presented by Hayouni et al. (2007) we can note that extracts from *J. phoenicea* obtained by mixture: acetone/water/acetic acid (90/9.5/0.5) showed more than 70% inhibition of scavenging of the ABTS radical, which is as strong as the synthetic BHA. This could be attributed to the higher levels of phenolics in *J. phoenicea* extracts. Juniperoside, a new 9-O(β-D-glucopyranoside)-3,4,5-trimethoxycinnamyl alcohol has been isolated along with the 9-O-(α-L-arabinofuranosyl-(1→6)-β-D-glucopyranoside)cinnamyl alcohol (rosarin) and coumarin 7-O-β-D-glucopyranoside (skimmin) from the acetone extract of the aerial parts of J. phoenicea (Comte et al. 1996).

## *Ephedraceae*

***Ephedra alata* Decaisne,** is a very effective sand-binder and resistant to desertification (Abdallah and Chaieb 2007). Most of the 50 *Ephedra* species worldwide (Stevenson 1993; Price 1996) are shrubs adapted to semiarid and desert conditions (Pearson 1929). A few lianoid species are found in temperate regions along the Mediterranean coastline and humid montane sites in North Africa and southwest Asia (Freitag and Maier-Stolte 1989). About 25 species of *Ephedra* are found in the drier regions of the Old World extending westwards from Central Asia across southwest Asia and into Mediterranean Europe and North Africa (Freitag and Maier-Stolte 1994). According to Boulos (1983), *E. alata* which grows wild in the Egyptian desert provides extracts used in folk medicine as depurative, hypotensive, anti-asthmatic, sympathomimetic and astringent agents. The branches are chewed for cephalalgia, used in miscarriage and as a bronchodilator.

An Asian *Ephedra* species has been used for thousands of years in Traditional Chinese Medicine herbal formulas designed to treat asthma and other respiratory diseases (Rotblatt and Ziment 2002). According to Chen and Schmidt (1930) and Bloedorn and Dickens (1928), *Ephedra* gained widespread medical use in the United States in the 1920s as a nasal decongestant, central nervous system stimulant, and asthma treatment, but use declined substantially in the following decade because of safety concerns and the availability of safer alternatives. Many Ephedra-containing products are claimed to be effective for weight-loss, as energy/performance boosters, euphorics or aphrodisiacs, claims that are especially appealing to young adults (Josefson 1995). Recently, *Ephedra* has become a common ingredient in many dietary supplements that are promoted to increase energy or assist with weight loss. Since herbs are regulated as dietary supplements in the United States, these ephedra-containing products that can be sold without approval by the U.S. Food and Drug Administration (FDA) (FDA 1994). Numerous case studies have documented adverse effects in persons using ephedra (Haller and Benowitz 2000; Vahedi et al. 2000). In the largest published case series, Haller and Benowitz (2000) reviewed 140 adverse event reports involving ephedra that were submitted to the FDA. They concluded that 43 cases were definitely or probably related to ephedra and another 44 were possibly related. However, there were certain limitations in determining causality from case reports, the foremost of which is the inability to determine relative risk.

Almost all commercial applications of Ephedra extracts derive from the ephedrine alkaloids found in the stems in many Eurasian species. The best-documented drug made from Ephedra is Ma-huang. It contains several *Ephedra* alkaloids, including the primary active constituent, ephedrine, as well as smaller amounts of pseudoephedrine, phenylpropanolamine, methylephedrine, methylpseudoephedrine, and norpseudoephedrine (Gurley et al. 1998), used in Chinese medicine for 5000 year as a treatment for fever, nasal congestion, and asthma (Zhu 1998). Ma-huang is also an effective respiratory sedative and cough remedy. These drugs have also sympathomimetic activity and can lead to various physiologic responses, including vasoconstriction; bronchodilation; and increases in blood pressure, heart rate, cardiac contractile force, and automaticity (Hardman and Limbird 2001). Herbal mixtures containing Ma-huang are sold in health food stores in the West as nutritional supplements under such names as Herbal Ecstasy and Escalation (Gurley et al. 1998) accompanied by dubious claims that they have energizing value or assist in dieting (White et al. 1997). The organs used in traditional medicine are the dried green stems, which are usually boiled in water and administered as a hot tea. The usual daily dose is 1.5–9 g of the decocted herb (Leung 1999). The aerial parts of different *Ephedra* species contain from 0.02% to 3.4% of 6 optically active alkaloids concentrated in the internodes (Leung and Foster 1996). (−)-Ephedrine (EPH) is the major isomer comprising 30–90% of the total alkaloids. It was the first alkaloid isolated from *Ephedra* by Nagai in 1887. (+)-Pseudoephedrine (PSE), the diastereomer of (−)-EPH, was subsequently isolated by Ladenburg and Ölschlägel in 1889 followed by the remaining isomers in the late 1920 (Ehab et al. 2003). In addition to the ephedrine-type alkaloids, other alkaloids and amino compounds have also been isolated from different species of *Ephedra* (Ehab et al. 2003).

The isolation of β-coumaric acid and (f)-2,6-bis-(3,5-dimethoxy-4-hydroxyphenyl)-3,7- dioxabicyclo(3,3,0)octane (syringaresinol, 2) from *E. alata* chloroform extract, the 2,3-digalloylglucopyranose (nilo- citin, 3) from *E. alata* ethanolic extract and the new alkaloid 7- methoxy-4-quinolone 2-carboxylic acid (ephedralone, 4) from *E. alata* aqueous extract of the same plant were reported (Nawwar et al. 1984). In addition to the known β-coumaric acid, the furanofuran lignan (f)-syringaresinol and the digalloyl-glucose, nilocitin, were obtained from the whole plant of *E. alata*. A new natural alkaloid, ephedralone, was also isolated (Nawwar et al. 1985). Flavonol glucosides have been identified in *E. alata*, namely, herbacetin methyl ether 3-0- glucoside-7-0-rutinoside and herbacetin 7-0-(6″-quinylglucoside). The known flavonoids vicenin II, lucenin III, kaempferol 3-rhamnoside, quercetin 3-rhamnoside and herbacetin 7-glucoside were also found. The structure of the isolated compounds was determined mostly by FABMS and HNMR spectroscopy (Nawwar et al. 1984).

The pharmacology of *Ephedra* is complex, but for both traditional and more recent popular uses, the established pharmacological effects appear to be attributable to its ephedrine-type alkaloids, mainly (−)-EPH and (+)-PSE. Pharmacological studies of *Ephedra* began shortly after ephedrine was first isolated and were focused on the pure compound rather than on the whole herb extract (Takahashi and Miura 1888).

## *Globulariaceae*

***Globularia alypum* L.,** is a perennial wild-growing shurb found throughout the mediterranean area locally named as "Ain Larneb". Its leaves are traditionally used as hypoglycaemic agent, laxative (Bellakhdar et al. 1991), cholagogue, stomachic, purgative and sudorific (Sijelmassi 1993). A recent ethnobotanical survey has demonstrated that *G. alypum* is one of the most used medicinal plants in Fez region (Jouad et al. 2001). Previous investigations of 624 patients suffering from diabetes and/or hypertension in oriental Morocco have demonstrated that *G. alypum* represented the second highest percentage of medicinal plant utilization (9.9%) after *Trigonella foenum graecum* (13.3%) (Ziyyat et al. 1997). Similar studies undertaken in the Wilaya of Marrakesh have shown that this plant was the fifth highest percentage (9.86%) of utilization (Jaouhari et al. 1999). Investigations were carried out to evaluate the hypoglycaemic activity of the infusion of *G. alypum* leaves. Oral and intraperitoneal administration of the plant (0.7 g/kg) produced a significant hypoglycaemic effect in normal as well as in hyperglycaemic rats. The infusion increased plasma insulin levels significantly in normal rats. It is suggested that the hypoglycaemic activity of this plant may be mediated through enhancement of peripheral metabolism of glucose and increase in insulin release (Skima et al. 1999). A significant antileukemic activity of an aqueous extract of *G. alypum* was also reported (Caldes et al. 1975).

The antioxidant activity of the *G. alypum* phytochemicals were evaluated by Es-Safi et al. (2007) for their capacity to scavenge the 1,1-diphenyl-2-dipicrylhydrazyl (DPPH) free radical and some structure–activity relationships were obtained. Assay guided fractionation led to the isolation of syringin, four phenylethanoids, four flavonoids and six iridoids as the main constituents of the extract and their antioxidant activity was determined. The obtained results showed that the activity towards the DPPH free radical was mainly due to the flavonoid and phenyl ethanoid constituents which were most active free radical scavengers than iridoids. Their high antioxidant activity could be attributed to the caffeoyl moieties contained in them, while iridoids showed moderate free radical scavenging activity. These results demonstrated that some of the isolated compounds play an important role for the antioxidant activity of *G. alypum* and give a scientific basis to the use of this plant in traditional medicine. The hydromethanolic extract of *G. alypum* could thus be considered as a source of potential antioxidants and will promote the reasonable use of this plant in food technology and processing as well as for medical use.

*G. alypum* is used in the treatment of cardiovascular and renal diseases as demonstrated by recent ethnobotanical surveys (Jouad et al. 2001; Jaouhari et al. 1999). Recently, methanol and dichloromethane extracts of *G. alypum* were also shown to reduce histamine and serotonin contraction *in vitro* (Bello et al. 2002). From the hydromethanolic extract of the aerial parts of *G. alypum* grown in Morocco, a new chlorinated iridoid glucoside, globularioside has been isolated beside 5 known iridoid glycosides, globularin, globularicisin, globularidin, globularinin and globularimin (Es-Safi et al. 2006).

## Poaceae

***Hyparrhenia hirta* (L.) Stapf** is a tufted perennial grass growing to 1 m in height that is distributed mainly in the shore regions of the Mediterranean and across the African continent (Clayton 1969). *H. hirta* Stapf occurs in Tunisia in higher densities in central and southern parts of the country (Chaieb and Boukhris 1998). *H. hirta* is a pasture grass that has become highly invasive in several parts of the world, including Australia where it has become a serious environmental weed in recent decades. In Tunisia the plant is known for its diuretic properties (Boukef 1986).

The phytochemistry of *H. hirta* has been investigated until relatively recently, when triterpenes, a β-ketone, aromatic aldehydes and some common flavonoids were described (Ben Salah et al. 2000 and Bouaziz et al. 2001). Leaves of *H. hirta* yielded the rare diastereoisomeric flavonolignans tricin 4′-O-(erythro-β-guaiacylglyceryl) ether and tricin 4′-O-(threo-β-guaiacylglyceryl) ether together with their 7-O-glucosides, which are the first flavonolignan glycosides to be isolated as natural products (Bouaziz et al. 2002). Although the overall flavonoid profile of *H. hirta* is typical of Poaceae in containing flavone C-glycosides and tricin 5-O-glucoside, the multitude of glycosides present in this species shows that the exact glycosidic pattern of species may be of chemotaxonomic importance in the family at lower levels of classification (Bouaziz et al. 2001). The methanol extract of *H. hirta* show an anti-inflammatory (Hoult and Moroney 1994) and anti-oxidant (Joyeux et al. 1995) activity, which could be attributed to the isolated flavonoids (Bouaziz et al. 2001).

## *Labiatae (Lamiaceae)*

*Rosmarinus officinalis* **L.** is a woody, perennial herb with fragrant evergreen needle-like leaves. It is native to the Mediterranean region. Forms range from upright to trailing; the upright forms can reach 1.5 m, rarely 2 m. The leaves are evergreen, 2–4 cm long and 2–5 mm broad, green above, and white below with dense short woolly hair. Flowering, very common in a mature and healthy specimen, usually appears in winter or spring and is variable in color, being white, pink, purple, or blue. *R. officinalis* is very common in Europe, Asia and Africa, but only in the areas around the Mediterraneean Sea and in many islands, particularly Sicily, Sardinia, Corsica, Baleari and Elba (Pintore et al. 2002). In Tunisia, *R. officinalis,* this is an interesting plant due to its essential oil, grows in mountains at an altitude of 400–500 m and on plains but it is rare or absent in the islands (Djerba, Kerkanah).

In Tunisia, *R. officinalis* is present with an endemic variety (*troglodytorum*), growing in the Matmatas region (southern Tunisia). It is characterized by its particular ecological adaptation to arid superior environmental factors (i.e. irregularity and insufficiency of rainfall, low soil quality, *etc.*). Because of its rusticity, it grows in every soil type, but it prefers a sandy, arid, calcareous, humus-poor soil. It can be distinguished from the other varieties by its dark purple flowers, long stamens and its greyish leaves and stem (Pottier-Alapetite 1981). *R. officinalis* is cultivated in many family gardens in Tunisia as ornamental plant and to use its leaves in traditional medicine and for flavoring food. It is also cultivated for the valuable oil which can be extracted from the flowering aerial tops, comprising leaves, twigs and flowers, collected from spring to late autumn. Many studies have been interested in the qualitative and quantitative variability of the composition, yield and biological characteristics of the essential oil, due to intrinsic (genetics and plant age) or extrinsic parameters such as climate and cultivation conditions or isolation methods (Bekkara et al. 2007; Stevanovits-Banyai et al. 2003).

The fresh and dried leaves are used frequently in traditional Mediterranean cuisine as food flavoring and for medicinal purposes (Le Floc'h 1983). It contains a great amount of essential oil (0.5–2.5% v/w), which is largely used in traditional medicine as a pulmonary antiseptic, a choloretic and a colagoguic. It has alsostomachic, antidiarrhoic and antirhumatic properties (Oury 1984). It is also utilized in the cosmetic industry producing various Cologne waters, bathing essence, hair lotions and shampoos (Burt 2004). *R. officinalis* is known by its antioxidant properties (Raul et al. 2005; Sacchetti et al. 2005; Wang et al. 2008). The biological characteristics of essential oil of *R. officinalis* have been studied by several authors. It has been demonstrated that this oil exhibited antibacterial, antioxidant, insecticidal and antimutagenic activities (Bannour et al. 2006; Gachkar et al. 2007; Moghtader and Afzali 2009; Wang et al. 2008; Isman et al. 2008; Miresmailli et al. 2006). Al-Sereitia et al. (1999) reported different pharmacological actions of *R. officinalis* such as effects on central nervous system, on the circulation and on smooth muscle, choleritic and antitumoriginic effect and antimycotic effect.

According to the geographic location of the species, in North Africa and in Europe, the analyses of chemical composition showed that the species is mainly represented by two major chemotypes, 1,8-cineole (over 40%) and α-pinene/camphor (with equal ratio, 20-30%). The species is cultivated extensively in several countries such as Portugal, Spain, Italy, in near orient, in Asia and in America.

This chemical composition of *R. officinalis* essential oil has been the subject of considerable study. The reported components were mostly monoterpenes, the majorones being α-pinene, 1,8-cineole and camphor, associated with variable amounts of camphene, myrcene, limonene, borneol, verbenone, bornyl-acetate, etc. The major types of *R. officinalis* can be distinguished with respect to these main constituents: oils with over 40% of 1,8-cineole (oils from Morocco, Tunisia, Turkey, Greece, Yugoslavia, Italy, France, Algeria) (Chalchat et al. 1993; Boutekedjiret et al. 1998; Boutekedjiret et al. 1999; Rezzoug et al. 1998; Zaouali et al. 2005) and oils with approximately equal ratios (20–30%) of 1,8-cineole, α-pinene and camphor (oils from France, Spain, Italy, Greece, Tunisia, Algeria, South Africa) (Domokos et al. 1997; Mangena and Muyima 1999; Mastelic and Kustrak 1997; Tomei et al. 1995). One other chemical composition could be defined according to the comparatively higher amount of myrcene in oils from Argentina, Portugal and Spain (Mizrahi et al. 1991; Dellacassa et al. 1999). The literature also revealed some other chemical compositions for rosemary oils. Oils dominated by camphor/1,8-cineole/borneol from Cubaand Morocco (Pino et al. 1998; Elamrani et al. 2000), by 1,8-cineole/borneol/p-cymene from Turkey (Perez-Alonso et al. 1995), by α-pinene/1,8-cineole/camphene from Iran and Morocoo (Elamrani et al. 2000; Jamshidi et al. 2009), by α-pinene/camphor/α--pinene from Algeria, by1,8-cineole/camphor/camphene/α-pinene from southern of Tunisia. Oils from Sardinia and Corsica (Mediterranean islands) were characterized by its relatively high content of verbenone (15.7–24.9%) with a considerable amount of α-pinene (13.7–24.6%) and camphor (2.9–14.1%). Oils from Izmir and Canakkale (Turkey) were also relatively rich in verbenone (4.4–45.2%) but with a considerable amount of 1, 8-cineole (12.1–34.3%) and camphor (9.9–24.1%). The essential oils extracted by Clavenger apparatus from leaves of *R. officinalis* cultivated in different areas in Djerba (Island in the southern of Tunisia) were evaluated for their antioxidant (DPPH method) and antibacterial activities (agar-well diffusion method) by (Akrout et al. 2010a, b). The three essential oils chemotypes exhibited moderate antioxidant activity with an $IC_{50}$ ranged from 4.186 mg/ml for chemotype I to 7.298 for chemotype II and showed strong to moderate antibacterial activity with a MIC ranged from 0.156 to 1.25 mg/mL (Chemotype I and chemotype II) and from 1.25 to 5 mg/mL (chemotype III).

The *Thymus* genus belonging to the Lamiaceae family includes approximately 350 species, existing mainly in Europe, Western Asia and the Mediterranean regions (Mabberley 1997). Many species of *Thymus* have been widely used in folk medicine in the world for their carminative, antispasmodic, emmenagogic and tonic properties (Bruneton 1993). In Tunisia, the genus *Thymus* includes four species: *T. capitatus* (L.) Hoffm. et Link., *T. numidicus* Poir., *T. algeriensis* Boiss. et Reut., and *T. vulgaris* L. The species can be sympatric in a wide part of their distribution area.

The species *Thymus algeriensis* (Hyphodromi Section) is an endemic plant of semi-arid and arid areas of Tunisia and Algeria (Pottier-Alapetite 1981; Le Floch and Boulos 2008). In Tunisia, *T. algeriensis* populations are distributed from the sub-humid to the lower arid climates at altitudes ranging from 120 to 1100 m. The species grows on poor fertile calcareous soils and occurs in scattered, small populations, showing different levels of destruction, mainly due to overharvesting and overgrazing. Close to 100 species have been identified in the kind of *Thymus* across the world (Richard et al. 1985). They are reported to possess some biological effects such as antispasmodic (Meister et al. 1999), antibacterial (Essawi and Srour 2000; Dob et al. 2006), antifungal (Soliman and Badeaa 2002), anti-tabagism (Carlini et al. 2006), giardicidal (Amaral et al. 2006), anticancer (Jaafari et al. 2007) and antioxidant activities (Tepe et al. 2005). The First Edition of Brazilian Official Pharmacopoeia, published in 1926, includes 1702 Monographs for medicinal products. Among those, two Monographs refer the thyme (Brandão et al. 2006).

Tunisian ***Thymus algeriensis* Boiss & Reuter** (Synonym *T. hirtus* Willd. subsp. *algeriensis* Boiss. & Reut.) is known under vernacular name of “Mazoukcha”. In traditional medicine, infusions, leaves and flowers decoctions were used as tonic, anti-inflammatory, antiparasitic, antitussive and carminative (Stahl-Biskup and Sáez 2002). It is an herbaceous fragrant plant largely used, fresh or dried, as a culinary herb (Pottier-Alapetite 1981; Le Floch and Boulos 2008). Furthermore, this plant is also widely used in folk medicine against illnesses of the digestive tube and antiabortion (Le Floc’h 1983). *T. algeriensis* is used fresh or dried as a culinary herb and in folk medicine for its antiseptic, antispasmodic and antifungal properties (Hazzit et al. 2009; Giordani et al. 2008). In Algeria, they have been used as astringent, expectorant and cicatrising agents (Baba Aïssa 1991). The thyme has been used, in Moroccan traditional medicine, in the treatment of diarrhoea, fever, cough, infected areas and wounds. It was also used as a tonic and stimulant (Bellakhdar 1996; Sijelmassi 1993) and, generally, for its anti-inflammatory properties after topical or oral administration (Ismaili et al. 2001, 2002, 2004).

Data on the essential oil composition of *T. vulgaris* of other continents have been presented in numerous publications. Thyme oil is among the world’s top ten essential oils also used as a preservative for food purposes (Alçiçek 2011; Ehivet et al. 2011) and (Stahl-Biskup and Sáez 2002). The demand for essential oils from these species is increasing for perfumery, cosmetic and medicinal uses (Hazzit et al. 2009). *T. algeriensis* are rich in essential oils and were characterized by a great variability of both morphology and chemotypes (Stahl-Biskup 1991). There is obviously a chemical polymorphism of essential oils within the plants belonging to the genus *Thymus*. In addition, the infraspecific variability of the essential oils in the genus *Thymus* was also observed (Stahl-Biskup 1991). These differences observed in the oil composition may be due to the collection time (development stage) and ecological factors (Dob et al. 2006). Depending on the geographical origin and the specific ecological sites from which it is collected, its quantitative and qualitative composition vary greatly between countries and the diversity of essential oils arises from the existence of many chemotypes (Hudaib et al. 2002; Owen et al. 2002; Goodner et al. 2006; Snoussi et al. 2008). Even in the same country, genetic and ecological factors, especially climatic, determine the distribution frequency of the chemotypes among populations such as in Spain: thymol (Rota et al. 2008), 1.8-cineole and eucalyptol (Goodner et al. 2006) and in France (geraniol, linalol, α-terpineol, thujanol-4, terpineol-4, thymol and carvacrol) (Granger and Passet 1973). Recently, investigations on the chemical composition of *T. algeriensis* essential oil wildly growing in Tunisia, allowed the identification of six chemotypes according to the main compounds. In fact, a high variation among populations for the majority of the compounds was shown (Ben El Hadj Ali et al. 2010). The variation of the essential-oil composition among 14 Tunisian natural populations of T. algeriensis was assessed by GC (RI) and GC/MS. The populations were collected from different geographical regions belonging to the sub-humid, upper semi-arid, mean semi-arid, lower semi-arid, and upper-arid bioclimates. A total of 47 constituents, representing 81.0–96.5% of the total oil, were identified. The main volatiles at the species level were 1,8-cineole (17.7%), α-pinene (15.5%), and camphor (8.2%). A high variation among populations for the majority of the compounds was shown. Camphor (0.2–14.0%), linalool (0.2–22.4%), borneol (<0.01–24.3%), caryophyllene oxide (<0.01–18.8%), thymol (<0.01–54.9%), γ-terpinene (0.4–6.5%), α-copaene (0.4–7.6%), linalyl acetate (<0.01–6.4%), and methyl eugenol (<0.01–6.9%) were the main constituents differentiating the populations (Ben El Hadj Ali et al. 2010).

Many studies on the antimicrobial activity (Zambonelli et al. 2004; Dorman and Deans 2004) and antioxidative activity (Dorman and Deans 2004; Youdim et al. 2002) of these oils have been reported. On the other hand, several extracts of these plants were tested for their pharmacological activity (Marti et al. 2005). Furthermore, the oil was tested for antimicrobial activity against four bacteria, two fungi and two yeasts. This oil exhibited a significant *in vitro* antimicrobial activity against *Bacillus subtilis* (MIC = 0.5 lL/mL), as well as against all yeast and all filamentous fungi tested (MIC = 0.5 and 1.0 lL/mL) (Dob et al. 2006). The antimicrobial properties of the oil could be associated with the high percentage of linalool, which is known to possess strong antimicrobial activities (Burt 2004; Pitarokili et al. 2002). In addition, it has been well known that the phenolic components of the essential oils showed the strongest antimicrobial activity (Burt 2004). The essential oil activity is then due either to minor constituents or a synergy between these and the major components (Zakarya et al. 1993). Thymes are well known for their antispasmodic, sedative, antioxidant (Mkaddem et al. 2011; Safaei-Ghomi et al. 2009) and antiaflatoxigenic (Razzaghi-Abyaneh et al. 2009) properties. The *T. algeriensis* essential oil was found to possess an interesting inhibitory activity towards ACE with an $IC_{50}$ value of 150 μg/ml. The obtained results also showed that this oil can act as radical scavengers ($IC_{50}$ = 0.8 mg/ml) and displayed a lipidperoxidation inhibitory activity ($IC_{50}$ = 0.5 mg/ml) as evaluated by 2,2-diphenyl-1-picrylhydrazyl and β-carotene bleaching methods, respectively. Previously isolated constituents flavonoids were: taxifolin, eriodictyol, 5,6-dihydroxy-7-3′,4′-trimethoxyflavone, and 5,6,4′-trihydroxy-7,3′-dimethoxyflavone (El-Domiaty et al. 1997).

*Thymus vulgaris* **L.,** locally known "zaatar" and member of the Lamiaceae family, is widely used in Mediterranean medicine for its biological activities (Biondi et al. 1993; Cosentino et al. 1999). *T. vulgaris*, also known as Spanish common thyme, has its flowering period between March and July. It can be normally located from sea level to 2000 m above sea level (Morales 1996). The buds of cloves known as a dental analgesic, carminative, stimulant and antiseptic agent, are also used in the traditional preparation of some Tunisian food dishes such us "couscous" due to their flavouring activity (Snoussi et al. 2008).

*T. vulgaris* oil exhibited a high level of antimicrobial activities against *Vibrio* spp. Strains compared to *Mentha longifolia*, *M. pulegium*, *Eugenia caryophyllata* and *Rosmarinus officinalis* oil (Snoussi et al. 2008). As a valuable medicinal plant, *T. vulgaris* possesses, antiseptic, antimicrobial, and antioxidative properties (Dorman and Deans 2000; Nguyen et al. 2000; Dapkevicius et al. 2002).

Totally, 61 constituents were identified in *T. vulgaris* essential oil by capillary GC and GC-MS. Thymol was the dominating compound in the all analyzed oils (44.4–58.1%), followed by α-cymene (9.1–18.5%), α-terpinene (6.9–18.9%), and carvacrol (2.4–4.2%). Phenolic compounds are the aroma principles in this chemotype of thyme. The quantitatively most important compounds are the phenols thymol (44.4–58.1%) and carvacrol (2.4–4.2%), which constituted almost half of the essential oil, followed by monoterpene hydrocarbons *p*-cymene (9.1–18.5%), α-terpinene (16.1–18.9%), myrcene (1.5–2.2%), and β-thujene (1.0–1.5%). (*E*)-Sabinene hydrate (0.7–1.2%), linalool (1.5–2.8%), and borneol (0.4–1.7%) were dominant among monoterpenic alcohols (Baranauskiene et al. 2003).

***Thymus capitatus* (L.) Hoffm. et Link (Synonym *Thymbra capitata* (L.) Cav. or *Coridothymus capitatus* Rch. f.)** (Figueiredo et al. 2008).

It is widespread in the Mediterranean area. In Tunisia, populations grow on sandy and often on rocky soils from the sub-humid to the upper arid bioclimates. The annual rainfall of sites ranged from 300 to 1000 mm/year. Altitudes varied from 150 to 500 m (Nabli 1995). Populations were mainly associated with *Ceratonia siliqua* L., *Fumana thymifolia* Spach, *Hyparrhenia hirta* Stapf., *Olea europea* L., *Quercus coccifera* L. and *Rosmarinus officinalis* L. It's a perennial, herbaceous shrub commonly used as a spicy herb and locally known under the common name "Zaâtar". *T. capitatus* is used as spicy herb for flavouring cheeses, soups, stews, stuffings, meats, fishes, dressings, sauces, and honey. The essential oil of *T. capitatus* are also used in the flavour and food industries. The oil is used in the flavouring of toothpaste, mouthwashes, cough suppressant medicines and in the manufacture of perfumes and cosmetics. As a medicinal plant, *T. capitatus* has traditionally been considered an anthelmintic, antispasmodic, carminative, emmenagogue, expectorant, rubefactient, sedative, stimulant, and tonic. The plant has been used as a folk medicine against asthma, arteriosclerosis, colic, bronchitis, coughs, diarrhea, and rheumatism (Le Floc'h 1983). Nowadays, essential oils and their components are gaining increasing interest because of their relatively safe status, their wide acceptance by consumers, and their exploitation for potential multi-purpose functional use (Ormancey et al. 2001; Sawamura 2000). The essential oils of *T. capitatus* have been investigated by many researchers whose reported that the major components of these oils were thymol, carvacrol, linalool, γ-terpinene with the presence of β-cymene, borneol and β-bisabolene in relatively lower amounts (Amirti et al. 2008a; Chaouch 2008a; Akrout 2004). *T. capitatus,* growing wild in the southern of Tunisia, oil was mainly composed of carvacrol (68.8%) and β-cymène (11.1%) (Bounatirou et al. 2007a). This chemical composition is in concordance with the carvacrol chemotype growing in Tunisia (Bounatirou et al. 2007a, b; Akrout 2004; Bounatirou et al. 2008), Morocco (El Ajjouri et al. 2008) and Greece (Karousou et al. 2005). The Sardinian *T. capitatus* oil differed from Tunisian oil by its low content of carvacrol (10.8%) and its high amount of thymol (29.3%) and β-cymene (26.4%) (Cosentino et al. 1999). The Turkish oil is dominated by carvacrol (35.6%) in addition with thymol (18.6%), β-cymene (26.4%) and γ-terpinene (12.3%) (Goren et al. 2003). *T. capitatus* essential oils collected from the southern of Tunisia exhibited antibacterial, antifungal and antioxidant activities (Bounatirou et al. 2007a, 2008; Amirti and Aberchane 2008; El Ajjouri et al. 2008). The determination of the antiradical activity by DPPH method showed that *T. capitatus* oil exerted the highest activity with (0.15 μl/ml), followed by the more commonly medicinal plant used by local population in the southern parts of Tunisia for several purposes, *Artemisia herba-alba* (1.0 μl/ml) and *Artemisia campestris* (2.09 μl/ml). The screening of the antibacterial activity against seven bacteria using the disc diffusion method showed that *T. capitatus* oil strongly inhibited the growth of all bacteria studied (20–30 mm) except *Pseudomonas aerogunosa* which was resistant to all oils. These results show and confirm that *T. capitatus* possesses strong antiradical and antibacterial activities, and therefore it could be used as a natural preservative ingredient in food and/or pharmaceutical industries (Akrout et al. 2010a, b).

**_Ajuga iva_ (L.) Schreber**

Several species of the genus *Ajuga* are used in African and Asian folk medicine. *Ajuga iva* (L.) Schreber is an annual herbaceous plant, commonly found in the different regions of Morocco, where it is known as "Chendgora" (Ziyyat et al. 1997). It is known for its numerous beneficial effects as a panacea (cure-all) (Hassar 1999), specifically for gastrointestinal disorders (Bellakhdar et al. 1991), hypertension (Ziyyat et al. 1997) diabetes (Ziyyat et al. 1997), and as an anthelmintic. In East Africa, the plants of the genus *Ajuga* have been used as a remedy for fever, toothache, dysentery, and high blood pressure (Kokwaro 1976), and in the traditional Chinese pharmacopoeia, they are known for their diuretic effect (Aliotta and Pollio 1994).

The plants of the genus *Ajuga* have been reported to have antifungal (Anonymous 2000; Kariba 2001), antibacterial (Chen et al. 1996; Anonymous 2000; Bennaghmouch et al. 2001), antimycobacterial (Cantrell et al. 1999), antihypertensive (Odek-Ogunde et al. 1993), antiplasmodial (Kuria et al. 2001, 2002), hypoglycaemic (El Hilaly and Lyoussi 2002), and larvae and insect antifeedant (Bondi et al. 2000; Ben et al. 2000, Ben Jannet et al. 2001) activity. The empirical use of *Ajuga* is corroborated by the isolation and identification of a number of the active compounds, including antileukemic sterol glycosides (Akbay et al. 2002), hypoglycaemicecdysteroids (Kutepova et al. 2001), antibacterial and insect antifeedant neo-clerodane diterpenoids (Bremner et al. 1998; Ben et al. 2000; Bondi et al. 2000), insect antifeedant diglyceride (Ben Jannet et al. 2001), vasoconstrictor 8-O-acetyharpagide (Breschi et al. 1992), and insect ecdysis inhibitors (Kubo et al. 1981). Some of the phytochemicals isolated from the *Ajuga* plantsare reported to be cardiotonic (Kuria and Muriuki 1984), renal stimulant (Aliotta and Pollio 1994), biliary sectretagogue (Syrov et al. 1986), antidote for liver toxicity (Syrov and Khushbaktova 2001) and erythropoiesis stimulator (Syrov et al. 1997).

*A. iva* aqueous extract improves the antioxidant status by reducing lipid peroxidation and enhancing the antioxidant enzymes activities in plasma, erythrocytes and tissues of diabetic rat (Taleb-Senouci et al. 2009). An aqueous extract of the *A. iva* whole plant showed hypolipidemic activity, in addition to its hypoglycemic effect in normoglycemic and diabetic rats. In view of the hypoglycemic and hypolipidemic activity, and its relatively non-toxic, *A. iva* may be a candidate for development as an anti-diabetic agent in human (El-Hilaly et al. 2006).

***Marrubium deserti* De Noé**

The genus *Marrubium* is represented by 97 species which are widely spread over the temperate and warm regions. *Marrubium* species are indigenous in Europe, the Mediterranean area and in Asia (Watt and Breyer-Brandwizk Watt and Breyer-Brandwijk 1962). *M. deserti* is a common plant in Tunisia and is used in traditional medicine in the form of a decoction as a remedy for asthma, diabetes and as a diuretic. Furthermore, it is frequently used in traditional medicine for treating digestive disorders as antispasmodic, and as local treatment against scorpion stings and allergy (Maiza et al. 1993; Ould El Hadj et al. 2003). Recent research demonstrated the antibacterial activity of *M. deserti* extracts (Edziri et al. 2007).

*Marrubium* species are reported in the special literature to be used in folk medicine. As an example: aqueous and hydroalcoholic extracts of flowered aerial parts of different *Marrubium* sp. are reported to treat cough and as choleretic, in digestive and biliary complaints (Newall et al. 1996 and Wichtl and Anton Wichtel and Anton 1999). Traditionally, they are also used for their neurosedative and antiinflammatory activities (Girre 2000).

The essential oil from aerial parts of *M. deserti*, obtained by hydrodistillation, was analyzed by gas chromatography (GC) and gas chromatography-mass spectrometry (GC-MS). Thirty-seven compounds were identified in the oil, with germacrene-D, as the major component (45.7%). This oil was characterized by a significant hydrocarbon fraction (78.1%) and the predominance of sesquiterpenes (67.4%) (Edziri et al. 2007).

The best known and first isolated diterpenoid was obtained from *M. vulgare,* in 1842. It was characterized as marrubiin. This furanic labdane diterpene is, in part, responsible for therapeutic properties of *Marrubium* sp. (Stulzer et al. 2006). The genus *Marrubium* is a rich source of diterpenoids, flavonoids and phenylethanoid glycosides (Takeda et al. 2000; Rigano et al. 2006; Piozzi et al. 2006; Dendougui et al. 2011).

Dichloromethane and methanolic extracts of *M. deserti* led to the isolation of terpenoid among which the two new labdane diterpenes were named marrulibacetal-A and desertine, respectively. There were also six known compounds (a mixture of the isomers cyllenin A and 15-epi-cyllenin A, marrubiin, marrulactone, marrulibacetal and β-stigmasterol) and seven known phenolic compounds were also isolated: apigenin and several 7-O-substituted derivatives (apigenin-7-O-β-neohesperi doside, apigenin-7-O-glucoside, terniflorin and apigenin-7-O-glucuronide) together with two phenylethanoid glucosides (acteoside and forsythoside B). The structures and relative configurations of the new compounds were elucidated by MS and a series of 1D and 2D NMR analyses. Some pure compounds have been evaluated for their antioxidant activities through different methods: DPPH and ABTS assays. Genotoxic and antigenotoxic activities of extracts and pure compounds were also evaluated *in vitro* on *Escherichia coli* PQ37 cells by the SOS Chromotest. Some of the isolated compounds like phenylethanoid derivatives showed stronger antioxidant capacity than trolox and were also able to significantly inhibit β-galactosidase induction caused by the mutagen agent nitrofurantoin (Zaabat et al. 2011).

The antiviral activity of petroleum ether, ethyl acetate, chloroform, butanol and methanol extracts of *M. deserti* was evaluated against human cytomegalovirus (HCMV) strain AD-169 (ATCC Ref. VR 538) and Coxsackie B virus type 3 (CoxB-3) using a cytopathic effect (CPE) reduction assay. The butanol, methanol and ethyl acetate extracts showed significant antiviral activity against Coxsackie B3 virus (Dziri et al. 2012).

***Marrubium alysson* L.**

*M. alysson* is a perinial herb of the Lamiaceae family which is commonly distributed in Europe, Mediterranean, Asia (Mabberly 1997) and in Tunisia. Its flowering aerial parts, as well as their aqueous and hydroalcoholic extracts are used in traditional medicine for treating cough but also as a laxative during digestive and biliary complaints (Wichtel and Anton 1999).

*M. alysson* witch had an interesting antibacterial activity with MIC varied from 128 to 2000 μg/ml against different *enterobacteriacea* and had good antifungal activity with MIC between 128 and 256 μg/ml (Edziri et al. 2007).

***Teucrium polium* L.**

*T. polium* L. is a wild-growing flowering plant belonging to the Labiatae family and is found abundantly in south western Asia, Europe and North Africa. *Teucrium* species have been used as medicinal herbs for over 2000 years as diuretic, diaphoretic, tonic, antipyretic, antispasmodic, an appetizer especially for children and cholagogic (Galati et al. 2000). An infusion of the leaves and flowers of the plant is consumed as a refreshing beverage (Facciola 1990). The biological activities of *T. polium* is widely reported and it has been shown to possess anti-inflammatory, antinociceptive, anti-bacterial, anti-hypertensive, hypolipidemic, and anti-rheumatoid effects (Rasekh et al. 2001; Abdollahi et al. 2003). The effect of ethanolic extract of *T. polium* on carrageenan-induced acute inflammation, cotton-pellet granuloma and some of the biochemical parameters were studied by Tariq et al. (1989). The ethanolic extract of *T. polium* at a dose of 500 mg/kg body weight produced significant inhibition of carrageenan-induced inflammation and cotton-pellet granuloma. The presence of flavonoids and sterols might be responsible for the anti-inflammatory activity of this plant.

Two known phenylethanoid glycosides, verbascoside and poliumoside have been reported from the aerial parts of *T. polium* (Bedir et al. 1999).

The crude extracts and isolated compounds were screened for their antioxidant and free radical scavenging activities using DPPH radical-scavenging, beta-carotene/linoleic acid and ammonium thiocyanate methods. Methanol extract, rutin and apigenin were found to be the most active fractions as radical-scavengers with $CI_{50}$ values of 20.1 ± 1.7, 23.7 ± 1.9 and 30.3 ± 2.1 mg/ml, respectively. The samples with the highest inhibition of oxidation of beta-carotene and lipid peroxidation in ammonium thiocyanate methods were also found to be methanol extract, rutin and apigenin. Methoxylated flavonoids exhibited a lesser antioxidant activity (Sharififar et al. 2009). *T. polium* has a folk reputation, as a hypoglycemic agent. The hypoglycemic activity of an aqueous decoction of plant aerial parts was tested in normoglycemic and streptozotocin-hyperglycemic rats. Results indicate that this extract caused significant reductions in blood glucose concentration 4 h after intravenous administration and 24 h after intraperitoneal administration (Gharaibeh et al. 1988). To assess the claimed hypoglycemic property of the crude drug and to get some-knowledge about its mechanism of action. This extract was administered orally to a group of streptozotocin diabetic rats for six consecutive weeks (0.5 g plant powder per kg body weight). A significant decrease (64%) in blood glucose concentration was observed in the treated animals compared to the untreated diabetic rats, without any measurable effects on the major biochemical factors. In addition, the crude extract significantly enhanced the blood insulin level by almost 160% compared to the untreated diabetic rats. The insulinotropic property of the *T. polium* extract was further assessed by an *in vitro* investigation using isolated pancreatic rat islets. The data indicated that *T. polium* crude extract is able to enhance insulin secretion by almost 135% after a single dose of plant extract (equivalent to 0.1 mg plant leaf powder per ml of the culture medium) at high glucose concentration (16 mmol/l) (Esmaeili and Yazdanparast 2004).

Some classes of constituents (diterpenoids, flavonoids, iridoids, sterols and terpenoids) was isolated (Carreiras et al. 1989; Hassan et al. 1979). Piozzi et al. (1987), have isolated the previously known neo-clerodane diterpenoids 19-acetylgnaphalin, eriocephalin, isoeriocephalin, and 3-deacetyl-20-epi-teulanigin (De la Torre et al. 1988), together with 5,3′,4′-trihydroxy-6,7-dimethoxyflavone (cirsiliol) (Harborne et al. 1975) and 5,7,4′-trihydroxyflavone (apigenin) (Harborne et al. 1975). In addition, three new diterpenoids, teuvincentins A, B and C, have also been isolated from the same source and their structures (2, 3 and 4, respectively) established by chemical and spectroscopic means and, in the case of teuvincentins A and B, by X-ray diffraction analyses of their respective acetyl derivatives 6 and 10. From the aerial parts of *T. polium*subsp. Vincentinum, three new neo-clerodane diterpenoids, teuvincentins A, B and C, have been isolated besides four already known diterpenes (19-acetylgnaphalin, eriocephalin, isoeriocephalin and 3-deacetyl-20-epi-teulanigin) and the flavones cirsiliol and apigenin. The structures of teuvincentin-A

((12S,20S)-20-0-acety1-19-acetoxy-18-ch1oro-15,16-epoxy-4~,7-dihydroxy-6-oxo-neo-c1erodane-7,13(16),14-triene-20,12-hemiacetal), –B ((12S,20S)-20-0-acetyl-4a,l8;15,16-diepoxy-6a-hydroxy-l7-neo-clerodane-13(16),14-diene-7a,19; 20,12-dihemiacetal) and C ((12S,20S)-20-0-acetyl-6a,l9-diacetoxy-4a,l8; 15,16-diepoxy-7-oxo-17-neo-clerodane-13(16),14-diene-20,12-hemiacetal) were established by chemical and spectroscopic means and, in the case of teuvincentins-A and -B, also confirmed by X-ray diffraction analyses of their acetyl derivatives (Carreiras et al. 1989).

## *Leguminosae/Papilionacées/Fabaceae*

### *Astragalus gombiformis* Pomel

*Astragalus*, genus of Fabaceae, comprises more than 1500 species that are distributed in the Orient and North Africa (Ozenda 1991). Many *Astragalus* species are toxic. In contrast, several therapeutic properties are assigned to other plants of this genus. In folk and modern medicine, *Astragalus* species such as *A. monghalicus* Bunge and *A. membranaceus* (Fish) Bugne are used against various diseases and for cancer therapy (Lei et al. 2003; Yesilada et al. 2005; Yin et al. 2006). Several species of *Astragalus* are reported from the Tunisian flora, including *A. gombiformis* Pomel, which grows in desert regions. In these areas, the plants normally synthesize several metabolites in order to adapt to drought stress and are therefore a promising source of active molecules *A. Gombiformis* is traditionally used against the bites of snakes and scorpions (El-Rhaffari and Zaid 2002).

In Tunisia, the flora containes many *Astragalus* species such as *A. epiglottis* L., *A. sesameus* L., *A. baeticus* L., and *A. caprinus* usually distributed in desert regions (Le Floc'h 1993). Except the studies by Semmar et al. (2001, 2002, 2005) on *A. caprinus*, no previous phytochemical investigations have been reported on *Astragalus* species of Tunisian flora. *A. gombiformis* Pomel, syn. *A. gombo* subsp. *gombiformis* (Pomel) Ott, has a narrow spread in the big oriental Erg of Tunisia under Saharan climate characterized by different stresses. Although, this plant has a significant biomass, it is only moderately edible for animals, probably due to its toxicity or its rankness (Teyeb et al. 2012).

The main chemical compounds of *A. gombiformis* are: 6,10,14-trimethylpentade can-2-one, hexadecanoic acid, phytol and dillapiole. Many volatiles of interest are identified such as butylated hydroxytoluene (BHT), an antioxidant compound. Other compounds which can be responsible for toxicity are identified with different amounts between organs such as caprolactam and benzencacetonitrile (Teyeb et al. 2010). *A. gombiformis* essential oils are complex mixtures of important natural compounds, which varied qualitatively and quantitatively between cultivated and wild plants and between phenological stages of development. All analysed oils are characterized by the constant presence of phytol, 6,10,14-trimethyl-2-pentadecanone, 4-terpinol, and γ-terpinene (Teyeb et al. 2011). The antibacterial and cytotoxic activities of extracts from the leaves of wild *A. gombiformis*were reported. The strongest cytotoxic activity against the human A549 lung epithelial carcinoma cell line was detected for dichloromethane extracts at $IC_{50} = 85 \pm 21.7$ ug/ml. The methanol extract was found to be the most active against *Pseudomonas aeruginosa* and *Salmonella typhimurium* (Teyeb et al. 2012). In Turkish folk medicine, the roots of *Astragalus* species are used for the treatment of leukemia and for the healing of wounds (Yesilada et al. 2005). Some *Astragalus* products, such as gum tragacanth, are widely in use in the preparation of pharmaceuticals and as thickening agents in certain foods (Zarre 2000). Some products may even have applications in controlling cancer cells (Somer and Calışkan 2007).

***Calicotome villosa* (Poir.) Link.** is a spiny shrub, with yellow papilionaceous flowers during the spring season (Pignatti 1982; Tutin 1972), very common in the Mediterranean area, where it grows near the sea (0–1200-m altitude). *C. villosa* was considered as an antitumoral agent (Hartwell 1982), while in the Sicilian folk medicine it was used for the treatment of foruncle, cutaneous abscess and chilblain (Lentini et al. 1993). According to Dessi et al. (2001), the methanol extract and essential oil of this species exibit an antioxidant, cytotoxic and antimicrobial activities. Previously isolated constituents in *C. villosa*, isoflavones (Harborne 1969), alkaloids and chrysin (Dessí et al. 2001). Two new acylated flavone glycosides, chrysin 7-(6″-*O*-acetyl)-*O*-β-d-glucopyranoside (1) and chrysin 7-(4″-O-acetyl)-O-β-D-glucopyranoside (2) were isolated from the aerial parts of C. villosa, along with 17 known flavonoids and one triterpene (Pistelli et al. 2003).

***Genista saharae* Coss. et Dar.,** growing in Sahara, is a spontaneous shrub legume. This evergreen legume is well adapted to drought stress and it is a goodcontributor to dune stabilization (Mahdhi et al. 2007). According to ethnobotanical investigation, arial parts of *G. saharae* are traditionally used for treating respiratory diseases and possess diuretic property (Mekkiou et al. 2005). A good antioxidant capacity (26.5 mg of Vitamin C Equivalent/g plant extract and antibacterial activity against: *Escherichia coli* (ATCC 25922), *Staphylococcus aureus* (ATCC 25923) were shown by ethyl acetate fractions (EAF) of *G. saharae* (Bouchouka et al. 2012). The ethyl acetate fraction (EAF) of *G. sahara* was found to have the highest content of both total phenolic compounds and total flavonoids with 459.28 mg GAE and 242 mg QuE per g of plant extract, respectively (Bouchouka et al. 2012).

Recently, the scientific community got interested in the studies involving G. saharae (Mekkiou et al. 2005; Boumaza et al. 2006). According to these authors, *G. saharae*, like other species of Fabaceae family, contains high levels of isoflavones, notably C-glycosylated isoflavones.

***Retama raetam* (Forsk.) Webb.** Locally named as "R'tm", is a wild plant common to the north and east Mediterranean regions (Mittler et al. 2001). The plant flowers from April to May. It is used for traditional treatment of some renal diseases and for treatment of hypertension (Kassem et al. 2000; Burits and Bucar 2000). Since it was reported to show significant diuretic activity (Caceres et al. 1987). This specieshas been used since ancient times as a herbal remedy for traditional treatment of wounds, bites of snakes and sheep scabies. Branches are used as febrifuge; powdered branches mixed with honey are emetic, given as a purgative and vermifuge, abortive in large doses. The plant is used for making eye wash for eye troubles. The antibacterial, antifungal, antioxidant and antiviral activities were also recent reprted by Edziri et al. (2008). Edziri et al. (2008) added that among the different extracts studied by the butanol extract of *R. raetam* showed high antimicrobial activity against Gram-positive bacteria than the other extracts. It has MICs of 0.256–0.512 mg/ml against *Bacillus subtilis*, *Enterococcus faecium*, *Streptococcus agalactiae*, *Streptococcus pyogenes*, *Corynebacterium* spp, MSSA, and MRSA. Ethyl acetate and butanol *R. raetam* extracts displayed more important antibacterial activity against Gram-positive bacteria than Gram-negative bacteria.

The screening of chemical groups in the methanol extract of the flowers of *R. raetam*collected from Tunisia revealed the presence of alkaloids, flavonoids, and tannins (Edziri et al. 2008). The total phenolic contents varied from 135.91 to 220.81 mg CAE/g in *R. raetam* butanolic extracts which exeded the total phenolic contents of other extracts (Edziri et al. 2008). Flavonoid content of the extracts of *R. raetam* flowers ranged from 41.58 to 54.97 mg CAE/g. The major components in the *R. raetam* oil detected were nonanal (35.75%), humulene (29.29%), acetaldehyde (7.84%), linalool (5.62%), myrcene (3.38%), tridecanal (2.21%), caryophyllene (1.79%), terpinyl acetate (1.46%), terpinolene (1.26%) and methyl anthranilate (1.06%) (Edziri et al. 2010). *Retama* species have been reported to contain alkaloids (Abdel Halim et al. 1997) and flavonoids such as daidzein, vicenin-2, naringenin, apigenin, kaempferol, quercetin and kaempferol-7-O-glucoside in the seeds, and daidzein 7, 4′-dimethyl ether, chrysoeriol 7-O-glucoside and orientin in the leaves (Abdalla and Saleh 1983). Kassem et al. (2000) have isolated two new flavonoids from the aerial part: luteolin 4′-O-Neohesperidoside and 5, 4′-dihydroxy-(3″, 4″-dihydroxy)-2″, 2″-dimethylpyrano-(5″, 6″: 7, 8)-flavone.

***Ceratonia siliqua* L.**

*Ceratonia siliqua* L., commonly known as carob, is distrubuted in the Mediterranean region (Polunin and Huxley 1972). The nonfleshy and bean-like fruits, called "carob pods", are a traditional part of the diet in the Mediterranean region, and the plant has been cultivated in the region for centuries for its edible fruits (Ayaz et al. 2007). Current world production of carob pod has been estimated at about 310,000 tons per year, produced from about 200,000 ha with very variable yields depending on the cultivar, region, and farming practices (Makris and Kefalas 2004). Its bark and leaves are used in Turkish folk medicine as an antidiarrheal and diuretic (Polunin and Huxley 1972; Baytop 1984). The fruits are traditionally used as an antitussive and against warts (Merzouki et al. 1997, Amico and Sorce 1997). Its hormone-like effects have been reported. Carob pods have been consumed traditionally as animal and human food (Binder et al. 1959). Industrially, carob is mainly used for locust bean gum production from the seeds. From the pulp, different human foods can be derived, such as sugar syrups or molasses, unroasted and roasted carob powder used as cocoa substitutes, or especially tannin-rich preparations as antidiarrheic products (Marakis 1996; Loeb et al. 1989). Carob pulp products are especially rich in insoluble dietary fiber (carob fiber) that has recently shown to have promising cholesterol-lowering properties both in animals (Pérez-Olleros et al. 1999a, b) and human trials

(Zunft et al. 2001; Zunft et al. 2003). Furthermore, carob fiber products, as well as water extracts from carob pods, showed a high antioxidative activity in *in vitro* tests (Haber 2002; Kumazawa et al. 2002). Results from the antimicrobial screening of ethanol extract of *C. siliqua* inhibited the growth of five out of the ten microorganisms studied and showed that ethyl-acetate and n-hexane extracts of *C. siliqua* were more active against *Enterococcus faecalis* ATCC 29212 than Ceftazidime.

Carob pods contain condensed tannins (proanthocyanidins), composed of flavan-3-ol groups and their galloyl esters (Bravo et al. 1994; Marakis et al. 1997), gallic acid (Corsi et al. 2002), (+)-catechin, (−)-epicatechin gallate, (−)-epigallocatechin gallate, and quercetin glycosides (Sakakibara et al. 2003). Carob pod meal contained high levels of carbohydrates (45%), appreciable amounts of protein (3%), and low levels of fat (0.6%). Germ and seedmeal contained more fat and less carbohydrates as compared to the pods. Carob pods contain a mean value of 19 mg of total polyphenols/g, 2.75 mg of condensed tannins (proanthocyanidins)/g, 0.95 mg of hydrolysable tannins (gallo- and ellagitannins)/g. Germs contained higher concentrations of total polyphenols (40.8 mg/g) and tannins (16.2 mg ofcondensed tannins/g and 2.98 mg of hydrolysable tannins/g) while only traces of these compounds were detected in carob seed (Avallone et al. 1997).

Carob meal contains, among total polyphenols, a high level of pro-anthocyanidins in comparison to hydrolysable ones represented by ellagitannins and gallotannins (Avallone et al. 1997). Extracts from pods and leaves of carob were tested for their abilityto inhibit cell proliferation of mouse hepatocellular carcinoma cell line (T1). The two extractsshowed a marked alteration of T1 cell proliferation in a dose-related fashion reaching themaximal effect at 1 mg/ml. Moreover, we demonstrated that leaf and pod extracts were able to induce apoptosis in T1 cell lines after 24-h treatment mediating a direct activation of the caspase 3 pathway (Corsi et al. 2002). HPLC analysis revealed the presence of gallic acid, (y) epigallocatechin-3-gallate and (y) epicatechin-3-gallate in pod and leaf extracts, compounds well known to exert antiproliferative effects. The discovery that carob pod and leaf extracts contained antiproliferative agents could be of practical importance in the development of functional foods and/or chemopreventive drugs (Corsi et al. 2002).

***Genista microcephala* Coss et Dur.** is an endemic plant from North Africa that flourishes from May to July (Quezel and Santa 1962). *G. microcephala* is distributed in Eastern Algeria (Aurès, Hauts-Plateaux) and Tunisia (Quezel and Santa 1962; Maire 1987). *G. microcephala* var. *tripolitana* (Bornm.) Maire occurs only in Libya (Tripolitania) and South-Eastern Tunisia (Maire 1987).

The *Genista* species show interesting biological properties such as hypoglycemic, antiinflammatory, anticuler, spasmolytic, antioxidant, estrogenic and cytotoxic activity against different human cancer cell lines (Scarpato et al. 2008; Rauter et al. 2009; Rigano et al. 2009). Phytochemical analysis of *G. microcephala* has revealed the presence of flavorous, isoflavones and alkaloids (Martins et al. 2005; Tosun et al. 2009). An acceptable yield, 0.2% is obtained 27 compounds were identified by Lograda et al. (2012) representing 98.9% of the total essential oil. The oils showed significant antibacterial activities against *Echerchia coli, Pseudomonas agreunisa* and *Staphylocus aureus.* A chemical investigation of the aerial parts of *G. micro-*

*cephala*, yielded two alkaloids, identified as lupanine (2-oxosparteine) and S-calycotomine. The complete 1H and $^{13}C$ chemical shifts assignments of Lupanine and Scalycotomine were determined using 1D and 2D NMR spectroscopy. The alkaloid extract of *G. microcephala* was tested against bacteria and fungi strains, which showed a strong sensitivity (Zellagui et al. 2004).

## *Poaceae*

***Cymbopogon schoenanthus* (L.) Spreng,** is an aromatic herb known in Tunisia by the name of "El bekhirai". Fresh young leaves are consumed in salads and are used to prepare traditional meat recipes. Due to its pleasant aroma and taste it is used to prepare an aromatic "tea" that is much appreciated and largely consumed in the north of Africa (IUCN 2005). Besides its culinary use, *C. schoenanthus* is also used in folk medicine. Its medicinal properties are known from the antiquity, being already described by "Pliny the Eldey" in his book Naturalis Historia (Pline L'Ancien 1848–1850). Le Floc'h (1983) reports its use for the treatment of rheumatism and fever. This author describes also its use as a diuretic, insecticide and a poultice to cure dromedary wounds.

*C. schoenanthus* or camel grass is a major essential oil bearing aromatic grass that grows well in arid areas. Essential oils obtained from the aerial parts of plants are used extensively in perfumery and pharmaceutical industries. *C. schoenanthus* oil has been reported to be severely toxic for some parasite species. The oil is used to reduce density of parasitoid populations and increase seed losses (Ketoh et al. 2002). The chemical composition, antioxidant activities and acetylcholinesterase inhibitory properties of the essential oils from fresh leaves, dried leaves and roots collected from three different locations in southern Tunisia, were evaluated by Khadri et al. (2008). Essential oils were analysed by GC–mass spectrometry and 13C NMR. The major components were limonene (10.5–27.3%), β-phellandrene (8.2–16.3%), α-terpinene (4.3–21.2%) and α -terpineol (6.8–11.0%) (Khadri et al. 2008). Antioxidant, acetylcholine-esterase activity of *C. schoenanthus* essential oil was reported by Khadri et al. (2008, 2010). *C. schoenanthus* essential oil was active against both termites (Koba et al. 2007) and the bruchid (*Callosobruchus maculatus*), which is a major pest of stored grains (Ketoh et al. 2002).

## *Polygonaceae*

*Calligonum* genus belongs to the Polygonaceae family (Wang 2007), with some 80 species distributed throughout Western Asia, Southern Europe and North Africa. Only *C. comosum* L'Hérit, *C. azel* Maire and *C. arich* Le Houérou are represented in the Tunisian arid zone and the last one is endemic (Le Houérou 1959). The three *Calligonum* species are dominant perennials in active sand dunes and stabilized sand fields in the southern desert of Tunisia and grow naturally in the eastern Great Erg (Le Houérou 1959). Several species have a C4-type of photosynthesis and can grow in mobile sand dunes, under extreme drought conditions (Ren 2001). Due to their high tolerance of water deficits, these speceies appear to be suitable for re-vegetating deserts (Dhief et al. 2009).

*Calligonum comsum* **Hérit,** is widely distributed, from the North African deserts to the desert sands of the Middle East and as far east as the Rajputana desert in western India (Le Houérou 1959).

In the Tunisian desert, *C. comosum* was used for the treatment of the scabies of the camel. The decoction of roots of this species was anthelmintic. In theUnited Arab Emirates, *C. comosum* is used by some healers to treat stomach ailments, the stems and leaves are chewed for curing toothache. The fresh aerial parts of this species, collected in Abu Dhabi Emirates, show an anti-inflammatory and antiulcer activity. Some compounds, extracted from Egyptian *C. comosum*, presented cytotoxic and antioxidant activity. *Calligonum* species are known to contain high levels and various types of polysaccharides, secondary metabolites, polyphenols, alkaloids, flavonoids, and tannins (Liu et al. 2001). Dhief et al. (2011) studied the essential oils of the three wild species of *Calligonum* growing in the Tunisian desert. The oils obtained by hydrodistillation were analyzed by GC-FID and GC-MS in order to compare their chemical composition. A total of 110 compounds, which accounted for 94.0–99.7% of the total composition of the essential oils, have been identified. The main constituents were viridiflorol (9.6%) in *C. azel*, hexadecanoic acid (20.1%) in *C. arich* and 9-octadecenoicacid (19.8%) in *C. comosum*. Based on their chemical composition, two chemotaxonomic groups are proposed: C. *azel* on one hand, *C. arich* and *C. comosum.*

*Polygonum equisetiforme* **Sibth. et Smis**

In the south of Tunisia, *Polygonum equisetiforme* Sibth. et Smis is used for the disinfection of wounds, healing and fortifying. It is also used to treat cough, cold and sore throat (Khafagi and Dewedar 2000). Crude extracts of *P. equisetiforme,* as well as the pure compounds isolated from them, were screened for antibacterial and antifungal activity against 20 bacterial and 17 fungal cultures. The crude extract of *P. equisetiforme* inhibited the growth of *Trichophyton mentagrophytes, M. canis* and *Allescheria* boydiiat an MIC of 450–500 μg/ml whereas the pure compounds inhibited the growth at MIC of 300–350 μl. Species of *Corynebacterium*, *Bacill*us and *Staphylococcus* aureus were found to be highly susceptible to the crude and the pure compounds. MIC values of both crude extracts for different organisms were found to be higher (200–350 μg/ml) than those of the pure compounds (150–170 μg/ml). The crude extract of *P. equisetiforme* did not inhibit the growth of *P. aeruginosa*, however, the pure compound was found to be bacteriostatic (Kazmi et al. 1991). Quercetin (a flavonol) has been isolated and identified as the major constituent of *P. equisetiforme* (Ghazal et al. 1992).

## *Rhamnaceae*

The genus *Ziziphus* belongs to the family Rhamnaceae and is represented by 135–170 species (Liu and Cheng 1994). These are evergreen or deciduous trees, or shrubs, usually armed with unequal stipular spines. Of these, only *Z. spina-christi* (L.) Willd, *Z. vulgaris* Lam. and *Z. lotus* (L.) Lam. are found in Tunisia (Laamouri and Zine El Abidine 2000). Although highly similar, *Z. lotus* differs from the other two species by its deciduous shrubby habit with intricately branched stems and smaller flowers and fruits (Jafri 1977). This species is indigenous to Tunisia. It grows under a variety of environmental conditions and has a wide ecological and geographical distribution in both climatic regions of Tunisia. It is a shrub that reaches 2–5 m in height and is found in depressions with deep sandy soil. Mounds composed of wind-borne sediments that accumulate around *Z. lotus* thorn scrub have long been reported from the Tunisian steppe regions (Tengberg and Chen 1998). This species is a dominant perennial shrub in both the active sand dunes and stabilized sand fields of the southern, arid zone of Tunisia. *Z. lotus* is dormant from October through March and mature plants flower in May–June and produce fruits in August.

*Ziziphus lotus* **(L.) Desf.** is abundantly present in the Mediterranean region, throughout Libya to Morocco, Algeria and southern European countries like Spain, Sicily, Greece and Cyprus (Gorai et al. 2010). The North African species of the genus *Zizyphus* are known with the vernacular names 'sedra', 'addhal', 'roubaidh', 'dhou achaouk', 'sder alberri' and 'cder, nabga' (Marie 1964; Boukef 1986). Fruits, leaves, seeds, roots and bark of jujube have been commonly used as a source of simple medicines in traditional medicine (Le Croueour et al. 2002; Abdel-Zaher et al. 2005; Li et al. 2005). Flowers of Chinese jujube have high-quality nectar, and the leaves are consumed as tea (Zhao et al. 2008). The fruit is the edible part of the plant by local population. Several parts of *Zizyphus* have been used by traditional and ancestral medicine, both in North Africa and Middle East, for the treatment of several pathologic conditions including digestive disorders, weakness, liver complaints, obesity, urinary troubles, diabetes, skin infections, fever, diarrhoea and fatigue, and distress, which cause insomnia (Boukef 1986; Adzu et al. 2003; Lahlou et al. 2002). *Z. lotus* are used is emollient (Le Floc'h 1983). The Ethnomedicinal evaluation of this plant in Tunisia by Le Floc'h (1983) indicates that seeds and flowers have sedative effect. They can contribute to the overall tonification therapy, normalize the function of stomach, calm the mind and treat eyes inflammation. They are reported to be effective in treating vitamin B deficiencies. The fruits are known to be beneficial to lose weight and as such, they are used in treating cases of obesity. Farther uses are: spleen tonic, aphrodisiac, blood cleaning, appetizer. The dried fruits have been reported to facilitate digestion (Boukef 1986). *Zizyphus* is also used for treatment of fever, skin problems and for fortifying hair (Le Floc'h 1983).

Jujube fruits are consumed fresh, dried and processed (jams, loaf, cakes, jelly, *etc.*) throughout the world (Pareek 2002). Fruits have a high sugar content and high levels of vitamin A, C and B complexes, phosphorus and calcium (Pareek 2002).

Alkaloids, flavonoids, sterols, tannins, saponin, and fatty acids have been detected in the different species of the genus *Ziziphus* (Le Croueour et al. 2002; Abdel-Zaher et al. 2005; Bhargava et al. 2005; Zhao et al. 2006). *Z. lotus* extracts contain anti-spasmodic constituents mediating their effect through cholinergic receptors and blockade $Ca^{2+}$ influx. This could explain the traditional use of *Z. lotus*in the treatment of the intestinal diseases (Borgi et al. 2007). The different extracts of *Z. lotus* exerted immune-suppressive on T-cell proliferation and IL-2 mRNA expression and antioxidant effects. Borgi et al. (2007) have shown that root barks of *Z. lotus*, given intraperitoneally, showed a significant and dose-dependent anti-inflammatory and analgesic activity in carrageenan-induced paw edema in the rat. Hence, the presence of flavonoids in the *Zizyphus* extracts was supposed to be responsible for these beneficial effects. Adhvaryu et al. (2007) have shown that *Zizyphus* extracts used with other plants stimulated neutrophil functions and exerted hepatotoxic and immunomodulatory effects in *guinea* pigs. Chan et al. (2006) assessed cell signaling mechanisms in T-cells and provided the evidence that a mixture of herbs containing *Zizyphus* extract induced the expression of mitogen-activated protein kinases (i.e. ERK, JNK and p38) in T-cells, indicating that the immunomodulatory effects of *Zizyphus* involve the activation of second messenger cascade. Since *Z. lotus* has been shown to modulate different disorders (Adzu et al. 2003; Han and Park 1986).

Four cyclopeptide alkaloids, lotusines B, C, E and F, have been isolated from the root bark of *Z. lotus*. Their structures were elucidated mainly by homo- and heteronuclear NMR techniques. Besides, several biologically active molecules, particularly cyclopeptide alkaloids, termed lotusiones (Ghedira et al. 1993; Le Croueour et al. 2002) and dammarane saponins have been isolated from this shrub (Renault et al. 1997). Cyclopeptides extracted from *Z. lotus* exhibited antibacterial and antifungal properties (Pandey et al. 1990). Five dammarane-type saponins were isolated by means of centrifugal partition chromatography from the leaves of *Z. lotus*. Their structures were elucidated using a combination of 1D and 2D 1H and 13C NMR spectra and mass spectroscopy. One of these glycosides is the known jujuboside B (**5**). Three are new jujubogenin glycosides, identified as 3-*O*-α-L-rhamnopyranosyl-(1_6)-*β*-D-glucopyranosyljujubogenin-20-*O*-(2,3,4-*O*-triacetyl)-α-L-rhamnopyranoside (**1**), 3-*O*- α -L-rhamnopyranosyl-(1_6)- *β* -D-glucopyranosyljujubogenin-20-*O*- α -L-rhamnopyranoside (**2**), and 3-*O*- α -L-rhamnopyranosyl-(1_2)-((4-sulfo)- *β* Dglucopyranosyl-(1_3))- α -L-arabinopyranosyljujubogenin (**3**). The last is a new sulfated derivative of jujubasaponine IV, identified as 3-*O*-R-L-rhamnopyranosyl-(1_2)-((4-sulfo)- *β* -D-glucopyranosyl-(1–3))- *β* -D-galactopyranosyl-(20*R*,22*R*)-16 *β*,22:16R,30-diepoxydammar-24-ene-3 *β*,20-diol (**4**) (Borgi et al. 2007). Catechin, caffeic acid, p-coumaric acid, ferulicacid, rutin, apigenin-7-glucoside, eriodictyol, quercetin, p-hydroxybenzoic acid, chlorogenic acid andsyringic acid were isolated from leaves were analyzed withan HPLC device equipped with a diode array detector, and the fatty acid methyl esters were analyzedwith QP 5050 GC/MS equipped with a CP-Wax 52 CB column. Rutin content was very high in the leaves of all the jujubeselections, ranging from 269.0 to 367.90 mg/100 g, followed by apigenin-7-glucoside (22.90–49.38 mg/100 g) and eriodictyol (5.06–6.27 mg/100 g). Seven phenolic compounds, catechin, caffeic acid, epicatechin, ferulic acid, rutin, p-hydroxybenzoic acid and chlorogenic acid, were isolated from fruits of jujube selections. Catechin level ranged from 2.46 to 3.74 mg/100 g, and rutin level ranged from 0.88 to 3.60 mg/100 g for fruits (Borgi et al. 2007).

The predominant fatty acids in all jujube selections were oleic acid, linoleic acid, palmitic acid andpalmitoleic acid. Unsaturated fatty acids comprised 68.54–72.44% of the total fat in jujube fruit.

## *Rutaceae*

***Ruta graveolens* L. (*chalepensis* L. or** *montana)*
The most common medicinal plant of this family is *Ruta graveolens*, known as rue and native to Europe. The plant is now available all over the world, though preferably grown in Mediterranean climates. This plant is in medicinal use for various clinical conditions from very ancient time but rationality of its use is still controversial. *R. graveolens* is a medicinal plant which has been traditionally used as a sedative and antihelmintic to relievemenstrual and gastrointestinal disorders (Guarrera 1999; Skidmore-Roth 2001), hypotensive, (Trovato et al. 2000) and antifertility (Al-Okbi et al. 2002). In homeopathy, rue is an important remedy for deep aching pain and rheumatism besides being used for eyestrain-induced headache. It has also been used as a remedy for gastric disorders, stiff neck, dizziness and headache (Conway and Slocumb 1979). *R. graveolens* L.has been notorious, since at least the first century of the Christian era, for its ability to produce erythema and pustular eruptions on human skin coming in contact with it. In the present century, the lesions were characterized as a photophytodermatitis induced by exposure to UV and due to the presence in this species of linear furanocoumarins, psoralen and some of its derivatives (Murray et al. 1982).

*R. graveolens* is a medicinal plant, used since time immemorial. Traditionally Rue is considered to be emmenagogue, ecbolic, anthelminthic and antispasmodic. It has been approved by Food and Drug Administration (FDA) as a flavouring agent. Antiinflammatory (Raghav et al. 2006) anti fungal (Oliva et al. 2003) antibacterial (Ojala et al. 2000) and hypotensive (Chiu and Fung 1997) activities of *R. graveolens* have been scientifically proven but at higher concentrations it has been reported to exhibit toxicity (El Agraa et al. 2002). Most of these biological properties that have beendenoted by preceding studies are likely related to some chemical constituents including flavonoids (Chen et al. 2001; Chiu and Fung 1997). The most important analyzed flavonoid in *R. graveolens* L. is rutin (quercetin-3-*O*- b-rutinoside) that belongs to flavonol glycoside. Quercetin is another major flavonoid found in *R. graveolens* L. and can also be obtained by rutin hydrolysis. The purpurted therapeutic effects of many traditional medicines have been attributed to flavonoids in particular as a result of their enzyme inhibitory and antioxidant activity (Havsteen 1983; Chiu and Fung 1997). *R. graveolens* plant contains approximately 2% of rutin. An interesting recent study reported the decrease in lipopolysaccharide-(LPS) induced nitric oxide production by rutin *in vivo* due to inhibition of nitric oxide synthase (iNOS) protein expression. *R. graveolens*, a related plant has been found to reduce LPS-induced nitric oxide level (measured nitrite level) without altering the cytokines (Iauk et al. 2004). The study of Raghav et al. (2006) is designed to investigate the effect of plant extract of *R. graveolens* on murine macrophage cells (J-774) challenged with lipopolysaccharide (LPS). The inhibition observed for the extract was significantly higher than that observed for rutin, a flavonoid constituent of the plant. At 40 μM rutin, a comparable concentration of this flavonoid in the highest concentration (500 μg/ml) of plant extract was used in this study; a 20% inhibition ($p = 0.058$) was observed. Inhibition in inducible nitric oxide synthase (inos) gene expression in the cells treated with the plant extract suggests an inhibition at the transcription level.

*R. graveolens* extract exhibited a high inhibition on aldehyde oxidase activity (89–96%) at 100 mg/ml which was comparable with 10 mM of menadione, a specific potent inhibitor of aldehyde oxidase (Saieed et al. 2006).

Extracts of rue have been proposed as topical pharmaceutical fungicides (Ali-Shtayeh and Abu Ghdeib 1999; Trovato et al. 2000). Antifungal compounds were found in callus cultures of *R. graveolens* (Wolters and Eilert 1981). Rue extracts and two constituents of rue, 5- and 8-methoxypsoralen, are active against fungal plant pathogens (*Rhizoctonia solanii*, *Fusarium* spp., *Pyrenochaeta lycopersici*, *Trichoderma viride*, *Penicillium* spp., *Thielaviopsis basicola*, and *Verticillium dahliae*) (Aliotta et al. 2000; Ojala et al. 2000; Oliva et al. 1999).

The constituents of *R. graveolens* were further studied not only because of interest in the chemistry of natural products, but also because of several biologically active compounds providing a base for the use of *R. graveolens* in folk medicine and to find some more biologically active compounds. Bioassay-directed isolation of antifungal compounds from an ethyl acetate extract of *R. graveolens* leaves yielded two furanocoumarins, one quinoline alkaloid, and four quinolone alkaloids, including a novel compound, 1-methyl-2-(6′-(3′′,4′′-methylenedioxyphenyl)hexyl)-4-quinolone. Four of the alkaloids had moderate activity against *Colletotrichum* species, including a benomyl-resistant Colletotrichum*acutatum*. These compounds and the furanocoumarins 5- and 8-methoxypsoralen had moderate activity against *Fusarium oxysporum*. The novel quinolone alkaloid was highly active against *Botrytis cinerea*. Phomopsis species were much more sensitive to most of the compounds, with *P. viticola* being highly sensitive to all of the compounds (Oliva et al. 2003).

The use of *R. graveolens* against intracranial cysticercosis has been reported (Banerji and Banerji 2001). Recently Pathak et al. (2003) showed that homeopathically potentiated *Ruta* induces cell death in brain cancer cells. Telomere dynamics, mitotic catastrophe and apoptosis have been shown to be the possible mechanisms. An extract of *R. graveolens* was found to be cytotoxic to Dalton's lymphoma ascites (DLA), Ehrlich ascites carcinoma (EAC) and L929 cells in culture (IC100 = 16mg/ml) and also to increase the lifespan of tumour bearing animals. The extract further decreased solid tumours developing from DLA and EAC cells when given simultaneously with elongation of the lifespan of tumour-bearing animals. A homeopathic preparation of R. *graveolens* (200c) was equally effective. Neither was effective for reducing already developed tumours. The R. graveolensextract was found to scavenge hydroxyl radicals and inhibit lipid peroxidation at low concentrations. However, at higher concentrations the extract acted as a prooxidant as inhibition of lipid peroxidation and scavenging of hydroxyl radical was minimal. These data indicate that the prooxidant activity of *R. graveolens* may be responsible for the cytocidal action of the extract and its ability to produce tumour reduction (Preethi et al. 2006).

The effects on the reproductive system and fertility using adult male albino rats with special emphasis on the aggressive behavior and sex behavior was recently study. The aqueous extract of *R. graveolens* solution was fed orally to male albino rats at a dose of 500 mg/kg body weight for 60 days. This dose induces a significant decrease in the weight of reproductive organs ($P < 0.01$) when compared to controls. The sperm motility and density in cauda epidydimides and testicular ducts were significantly decreased ($P < 0.01$). A significant decreased ($P < 0.001$) in spermatogenesis activity is observed in somniferous tubule. Treated rats testicular cell population showed a decrease in number of spermatocytes and spermatids ($P < 0.001$) when compared to controls. Serum hormonal assay indicated a decrease in Testosterone and Follicular Stimulating Hormone levels in treated rats. A decrease in the number female rats impregnated by males receiving treatment was observed and demonstrated by a decrease inthe implantation sites and viable fetuses number ($P < 0.01$). The ingested extract also suppresses the sexual behavior in adult male rats expressed by a prolongation of first mount time, increase in intromission latency, decrease in intromissions number, and prolongation of the post-ejaculatory interval. This led to reduce the ejaculation time and increase the post ejaculatory intervals. Ingestion of *R. graveolens* markedly abolished aggressive behavior parameters in adult male treated rats namely, a suppression in lateralization, boxing bouts and ventral presenting postures (Khouri and EL-Akawi 2005).

Nineteen compounds have been isolated from the methanol extract of the root and aerial parts of *R. graveolens* by (Wu et al. 2003). Their antiplatelet aggregation andcytotoxic activities were examined to find potent antiplatelet aggregation and cytotoxic compounds from natural resources. Among them, dictamine, skimmianine, psoralen, chalepensin, clausindinand graveolinine showed significant inhibition of platelet aggregation, induced by arachidonic acid and collagen. Arborinine, dictamine, isopimpinellin, clausindin, and graveoline exhibited cytotoxic activity against KB, Hela, DLD, NCI and Hepa tumor cell lines (Wu et al. 2003).

The essential oil of aerial parts of *R. graveolens* was obtained by hydrodistillation with a 0.74% yield on a dry weight basis. Thirty-eight components were identified by GC and GC--MS analyses. 2-Ketones were predominant in the essential oil, with undecan-2-one (46.8%) and nonan-2-one (18.8%) as the main constituents (De Feo et al. 2002). Subcritical ($CO_2$) extraction, carried out in a J&W Scientific High Pressure Soxhlet Extractor, was used to isolate secondary metabolites from leaves, flowers, stems and roots of Colombian rue (*R. graveolens*). The various extracts were analyzed by capillary chromatography, on an HP-5 (30 m) column, using nitrogen-phosphorus, flame ionization, and mass selective detection systems. The extracts from the various parts of rue had different compositions. The number of compounds detected at concentrations above 0.01% (w/w) in the extracts from leaves, flowers, stems and roots, was 78, 45, 25 and 24, respectively. 2-Nonanone (8.9%), 2-undecanone (13.4%), chalepensin (13.0%), and geijerene (19.3%) were the main constituents found in the extracts from rue leaves, flowers, stems and roots, respectively. Furanocoumarins, furoquinolines, hydrocarbons and benzodioxol derivatives were the main compound families found in all extracts, at total concentrations between 3.7 and 33.9%, depending on the part of the plant (Stashenko et al. 2000).

## *Thymelaeaceae*

***Thymelaea hirsuta* (L.) Endel.**

*Thymelaea* is a genus comprising about 30 species of evergreen shrubs of the flowering plant family Thymelaeaceae, native to the Canary Island, the Mediterranean region, north of central Europe, and east of central Asia. In this report, we focused on *T. hirsuta*, which is native to North Africa.

*Thymelaea hirsuta* (L.) Endel. is comonly known as "Methnane". In Tunisia, this plant has been used traditionally as an antiseptic and anti-inflammatory agent and for the treatment of hypertension by external application. Murine melanoma B16 cells were treated with *T. hirsuta* extract, and then cell viability and synthesized melanin content were measured. We found that the T. hirsuta extract decreased the synthesized melanin content in B16 cells without cytotoxicity. Tyrosinase is a key enzyme of melanogenesis and extracellular signal-regulated kinase (ERK)-1/2 phosphorylation is known to be related to melanogenesis inhibition. To clarify its mechanism, authors also determined ERK1/2 phosphorylation and tyrosinase expression level. ERK1/2 was immediately phosphorylated in cells just after treatment with the extract. The tyrosinase expression was inhibited after 24 h of stimulation with the extract. The same authors added that the fraction contains daphnanes as the main component (Kawano et al. 2007).

A study on the effect of oral administration of the *T. hirsuta* aqueous extract on blood glucose levels in normal, glucose-hyperglycemic and streptozotocin (STZ)-induced diabetic rats showed that in normal rats, single oral administration of *T. hirsuta* lowered blood glucose levels significantly. It omitted significantly reduced the fasting glucose level in rats with oral glucose (2 g/kg) induced hyperglycemia. In STZ-induced diabetic rats, single oral administration of *T. hirsuta* also produced a significant decrease of blood glucose levels. The aqueous extract of *T. hirsuta* possesses both hypoglycaemic and antidiabetic effects in normoglycaemic and streptozotocin-induced diabetic rats. This may indicate the ethnopharmacological reason of the use of *T. hirsuta* in traditional medicine for treating diabetes (El-Amran et al. El Amrani et al. 2009). The bioassay-guided fractionation of this plant implied that the major bioactive components are daphnane diterpenoids. It was previously reported that five daphnane diterpenoids were isolated from *T. hirsuta* (Brooks et al. 1990). These compounds are known to have various bioactivities, such as antileukemic and neurotrophic effects (Park et al. 2007). The structures of two new daphnane diterpenoids, hirseins A and B, from *T. hirsuta*, and their antimelanogenesis activities were reported.

## *Apiaceae (Umbelliferae)*

***Critimum maritmum* L.** called sea fennel, crest marine, marine fennerl, sampler and also rock samphier, is a fleshy aromatic, perennial plant which grows wild on maritime rocks, piers, and breakwaters, more rarely on sandy beaches along the Mediterranean countries, pacific and Atlantic coasts. It grows on all the world's coastlines but is especially abundant in France, Italy, Portugal, Spain and Turkey coastal regions (Barroso et al. 1991, 1992; Coiffard et al. 1993; Cunsolo et al. 1993). *C. maritimum* is a facultative halophyte growing on maritime cliffs and sometimes in sand. Several uses of this plant are known for culinary purposes. Indeed, *C. maritimum* leaves are rich in several compounds such as vitamin C, carotenoids, flavonoids, as well as bioactive substances which could be used for aromatic and medicinal purposes (Ozcan 2000; Ruberto and Baratta 2000; Kumarasamy et al. 2002). Sea fennel contains fragrant oil other aromatic substances that are widely used in cosmetics due to claimed slimming properties and medicine (Barroso et al. 1991; Senatore and De Feo 1994).

Its fresh leaves and young branches are pickled in vinegar and used as appetizers and condiments (Ozcan 2000; Akgül 1993). The leaves are also used in flok medicine as antiscorbutic, tonic, carminative, diuretic, depurative and vermifuge (Barroso et al. 1991; Senatore and De Feo 1994; Ruberto et al. 1991).

Same results indicate that sea fennel can be considered as a valuable source of antioxidant products, especially of chlorogenic acid (Meot-Duros and Magné 2009). According to Guenther (1950), early studies on *C. maritmum* oil showed that all parts of the plant contain volatiles compounds, but the yield and the chemical composition of the oils varied considerably depending on the site of growth and soil conditions. The essential oil of *C. maritmum* showed major fluctuations in the relative amounts of several components, mainly sabinene (7–42%) and gamma terpinene (26–55%) (Barroso et al. 1992). More recently, Ozcan et al. (2006) showed that the essential oil obtained from young leaves and branches of *C. maritmum* collected from Turkey contained mainly monoterpens, amounting to 89 % and 99, respectively of the oil, with γ-terpinene (36% and 32%), β-phellandrene (21% and 22%), and sabinene (13% and 9%) as the main components.

***Thapsia garganica* L.**

The genus *Thapsia* family Apiaceae (Umbelliferae) is widely distributed throughout the Mediterraean area. Umbelliferous plants are popular as drugs and spices as they contain useful secondary compounds such as essential oils, coumarins and sequiterpenes (Hetwood 1971). The resin of *Thapsia garganica* causes contact dermatitis and is extensively used by the Northern African Arabs for medicinal purposes (Jager et al. 1993).

The major bioactive compounds synthesized by *T. garanica* are Thapsigargin, Thapsigargicin, notrilobolid and thapsivillosin (Rasmussen et al. 1981). *T. garganica,* especially the root, contains potent skin irritants (Rasmussen et al. 1978). Because of this property, described by Hippokrates about 400 B.C., drugs prepared from the plant have been recorded in several pharmacopoieas, most recently the 1937 edition of the French pharmacopoiea. Two very potent skin irritants, named thapsigargin (yield 0.1% of fresh material) and thapsigargicin (yield 0.02%), have been isolated from an ethanolic extract of the root. The drugs are still used as ingredients of rheumatic pain releasing ointments in Arabian folk medicine (Rasmussen 1978). The most famous compounds from either biosynthetical or pharmacological point of view, have been isolated from *T. garganica*, which has become an indispensable "tool" for the study of the sarco/endoplasmic calcium pump (SERCA) (Olesen et al. 2004) and might offer a new type of chemotherapeutics for treatment of prostate cancer (Denmeade and Isaacs 2005; Denmeade et al. 2003).

***Pituranthos chloranthus* L.**

*Pituranthos* is the largest genus of the Apiaceae family and includes about 2500 species, which are spread throughout North Africa (Touil et al. 2006).

In Tunisia three species were mentioned by Pottier-Alapetite (1979), *P. chloranthus*; *P. scoparius* (Coss. et Dur.) and *P. tortuosus* (Coss. and Dur.). Furher, *Pituranthos* species are used in traditional medicine for the treatment of asthma, rheumatism, postpartum care, spasms, pains, fevers, diabetes, lice (head and pubis), hepatitis, digestives difficulties, urinary infections and scorpions stings (Verite et al. 2004; Yangui et al. 2009). Stems of *P. chloranthus* have been, traditionally, used as straw for farmers to dry figs and grapes. This plant has a double advantage; first, it is used for its aroma and distinctive taste that adhere to the dry fruits. Second, it has an insecticidal effect. In the South of Tunisian, a tuft of *P. chloranthus* was traditionally suspended on the surface of drinking water to disinfect the underground cisterns of the rainwater storage.

As a consequence, some species of *Pituranthos*, for example *P. tortuosus*, have been investigated for biological activity (Abdelwahed et al. 2006). *Pituranthos* essential oil and its main constituents are considered as a potential source of biologically active compounds (Le Floc'h 1983). The *P. chloranthus* essentials oils reduced significantly Benzo (a) pyrene (B(a) P) and sodium-azide-induced mutagenicity. The antioxidant activity was also showed. More recently Neffati et al. (2009) showed that essential oils of *P. chloranthus* could notably protect against damage induced by radicals. Dahia et al. (2007) identified 54 volatile constituents in the essential oil of *P. chloranthus* harvested in Algeria at the flowering stage. One hundred and fifty compounds were identified in which α-pinene, β-pinene, α-phellandrene, β-myrcene, β-phellandrene, p-cymene, 8-methyldecanal, exo-2-hydroxycineole acetate and carvacrol could reach more than 10 % of the total amount of Tunisian *P. chloranthus* (Neffati et al. 2009).

***Pituranthos tortuosus* (Coss.) Maire**

*P. tortuosus*, known in Arabic as "Guezzah", is a small shrub without leaves. This aromatic plant, which grows naturally in North Africa, is widespread in central and southern Tunisia (Pottier-Alapetite 1981). In Tunisia, *P. tortuosus* is used traditionally as an anti-asthmatic and against scorpion stings (Boukef et al. 1982). This species is used by the Egyptians for the preparation of a carminative drink and is occasionally eaten by grazing animals (Ahmed et al. 1962). It is also used for relief of stomach pains, against intestinal parasites, when blood is excreted in the urine or when coughing blood, and for the regulation of menstruation (Ashkenazy et al. 1984).

Plants of the family Apiaceae are generally known to be rich in essential oils showing antimicrobial activity against bacteria and fungi (Al-Gaby and Allam 2000). The essential oil of *P. tortuosus* from Egypt is a mixture of a number of volatile components, mainly terpenoids, with camphene (31%) as the major constituent, of the oil. This oil was ineffective against Gram-positive bacteria but exhibited antifungal activity at high concentrations (Al-Gaby and Allam 2000). The aerial parts of *P. tortuosus*, collected during November and April from Tunisia, were analysed by GC and GC–MS. In total, 56 compounds were identified by their retention indices. Antimicrobial assays showed that the essential oil obtained from plants in November is more effective against the Gram-positive bacteriathan the oil of April. *Enterococcus faecalis* and *Staphylococcus aureus*. The essential oil obtained in April displayed the highest activity against *Staphylococcus aureus* (Abdelwahed et al. 2006).

Mutagenic and antimutagenic activities against direct acting mutagens, nifuroxazide (NF) and sodium azide (SA), and indirect acting mutagen aflatoxin B1 (AFB1) of extracts prepared from aerial parts of *P. tortuosus*were investigated in bacterial assay systems (i.e., the Ames test with *Salmonella typhimurium* TA100, TA98, TA1538, TA1535, and the SOS chromotest with *Escherichia coli* PQ 37). It was found that all extracts obtained from *P. tortuosus* decreased the mutagenicity induced by AFB1 (10 μg/assay), SA (1.5 μg/assay), and NF (20 μg/assay). Ethyl acetate, acetone, methanol, and total oligomer flavenoid extracts exhibited the highest inhibition level of mutagenicity induced by the indirect mutagen AFB1 (Abdelwahed et al. 2008). In addition, antiproliferative and apoptotic properties of these extracts have also been reported using two leukemia cell lines, L1210 and K562. The results revealed that all extracts showed a significant cytotoxic effect on these cell lines, and the effect was greater in the presence of human K562 chronic myelogenous leukemia cells, whereas they do not induce apoptosis (Abdelwahed et al. 2008).

Using different spectroscopic techniques, the isolated compounds from the $CHCl_3$ soluble fraction of the roots of *P. tortuosus* were identified as bergapten, graveolone, xanthotoxin, isopimpinellin, aesculetin dimethyl ether, stigmasterol glucoside, in addition to the new ester *4*-methoxyphenylumbellate (Abdel-Kader 2003).

## *Zygophyllaceae*

*Nitraria retusa* **(Forsk) Asch.** is native to the desert areas of Northern Africa. *N. retusa* (Forssk.) Asch is one of the leading shrubs in steppes, deserts and saline soils. It belongs to the family Nitrariaceae where it is the only genus of this family (Täckholm 1974; Boulos 2000) or to the family Zygophyllaceae (Zohary 1972). It has a fairly continuous range of distribution extending from Arabia in the north to Somalia in the south, and from Palestine in the east to Senegambia in the west (Zohary 1973). It is definitely a Saharo-Arabian species with some trends towards Sudanian territories. It has tiny, white to green, fragrant flowers and small edible red fruits. Along with its culinary use, *N. retusa* is also used in folk medicine. *N. retusa* aqueous extracts showed significant anti-inflammatory activities.

***Peganum harmala*** **(L.)**
*Peganum harmala* L. is a perennial herbaceous, glabrous plant, which may grow up to 30–100 cm. Its normal habitat is semi-arid rangeland, steppe areas and sandy soils. The plant is widely distributed in the Central Asia, North Africa and Middle East and has been introduced in America and Australia. This plant is known as "Espand" in Iran, "Harmel" in North Africa and "African Rue", "Mexican Rue" or "Turkish Rue" in United States. It has alternate thong-like leaves, which have a strong deterrent odor when rumpled. Opposite to the leaves are solitary white flowers with green veins (Mahmoudian et al. 2002).

Seeds of *P. harmala*, a widespread species growing wild in Turkey are used as an antiheamoroidal, helmicide, and central nervous system (CNS) stimulating agent in folk medicine (Baytop 1999 and Kusmenoglu 1996). *P. harmala* has been used in Turkey for ritual purification for drift away the evil spirits by burning the seeds. It is used for dyes in Turkish rugs, it was source of the richly coloured textile dye (Baytop 1999). In traditional medicine in the Middle East and North Africa, this plant is used as an anthelmintic, lactogogue, antispasmodic, antipyretic, abortifacient, emetic and emmenagogue (Chopra et al. 1958; Nadkarni 1954). Its ethnobotanical preparation is used in the treatment of cancer by the late Hakim Analoui, a local healer in Iran, is a mixture of the extracts of *P. harmala* L. (Zygophyllaceae) and *Dracocephalum kotshyi* Boiss. (Labiatae). Eighty percent of this local remedy is *P. harmala* seeds extract. Its cytotoxic effects on cancerous cell-lines have been demonstrated in previous studies (Lamchouri et al. 1999; Lamchouri et al. 2000; Sobhani et al. 2002).

*P. harmala* alkaloids (Munir et al. 1993) are known for their antimicrobial (Ahmed et al. 1992), hypothermic (Abdelfattah et al. 1995) and hallucinogenic (Bruinvels 1969; Stratton and Lorden 1991; O'Hearn and Molliver 1993) properties. In Moroccan traditional medicine, seeds of P.harmala have been used for the empiric treatment of cancers (Bellakhdar 1997). The alkaloidic fraction of the methanol extract of *P. harmala* seeds was tested *in vitro* on three tumoral cell-lines: UCP-Med and Med-mek Carcinoma, and UCP-Med Sarcoma. Proliferation was significantly reduced at all tested concentrations (20–120 μg/ml) during the first 24 h of contact. A cell lysis effect occurred after 24 h and increased thereafter to complete cell death within 48–72 h, depending on tested concentration (Lamchouri et al. 2000). *P. harmala* is a poisonous plant that grows in Central Asia, North Africa and Middle East

The pharmacologically active compounds of *P. harmala* are several alkaloids, which are contained found especially by the seeds and the roots. These include β-carbolines such as: harmine, harmaline (identical with harmidine), harmalol and harman and quinazoline derivatives: vasicine and vasicinone. The alka-loidal content of the unripe seeds is less than the ripeones (Kamel et al. 1970). Harmaline (harmidine, $C_{13}H_{15}ON_2$), first isolated by Göbel (Budavari and O'Neil 1996) from the seeds and roots of *P. harmala*, this is the major alkaloid of this plant. Harmaline is almost twice as toxic as harmine and in moderate doses causes tremors and clonic convulsions but with no increase in spinal reflex excitability (Glasby 1978). Lethal doses bring about convulsions, which are soon followed by motor paralysis due to the marked depressive action upon the central nervous system. Respiration is paralyzed and a decrease in body temperature occurs. The perfused heart is arrested in diastolic phase and the con-tractions of smooth muscle are diminished with the exception of the uterus, which may be made to contract more powerfully. Over a wide range of doses there is a reduction in blood pressure due to a pronounced weakening of the heart muscle.

**Harmine (banisterine, $C_{13}H_{12}ON_2$) is present** in *P. harmala.* Pharmacologically, harmine resembles harmaline in its actions but is less toxic. The hydrochlo-ride has been found to be highly active against Myco-bacterium tuberculosis (Glasby 1978).

**Harmalol** ($C_{12}H_{12}ON_2$) which occurs in *P. harmala* crystallizes from water as the trihydrate. It is freely soluble in hot water, acetone or chloroform but only sparingly soluble in benzene. The alkaloid is unstable when exposed to air. Its methyl ether is harmaline (Glasby 1978).

**Vasicine (peganine,** $C_{13}H_{15}ON_2$), discovered in *P. harmala* under the name of Peganine. The salt are readily obtained as crystals and is used in India as a remedy for asthma and the pure alkaloid acts as a bronchodilator (Puzii et al. 1980).

Vasicinone (C11H10O2N2) a further alkaloid present in P. harmala. The base forms colorless crystals from 95% alcohol. It has (α) 22–100° (c = 0.5 in CHCl3) and the UV spectrum has absorption maxima at 227, 272, 302 and 315 nm. The alkaloid yields crystalline salts with mineral acids. It is an active bronchodilator (Puzii et al. 1980). In addition to their effects on CNS (Airaksinen and Kari 1981) and cardiovascular system (Aarons et al. 1977), there are reports of antimicrobial (Ahmed et al. 1992) and cytotoxic activities (Sobhani et al. 2002; Ahmed et al. 1992) of β-carbolines.

#### *Zygophyllum album* L.

*Zygophyllum album* L. is a succulent plant belonging to the Zygophyllaceae family that adapts to dry environments such as rock walls, the warm and temperate regions. Folk medicine uses a variety of species of *Zygophyllum*, including *Z. simplex* L., *Z. gaetulum*, *Z. cornutum* Coss. and *Z. album* L. for therapeutic treatments. It was reported that different parts of *Zygophyllum* are used against diabetes, and as a drug active against rheumatism, gout, asthma and hypertension. It is also used as diuretic, local anaesthetic, antihistaminic, anti-inflammatofry, molluscicidal and expectorant activity (Pöllmann et al. 1997; Safir and Fkih-Tetouani 1998; Jaouhari et al. 1999; Jaouhari and Lazrek 2000; Rimbau et al. 1999). Local surveys show that *Z. album* L. is used against aches and for calm thirst. It is also used for wound care, treatment of dental caries, and washing clothes and hair (Grim and Bounaga 1996).

The essential oil of *Z. album* was extracted, by hydrodistillation (HD) and analysis performed by gas chromatography and gas chromatography coupled to mass spectrometry which showed that (E)-β-damascenone was the major component of the oil (Tigrine-Kordjani et al. 2006a). Employing microwave accelerated distillation (MAD) extraction (100 °C, 30 min), *Z. album* essential oil contained mainly oxygenated monoterpenes with major constituents: carvone and a-terpineol. Remarkably, most of the compounds present in the hydrodistilled volatile fraction were not terpene species, with β-damascenone as a major constituent (Tigrine-Kordjani et al. 2010). Further investigations into the leaves and stems of *Z. album* yielded four triterpene glycosides which were characterized as 14-decarboxyquinovic acid-3β-O-β-D-quinovopyranosyl (1→4)-quinovopyranoside, quinovic acid 28-O-β-D-glucopyranosyl (2→1)β-D-glucopyranosyl ester, quinovic acid 27-β-D-glucopy ranosyl(2ster-D-glucopyranosyl ester and quinovic acid-3-β-O-glucopyranosyl(2ster and quinovic acid-3-β-uinovic acid 28-le fraction were not terpenes (Hassanean et al. 1993). Besides the two known saponins, quinovic acid-3β-O- β -D-quinovoside and 3β-O- β -D quinovopyranosyl quinovic acid (28→l) β-D-glycopyranosyl ester, two new glycosides having the structure quiaovic β acid 3β-O- β -D quinovopyranosyl(3→1) β-D-xylopyranoside and 3β-O- β -D quinovopyranosylquinovic acid (28→ l)quinovopyra-nosy1 ester were isolated from the aerial parts of *Z. album* growing in Egypt (Hassanean et al. 1993).

## References

Aarons DH, Rossi GV, Orzechowski RF (1977) Cardiovascular actions of three *harmala* alkaloids: harmine, harmaline, and harmalol. J Pharm Sci 66:1244–1248

Abdallah L, Chaieb M (2007) Water status and growing phonology of a Saharan Shrub in North Africa. Afr J Ecol 1(45):80–85

Abdalla MF, Saleh NAM (1983) Flavonoids of Retama raetam. J Nat Prod 46:755–756.

Abdel Halim OB, Abdel Fattah H, Halim AF, Murakoshi I (1997) Comparative chemical and biological studies of the alkaloidal content of Lygos species and varieties growing in Egypt. Acta Pharm Hung 67:241–247

Abdel-Kader MS (2003) New ester and furanocoumarins from the roots of *Pituranthos tortuosus*. J Braz Chem Soc 14:48–51

Abdel-Fattah AFM, Matsumoto K, Gammaz HAK, Watanabe H (1995) Hypothermic effect of harmala alkaloid in rats: involvement of serotonergic mechanism. Pharmacol Biochem Behav 52:421–426

Abdelwahed MSA, Amin AA, El-Rashad SM (2006) Physiological effect of some bioregulators on vegetative growth, yield and chemical constituents of yellow maize plants. World J Agric Sci 2:149–155

Abdelwahed A, Skandrani I, Kilani S, Neffati A, Sghaier MB, Boubaker J, Ammar RB, Mahmoud A, Ghedira K, Chekir-Ghedira L (2008) Mutagenic, antimutagenic, cytotoxic and apoptotic activities of extracts from *Pituranthos tortuosus*. Drug Chem Toxicol 31(1):37–60

Abdel-Zaher AO, Salim SY, Assaf MH, Abdel-Hady RH (2005) Antidiabetic activity and toxicity of Zizyphus spina-christi leaves. J Ethnopharmacol 101:129–138

Abdollahi M, Fooladian F, Emami B, Zafari K, Bahreini-Moghadam A (2003) Protection by sildenafil and theophylline of lead acetate-induced oxidative stress in rat submandibular gland and saliva. Hum Exp Toxicol 22:587–592

Adam SEI (1978) C/in. 7bxicol 13:269

Adams RP (1999) Systematics of multi-seeded eastern hemisphere *Juniperus* based on leaf essential oils and RAPD DNA fingerprinting. Biochem Syst Ecol 27:709–725

Adams RP (2000) Systematics of Juniperus section Juniperus based on leaf essential oils and random amplified polymorphic DNAs (RAPDs). Biochem Syst Ecol 28:515–528

Adams RP, Barrero AF, Lara A (1996) Comparisons of the leaf essential oils of *Juniperus phoenicea, J. phoenicea* subsp. eu-mediterranea Lebr. & Thiv. and *J. phoenicea* var. *turbinata* (Guss) Parl. J Essent Oil Res 8:367–371

Adhvaryu MR, Reddy N, Parabia MH (2007) Effects of four Indian medicinal herbs on Isoniazid-, Rifampicin- and Pyrazinamide-induced hepatic injury and immunosuppression in *guinea pigs*. World J Gastroenterol 13:3199–3205

Adzu B, Amos S, Amizan MB, Gamaniel K (2003) Evaluation of the antidiarrhoeal effects of *Zizyphus spina-christi* stem bark in rats. Acta Trop 87(2):245–250

Afsharypuor S, Jeiran K, Jazy AA (1998) First investigation of the flavor profiles of leaf, ripe fruit and root of *Capparis spinosa* var. *mucronifolia* from Iran. Pharm Acta Helv 72:307–309

Ahmed A, Cameron IT, Ferriani RA (1992) Activation of phospholipase A2 and phospholipase C by endothelin-1 in human endometrium. J Endocrinol 135:383–390

Ahmed MS, Galal AM, Ross SA, Ferreira D, El Sohly MA, Ibrahim ARS, Mossa JS, El-Ferly FS (2001) A weakly antimalarial biflavonone from *Rhus retinorrhoea*. Phtochemistry 58:599–602

Ahmed ZF, Wassel GM, Abdel-Moneim FM (1969) A preliminary phytochemical investigation of pituranthostortuosus (Desf.) Benth and Hook. J Pharm Sci U A R 10(1):31–36

Airaksinen MM, Kari I (1981) b-Carbolines, psychoactive compounds in the mammalian body. Effects Med Biol 59:190–211

Akbay P, Gertsch J, Çalis I, Heilmann J, Zerbe O, Sticher O (2002) Novel antileukemic sterol glycosides from *Ajuga salicifolia*. Helv Chim Acta 85:1930–1942

Akgül A (1993) Spice science and technology (Baharat Bilimi ve Teknolojisi; in Turkish) Turkish. *Association of Food Technologists*, Publ. No. 15, Ankara

Akrout A (2004) The study of chemical compositions of essential oils of three pastoral plants from Matmata (south Tunisia) (in French). Cah Options Méditerr 62:289–292

Akrout A, Chemli R, Chrief I, Hammami M (2001) Analysis of the essential oil of *Artemisia campestris* L. J Flavour Fragr 16:337–339

Akrout A, Chemli R, Simmonds M, Kite G, Hammami M, Cherif I (2003) Seasonal variation of the essential oil of Artemisia campestris L. J Essent Oil Res 15(5):333–336

Akrout A, El-Jani H, Amouri A, Neffati M (2010a) Screening of antiradical and antibacterial activities of essential oils of *Artemisia campestris* L., *Artemisia herba-alb* Asso. & *Thymus capitatus* Off. et Link. Growing wild in the southern of Tunisia. Recent Res Sci Technol 2(1):29–39

Akrout A, Hajlaoui H, Mighri H, Najjaa H, El Jani H, Zaidi S, Neffati M (2010b) Chemical and Biological Characteristics of Essential Oil of *Rosmarinus officinalis* Cultivated in Djerba. Jeobp 13(4):398–411

Alçiçek Z (2011) The effects of thyme (*Thymus vulgaris* L.) oil concentration on liquidsmoked vacuum-packed rainbow trout (Oncorhynchus mykiss Walbaum, 1792) fillets during chilled storage. Food Chem 128:683–688

Al-Gaby AM, Allam RR (2000) Chemical analysis, antimicrobial activity and the essential oils from some wild herbs in Egypt. J Herbs Species Med Plant 7(1):15–23

Alghazeer R, El-Saltani H, Saleh N, Al-Najjar A, Hebail F (2012) Antioxidant and antimicrobial properties of five medicinal Libyan plants extracts. Nat Sci 4(5):324–335

Aliotta G, Pollio A (1994) Useful plants in renal therapy according to Pliny the elder. Am J Nephrol 14:341–399

Aliotta G, Caiero G, De Feo V, Di Blasio B, Iacovino R, Oliva A (2000) Allelochemicals from rue (*Ruta graveolens* L.) and olive (*Olea europaea* L.) oil mill waste waters as potential natural pesticides. Curr Top Phytochem 3:167–177

Al-Mustafa AH, Al-Thunibat OY (2008) Antioxidant activity of some Jordanian Medicinal plants used traditionally for treatment of diabetes, Pakistan. Aust J Biol Sci 1(3):351–358

Al-Okbi SY, El-Sayed EM, Ammar NM, El-Sayed NK, Abou-El Kassem LT (2002) Effect of *Ruta graveolens* L. and *Euphorbia peplus* L. anti-inflammatory extracts on nutritional status of rats and the safety of their use. Indian J Exp Biol 40(1):45–48

Al-Quran S (2008) Taxonomical and pharmacological survey of therapeutic plants in Jordan. J Nat Prod 1:10–26

Al-Rawi A, Chakravarty HL (1964) Medicinal plants of Iraq. Ministry of Agriculture *Technical Bulletin No. 15,* Govt. Press, Baghdad.

Al-Said MS, Abdelsatter EA, Khalifa SI, El-Ferly FS (1988) Isolation and identification of an anti-inflammatory principle from *Capparis spinosa*. Pharmazie 43:640–641

Al-Sereitia MR, Abu-Amerb KM, Sena P (1999) Pharmacology of rosemary (*Rosmarinus officinalis* Linn.) and its therapeutic potentials. Indian J Exp Biol 37:124–131

Ali-Shtayeh MS, Abu Ghdeib SI (1999) Antifungal activity of plant extracts against dermatophytes. Mycoses 42:665–672

Amaral FMM, Ribeiro MNS, Barbosa-Filho JM, Reis AS, Nascimento FRF, Macedo RO (2006) Plants and chemical constituents with giardicidal activity. Rev Bras Farmacogn 16(Supl.):696–720

Amer A, Mehlhorn H (2006a) Larvicidal effects of various essential oils against *Aedes anopheles* and *Culex larvae* (Diptera, Culicidae). Parasitol Res 99:466–472

Amer A, Mehlhorn H (2006b) Repellency effects of forty-one essential oils *against Aedes anopheles* and *Culex mosquitoes*. Parasitol Res 99:478–490

Amico FP, Sorce EG (1997) Medical plants and phytotherapy in Mussomeli area (Caltanisseta, Sicily, Italy). Fitoterapia 68:143–159

Amirti F, Aberchane M (2008) Activité antifongique des huiles essentielles de *Thymus bleicherianus* Pomel et *Thymus capitatus* (L.) Hoff. & Link contre les champignons de pourriture du bois d'eouvre. BASE 12(4):345–351

Amirti F, Satrani B, Aafi A, Ghanmi M, Farah A, Aberchane M, El Ajjouri M, Al Antry S, Chaouch A (2008a) Chemical composition and antimicrobial activity of the essential oils of Moroccan *Thymus capitatus* and *Thymus bleicherianus*. Phytothérapie 6(6):342–347

Angelini G, Vena GA, Filotico R, Foti C, Grandolfo M (1991) Allergic contact dermatitis from *Capparis spinosa* L. applied as wet compresses. Contact Dermatitis 24(5):382–383

Aniya Y, Shimabukuro M, Shimoji M, Kohatsu M, Gyamfi MA, Miyagi C, Kunii D, Takayama F, Egashira T (2000) Antioxidant and hepatoprotective actions of the medicinal herb Artemisia campestris from the Okinawa Islands. J Biol Pharm Bull 23(3):309–312

Anonymous (2000) Studies on the cytotoxicity and antimicrobial effects of the extract of *Ajuga multiflora* Bunge. Korean J Pharmacogn 31:72–76

Arena A, Bisignano G, Pavone B, Tomaino A, Bonina FP, Saija A (2008) Antiviral and immuno modulatory effect of a lyophilized extract of *Capparis spinosa* L. buds. Phytother Res 22:313–317

Arnault I, Haffner T, Siess MH, Vollmard A, Kahane R, Auger J (2005) J Pharm Biomed Anal 37:963–970

Ashkenazy D, Friedman J, Kashman Y, Egyed MN, Shlosberg A, Nyska A (1984) Photosensitization in ducklings induced by Pituranthos triradiatus. Vet Hum Toxicol 26(2):118–120

Askri M, Mighri Z, Bui AM, Das BC, Hylands PJ (1989) Medicinal plants of Tunisia. The structure of periplocadiol, a new elemane-type sesquiterpene isolated from the roots of *Periploca laevigata*. J Nat Prod 52:792–796

Auger J, Yang W, Arnault I, Pannier F, Potin-Gautier M (2004) J Chromatogr A 1032:103–107

Avallone R, Plessi M, Baraldi M, Monzani A (1997) Determination of chemical composition of carob (*Ceratonia siliqua*): protein, fat, carbohydrates and tannins. J Food Compos Anal 10:166–172

Ayaz FA, Torun H, Ayaz S, Correia PJ, Alaiz M, Sanz C, Gruz J, Strnad M (2007) Determination of chemical composition of Anatolian carob pod (*Ceratonia siliqua* L.): sugars, amino and organic acids, minerals and phenolic compounds. J Food Qual 30(6):1040–1055

Baba Aïssa F (1991) Les Plantes Me´dicinales en Algérie. Coédition Bouchène et Ad-Diwan, Alger

Bakkali F, Averbeck S, Averbeck D, Idaomar M (2008) Biological effects of essential oils – a review. Food Chem Toxicol 46:446–475

Bakri LM, Douglas CWI (2005) Inhibitory effect of garlic extract on oral bacteria. Arch Oral Biol 50(7):645–651

Banarjee SP, Dandiya PC (1967) smooth muscle and cardiovascular pharmacology of α-elaterin-2-D-glucopyranoside glycoside of Citrullus colocynthis. J Pharm Sci 56(12):1665–1667

Banerji P, Banerji P (2001) Intracranial cysticercosis: an effective treatment with alternative medicines. In Vivo 15:181–184

Bannour F, Bel Haj Saleh K, Chraief I, Hammami M, El Gazzah M, Chemli R, Aouni M (2006) Variability in essential oil composition and antibacterial activity of Tunisian *Rosmarinus officinalis*. Microbiol Hyg Alim 18(51):59–64

Banthorpe DV, Davies HS, Gatford C, Williams SR (1973) Monoterpene patterns in *Juniperus* and Thuja species. Planta Med 23:64–69

Baranauskiene R, Venskutonis PR, Viskelis P, Dambrauskiene E (2003) Influence of nitrogen fertilizers on the yield and composition of thyme (*Thymus vulgaris*). J Agric Food Chem 51(26):7751–7758

Barroso JG, Pedro LG, Pais MSS, Scheffer JJC (1991) Analysis of the essential oil of *Crithmum maritimum* L. J Essent Oil Res 3:313–316

Barroso JG, Pedro LG, Figueiredo AC, Salomé M, Pais S (1992) Seasonal variation in the composition of the essential oil of *Crithmum maritimum* L. Flavour Fragr J 7:147–150

Baytop T (1984) Therapy with medicinal plants (past and present). Istanbul University publications, Istanbul, pp 268–269

Baytop T (1999) Turkçe Bitkiler ile Tedavi-Geçmisten Bugüne (Therapy with Medicinal Plants in Turkey-Past and Present), 2nd edn. Nobel Tip Basimevi, Istanbul, pp 348–349

Bedir E, Tasdemir D, Çalis I, Zerbe O, Sticher O (1999) Neo-clerodane diterpenoids from *Teucrium polium*. Phytochemistry 51:921–925

Bekkara FA, Bousmaha L, Bendiab SAT, Boti JB, Casanova J (2007) Chemical composition of essential oil cultivated and wild growing *Rosmarinus officinalis* L. from Tlemcen (Algeria). Biology & Health 7(1):6–11. (in french)

Bellakhdar J (1996) La Pharmacopée Marocaine Traditionnelle. Ibiss Press, p 358

Bellakhder J (1997) La Pharmacopée Marocaine Traditionnelle. Ibis Press, Paris, pp 271–272

Bellakhdar J, Claisse R, Fleurentin J, Younos C (1991) J Ethnopharmacol 35:123–143

Bellei B, Maresca V, Flori E, Pitisci A, Larue L, Picardo M (2010) Regulates pigmentation via proteasomal degradation of tyrosinase. J Biol Chem 285:7288–7299

Bello R, Moreno L, Primo-Yufera E, Esplugues J (2002) Phytother Res 16:389–392

Ben El Hadj Ali I, Zaouali Y, Bejaoui A, Boussaid M (2010) Variation of the chemical composition of essential oils in Tunisian populations of *Thymus algeriensis* Boiss. et Reut (Lamiaceae) and implication for conservation. Chem Biodivers 7:1276–1289

Ben Jannet H, Harzallah-Skhiri F, Mighri Z, Simmonds MSJ, Blaney WM (2001) Antifeedant activity of plant extracts and of new natural diglyceride compounds isolated from *Ajuga pseudo-iva* leaves against *Spodoptera littoralis*. Ind Crop Prod 14:213–222

Ben Jannet H, Harzallah-Skhiri F, Mighri Z, Simmonds MSJ, Blaney WM (2000) Responses of *Spodoptera littoralis* larvae to Tunisian plant extracts and to neo-clerodane diterpenoids isolated from *Ajuga pseudoiva* leaves. Fitoterapia 71:105–112

Ben Salah H, Bouaziz M, Bahroun A, Damak M, McKillop M, Simmonds MSJ (2000) Contribution à l'étude chimique *Hyparrhenia hirta*. J Soc Chim Tunis 5(7):647–652

Ben Salah H, Jarraya R, Martin MT, Veitch NC, Grayer RJ, Simmons MSJ, Damak M (2002) Flavonol triglycosides from the leaves of *Hammada scoparia* (POMEL) ILJIN. Chem Pharm Bull 50(9):1268–1270

Benkrief R, Brum-Bousquet M, Tillequin F, Koch M (1990) Alkaloids and flavonoid from aerial parts of *Hammada articulata* spp *scoparia*. Ann Pharm Fr 48:219–224

Bennaghmouch L, Hajjaji N, Zellou A, Cherrah Y (2001) Pharmacological study of Ajuga iva. Ann Pharm Fr 59:284

Bhargava R, Shukla AK, Chauhan N, Vashishtha BB, Dhardar DG (2005) Impact of hybridity on flavonoid spectrum of ber (*Ziziphus mauritiana* Lamk.) Environ Exp Bot 53:135–138

Binder RJ, Coit JE, Williams KT, Brekke JE (1959) Carob varieties and composition. Food Technol 13:213–216

Biondi D, Cianci P, Geraci C (1993) Antimicrobial activity and chemical composition of essential oils from Sicilian aromatic plants. J Flavour Fragr 8:331–337

Bloedorn WA, Dickens PF (1928) Danger of the administration of ephedrine in heart failure. Arch Intern Med 42:322–330

Bondi ML, Al-Hillo MRY, Ladjel S, Bruno B, Piozzi F, Simmonds MSJ (2000) Occurrence of the antifeedant 14,15-dihydroajugaritin in the aerial parts of *Ajuga iva* from Algeria. Biochem Syst Ecol 28(10):1023–1025

Bonnier G, Douin R (1990) La grande flore. Belin, Paris

Borgi W, Bouraoui A, Chouchane N (2007) Antiulcerogenic activity of *Zizyphus lotus* (L.) extracts. J Ethnopharmacol 112:228–231

Borkowski B, Pasich B (1958) Die isolation der flavone aus herba Herniariae glabrae L. Planta Med 6(3):225–237

Bouaziz M, Simmonds MSJ, Grayer RJ, Kite GC, Damak M (2001) Flavonoids from *Hyparrhenia hirta* Stapf (Poaceae) growing in Tunisia. Biochem Syst Ecol 29:849–851

Bouaziz M, Veitch NC, Grayer RJ, Simmonds MSJ, Damak M (2002) Flavonolignans from *Hyparrhenia hirta*. Phytochemistry 60:515–520

Bouchouka EL, Djilani A, Bekkouche A (2012) Antibacterial and antioxidant activities of three endemic plants from Algerian Sahara. Acta Sci Pol Technol Aliment 11(1):61–65

Boukef MK (1986) Les Plantes dans la médicine traditionnelle Tunisienne, Agence de Coopération Culturelle et Technique, pp 82–83

Boukef MK, Souissi HR, Ballansard G (1982) Contribution à l'étude des plantes utilisées en médicine traditionnelle tunisienne. Plants Med Phyto 16:260–279

Boukrich F, Zouari S, Neffati M, Abdelly C, Liu K, Casanova J, Tomi F (2010) J Essent Oil Res 22:331

Boulos L (1983) Medicinal plants of North Africa. Reference Publications, Algonac

Boulos L (2000) Flora of Egypt. Vol. II (Geraniaceae- Boraginaceae). Al-Hadara Publishing, Cairo, p 352

Boumaza O, Mekkiou R, Seghiri R, Sarri D, Benayache S, Garcia VP, Benayache F (2006) Flavonoids and isoflavonoids from *Genista tricuspidata*. Chem Nat Compd 42(6):730–731

Bounatirou S, Smiti S, Miguel MG, Rejeb MN, Neffati M, Costa MM, Faleiro L, Figueiredo AC, Barroso JG, Pedro LG (2007a) Chemical composition, antioxidant and antibacterial activities of the essential oils isolated from Tunisian *Thymus capitatus* Hoff. Et Link. Food Chem 105:146–155

Bounatirou SG, Zouari S, Figueiredo ACS, Barroso JG, Pedro LG, Neffati M, Rejeb MN, Smiti S (2007b) Chemical homogeneity of *Thymus capitatus* Hoff. et Link. essential oils from Tunisia. Revue des Régions Arides 2:679–686

Bounatirou S, Smiti S, Miguel MG, Rejeb MN, Neffati M, Costa MM, Faleiro L, Figueiredo AC, Barroso JG, Pedro LG (2008) *Tymus capitatus* grown in Tunisia: antioxidant ability of the essential oils on lilnoleic acid evaluated by different methods. Acta Horticulturae (ISHS) 765:315–324

Bourogaa E, Bertrand J, Despeaux M (2011) *Hammada scoparia* flavonoids and rutin kill adherent and chemoresistant leukemic cells. *Leuk Res* 35:1093–1101

Bourogaa E, Nciri R, Sulta CR, Damak M, Elfeki A (2012) Antioxidant activity and hepatoprotective potential of Hammada scopariaagainst ethanol-induced liver injury in rats. J Physiol Biochem

Boutekedjiret C, Bentahar F, Belabbes R, Bessière JM (1998) The essential oils of *Rosmarinus officinalis* from Algeria. J Esent Oil Res 10:680–682
Boutekedjiret C, Belabbes R, Bentahar F, Bessière JM (1999) Study of *Rosmarinus officinalis* L. essential oil as a function of the plant life cycle. J Esent Oil Res 11:238–240
Brandão MGL, Cosenza GP, Moreira RA, Monte-Mor RLM (2006) Medicinal plants and other botanical products from the Brazilian Official Pharmacopoeia. Rev Bras Farm 16:408–420
Bravo L, Grados N, Saura-Calixto F (1994) Composition and potential uses of mesquite pods (*Prosopis pallida* L.). Comparison with carob pods (*Ceratonia siliqua* L.) J Sci Food Agric 65:303–306
Bremner PD, Simmonds MSJ, Blaney WM, Veitch NC (1998) Neo-clerodane diterpenoid insect antifeedants from *Ajuga reptans* ev Catlins Giant. Phytochemistry 47:1227–1232
Breschi MC, Martinotti E, Catalano S, Flamini G, Morelli I, Pagni AM (1992) Vasoconstrictor activity of 8-Oacetylharpagide from *Ajuga reptans*. J Nat Prod 55:1145–1148
Brevard H, Brambilla M, Chaintreau A, Marion JP (1992) Occurrence of elemental sulphur in capers (*Capparis spinosa* L.) and first investigation of the flavour profile. Flavour Fragr J 7:313–321
Brigelius-Flohé R, Traber MG (1999) Vitamin E: function and metabolism. FASEB J 13:1145–1155
Brooks G, Evans AT, Aitken A, Evans FJ, Rizk AFM, Hammouda FM, El-Missiry MM, Ismail SE (1990) Daphnane diterpenes of *Thymelaea hirsuta*. Phytochemistry 29(7):2235–2237
Bruinvels J (1969) Comparison of intracisternally and intraperitoneally injected harmaline on body temperature and tremor in the rat. J Pharm Pharmacol 21:506–508
Bruneton J (1993) Pharmacognosie: Phytochimie, Plantes Me´dicinales, 2nd edn. Technique et Documentation, Lavoisier, Paris
Budavari S, O'Neil MJ (1996) The Merck Index, 12nd edn. CRC Press, pp 4644–4645
Burits M, Bucar F (2000) Antioxidant activity of Nigella sativa essential oil. Phytother Res 14:323–328
Burt S (2004) Essential oils: their antibacterial and potential application in foods- A review. Int J Food Microbiol 94:223–253
Caceres A, Giron LM, Martinez AM (1987) Diuretic activity of plants used for the treatments of urinary ailments in Guatemaela. J Ethnopharmacol 43:197–201
Caldes G, Prescott B, King JR (1975) Planta Med 27:72–76
Calis I, Kuruözüm-Uz A, Lorenzetto PA, Rüedi P (2002) (6S)-Hydroxy-3-oxo-α-ionol glucosides from *Capparis spinosa* fruits. Phytochemistry 59(4):451–457
Cao YL, Li X, Zheng M (2008) Effect of *Capparis spinosa* on fibroblast proliferation and type II collagen production in progressive systematic sclerosis. Zhongguo Zhong Yao Za Zhi 33:560–563
Cantrell CL, Rajab MS, Franzblau SG, Fronczek FR, Fisher NH (1999) Antimycobacterial ergosterol-5,8-endoperoxide from Ajuga remota. Planta Med 65:732–734
Carling C, Sandberg F (1970) Alkaloids of *Haloxylon articulatum*. Acta Pharm Suec 7:285–288
Carlini EA, Rodrigues E, Mendes FR, Tabach R, Gianfratti B (2006) Treatment of drug dependence with Brazilian herbal medicines. Rev Bras Farm 16:690–695
Carreiras MC, Rodríguez B, Piozzi F, Savona G, Torres MR, Perales A (1989) A chlorine-containing and two 17β-*neo*-clerodane diterpenoids from *Teucrium polium* subsp. *Vincentinum*. Phytochemistry 28(5):1453–1461
Chaieb M, Boukhris M (1998) Flore succinte et illustrée des zones arides et désertiques de Tunisie. Association pour la protection de la nature et de l'environnement. Sfax, 290 pages
Chalchat JC, Carry RP, Michet A, Benjilali B, Chabart JL (1993) Essential oil of Rosemary (*Rosmarinus officinalis* L.), the chemical composition of oil from various origins. J Esent Oil Res 5:613–618
Chan AS, Yip EC, Yung LY, Pang H, Luk SC, Pang SF, Wong YH (2006) Ckbm stimulates MAPKs but inhibits LPS-induced IFN-gamma in lymphocytes. Phytother Res 20:725–731
Chao HC, Najjaa H, Villarcal MO, Ksouri R, Han J, Neffati M, Isoda H (2013) *Arthrophytum scoparium* inhibits melanogenesis through the down-regulation of tyrosinase and melanogenic gene expressions in B16 melanoma cells. Exp Dermatol 22(2):131–136

Chaouch A (2008a) Chemical composition and antimicrobial activity of the essential oils of Moroccan *Thymus capitatus* and *Thymus bleicherianus*. Phytothérapie 6(6):342–347
Chen KK, Schmidt CF (1930) Ephedrine and related substances. Medicine 9:91–117
Chen CC, Huang YL, Huang FI, Wang CW, Ou JC (2001) Water-soluble glycosides from *Ruta graveolens*. J Nat Prod 64(7):990–200
Chen H, Tan RX, Liu ZL, Zhang Y (1996) Antibacterial neoclerodane diterpenoids from Ajuga lupulina. J Nat Prod 59:668–670
Chevalier A (1935) Les iles du Cap Vert. Flore de l'archipel. Rev Bot Appl Agric Trop 15:733–1090
Chevallier A (1996) The encyclopedia of medicinal plants. Dorling Kindersley, London
Chiu KW, Fung AY (1997) The cardiovascular effects of green beans (*Phaseolus aureus*), common rue (*Ruta graveolens*) and kelp (Laminaria japonica) in rats. Gen Pharmacol 29:859–862
Choi W, Jung H, Kim K, Lee S, Yoon S, Park J, Kim S, Cheon S, Eo W, Lee S (2012) *Rhus verniciflua* stokes against advanced cancer: a perspective from the Korean integrative cancer Center. J Biomed Biotechnol 2012:181–191
Chopra RN, Chopra IC, Handa KL, Kapur LD (1958) Indigenous drugs of India. Academic Publishers, Calcutta
Chopra RN, Nayar SL, Chopra IC (1986) Glossary of Indian Medicinal Plants (including the supplement) Council of Scientific and Industrial Research, New Delhi
Clayton WD (1969) A revision of the Genus *Hyparrhenia*. Kew Bulletin Additional Series II. HMSO, London
Coiffard L, Piron-Frenet M, Amicel L (1993) Geographical variations of the constituents of the essential oil of *Crithmum maritimum* L. Apiaceae. Int J Cosm Sei 15:15–21
Comte G, Chulia AJ, Vercauteren J, Allais DP (1996) Phenylpropane glycosides from *Juniperus phœnicea*. Planta Med 62(1):88–89
Conway GA, Slocumb JC (1979) Plants used as abortifacients and emmenagogues by Spanish New Mexicans. J Ethnopharmacol 1(3):241–261
Corsi L, Avallone R, Cosenza F, Farina F, Baraldi C, Baraldi M (2002) Antiproliferative effects of *Ceratonia siliqua* L. on mouse hepatocellular carcinoma cell line. Fitoterapia 73(7–8):674–684
Cosentino S, Tuberoso CIG, Pisano B, Mascia E, Arzedi E, Palmas F (1999) *In-vitro* antimicrobial activity and chemical composition of Sardenian *Thymus* essential oils. Lett Appl Microbiol 29:130–135
Cuénod A (1954) Flore analytique et synoptique de la Tunisie. Cryptogames vasculaires, Gymnospermes et Monocotylédones, S.E.F.A.N. (Eds). Tunis, p 287
Cunsolo F, Ruberto G, Amico V, Piattelli M (1993) Bioactive metabolites from Sicilian marine fennel, *Crithmum maritimum*. J Nat Prod 56:1598–1600
Dahia M, Laouer H, Chake AN, Prado S, Meierhenrich UJ, Baldovini N (2007) Chemical composition and antibacterial activity of *Pituranthos chloranthus* volatile oil. Nat Prod Commun 2(11):1159–1162
Dapkevicius A, Van Beek VTA, Lelyveld GP, Van Veldhuizen A, DeGroot A, Linssen JPH, Venskutonis R (2002) Isolation and structural elucidation of radical scavengers from *Thymus vulgaris* leaves. J Nat Prod 65(6):892–896
Darwish-Sayed M, Balbaa SI, Afifi MS (1974) The glycosidal content of the different organs of *Citrullus colocynthis*. Planta Med 26(3):293–298
Daxenbichler ME, Spencer GF, Carlson DG, Rose GB, Brinker AM, Powel RG (1991) Glucosinolate composition of seeds from 297 species of wild plants. Phytochemistry 30:2623–2638
De Feo V, De Simone F, Senatore F (2002) Potential allelochemicals from the essential oil of *Ruta graveolens*. Phytochemistry 61:573–578
De la Torre M, Rodriguez B, Bruno M, Piozzi F (1988) Montanin C from *Teucrium polium* var. *album*. Fitoterapia 59:129–130
Dellacassa E, Lorenzo D, Moyna P, Frisso CD, Serafini A, Dugo P (1999) *Rosmarinus officinalis* L. (Labiatae) essential oils from the south of Brazil and Uruguay. J Esent Oil Res 11(1):27–30
Dendougui H, Seghir S, Belloum Z, Benayache F, Leon F, Brouard I, Bermejo J, Benayache S (2011) A new labdane diterpene and other constituents from *Marrubium deserti* Noe ex coss. Rec Nat Prod 5(4):300–304

Denmeade SR, Isaacs JT (2005) The SERCA pump as a therapeutic target: making a "smart bomb" for prostate cancer. Cancer Biol Ther 4:14–22
Denmeade SR, Jakobsen CM, Janssen S, Khan SR, Garrett ES, Lilja H, Christensen SB, Isaacs JT (2003) Prostate-specific antigen-activated thapsigargin prodrug as targeted therapy for prostate cancer. J Natl Cancer Inst 95:990–1000
Dessí MA, Deiana M, Rosa A, Piredda M, Cottiglia F, Bonsignore L, Deidda D, Pompei R, Corongiu FP (2001) Antioxidant activity of extracts from plants growing in Sardinia. Phytother Res 15(6):511–518
Dhief A, Gorai M, Aschi-Smiti S, Neffati M (2009) Comparative phenological and water potential patterns of three *Calligonum* species in the eastern great Erg of Tunisia. Flora – Morphol Distrib Funct Ecol Plant 204(8):581–592
Dhief A, Zouari S, Abdellaoui R, Aschi-Smiti S, Neffati M (2011) Comparative study of chemical composition of the essential oils from three Calligonum species growing wild in Tunisian desert. Jeobp 14(1):11–22
Djeridane A, Yousfi M, Nadjemi B, Boutassouna D, Stocker P, Vidal N (2006) Antioxidant activity of some Algerian medicinal plants extracts containing phenolic compounds. Food Chem 97:654–660
Dob T, Dahmane D, Benabdelkader T, Chelghoum C (2006) Studies on the essential oil composition and antimicrobial activity of *Thymus algeriensis* Boiss.et Reut. Int J Aromather 16:95–100
Domokos J, Hethelyi E, Palinkas J, Szirmai S, Tulok HM (1997) Essential oil of rosemary (*Rosmarinus officinalis* L) of Hungarian origin. J Esent Oil Res 9:41–45
Dorman HJD, Deans SG (2000) Antimicrobial agents from plants: antibacterial activity of plant volatile oils. J Appl Microbiol 88:308–316
Dorman HJD, Deans SG (2004) Chemical composition, antimicrobial and in vitro antioxidant properties of *Monaranda citriodora* var. *citriodora*, *Myristica fragrans*, *Origanum vulgare* ssp. *hirutum*, *pelargonium* sp. and *Thymus zygis* oils. J Essent Oil Res 16:145–150
Durenkamp M, Dekok LJ (2004) J Exp Bot 55(404):1821–1830
Dziri S, Hassen I, Fatnassi S, Mrabet Y, Casabianca H, Hanchi B, Hosni K (2012) Phenolic constituents, antioxidant and antimicrobial activities of rosy garlic (*Allium roseum* var. *odoratissimum*). J Funct Foods 4(2):423–432
Eddouks M, Maghrani M, Lemhadri A, Ouahidi ML, Jouad H (2002) Ethnopharmacological survey of medicinal plants used for the treatment of diabetes mellitus, hypertension and cardiac diseases in the south-east region of Morocco (Tafilalet). J Ethnopharmacol 82:97–103
Eddouks M, Lemhadri A, Michel JB (2004) Caraway and caper: potential antihyperglycaemic plants in diabetic rats. J Ethnopharmacol 94:143–148
Eddouks M, Lemhadri A, Michel JB (2005) Hypolipidemic activity of aqueous extract of *Capparis spinosa* L. in normal and diabetic rats. J Ethnopharmacol 98:345–350
Edziri H, Ammar S, Groh P, Mahjoub MA, Mastouri M, Gutmann L, Zine M, Aouni M (2007) Antimicrobial and cytotoxic activity of *Marrubium alysson* and *Retama retama* in Tunisia. Pak J Biol Sci 10:1759–1762
Edziri H, Mastouri M, Ammar S, Matieu M, Patrich G, Hiar R, Mahjoub MA, Ali SM, Laurent G, Mighri Z (2008) World. J Microbiol Biotechnol 24(12):2933–2940
Edziri H, Mastouri M, Chéraif I, Aouni M (2010) Chemical composition and antibacterial, antifungal and antioxidant activities of the flower oil of *Retama raetam* (Forssk.). Webb Tunisia Nat Prod Res Form Nat Prod Lett 24(9)
Ehab AA, El-Alfy AT, Khan IA, Walker L (2003) *Ephedra* in perspective – a current review. Phytother Res 17:703–712
Ehivet FE, Min B, Park MK, Oh JH (2011) Characterization and antimicrobial activity of sweet potato starch based edible film containing origanum (*Thymus capitatus*) oil. J Food Sci 76:178–184
Elamrani A, Zrira S, Benjilali B, Berrada MA (2000) Study of Moroccan Rosemary oils. J Esent Oil Res 12:487–495

El Agraa SE, El Balwi SM, Adam SE (2002) Preliminary observations on experimental *Ruta graveolens* toxicosis in Nubian goats. Trop Animal Health Proc 34:271–281
El Ajjouri M, Satrani B, Ghanmi M, Aafi A, Farah A, Rahouti M, Amarti F, Aberchane M (2008) Activité antifongique des huiles essentielles de *Thymus bleicherianus* Pomel et *Thymus capitatus* (L.) Hoff.& Link contre les champignons de pourriture du bois d'eouvre. BASE 12(4):345–351
El Amrani F, Rhallab A, Alaoui T, El Badaoui K, Chakir S (2009) Hypoglycaemic effect of *T. hirsuta* in normal and streptozotocininduced diabetic rats. J Med Plant Res 3:625–629
El-Demerdash FM, Yousef MI, Abou El-Naga NI (2005) Food Chem Toxicol 43(1):57–63
El-Domiaty MM, El-hafae AM, Abdel-Aal MM (1997) Alexandria J Pharmaceut Sci 11:13
El Hilaly J, Lyoussi B (2002) Hypoglycemic effect of the lyophilized aqueous extract of *Ajuga iva* in normal and streptozotocin diabetic rats. J Ethnopharmacol 80:109–113
El-Hilaly J, Tahraoui A, Israili ZH, Lyoussi B (2006) Hypolipidemic effects of acute and subchronic adminis-tration of an aqueous extract of Ajuga iva L. whole plant in normal and diabetic rats. J Ethnopharmacol 105:441–448
El-Rhaffari L, Zaid A (2002) Pratique de la phytothérapie dans le sud-est du Maroc (Tafilalet): Un savoir empirique pour une pharmacopée rénovée. Des sources du savoir aux médicaments du futur 1:293–318
Esmaeili MA, Yazdanparast R (2004) Hypoglycaemic effect of *Teucrium polium*: studies with rat pancreatic islets. J Ethnopharmacol 95:27–30
Es-Safi NE, Khlifi S, Kollmann A, Kerhoas L, El Abbouyi A, Ducrot PH (2006) Iridoid glucosides from the aerial parts of *Globularia alypum* L. (Globulariaceae). Chem Pharm Bull 54:85–88
Es-Safia NE, Kollmann A, Khlifi S, Ducrot PH (2007) Antioxidative effect of compounds isolated from *Globularia alypum* L. structure – activity relationship. LWT 40:1246–1252
Essawi T, Srour M (2000) Screening of some Palestinian medicinal plants for antibacterial activity. J Ethnopharmacol 70:343–349
Facciola S (1990) Cornucopia – a source book of edible plants. Kampong Publications, Vista
Faust RE, Cwalina G, Ramstad E (1958) The antineoplastic action of chemical fractions of the fruit of *Citrullus colocynthis* on sarcoma-37. J Am Pharm Assoc 47(1):1–5
Feuerstein I, Müller D, Hobert K, Danin A, Segal R (1986) The constitution of essential oils from *Artemisia herba-alba* populations of Israel and Sinaï. Phytochemistry 25:2343–2347
Fici S (2001) Intraspecific variation and evolutionary trends in *Capparis spinosa* L. (Capparaceae). Plant Syst Evol 228(3–4):123–141
Figliuolo G, Mang SM (2010) Characterization and discrimination of Mediterranean bulb-producing garlic. In: Acurar MP, Krejci G (eds) *Garlic consumption and health*. Nova Science Publishers, Inc., New York, pp 181–197
Figueiredo AC, Barroso JG, Pedro LG (2008) *Tymus capitatus* grown in Tunisia: antioxidant ability of the essential oils on lilnoleic acid evaluated by different methods. Acta Hortic (ISHS) 765:315–324
Food and Drug Administration (FDA) Center for Food Safety and Applied Nutrition (1994) Adverse events with Ephedra and other botanical dietary supplements
Fragiska M (2005) Wild and cultivated vegetables, herbs and spices in Greek antiquity. Environ Archaeol 10(1):73–82
Freitag H, Maier-Stolte M (1989) The *Ephedra* species of P. Forsskal: identity and typification. Taxon 38:545–556
Freitag H, Maier-Stolte M (1994) Ephedraceae. In: Browicz K (ed) Chorology of trees and shrubs in south-west Asia and adjacent regions. Bogucki Publishers, Poznan
Gachkar L, Yadegari D, Rezaei MB, Taghizadeh M, Astaneh SA, Rasooli I (2007) Chemical and biological characteristics of *Cuminum cyminum* and *Rosmarinus officinalis* essential oils. Food Chem 102:898–904
Gadgoli C, Mishra SH (1999) Antihepatotoxic activity of p-methoxy benzoic acid from *Capparis spinosa*. J Ethnopharmacol 66:187–192

Galati EM, Mondello MRD, Aquino A, Miceli N, Sanogo R, Tzakou O (2000) Effects of *Teucrium divaricatum* Heldr. ssp. *divaricatum* decoction on experimental ulcer in rats. J Ethnopharmacol 72:337–342

Gao R, Gao C, Tian X, Yu X, Di X, Xiao H, Zhang X (2004) Insecticidal activity of deoxypodophyllotoxin, isolated from *Juniperus sabina* L, and related lignans against larvae of Pieris rapae L. Pest Manag Sci 60:1131–1136

Germano MP, De Pasquale R, D'Angelo V, Catania S, Silvaria V, Costa C (2002) Evaluation of extracts and isolated fraction from *Capparis spinosa* L. buds as an antioxidant source. J Agric Food Chem 50:1168–1171

Gharaibeh MN, Elayan HH, Salhab AS (1988) Hypoglycemic effects of Teucrium polium. J Ethnopharmacol 24(1):93–99

Ghazal S, Abuzarqa MA, Mahasneh MA (1992) Antimicrobial activity of *Polygonum equisetiforme* extracts and flavonoids. Phytother Res 6(5):265–269

Ghedira K, Chemli R, Richard B, Nuzillard JM, Zeches M (1993) Two cyclopeptide alkaloids from *Zizyphus lotus*. Phytochemistry 32(6):1591–1594

Giancarlo S, Rosa LM, Nadjafi F, Francesco M (2006) Hypoglycaemic activity of two spices extracts: *Rhus coriaria* L. and *Bunium persicum* Boiss. Nat Prod Res 20:882–886

Gil de Meister E, Hoffman FR (1956) Die Atherischen Oll le, vol 4. Akademie-Verlag, Berlin, pp 296–297

Giordani R, Hadef Y, Kaloustian J (2008) Compositions and antifungal activities of essential oils of some Algerian aromatic plants. Fitoterapia 79:199–203

Girre L (2000) Infusions et Plantes de Santé en France. Ed. Tours, Ouest-France, p 60

Glasby JS (1978) Encyclopedia of the alkaloids. Plenum Press, London, pp 58–661

Gomes I, Leyens T, Da Luz B, Costa J, Gonçalves F (1999) New data on the distribution and conservation status of some angiosperms of the Cape Verde Islands, W Africa. Willdenowia 29:105–114

Goodner KL, Mahattanatawee K, Plotto A, Sotomayor JA (2006) Aromatic profiles of *Thymus hyemalis* and Spanish *T. vulgaris* essential oils by GC-MS/GC-O. Ind Crop Prod 24(3):264–268

Gorai M, Maraghni M, Neffati M (2010) The relationship between phenological traits and water potential patterns of the wild jujube *Ziziphus lotus* in southern Tunisia. Plant Ecol Divers 3:273–280

Goren AC, Bilsela G, Bilsela M, Demira H, Kocabas EE (2003) Analysis of essential oil of *coridothymus capitatus* (L.) and its antibacterial and antifungal activity. Z Naturfo B 58C:687–690

Granger R, Passet J (1973) *Thymus vulgaris* spontané de France: Races chimiques et chemotaxonomie. Phytochemistry 12(7):1683–1691

Grim F, Bounaga N (1996) Flavonic polymorphism and chemosystematics of the genus *Zygophyllum* studied in the Algerian Sahara. Acta Pharma 46:187–194

Guarrera PM (1999) Traditional anthielmintic, antiparasitic and repellent uses of plants in Central Italy. J Ethnopharmacol, 68(1–3):183–192.

Guenaoui C, Mang S, Figliuolo G, Neffati M (2012) Diversity in *Allium ampeloprasum*: from small and wild to large and cultivated. Genet Resour Crop Evol 53:19–25

Guenther E (1950) The essential oils, 3rd edn. D. Van Nostrand, New York

Guetat A, Vilatersana R, Neffati M, Bousaid M (2010) Genetic diversity in Tunisian rosy garlic populations (*Allium roseum* L.) as evidenced by chloroplastic DNA analysis: Sequence variation of non-coding region and intergenic spacers. Biochem Syst Ecol 38(4):502–509

Gurley BJ, Wang P, Gardner SF (1998) Ephedrine-type alkaloid content of nutritional supplements containing *Ephedra sinica* (Ma-huang) as determined by high performance liquid chromatography. J Pharm Sci 87:1547–1553

Haber B (2002) Carob fibre benefits and applications. Cereal Foods World 47:365–369

Hacıseferoğullari H, Őzcan M, Demir F, Çalışır S (2005) Some nutritional and technological properties of garlic (*Allium sativum* L.) J Food Eng 68:463–469

Hajji M, Masmoudi O, Ellouz-Triki Y, Siala R, Gharsallah N, Nasri M (2009) Chemical composition and antioxidant and radical-scavenging activities of *Periploca laevigata* root bark extracts. J Sci Food Agric 89:897–905

Hajji M, Masmoudi O, Triki Y, Souissi N, Kammoun S, Nasri M (2010) Chemical composition, angiotensin I-converting enzyme (ACE) inhibitory, antioxidant and antimicrobial activities of the essential oil from *Periploca laevigata* root barks. Food Chem 121(3):724–731

Haller CA, Benowitz NL (2000) Adverse cardiovascular and central nervous system events associated with dietary supplements containing ephedra alkaloids. N Engl J Med:3431833–3431838

Hamilton AC, Hamilton PB (2006) Plant conservation: an ecosystem approach. Earthscan, London

Hammami R, Zouhir A, Ben Hamida J, Neffati M, Vergoten G, Naghmouchi K, Fliss I (2009) Antimicrobial properties of aqueous extracts from three medicinal plants growing wild in arid regions of Tunisia. Pharm Biol 47(5):452–457

Han BH, Park MH (1986) Folk medicine: the art and science. The American Chemical Society, Washington, p 205

Harborne JB (1969) Chemosystematics of the leguminosae. Flavonoid and Isoflavonoid patterns in the tribe genisteae. Phytochemistry 8(8):1449–1456

Harborne JB, Mabry TJ, Mabry H (1975) The Flauonoids. Chapman & Hall, London

Hardman JG, Limbird LE (2001) The pharmacological basis of therapeutics, 10th edn. McGraw-Hill, New York

Hartwell JL (1982) In plant used against cancer a survey. Quarterman Publication, Inc., Lawrence

Hashem FA, Saleh MM (1999) Antimicrobial components of some cruciferae plants (*Diplotaxis harra* Forsk. and *Erucaria microcarpa* Boiss.) Phytother Res 13:329–332

Hassan MM, Muhtadi FJ, Al-Badr AA (1979) Antispasmodic GLC-mass spectrometry of *Teucrium polium* oil. J Pharm Sci 68:800–801

Hassanean HA, El-Hamouly MMA, El-Moghazy SA, Bishay DW (1993) 14-decarboxyquinovic and quinovic acid glycosides from *Zygophyllum album*. Phytochemistry 33:667–670

Hassanien Bey AM (1925) The lost oases. Century Co., New York/London. pp xxiii, 363

Hassar M (1999) La phytothérapie au Maroc. Espérance Médicale 47:83–85

Hatimi S, Boudouma M, Bichichi M, Chaib N, Idrissi NG (2000) Evaluation *in vitro* of antileishmanien activity of *Artemisia herba-alba* Asso. (in French). Presented at Franco-African meeting of pediatrics N°14, Paris, France. Exotic pathology Society, Paris, pp 57–70

Hatimi S, Boudouma M, Bichichi M, Chaib N, Idrissi NG (2001) Bull Soc Pathol Exot 94:29

Havey MJ (1999) Advances in new *Alliums*. In: Janick J (ed) *Perspectives on new crops and new uses*. ASHS Press, Alexandria

Havsteen B (1983) Flavonoids, a class of natural products of high pharmacological potency. Biochem Pharmacol 32:1141–1148

Hayouni E, Abedrabba M, Bouix M, Hamd M (2007) The effects of solvents and extraction method on the phenolic contents and biological activities in vitro of Tunisian Quercus *coccifera* L. and *Juniperus phoenicea* L. fruit extracts. Food Chem 105:1126–1134

Hazzit M, Baaliouamer A, Verıssimo AR, Faleiro ML, Miguel MG (2009) Chemical composition and biological activities of Algerian *Thymus* oils. Food Chem 16:714–721

Hetwood VH (1971) The biology and chemistry of the Umbelliferae. Academic Press, New York

Hooper D (1937) Useful plants and drugs of Iran and Iraq. Botanical series, Publication 387, Field Museum of Natural History, Chicago, p 94

Horbowicz M, Kotlińska T (2000) Level of flavonols in wild and cultivated *Allium* species. Acta Hortic 517:375–380

Hoult JRS, Moroney MA (1994) Paya, M. Methods Enzymol 234:443

Hudaib M, Aburjai T (2006) Composition of the Essential Oil from *Artemisia herba-alba* Grown in Jordan. J Essent Oil Res 18:301–304

Hudaib M, Speroni E, Di Pietra AM, Cavrini V (2002) GC/MS evaluation of thyme (*Thymus vulgaris* L.) oil composition and variations during the vegetative cycle. *J. Pharmaceut. Biomed.* Analysis 29(4):691–700

Iauk L, Mangano K, Rapisarda A, Maiolino L, Musumeci R, Costanzo R, Serra A (2004) Protection against murine endotoxemia by treatment with *Ruta chalepensis* L., a plant with anti-inflammatory properties. J Ethopharmacol 90:267–272

Infantino A, Tomassoli L, Peri E, Colazza S (2007) Viruses, fungi and insect pests affecting caper. Eur J Plant Sci Biotechnol 1(2):170–179

Inocencio C, Rivera D, Alcaraz F, Tomas-Barberan FA (2000) Flavonoid content of commercial capers (*Capparis spinosa*, *C. sicula*, and *C. orientalis*) produced in Mediterranean countries. *Eur. Food Res*. Dent Tech 212:70–74

Inocencio C, Alcaraz F, Calderón F, Obón C, Rivera D (2002) The use of floral characters in Capparis sect. Capparis' to determine the botanical and geographical origin of capers. Eur Food Res Technol 214(4):335–339

Inocenio C, Rivera D, Obon MC, Alcaraz F, Barrena JA (2006) A systematic revision of *Capparis* section *Capparis* (Capparaceae). Ann Mo Bot Gard 93:122–149

Ismaili H, Tortora S, Sosa S, Fkih-Tetouani S, Ilidrissi A, Della Loggia R, Tubaro A, Aquino R (2001) Topical anti-inflammatory activity of *Thymus willdenowii*. J Pharm Pharmacol 53:1645–1652

Ismaili H, Sosa S, Brkic D, Fkih-Tetouani S, Ilidrissi A, Touati D, Aquino R, Tubaro A (2002) Topical anti-inflammatory activity of extracts and compounds from *Thymus broussonettii*. J Pharm Pharmacol 54:1137–1140

Ismaili H, Milella L, Fkih-Tetouani S, Ilidrissi A, Camporese A, Sosa S, Altinier G, Della Loggia Aquino R (2004) *In vivo* topical anti-inflammatory and in vitro antioxidant activities of two extracts of *Thymus satureioides* leaves. J Ethnopharmacol 91:31–36

Isman MB, Wilson JA, Bradbury R (2008) Insecticidal activities of commercial rosemary oils (*Rosmarinus officinalis* L.) against *Trichoplusiani* in relation to their chemical compositions. Pharm Biol 46(1–2):82–87

Itokawa H, Xu JP, Takeya K (1988) Studies on chemical constituents of antitumor fraction from *Periploca sepium*. IV. Structures of new pregnane glycosides, periplocosides D, E, L, and M. Chem Pharm Bull 36:2084–2089

IUCN (2005) A guide to medicinal plants in North Africa. ISBN: 2, 8317-0893-1. p 256

Jaafari A, Ait Mouse H, El Rakib M, Ait L, Barek M, Tilaoui M, Benbakhta C, Boulli A, Abbad A, Zyad A (2007) Chemical composition and antitumor activity of different wild varieties of Moroccan *thyme*. Rev Bras Farm 17(4)

Jafri SMH (1977) Rhamnaceae. In: Jafri SMH, EL-Gadi A (eds) Flora of Libya vol 30. Al-Faateh University, Department of Botany, Tripoli, pp 1–13

Jager AK, Schottlander S, Smitt UW, Nyman U (1993) Somatic embryogenesis in cell cultures of *Thapsia garganica*: correlation between the state of differentiation and the content of thapsigargins. Plant Cell Rep 12:517–520

Jagtap SD, Deokule SS, Bhosle SV (2006) Ethnopharmacological communication some unique ethnomedicinal uses of plants used by the Korku tribe of Amravati district of Maharashtra. India J Ethnopharmacol 107(03):463–469

Jamshidi R, Afzali Z, Afzali D (2009) Chemical composition of hydrodistillation essential oil of rosemary in different origins in Iran and comparaison with other countries. American-Eurasian J Agr Envriron Sci 5(1):78–81

Janick J, Paull RE (2006) The encyclopedia of fruits and nuts. Edited by Janick J, Purdue University, USA, and Paull RE., University of Hawaii at Manoa, USA.; pp 228–231

Janssen MA, Scheffer JJC, Svendsen AB (1987) Antimicrobial activities of essential oils: a 1976–1986 literature review on possible applications. Pharmaceutische Weekblad (Scient Edn) 9:193–197

Jaouhari JT, Lazrek HB (2000) Hypoglycemic activity of *Zygophyllum gaetulum* extracts in alloxan induced hyperglycemic rats. J Ethnopharmacol 69(1):17–20

Jaouhari JT, Lazrek HB, Seddik A, Jana M (1999) Hypoglycaemic response to *Zygophyllum gaetulum* extracts in patients with non-insulin-dependent diabetes mellitus. J Ethnopharmacol 64(3):211–217

Jarraya R, Damak M (2001) Alcaloïdes des feuilles de *Hammada scoparia* (pomel) Iljin. J de la Société Chimique de Tunisie 4(9):941–948

Jarraya R, Chaieb M, Damak M (1993) Screening des plantes à alcaloïdes au sein de la flore Tunisienne. Plant Med Phytother 26:177–189

Jarraya MR, Bouaziz A, Hamdi B, Ben Salah A, Damak M (2008) *N*-methylisosalsoline from *Hammada scoparia*. Acta Cryst E 64:1710–1714

Jones MG, Hughes J, Tregova A, Mline J, Tomsett AB, Collin HA (2004) J Exp Bot 55(404):1903–1918

Josefson D (1995) Herbal stimulant causes US deaths. Br Med J 312:1378–1379

Jouad H, Haloui M, Rhiouani H, ElHilaly J, Eddouks M (2001) Ethnobotanical survey of medicinal plants used for the treatment of diabetes, cardiac and renal diseases in the North centre region of Morocco (Fez-Boulemane). J Ethnopharmacol 77:175–182

Joyeux M, Lobsrein A, Anton R, Mortier F (1995) Planta Med 61:126

Kamel S, Ibrahim L, Afifi A, Hamza (1970) Major alkaloidal constituents of the Egyptian plant. *Peganum harmala*. UARJ Vet Sci 7:71–86

Kanat M, Alma MH (2004) Insecticidal effects of essential oils from various plants against larvae of pine processionary moth (*Thaumetopoea pityocampa* Schiff) (Lepidoptera: Thaumetopoeidae). Pest Manag Sci 60:173–177

Karaman I, Sahin F, Gulluce M, Ogutcu H, Sengul M, Adiguzel A (2003) Antimicrobial activity of aqueous and methanol extracts of *Juniperus oxycedrus* L. J Ethnopharmacol 85:231–235

Kariba RM (2001) Antifungal activity of *Ajuga remota*. Fitoterapia 72:177–178

Karousou R, Koureas DN, Kokkini S (2005) Essential composition is related to the natural habitats: *Coridothymus capitatus* and *satureja thymbra* in NATURA 2000 sites of Crete. Phytochemistry 66:2668–2673

Kassem M, Mosharrafa S, Saleh N (2000) Two new flavonoids from *Retama raetam*. Fitoterapia 71:649–654

Kawano M, Matsuyama K, Miyamae Y, Shinmoto H, Kchouk ME, Morio T, Shigemori H, Isoda H (2007) Antimelanogenesis effect of Tunisian herb *Thymelaea hirsuta* extract on B16 murine melanoma cells. Exp Dermatol 16(12):977–984

Kazmi SU, Ali SN, Jamal SA, Attaur R (1991) New bioactive compounds of plant origin. Pak J Pharm Sci 4(2):113–123

Ketoh GK, Glitho AI, Huignard J (2002) Susceptibility of the bruchid *Callosobruchus maculates* (Coleoptera: Bruchidae) and its parasitoid *Dinarmus basalis* (Hymenoptera: Pteromalidae) to three essential oils. J Econ Entomol 95(1):174–182

Keusgen M, Jinger M, Krest I, Schoning MJ (2003) Sensors Actuators B 95:297–302

Khadri A, Serralheiro MLM, Nogueira JMF, Neffati M, Smiti S, Araujo MEM (2008) Antioxidant and antiacetylcholinesterase activities of essential oils from *Cymbopogon schoenanthus* L. Spreng. Determination of chemical composition by GC–mass spectrometry and C NMR13. Food Chem 109:630–637

Khadri A, Neffati M, Smiti S, Fale P, Rosa A, Lino L, Luisa M, Serralheiro M, Eduarda M, Araujoc M (2010) Antioxidant, antiacetylcholinesterase and antimicrobial activities of *Cymbopogon schoenanthus* L. Spreng (lemon grass) from Tunisia. LWT Food Sci Technol 43:331–336

Khafagi IK, Dewedar A (2000) The efficiency of random versus ethno-directed research in the evaluation of Sinai medicinal plants for bioactive compounds. J Ethnopharmacol 71:365–376

Khakberdyev MM, Mansurov MM, Eshchanov TB (1968) Russian Medskii Zh. Uzbek 12:47–48 (Biol Abstr 50:109453)

Khouri NA, El-Akawi Z (2005) Anti-androgenic activity of *Ruta graveolens* L. in male albino rats with emphasis on sexual and aggressive behavior. Neuroendocrinol Lett 26:823–829

Kiritikar KR, Basu BD (1987) Indian medicinal plants, vol 1. International Book Distributors, Dehradun, pp 195–197

Kiritikar KR, Basu BD (1991) Indian medicinal plants. Periodical Expert Book Agency, Delhi, pp 197–198

Kjaer A, Thomsen H (1963) Isothiocyanate-producing glucosides in species of Capparidaceae. Phytochemistry 2:29–32

Koba K, Poutouli PW, Nenonene YA, Songai MS, Raynaud C, Sanda K (2007) Chemical composition and anti-termite activity of three tropical essential oils against termite species *Trinervitermes geminates* (Wasmann). J Sci Technol 5(2):39–46

Kokwaro JO (1976) Medicinal plants of East Africa. East African Literature Bureau, Nairobi, p 106

Kubo I, Klocke JA, Asano S (1981) Insect ecdysis inhibitors from the East African medicinal plant, Ajuga remota (Labiatae). Agric Biol Chem 45:1925–1927

Kumarasamy Y, Cox PJ, Jaspars M, Nahar L, Sarker SD (2002) Screening seeds of Scottish plants for antibacterial activity. J Ethnopharmacol 83:73–77

Kumazawa S, Taniguchi M, Susuki Y, Shimura M, Kwon MS, Nakayama T (2002) Antioxidant activity of polyphenols in carob pods. J Agric Food Chem 50:373–377

Kuria KAM, Muriuki G (1984) A new cardiotonic agent from Ajuga remota benth (Labiatae). East Afr Med J 61:533–538

Kuria KAM, Chepkwony H, Govaerts C, Roets E, Busson R, de Witte P, Zupko I, Hoornaert G, Quirynen L, Maes L (2002) The antiplasmodial activity of isolates from *Ajuga remota*. J Nat Prod 65:789–793

Kuria KAM, De Coster S, Muriuki G (2001) Antimalarial activity of *Ajuga remota* benth (Labiatae) and *Caesalpinia volkensii* Harms (Caesalpiniaceae): in vitro confirmation of ethnopharmacological use. J Ethnopharmacol 74:141–148

Kusmenoglu S (1996) The plant *Peganum harmala* L. and its biologically active constitutents. FABAD J Pharm Sci 21:71–75

Kutepova TA, Syrov VN, Khushbaktova ZA, Saatov Z (2001) Hypoglycemic activity of the total ecdysteroid extract from *Ajuga turkestanica*. Pharm Chem J 35:608–609

Laamouri A, Zine El Abidine A (2000) Multiplication des jujubiers en Tunisie. Annales de la Recherche Forestière au Maroc 33:37–49

Lahlou M, El Mahi M, Hamamouchi J (2002) Évaluation des activités antifongique et molluscicide de *Zizyphus lotus* (L.) Desf. du Maroc. Ann Pharm Fr 60(6):410–414

Lam SK, Ng TB (2009) A protein with antiproliferative, antifungal and HIV-1 reverse transcriptase inhibitory activities from caper (*Capparis spinosa*) seeds. Phytomedicine 16:444–445

Lam SK, Han QF, Ng TB (2009) Isolation and characterization of a lectin with potentially exploitable activities from caper (*Capparis spinosa*) seeds. Biosci Rep 29(5):293–299

Lamchouri F, Settaf A, Cherrah Y, Zemzami M, Lyoussi B, Zaid A, Hassar M (1999) Antitumor principles from *Peganum harmala* seeds. Therapie 54:753–758

Lamchouri F, Settaf A, Cherrah Y, Hassar M, Zemzami M, Atif N, Lyoussi B (2000) In vitro cell-toxicity of *Peganum harmala* alkaloids on cancerous cell-lines. Fitoterapia 71:50

Lamiri A, Belanger A, Berrada M, Ismaïli-Alaoui MM, Benjilali B (1997a) Origin of chemical polymorphism of Moroccan *Artemisia herba-alba* Asso (in French). Rabat, Morocco, pp 81–92

Lamiri A, Belanger A, Berrada M, Zrira S, Benjilali B (1997b) Chemical polymorphism of *Artemisia herba-alba* Asso from Morocco (in French). Rabat, Morocco, pp 69–79

Lang D (2008) Trade in plant material for medicinal and other purposes. TRAFFIC Bull 17:21–32

Larsen HO, Smith LK (2004) Alpine medicinal plant trade and mountain livelihood strategies. Geogr J 169(3):243–254

Lawrence BM (2003) Progress in essential oils. Perfum Flav 28(2):62

Le Croueour G, Thepenier P, Richard B, Petermann C, Ghedira K, Zeches-Hanrot M (2002) Lotusine G: a new cyclopeptide alkaloid from *Zizyphus lotus*. Fitoterapia 73:63–68

Le Floc'h E (1983) Contribution à une étude ethnobotanique de la flore tunisienne. Programme Flore et Végétation tunisienne. Ed., Ministère de l'Enseignement Supérieur et de la Recherche Scientifique, Tunis, pp 83–85

Le Floch E, Boulos L (2008) Flore de la Tunisie. Catalogue Synonymique Commente´, Montepellier, p 461

Le Houérou HN (1959) Recherches écologiques et floristiques sur la végétation de la Tunisie méridionale. Institut des Recherches Sahariennes, Algérie

Le Houerou HN (1969) La vegetation de la Ttunisie steppique (avec references au Maroc, en Algérie et la Libye). Ann Inst Nat Rech Agric Tunisie 42:642p

Lee JC, Lee KY, Kim JN (2002) Identification of *Rhus verniciflua* Stokes compounds that exhibit free radical scavenging and antiapoptotic properties. Biochim Biophys Acta 1570:181–191

Lee JC, Lee KY, Kim JN (2004) extract from *Rhus verniciflua* Stokes is capable of inhibiting the growth of human lymphoma Murray cells. Food Chem Toxicol 9:1383–1388

Lei H, Wang B, Li WP, Yang Y, Zhou AW, Chen MZ (2003) Anti-ageing effect of astragalo-sides and its mechanism of action. Acta Pharmacol Sin 24:230–234

Lentini F, Aleo M, Amenta R (1993) L'uso popolare delle piantenelle Isole Egadi (Sicilia). G Bot Ital 127(3):702
Leporatti ML, Ghedira K (2009) Comparative analysis of medicinal plants used in traditional medicine in Italy and Tunisia. J Ethnobiol Ethnomed 5:31
Leung AY, Foster S (1996) Encyclopedia of common natural ingredients. John Wiley & Sons, New York
Leung AY (1999) Ephedrine, Ephedra, Mahuang, Mahuanggen what are they? Presented at AHPA Ephedra international symposium, Arlington, VA
Levizon E, Drilias P, Kyparissis A (2004) Exceptional photosynthetic performance of *Capparis spinosa* L. under adverse conditions of Mediterranean summer. Photosynthetica 42:229–235
Li JW, Ding SD, Ding XL (2005) Comparison of antioxidant capacities of extracts from five cultivars of Chinese jujube. Process Biochem 40(11):3607–3613
Lin YM, Flavin MT, Schure R, Chen FC, Sidwell R, Barnard DL (1999) Antiviral activities of bioflavonoids. Planta Med 65:120–125
Liu MJ, Cheng CY (1994) A taxonomic study on Chinese jujube and wild jujube. J Agr Univ Hebei 17:251–263
Liu XM, Zakaria MN, Islam MW, Radhakrishnan R, Ismail A, Chen HB, Chan K, Al-Attas A (2001) Anti-inflammatory and anti-ulcer activity of *Calligonum comosum* in rats. Fitoterapia 72(5):487–491
Loeb H, Vandenplas Y, Wursch P, Guesry P (1989) Tannin-rich carob pod for the treatment of acute-onset diarrhea. J Paediatr Gastroenterol Nutr 8(4):480–485
Lograda T, Ramdani M, Figueredo G, Chalchat JC, Chalard P (2012) Essential oil composition and antimicrobial activity of Genista microcephala Coss. et Dur. Int J Med Arom Plants 2(1):75–79
Mabberley DJ (1997) The plant-book. Cambridge University Press, Cambridge
Mahdhi M, Nzoué A, Gueye F, Merabet C, De Lajudie P, Mars M (2007) Phenotypic and genotypic diversity of *Genista saharae* microsymbionts from the infra-arid region of Tunisia. Lett Appl Microbiol 45(6):604–609
Mahjoub MA, Ammar S, Edziri H, Mighri N, Bouraoui A, Mighri Z (2004) Anti-inflammatory and antioxidant activities of some extracts and pure natural products isolated from Rhus tripartitum. J Soc Alger Chim 14:293–297
Mahjoub MA, Ammar S, Mighri Z (2005) A new biflavonoid and an isobiflavonoid from *Rhus tripartitum.* Nat Prod Res Form Nat Prod Lett 19(8)
Mahjoub MA, Ammar S, Edziri H, Mighri N, Bouraoui A, Mighri Z (2010) Anti-inflammatory and antioxidant activities of some extracts and pure natural products isolated from *Rhus tripartitum* (Ucria). Med Chem Res 19(3):271–282
Mahmoudian M, Jalilpour H, Salehian P (2002) Toxicity of *Peganum harmala*: review and a case report. Iran J Pharmacol Ther 1:1–4
Maire R (1962) Flore de l'Afrique du Nord, vol 8, ed. Paul Lechevalier, Paris, pp 161–164
Maire R (1987) Flore de l'Afrique du Nord, Dicotyledonae, ed. Lechevalier, vol 16. Paris, p 302
Maiza K, Brac de la Perrière RA, Hammiche V (1993) Traditional Saharian pharmacopoeia. Acta Hortic 332:37–42
Makris DP, Kefalas P (2004) Carob pods (*Ceratonia siliqua* L.) as a source of polyphenolic antioxidants. Food Technol Biotechnol 42:105–108
Mangena T, Muyima NYO (1999) Comparative evaluation of the antimicrobial activities of essential oils of *Artemisia afra*, *Pteronia incana* and *Rosmarinus officinalis* on selected bacteria and yeast strains. Lett Appl Microbiol 28:291–296
Marakis S (1996) Carob bean in food and fee: current status and future potentials- a critical appraisal. J Food Sci Technol 33:365–383
Marakis S, Marakis G, Lambraki M (1997) Tannins of eight carob varieties from the island of Lefkada, Greece, Chimica Chronika. New Series 26:57–66
Marie R (1964) Flore de l'Afrique du Nord (Maroc, Algérie, Tunisie, Tripolitaine cyrénaïque et sahara). Paris XIéme, p 156

Marrif HI, Ali BH, Hassan KM (1995) Some pharmacological studies on *Artemisia herba-alba* (Asso.) in rabbits and mice. J Ethnopharmacol 49(1):51–55

Marti D, Villagrasa V, Martinez-Solis I, Blanquer A, Castillo E, Moreno Royo L (2005) Hystological and pharmacological study of *Thymus piperella* (L.) Phytother Res 19:298–302

Martins A, Wink M, Tei A, Brum-Bousquet M, Tillequin F, Rauter P (2005) A phyto-chemical study of the quinolizidine alkaloids from Genista tenera by gas chromatography–mass spectrometry. Phytochem Anal 16:264–266

Mason M (1936) The paradise of fools. Hodder and Stoughton, London, 282 pp

Mastelic J, Kustrak D (1997) Essential oil and glycosidically bound volatiles in aromatic plants. II. Rosemary (*Rosmarinus officinalis* L., Lamiaceae). Acta Pharmaceut 47(2):139–142

Matsuyama K, Villareal MO, El Omri A, Han J, Kchouk ME, Isoda H (2009) Effect of Tunisian *Capparis spinosa* L. extract on melanogenesis in B16 murine melanoma cells. J Nat Med 63:468–472

Mbark AN, Charrouf Z, Wieruszeki JM, Leroy Y, Kol O (1995) Herniaria Saponin A, A Novel Saponin from *Herniaria Fontanesii*. Nat Prod Lett 6(3)

Mbark AN, Guillaume D, Cherrah Y, Wieruszeki JM, Ricart G, Charrouf Z (1996) An acylated isorhamentin glycoside from *Herniaria fontanesii*. Phytochemistry 41(3):989–990

Mbark AN, Charrouf Z, Guillaume D, Kol O (1999) New glycosidesfrom *Herniaria fontanesii*. Stud Plant Sci 6:314–319. Adv Plant Glycos Chem Biol. Proceedings of the international symposium in plant glycosides

McCutcheon AR, Ellis SM, Hancock RE, Towers GH (1992) Antibiotic screening of medicinal plants of the British Columbian native peoples. J Ethnopharmacol 37:213–223

Meister A, Bernhardt G, Christoffel V, Buschauer A (1999) Antispasmodic activity of *Thymus vulgaris* extract on the isolated guinea-pig trachea: discrimination between drug and ethanol effects. Planta Med 65:512–516

Mekkiou R, Touahar H, Dijoux-Franca MG, Mariotte AM, Benayache S, Benayache F (2005) A new isoflavone from Genista saharae (Fabaceae). Biochem Syst Ecol 33:635–638

Meot-Duros L, Magné C (2009) Antioxidant activity and phenol content of *Crithmum maritimum* L. leaves. Plant Physiol Biochem 47:37–41

Merzouki A, Derfoufi A, El-Aallau A, Molero-mesa J (1997) Wild medicinal plants used by local Bouhmed population (Morocco). Fitoterapia 68:444–460

Mighri H, Akrout A, El-jeni H, Zaidi S, Tomi F, Casanova J, Neffati M (2010a) Composition and intraspecific chemical variability of the essential oil from *Artemisia herba alba* growing wild in a Tunisian arid zone. Chem Biodivers 7:2709–2717

Mighri H, Hajlaoui H, Akrout A, Najjaa H, Neffati M (2010b) Antimicrobial and antioxidant activities of *Artemisia herba-alba* essential oil cultivated in Tunisian arid zone. C R Chim 13:380–386

Milos M, Radonic A (2000) Gas chromatography mass spectral analysis of free and glycosidically bound volatile compounds from *Juniperus oxycedrus* L. growing wild in Croatia. Food Chem 68:333–338

Miresmailli S, Bradbury R, Isman MB (2006) Comparative toxicity of *Rosmarinus officinalis* L. essential oil and blends of its major constituents against *Tetranychus urticae* Koch (Acari: Tetranychidae) on two different host plants. Pest Manag Sci 62:366–371

Mittler R, Merquiol E, Hallak-Herr E, Rachmilevitch S, Kaplan A, Cohen M (2001) Living under a "dormant" canopy: a molecular acclimation mechanism of the desert plant *Retama raetam*. Plant J 25:407–416

Mizrahi I, Juarez MA, Bandoni A (1991) The essential oils of *Rosmarinus officinalis* growing in Argentina. J Esent Oil Res 3:11–15

Mkaddem MG, Romdhane M, Ibrahim H, Ennajar M, Lebrihi A, Mathieu F, Bouajila J (2011) Essential oil of *Thymus capitatus* Hoff. et Link. from Matmata Tunisia: gas chromatography–mass spectrometry analysis and antimicrobial and antioxidant activities. J Med Food 13:1500–1504

Moghtader M, Afzali D (2009) Study of the antimicrobial properties of essential oil of rosemary. American-Eurasian J Agr Envriron Sci 5(3):393–397

Morales R (1996) Studies on the genus *Thymus* L. Lamiales. Newsletter 4:6–8
Morales R (1986) Taxonom´ıa de los g´eneros Thymus (excluida la secci´on Serpyllum) y Thymbra en la Pen ´ınsula Ib´erica. Ruizia, 3,Monograf´ıas Del Real Jard´ın Bot´anico, Madrid
Munir C, Zaidi MI, Ahmad N, Rahman AU (1993) An easy rapid metal mediated method of isolation of harmine and harmaline from *Peganum harmala*. Fitoterapia 1:73–76
Munoz-Reinoso JC (2004) Diversity of maritime juniper woodlands. Forest Ecol Manag 192:267–276
Murray RDH, Mtndez J, Brown SA (1982) The natural coumarins: occurrence, chemistry and biochemistry, Wiley, Chichester
Nabli A (1989) Essai de synthèse sur la végétation et la phyto-écologie tunisienne. Tunis Imp. Off. Rep, Tunisia, p 186
Nabli MA (1995) Essai de synthèse sur la végétation et la phyto-écologie tunisiennes. II et III. Le milieu physique et la végétation. Ecologie végétale appliquée, No. 5, 6 I.O.R., Tunisia
Nadkarni AK (1954) *Peganum harmala* Linn. Indian Materia Medica, 3rd Edition, vol 1. G. R. Bhatkal and Co., Bombay Popular Book Depot, Bombay, pp 927–929, 1319 pp
Najjaa H, Neffati M, Zouari S, Emmar E (2007) Essential oil composition and antibacterial activity of different extracts of *Allium roseum* L., a North African endemic species. CR Chimie 10:820–826
Najjaa H, Zouari S, Ammar E, Neffati M (2010) Phytochemical screening and antibacterial properties of *Allium roseum* L. a wild edible species in North Africa. J Food Biochem 35:699–714
Najjaa H, Zouari S, Arnault I, Auger J, Ammar E, Neffati M (2011a) Différences et similitudes des métabolites secondaires chez deux espèces du genre *Allium, Allium roseum* L. et *Allium ampeloprasum* L. Acta Bot Gallica 158(1):111–123
Najjaa H, Zria K, Fattouch S, Ammar E, Neffati M (2011b) Antioxidant and antimicrobial activities of *Allium roseum* L. "Lazoul," a wild edible endemic species in North Africa. Prop Sci Food 14:371–380
Nawwar MAM, El Sissi HI, Barakat HH (1984) Flavonoid constituents of *Ephedra alata*. Phytochemistry 23:2937–2939
Nawwar MAM, Barakat HH, Buddrus J, Linscheid M (1985) Alkaloidal, lignan and phenlic constituents of *Epherdra alata*. Phytochemistry 24(4):818–819
Neffati A, Skandrani I, Sghaier MB, Bouhlel I, Kilani S, Ghedira K, Neffati M, Cherif I, Hammami M, Chekir-Ghedira L (2008) Chemical composition, mutagenic and antimutagenic activities of essential oils from (Tunisian) *Artemisia campestris* and *Artemisia herba-alba*. J Essent Oil Res 20(5):471–477
Neffati A, Bouhlela I, Ben Sghaiera M, Boubakera J, Limema I, Kilani S, Skandrania I, Bhouria W, Le Dauphinc J, Barillier D, Mosrati R, Chekir-Ghediraa L, Ghediraa K (2009) Antigenotoxic and antioxidant activities of *Pituranthos chloranthus* essential oils. Environ Toxicol Pharmacol 27:187–194
Newall CA, Anderson LA, Phillipson JD (1996) Herbal medicines. The Pharmaceutical Press, London, p 165
Nguyen DV, Takascova M, Jakubik T, Dang MN (2000) Antioxidant effect of thyme in rape-seed oil. Biologia 55:277–281
Odek-Ogunde M, Rajab MS, Migwi GJ, Ndegwa JM (1993) Blood pressure responses to an extract of *Ajuga remota* in experimentally hypertensive rats. Planta Med 59:573–574
O'Hearn E, Molliver ME (1993) Degeneration of purkinje cells in parasagittal zones of the cerebellar vermis after ibogaine or harmaline. Neuroscience 55:303–310
Ojala T, Remes S, Pasi Haansuu P, Vuorela H, Hiltunen R, Haahtela K, Vuorela P (2000) Antimicrobial activity of some coumarin containing herbal plants growing in Finland. J Ethnopharmacol 73(1–2):299–305
Ojewole JAO (2008) Anticonvulsant effect of *Rhus chirindensis* (Baker F.) (Anacardiaceae) stem-bark aqueous extract in mice. J Ethnopharmacol 117(1):130–135
Olchowik E, Sciepuk A, Mavlyanov S, Abdullajanova N, Zamaraeva M (2012) Antioxidant capacities of polyphenols from Sumac (*Rhus typhina* L.) leaves in protection of erythrocytes against oxidative damage. Biomed Prev Nutr 2(2):99–105

Olesen C, Sorensen TLM, Nielsen RC, Moller JV, Nissen P (2004) Dephosphorylation of the calcium pump coupled to counterion occlusion. Science 306:2251–2255
Oliva A, Lahoz E, Contillo R, Aliotta G (1999) Fungistatic activity of Ruta graveolens extract and its allelochemicals. J Chem Ecol 25:519–526
Oliva A, Meepagala KM, Wedge DE, Harries D, Hale AL, Aliotta G, Duke SO (2003) Natural fungicides from *Ruta graveolens* L. leaves, including a new Quinolone Alkaloid. J Agric Food Chem 51(4):890–896
Ormancey X, Sisalli S, Coutiere P (2001) Formulation of essential oils in functional perfumery. Parfums, Cosmetiques, Actualites 157:30–40
Ould El Hadj MD, Hadj-Mahammed M, Zabeirou H (2003) Place des plantes spontanées dans la médicine traditionnelle de la région d'Ouargla (Sahara septentrional est). Courrier du Savoir 3:47–51
Oury P (1984) Encyclopedia of medicinal plants and flowers, vol IV. Eds Alliance, Paris. (in French)
Owen SM, Harley P, Guenther A, Hewitt CN (2002) Light dependency of VOC emissions from selected Mediterranean plant species. Atmos Environ 36(19):3147–3159
Ozcan M (2000) The use of yogurt as starter in rock sapphire (*Crithmum maritimum* L.) fermentation. Z Eur Food Res Technol 210:424–426
Ozcan M (2005) Mineral composition of different parts of *Capparis ovata* Desf. Growing wild in Turkey. J Med Food 8:405–407
Ozcan M, Akgul A (1998) Influence of species harvest date and size on composition of capers (*Capparis* spp.) flower buds. Nahrung 42:102–105
Ozcan MM, Pedro LG, Figueiredo AC, Barroso JG (2006) Constituents of the essential oil of sea fennel (*Crithmum maritimum* L.) growing wild in Turkey. J Med Food 9(1):128–130
Ozenda P (1991) Flore et végétation du Sahara, CNRS éditions
Pandey RC, Pathak RA, Pathak RK (1990) Physico-chemical changes associated with growth and development of fruits in ber (*Zizyphus mauritiana* Lamk). Indian J Hortic 47(3):286–290
Pareek OP (2002) Ber-ziziphus Mauritania. Available by International centre for underutilized crops (online). Available: http://www.civil.soton.ac.uk/icuc/cd-icuc-ber-tamarind/content/er/ber-book-html via the INTERNET. Accesed 6 Apr 2004
Park Y, Moon B, Yang H, Lee Y, Lee E, Lim Y (2007) Complete assignments of NMR data of 13 hydroxymethoxyflavones. Magn Reson Chem 45:1072–1075
Pathak S, Multani AS, Banerji P, Banerji P (2003) Ruta 6 selectively induces cell death in brain cancer cells but proliferation in normal peripheral blood lymphocytes: a novel treatment for human brain cancer. Int J Oncol 23:975–982
Pearson HHW (1929) Gnetales. Cambridge University Press, Cambridge
Perez-Alonso MJ, Valasco NA, Emin DM, Harmandar M, Esteben JL (1995) Composition of the essential oils of *Ocimum basilicum* var. *glabratum* and *Rosmarinus officinalis* from Turkey. J Esent Oil Res 7(1):73–75
Pérez-Olleros L, Garcia-Cuevas M, Ruiz-Roso B (1999a) Influence of pulp and natural carob fiber on some aspects of nutritional utilization and lipidaemia. Food Sci Technol Int 5:425–430
Pérez-Olleros L, Garcia-Cuevas M, Ruiz-Roso B, Requejo A (1999b) Comparative study of natural carob fibre and psyllium husk in rats. Influence on Some Aspects of Nutritional Utilization and Lipidaemia. J Sci Food Agric 79:173–178
Perfumi M, Valentini G, Bellomaria B, Biondi E (1999) J Essent Oil Res 11:223
Pignatti S (1982) Flora d'Italia. Edagricole, Bologna 1:636–644
Pino JA, Estraron M, Fuentes C (1998) Essential oil of Rosemary (*Rosmarinus officinalis* L.) from Cuba. J Esent Oil Res 10(1):111–112
Pintore G, Usai M, Bradesi P, Juliano C, Boatto G, Tomi F, Chessa M, Cerri R, Casanova J (2002) Chemical composition and antimicrobial activity of *Rosmarinus officinalis* L. oils from Sardinia and Corsica. Flavour Fragr J 17:15–19
Piozzi F, Rodriguez B, Savona G (1987) Advances in the chemistry of the furanoditerpenoids from *Teucrium* species. Heterocycles 25:807–841

Piozzi F, Bruno M, Rosselli S, Maggio A (2006) The diterpenoids of the genus *Marrubium* (Lamiaceae). Nat Prod Commun 11(7):585–592

Pirzada AJ, Shaikh W, Kazi TG (2007) Isolation of essential elements and inhibition activity of medicinal plant *Rhazya stricta* Dcne. against dermatophytic fungi. Pak J Agri Agril Eng Vet Sci 23(1):34–38

Pistelli L, Fiumi C, Morelli I (2003) Flavonoids from Calicotom villosa. Fitoterapia 74:417–419

Pitarokili D, Couladis M, Petsikos-Panayotarou N, Tzakou O (2002) Composition and antifungal activity on soil-borne pathogens of the essential oil of *Salvia sclarea* from Greece. J Agric Food Chem 50:6688–6691

Pöllmann K, Gagel S, Elgamal MHA, Shaker KH, Seifert K (1997) Triterpenoid saponins from the roots of *Zygophyllum* species. Phytochemistry 44:485–489

Polunin O, Huxley A (1972) Flowers of the Mediterranean. Publications of Chatto and Windos Ltd, London

Pottier-Alapetite G (1979) Flore de la Tunisie. Angiosperme–Dicotylédones. Vol. I: Apétales–Dialypétales. Ministère de l'Enseignement Supérieur et de la Recherche Scentifique et le Ministère de l'Agriculture, Tunis, Tunisie

Pottier-Alapetite G (1981) Flore de la Tunisie. Angiosperme–Dicotylédones. Vol. II: Gamopétales. Ministère de l'Enseignement Supérieur et de la Recherche Scentifique et le Ministère de l'Agriculture, Tunis, Tunisie

Prajapati V, Tripathi AK, Aggarwal KK, Khanuja SPS (2005) Insecticidal, repellent and oviposition-deterrent activity of selected essential oils against *Anopheles stephensi*, *Aedes aegypti* and *Culex quinquefasciatus*. Bioresour Technol 96:1749–1757

Preethi KC, Kuttan G, Kuttan R (2006) Antitumor activity of *Ruta graveolens* extract. Asian Pac J Cancer Prev 7:439–443

Price RA (1996) Systematics of the Gnetales: a review of morphological and molecular evidence. Int J Plant Sci 157(Suppl. 6):S40–S49

Puzii AD, Vecherkin SS, Tribunskii MP, Romakhov VG (1980) Toxicity of the combined alkaloids of harmala *(Peganum harmala,* Zygophyllaceae). Veterinariia 4:57–58

Quezel P, Santa S (1962) Nouvelle Flore de l'Algérie et des Régions Désertiques, Méridionales, tome II. Edition CNRS, Paris

Raghav SK, Gupta B, Agrawal C, Goswami K, Das H (2006) Antiinflammatory effect of *Ruta graveolens* L. in murine macrophage cells. J Ethnopharmacol 104:234–239

Rasekh HR, Khoshnood-Mansourkhani MJ, Kamalinejad M (2001) Hypolipidemic effects of *Teucrium polium* in rats. Fitoterapia 72:937–939

Rasmussen U, Christensen SB, Sandberg F (1978) Thapsigargin and Thapsigargicin, two new Histamine Liberators from *Thapsia garganica* L. Acta Pharm Suee 15:133

Rasmussen U, Christensen SB, Sandberg F (1981) Phytochemistry of the genus *Thapsia*. Planta Med 43:336–341

Raul NC Jr, Lucinewton SM, Paulo TVR, Meireles MAA (2005) Supercritical fluid extraction from rosemary (*Rosmarinus officinalis*): kinetic data, extract's global yield, composition, and antioxidant activity. J Supercrit Fluids 35:197–204

Rauter AP, Martins A, Lopes R, Ferreira J, Serralheiro LM, Araujo ME, Borges C, Justino J, Silva FV, Goulart M, Thomas-Oates J, Rodrigues JA, Edwards E, Noronha JP, Pinto R, Mota-Filip H (2009) Bioactivity studies and chemical profile of the antidiabetic plant *Genista tenera*. J Ethnopharmacol 122:384–393

Razzaghi-Abyaneh M, Shams-Ghahfarokhi M, Rezaee MB, Jaimand K, Alinezhad S, Saberi R, Yoshinari T (2009) Chemical composition and antiaflatoxigenic activity of Carum carvi L., *Thymus vulgaris* and *Citrus aurantifolia* essential oils. Food Control 20:1018–1024

Reische DW, Lillard DA, Eintenmiller RR (1998) Antioxidants in food lipids. In: Ahoh CC, Min DB (eds) Chemistry, nutrition and biotechnology. Marcel Dekker, New York, pp 423–448

Ren J (2001) The study on soil seed banks of desert vegetation and germination ecology of *Calligonum* plants. Ph.D. thesis, Environment and Engineering Institute of Cold and Arid Regions, The Chinese Academy of Sciences, P. R. China (in Chinese with English abstract)

Renault JH, Ghedira K, Thepenier P, Lavand C, Zeches-Hanrot M, Le Men-Olivier L (1997) Dammarane saponins from *Zizyphus lotus*. Phytochemistry 44:1321–1327

Rezzi S, Cavaleiro C, Bighelli A, Salgueiro L, Hnio A, Da Cunha P, Casanova J (2001) Intraspecic chemical variability of the leaf essential oil of *Juniperus phoenicea* subsp. *turbinata* from Corsica. Biochem Syst Ecol 29:179–188

Rezzoug SA, Baghdadi MW, Isouka N, Boutekedjiret C, Allaf K (1998) Study of a new extraction process: controlled instantaneous decompression. Application to the extraction of essential oil from rosemary leaves. Flavour Fragr J 13(4):251–258

Richard H, Benjilali B, Banquour N, Baritaux O (1985) Etude de diverses huiles essentielles de thym du Maroc. Lebensm-Wiss Technol 18:105–110

Rigano D, Formisano C, Basile A, Lavitola A, Senatore F, Rosselli S, Bruno M (2006) Antibacterial activity of flavonoids and phenylpropanoids from *Marrubium globosum* ssp. *libanoticum*. Phytother Res 21:395–397

Rigano D, Venera C, Formisano C, Maldini MT, Piacente S, Bevilacqua J, Russo A, Senatore F (2009) *Genista sessilifolia* DC. and *Genista tinctoria* L. inhibit UV light and nitric oxide – induced DNA damage and human melanoma cell growth. Chem Biol Interact 180:211–219

Rima K, Hao Z, Wei C (2011) Antimicrobial and antioxidant activities of Chinese sumac (*Rhus typhina* L.) fruit extract. Food Control 22(1):128–132

Rimbau V, Cerdan C, Vila R, Iglesias J (1999) Antiinflammatory activity of some extracts from plants used in the traditional medicine of north-African countries (II). Phytother Res 13(2):128–132

Rivera D, Inocencio C, Obon C, Alcaraz F (2003) Review of food and medicinal uses of *Capparis* L. subgenus *Capparis* (Capparidaceae). Econ Bot 57:515–534

Rodrigo M, Lazaro MJ, Alvarruiz A, Giner V (1992) Composition of capers (*Capparis spinosa*): influence of cultivar, size and harvest date. J Food Sci 57:1152–1154

Romeo V, Ziino M, Giuffrida D, Condurso C, Verzera A (2007) Flavour profile of capers (*Capparis spinosa* L.) from the Eolian archipelago by HS-SPME/GC-MS. Food Chem 101:1272–1278

Rota C, Carraminana JJ, Burillo J, Herrera A (2004) *In vitro* antimicrobial activity of essential oils from aromatic plants against selected foodborne pathogens. J Food Prot 67:1252–1256

Rota MC, Herrera A, Martínez RM, Sotomayor JA, Jordán MJ (2008) Antimicrobial activity and chemical composition of *Thymus vulgaris*, *Thymus zygis* and *Thymus hyemalis* essential oils. Food Control 19(7):681–687

Rotblatt M, Ziment I (2002) Evidence-based herbal medicine. Philadelphia Hanley & Belfus

Ruberto G, Biondi D, Piattelli M (1991) Composition of the volatile oil of *Crithmum maritimum* L. Flavour Fragr J 6(2):121–123

Ruberto G, Baratta MMT (2000) Antioxidant activity of selected essential oil components in two lipid model systems. Food Chem 69:167–171

Saadaoui E, Khaldi A, Khouja ML, El-Gazzah M (2007) Etude de la variabilité morphologique du câprier (*Capparis* spp.) en Tunisie. Revue des régions arides 2:523–527

Sacchetti G, Maietti S, Muzzoli M, Scaglianti M, Manfredini S, Radice M, Bruni R (2005) Comparative evaluation of 11 essential oils of different origin as functional antioxidants, anti-radicals and antimicrobials in foods. Food Chem 91:621–632

Safaei-Ghomi J, Ebrahimabadi AH, Djafari-Bidgoli Z, Batooli H (2009) GC/MS analysis and *in vitro* antioxidant activity of essential oil and methanol extracts of *Thymus caramanicus* Jalas and its main constituent carvacrol. Food Chem 115:1524–1528

Safir O, Fkih-Tetouani S (1998) Saponins from *Zygophyllum gaetulum*. J Nat Prod 61:130–134

Saha JC, Savini EC, Kasinathan S (1961) Indian J Med Rex 49:130

Said O, Khalil K, Fulder S, Azaizeh H (2002) J Ethnopharmacol 83:251

Saieed P, Reza RM, Abbas D, Seyyedali R, Aliasghar H (2006) Inhibitory effects of *Ruta graveolens* L. extract on guinea pig liver aldehyde oxidase. Chem Pharm Bull 54:9–13

Sakakibara K, Nishiyama T, Sumikawa N, Kofuji R, Murata T, Hasebe M (2003) Involvement of auxin and a homeodomain-leucine zipper I gene in rhizoid development of the moss Physcomitrella patens. Development 130:4835–4846

Salido S, Altarejos J, Nogueras M, Shanchez A (2001) Chemical composition of essential oil of *Artemisia herba alba* Ass. Ssp. *valentine* (Lam.) Marc. J Essent Oil Res 13:221–224
Salido S, Valenzuela LR, Altarejos J, Nogueras M, Sanchez A, Cano E (2004) Composition and infraspecific variability of *Artemisia herba-alba* from southern Spain. Biochem Syst Ecol 32:265–277
Sawamura M (2000) Aroma and functional properties of Japanese yuzu (*Citrus junos* Tanaka) essential oil. Aroma Res 1:14–19
Scarpato R, Paganucci L, Bertoli A, Fiore L, Pistelli L, Federico G (2008) Licoflavone C attenuates the genotoxicity of cancer drugs in human peripheral lymphocytes. Phytother Res 22:1650–1654
Schraudolf H (1989) Indole glucosinolates of *Capparis spinosa*. Phytochemistry 88:259–260
Senatore F, De Feo V (1994) Essential oil of a possible new chemotype of *Crithmum maritimum* L. growing in Campania (southern Italy). Flavour Fragr J 9:305–307
Semmar N, Jay M, Chemli R (2001) Chemical diversification trends in *Astragalus caprinus* (Leguminosae), based on the flavonoid pathway. Biochem Syst Ecol 29:727–738
Semmar N, Fenet B, Gluchoff-Fiasson K, Comte G, Jay M (2002) New flavonol tetraglycosides from *Astragalus caprinus*. Chem Pharm Bull 50:981–984
Semmar N, Jay M, Farman M, Chemli R (2005) Chemotaxonomic analysis of *Astragalus caprinus* (Fabaceae) based on the flavonic patterns. Biochem Syst Ecol 33:187–200
Sharififar F, Dehghn-Nudeh GH, Mitrajaldini M (2009) Major flavonoids with antioxidant activity from *Teucrium polium* L. Food Chem 112:885–888
Sher H, Alyemeni MN (2010) Ethnobotanical and pharmaceutical evaluation of *Capparis spinosa* L, validity of local folk and Unani system of medicine. J Med Plant Res 4(17):1751–1756
Sher H, Hussain F (2009) Ethnobotanical evaluation of some plant resources in Northern part of Pakistan. Afr J Biotechnol 8(17):4066–4076
Sher H, Al-Yemeni MN, Sher H (2010a) Forest Resource utilization assessment for economic development of rural community, Northern parts of Pakistan. J Med Plant Res 4(12):1197–1208
Sijelmassi A (1993) Les plantes médicinales du Maroc. Editions Le Fennec, Casablanca, 286 pp
Skidmore-Roth L (2001) Mosby's handbook of herbs and natural supplements, 1st edn. Mosby, New York, pp 741–744
Skima F, Kaaya A, Jaouhari JT, Lazrek HB, Jana M, El Amri H (1999) Hypoglycaemic activity of *Globularia alypum* leaves in rats. Fitoterapia 70:382–389
Snoussi M, Hajlaoui H, Noumi E, Usai D, Sechi LA, Zanetti S, Bakhrouf A (2008) *In vitro* anti-Vibrio spp. activity and chemical composition of some Tunisian aromatic plants. World J Microbiol Biotechnol 24:3071–3076
Sobhani AM, Ebrahimi SA, Hoormand M, Rahbar N, Mahmoudian M (2002) Cytotoxicity of *Peganum harmala* L. seeds extract and its relationship with contents of β-carboline alkaloids. J Iran Univ Med Sci 8:432–438
Soliman KM, Badeaa RI (2002) Effect of oil extracted from some medicinal plants on different mycotoxigenic fungi. Food Chem Toxicol 40:1669–1675
Somer G, Calışkan AC (2007) Selenium and trace element distribution in *Astragalus* plants: developing a differential pulse polarographic method for their determination. Turk J Chem 31:411–422
Sommer A, Davidson FR (2002) Assessment and control of vitamin A deficiency: the annecy accords. J Nutr 132:2845S–2850S
Spera D, Siciliano T, De Tommasi N, Braca A, Vessières A (2007) Antiproliferative cardenolides from *Periploca graeca*. Planta Med 73:384–387
Stahl-Biskup E (1991) The chemical composition of *Thymus* oils: a review of the literature 1960–1989. J Essent Oil Res 3:61–82
Stahl-Biskup E, Sáez F (2002) Thyme: the genus *Thymus*. Taylor and Francis, London
Stashenko EE, Acosta R, Martínez JR (2000) High-resolution gas-chromatographic analysis of the secondary metabolites obtained by subcritical-fluid extraction from Colombian rue (*Ruta graveolens* L.) J Biochem Biophys Methods 43(1–3):379–390

Stassi V, Verykokidou A, Loukis A (1996) The antimicrobial activity of the essential oils of four *Juniperus* species growing wild in Greece. Flavour Fragr J 11:71–74

Stevanovits-Banyai E, Tulok MH, Hegedus A, Renner C, Varga IS (2003) Antioxidant effect of various rosemary (*Rosmarinus officinalis* L.) clones. Acta Biol Szegediensis 47(1–4):111–113

Stevenson DW (1993) Ephedraceae. In: Flora of North America Editorial Committee (eds) Flora of North America, vol 2. Oxford University Press, New York, pp , 428–434

Stratton SE, Lorden JF (1991) Effect of harmaline on cells of the inferior olive in the absence of tremor: differential response of genetically dystonic and harmaline-tolerant rats. Neuroscience 41:543–549

Stulzer HK, Tagliari MP, Zampirolo JA, Cechinel-Filho V, Schlemper V (2006) Antioedematogenic effect of marrubiin obtained from *Marrubium vulgare*. J Ethnopharmacol 108(3):379–392

Sushila R, Permender R, Dharmender R, Deepti R, Vikash K (2010) Phytochemical and pharmacological Potential of Kair (*Capparis Decidua*). Int J Phytomed 2:10–17

Syrov VN, Khushbaktova ZA (2001) Experimental study of pharmacotherapeutic effect of phytoecdysteroids and nerobol in toxic liver damage. Eksp Klin Farmakol 64(4):56–58

Syrov VN, Nabiev AN, Sultanov MB (1986) Action of phytoecdysteroids on the bile-secretory function of the normal liver and in experimental hepatitis. Farmakol Toksikol 49:100–103

Syrov VN, Nasyrova SS, Khushbaktova ZA (1997) The results of experimental study of phytoecdysteroids as erythropoiesis stimulators in laboratory animals. Eksp Klin Farmakol 60:41–44

Täckholm V (1974) Students' flora of Egypt, 2nd edn. Cairo University, Cairo, p 888

Tahraoui A, El-Hilalya J, Israili ZH, Lyoussi B (2007) J Ethnopharmacol 110:105

Takahashi D, Miura K (1888) Mitt Med Fak Tokyo 1:255

Takeda Y, Yanagihara K, Masuda T, Otsuka H, Honda G, Takaishi Y, Sezik E, Yesilada E (2000) Labdane diterpenoids from *Marrubium globosum* ssp. *globosum*. Chem Pharm Bull (Tokyo) 48(8):1234–1235

Taleb-Senouci D, Ghomari H, Krouf D, Bouderbala S, Prost J, Lacaille-Dubois MA, Bouchenak M (2009) Antioxidant effect of *Ajuga iva* aqueous extract in streptozotocin-induced diabetic rats. Phytomedicine 16:623–631

Tariq M, Ageel AM, Al-Yahya MA, Mossa JS, Al-Said MS (1989) Anti-inflammatory activity of *Teucrium polium*. Int J Tissue React 11(4):185–188

Tengberg A, Chen D (1998) A comparative analysis of nebkhas in central Tunisia and northern Burkina Faso. Geomorphology 22:181–192

Tepe B, Sokmen M, Akpulat HA, Daferera D, Polissiou M, Sokmen A (2005) Antioxidative activity of the essential oils of *Thymus sipyleus* subsp. *sipyleus* var. sipyleus and *Thymus sipyleus* subsp. *sipyleus* var. rosulans. J Food Eng 66:447–454

Teyeb H, Zouari S, Douki W, Najjar MF, Neffati M (2010) Essentiel oils of leaves, flowers and fruits of Atragalus gombiformis Pomel (Fabaceae). Acta Hort 853:ISHS

Teyeb H, Zouari S, Douki W, Najjar MF, Neffati M (2011) Variation in volatiles of *Atragalus gombiformis* Pomel. Verlag der Zeischrift fùr Naturforschung. Tùbingen:0939–5075

Teyeb H, Zanina N, Neffati M, Douki W, Najjar MF (2012) Cytotoxic and antibacterial activities of leaf extracts of *A. gombiformis* Pomel (Fabaceae) growing wild in Tunisia. Turk J Biol 36:53–58

Tharib SM, Gnan SO, Veitch GBA (1983) Antimicrobial activity of compounds from *Artemisia campestris*. J Food Prot 46:185–187

Tigrine-Kordjani N, Meklati BY, Chemat F (2006a) Analysis by gas chromatography–mass spectrometry of the essential oil of *Zygophyllum album* L., an aromatic and medicinal plant growing in Algeria. Int J Aromather 6(3–4):187–191

Tigrine-Kordjani N, Meklati BY, Chemat F (2010) Contribution of microwave accelerated distillation in the extraction of the essential oil of *Zygophyllum album* L. Phytochem Anal 22:1–9

Tlig T, Gorai M, Neffati M (2008) Germination responses of *Diplotaxis harra* to temperature and salinity. Flora 203:421–428

Tlili N, Nasri N, Saadaoui E, Khaldi A, Triki S (2009) Carotenoid and tocopherol composition of leaves, buds, and flowers of Capparis spinosa grown wild in Tunisia. J Agric Food Chem 57:5381–5385

Tlili N, Khalidi A, Triki S, Munné-Bosch S (2010) Phenolic compounds and vitamin antioxidants of Caper (*Capparis spinosa*). Plant Foods Hum Nutr 65:260–265
Tomei PE, Cioni PL, Flamini G, Stefani A (1995) Evaluation of the chemical composition of the essential oils of some Lamiaceae from Serrania de Ronda (Andalucia, Spain). J Esent Oil Res 7(3):279–282
Tosun F, Akyuz-Kizilay C, Tosun AU (2009) Flavonoïds and isoflavonoïds from *Genista sessilifolia* DC. Growing in Turkey. Chem Nat Compd 45(1):83–84
Touil A, Rhouati S, Creche J (2006) Flavonoid glycosides from *Pituranthos chloranthus*. Chem Nat Compd 42(1):104–105
Tregenza LA (1955) The Red Sea mountains of Egypt. Oxford University Press, London, pp viii, 247
Trovato A, Monforte MT, Forestieri AM, Pizzimenti F (2000) *In vitro* anti-mycotic activity of some medicinal plants containing flavonoids. Boll Chim Farm 139(5):225–227
Tutin TG (1972) Flora Europea. University Press, Cambridge
Uphof JCT (1968) Dictionary of economic plants. Verlag von Cramer, Germany, p 290
Vahedi K, Domigo V, Amarenco P, Bousser MG (2000) Ischaemic stroke in a sportsman who consumed MaHuang extract and creatine monohydrate for body building [Letter]. J Neurol Neurosurg Psychiatry 2000:68112–68113
Verite P, Nacer A, Kabouche Z, Seguin E (2004) Composition of seeds and stems essential oil of *Pituranthos scoparius* (Coss.& Dur.) Schinz. Flavour Fragr J 19:562–564
Vernin G, Merad O, Vernin GMF, Zamkotsian RM, Parkanyi C (1995) GC-MS analysis of *Artemisia herba-alba* Asso essential oils from Algeria. Dev Food Sci 37A:147–205
Vidrich V, Michelozzi M (1993) Sulla resae compozione stagionale di olio essenziale di galbule e foglie di *Juniperus phoenicea* L. Italia Forestale e Montana 48:133–140
Vourlioti-Arapi F, Michaelakis A, Evergetis E, Koliopoulos G, Haroutounian SA (2012) Essential oils of indigenous in Greece six *Juniperus taxa*: Chemical composition and larvicidal activity against the West Nile virus vector *Culex pipiens*. Parasitol Res 110:1829–1839
Wang RZ (2007) C4 plants in the deserts of China: occurrence of C4 photosynthesis and its morphological functional types. Photosynthetica 45:167–171
Wang W, Wu N, Zu YG, Fu YJ (2008) Antioxidative activity of *Rosmarinus officinalis* L. essential oil compared to its main components. Food Chem 108:1019–1022
Watt OM, Breyer-Brandwijk MG (1962) The medicinal and poisonus plants of Southern and Eastern Africa. E and S Livingstone, Edinburgh, p 841
Wealth of India (1992) A dictionary of Indian raw material and industrial products, vol 3. CSIR, New Delhi, pp 213–214
White LM, Gardner SF, Gurley BJ, Marx MA, Wang P-L, Estes M (1997) Pharmacokinetics and cardiovascular effects of MaHuang (*Ephedra sinica*) in normotensive adults. J Clin Pharmacol 37:116–122
Wichtel M, Anton R (1999) Plantes Thérapeutiques. Tradition, pratique, officinale, science et thérapeutique. Tec and Doc (Eds)/EM inter, Paris, p 341
Wolters B, Eilert U (1981) Antimicrobial substances in callus cultures of *Ruta graveolens*. Planta Med 43:166–174
Wu T-S, Shi LS, Wang JJ, Iou SC, Chang HC, Chen YP, Kuo YH, Chang YL, Tenge CM (2003) Cytotoxic and antiplatelet aggregation principles of *Ruta Graveolens*. J Chin Chem Soc 50(1):171–178
Yang J, Du Y (2003) Chemical modification, characterization and bioactivity of Chinese lacquer polysaccharides from lac tree *Rhus vernicifera* against leukopenia induced by cyclophosphamide. Carbohydr Polym 52:405–410
Yangui T, Bouaziz M, Dhouib A, Sayadi S (2009) Potential use of Tunisian Pituranthos chloranthus essential oils as natural disinfectant. Lett Appl Microbiol 48:112–117
Yesilada E, Bedir E, Caliş I, Takaishi Y, Ohmoto Y (2005) Effects of triterpene saponins from *Astragalus* species on in vitro cytokine release. J Ethnopharmacol 96(1–2):71–77
Yin G, Jeney G, Racz T, Xu P, Jun X, Jeney Z (2006) Effect of two Chinese herbs (*Astragalus radix* and *Scutellaria radix*) on non-specific immune response of tilapia, *Oreochromis niloticus*. Aquaculture 253:39–47

Youdim KA, Deans SG, Finlayson HJ (2002) The antioxidant properties of thyme (*Thymus zygis* L.) essential oil: an inhibitor of lipid peroxidation and a free radical scavenger. J Essent Oil Res 14:210–215

Yue-lan C, Xin L, Min Z (2010) *Capparis spinosa* protects against oxidative stress in systemic sclerosis dermal fibroblasts. Arch Dermatol Res 302:349–355

Zaabat N, Hay AE, Michalet S, Darbour N, Bayet C, Skandrani I (2011) Antioxidant and antigenotoxic properties of compounds isolated from *Marrubium deserti* de Noe. Food Chem Toxicol 49(12):3328–3335

Zakarya D, Fkih-Tetouani S, Hajji F (1993) Chemical compositionantimicrobial activity relationships of Eucalyptus essential oils. Plantes Med Phytother 26:319–331

Zambonelli A, D'aulerio AZ, Severi A, Benvenuti S, Maggi L, Bianchi A (2004) Chemical composition and fungicidal activity of commercial essential oils of *Thymus vulgaris* L. J Essent Oil Res 16:69–74

Zaouali Y, Messaoud C, Ben Salah A, Boussaid M (2005) Oil compositon variability among populations in relationship with their ecological areas in Tunisia. Flavour Fragr J 20(5):512–520

Zargham R, Zargham H (2008) Tannin extracted from Sumac inhibits vascular smooth muscle cell migration. McGill J Med 11:119–123

Zarre S (2000) Systematic revision of *Astragalus* sect. Adiaspastus, sect. Macrophyllium and sect. Pterophorus. Englera 18:1–219

Zellagui MA, Rhouti S, Creche J, Toth G, Ahmed AA, Pare PW (2004) Antimicrobial activity of the alkaloid extract of *Genista microcephala*: isolation and complete 1H and 13C. Chemical shifts assignments of lupanine and S-balycotomine. Revista Latinoamer Quím 32(3):109–114

Zhang T, Tan DY (2009) An examination of the function of male flowers in an andromonoecious shrub *Capparis spinosa*. J Integr Plant Biol 51(3):316–324

Zhao J, Li SP, Yang FQ, Li P, Wang YT (2006) Simultaneous determination of saponins and fatty acids in *Ziziphus jujuba* (Suanzaoren) by high performance liquid chromatography-evaporative light scattering detection and pressurized liquid extraction. J Chromatogr A 1108:188–194

Zhao Z, Liu M, Tu P (2008) Characterization of water soluble polysaccharides from organs of Chinese Jujube (*Ziziphus jujuba* Mill. cv. Dongzao). Eur Food Res Technol 226:985–989

Zhu YP (1998) Chinese Materia Medica: chemistry, pharmacology and applications. Harwood Academic, Amsterdam

Ziyyat A, Legssyer H, Mekhfi H, Dassouli A, Serhouchni M, Benjelloun W (1997) Phytotherapy of hypertension and diabetes in oriental Morocco. J Ethnopharmacol 58:45–54

Zohary M (1972) Flora of Palaestina, vol 2. p 25

Zohary M (1973) The geobotanical foundations of the middle East, Fischer G (ed). Stuttgart, 738p

Zunft HJF, Lueder W, Harde A, Haber B, Graubaum HJ, Gruenwald J (2001) Carob pulp preparation for treatment of hypercholesterolemia. Adv Ther 18:230–236

Zunft HJF, Lueder W, Harde A, Haber B, Graubaum HJ, Koebnick C, Gruenwald J (2003) Carob pulp preparation rich in insoluble fibre lowers total and LDL cholesterol in hypercholesterolemic patients. Eur J Nutr 42:235–242

# Promising Indigenous and Endemic Medicinal Plants from Mauritius

**Mohamad Fawzi Mahomoodally and Muhammad Zakariyyah Aumeeruddy**

**Abstract** The Republic of Mauritius is part of Sub-Saharan Africa and one of the three islands that constitute the Mascarene archipelago. Due to its volcanic origin, the tropical island of Mauritius has a diverse flora and has been endorsed by international authorities as one of the Indian Ocean islands biodiversity hotspots. The local flora bears its essence from the floristic heritage of the Mascarenes and harbor a wide array of endemic or highly specialized species that has evolved as a result of the variation in topography, climate and the geographical isolation of the islands. Indeed, a major portion of this flora is exploited for its medicinal properties by the local people to treat and/or manage common ailments. Nonetheless, there is still a dearth of scientific validation and compilation of some indigenous plants in common use. The present chapter endeavors to highlight the botanical description, ethnopharmacological uses, and the main therapeutic benefits of common indigenous and endemic medicinal plants of Mauritius. Special emphasis has been geared towards recent *in vitro* and *in vivo* data which tend to support ethnopharmacological uses of these medicinal plants in traditional medicine of Mauritius.

**Keywords** Mauritius • Indian Ocean • Medicinal plants • Traditional medicines • Ethnopharmacology

M.F. Mahomoodally (✉)
Department of Health Sciences, Faculty of Science, University of Mauritius, 230 Réduit, Mauritius
e-mail: f.mahomoodally@uom.ac.mu

M.Z. Aumeeruddy
Department of Agricultural and Food Science, Faculty of Agriculture, University of Mauritius, 230 Réduit, Mauritius

 
M. Neffati et al. (eds.), *Medicinal and Aromatic Plants of the World - Africa Volume 3*, Medicinal and Aromatic Plants of the World,
DOI 10.1007/978-94-024-1120-1_9

# 1 Introduction

The tropical island of Mauritius forms part of the African continent and lies in the southern hemisphere in the middle of the Indian Ocean with latitude and longitude 20.1625°S, 58.2903°E. The Republic of Mauritius is composed of the several islands including of Mauritius, Rodrigues, Agalega, St. Brandon and a number of outlying smaller islands, all located in the south of the Indian Ocean. Mauritius is the principal island some 800 km from the south east of Madagascar. Mauritius has an area of 1865 $km^2$. Mauritius enjoys a mild tropical maritime climate throughout the year, characterized by a warm humid summer extending between November and April and a relatively cool dry winter between June and September (Sreekeeson and Mahomoodally 2014).

As at 01st July 2014 the population of the island stood at 1.219.265 individuals with 603.473 males and 615.792 females (CSO 2014). About half of the population is concentrated in the urban areas which lie along the axis from the capital Port-Louis to the city of Curepipe. The standards of health, nutrition and education are high compared to other countries in Africa. The adult literacy rate is 83% and according to the most recent WHO bulletin published in 2013, the life expectancy at birth is about 70.4 for males and 78 for females (World Health Rankings 2014).

The population comprises of Indo-Mauritians, people of mixed European and African origin and Sino-Mauritians. The main ethnic groups are the Bhojpuri speaking Hindus, constituting 40.2% of the total population. The Tamils are the second largest ethnic community (13.9%), while Telugus (5.6%) and Marathis (4%) represent smaller minorities within the overall Hindu population. The Hindus have a common language (Bhojpuri), the same regional origin (Uttar Pradesh and Bihar), religious practices, and rituals (Hollup 1994, 2012).

Since the past years, Mauritius economy has switched from sugar and textiles, into financial services and information and communication technology, as well as providing a seafood hub for the region and becoming a destination for medical tourism and a centre for academic excellence (Joseph and Troester 2012).

Mauritius has a rich heritage of indigenous and endemic plants and possesses 58 families of plant species consisting of both indigenous and endemic plants. A total of 711 indigenous plant species has been recorded in Mauritius, including 246 endemic plant species (MOAIFS 2012). Mauritius is characterized by considerable endemism rate and species diversity.

Nonetheless, the surge in population and economic growth has led to loss, degradation and decline of ecosystems, species and genetic diversity. Economic development has stimulated forest clearance for settlement, industries and agriculture as well as intrusion on environmentally sensitive areas, including mountains and forested areas, hence resulting in biodiversity erosion. Natural disasters such as cyclones and droughts are significant threats to native biodiversity as it is highly fragmented and the populations are small, resulting in diminished adaptability to weather variations. Biodiversity consequently remains one of the top national priorities of the local authority. According to the National Biodiversity Strategy and

Action Plan, it is estimated that Rs. 200 million is spent annually, in addition to external funding, on the conservation of forests and terrestrial biodiversity. A number of policies and strategies have been adopted by Government for conservation and sustainable use of biodiversity. These comprise of the National Environment Policy (2007), National Biodiversity Strategy and Action Plan (2006–2015), National Invasive Alien Species Strategy and Action Plan (2010), Non-Sugar Sector Strategic Plan (2003–2007), Strategic Options in Crop and Livestock Sector102 (2007–2015) and Study on Environmentally Sensitive Areas.

## 2 Medicinal Plants of Mauritius

Mauritius is famous for its floricultural diversity and for the panoply of traditions practiced across the island by various ethnic groups. Moreover, the island is renowned for the extraordinary richness of its flora, which harbors an array of medicinal plant species. Traditional medicine is omnipresent in the Mauritian community whereby Mauritians still use traditional medicine including mainly medicinal plants for the treatment and/or management of various ailments. Indeed, the local population has a long-standing tradition in the use of herbal medicine. Many indigenous and endemic plant species of Mauritius have been used in folkloric medicine to treat various ailments of man including chronic diseases communicable and non-communicable diseases.

Available reports tend to highlight that indigenous and native folk-medicinal plant conservation and study is of uttermost importance because such plants are fully adapted to local environments (e.g. tropical climates, altitudes, humidity and temperature) compared to exotic species. Pharmacologically active compounds and phytochemicals isolated from endemic and indigenous plants used in folk medicine in Mauritius have been the areas of interest (Mahomoodally 2012a, b, c; Narod et al. 2004). Currently, several kinds of extracts from various exotic, endemic and indigenous plants are sold as decoctions or "tisanes" in several markets across Mauritius to treat minor ailments (Mahomoodally 2012a, b, c). This native herbal folk medicinal practice forms an essential part of the heritage of the local pharmacopoeia of Mauritius. Nonetheless, there is still a dearth of scientific validation and compilation of some indigenous plants in common use.

The present chapter endeavours to highlight the botanical description, ethnopharmacological uses, and main therapeutic benefits of common indigenous and endemic medicinal plants of Mauritius. Special emphasis has been geared towards recent *in vitro* and *in vivo* data which tend to support ethnopharmacological uses in the traditional medicine of Mauritius.

## Acalypha integrifolia *Willd. subsp.* integrifolia *var.* integrifolia *(Euphorbiaceae)*

*Acalypha integrifolia* is an indigenous species of the Mascarenes islands. It is commonly known as 'bois queue de rat', 'bois de crève' and 'bois de Charles' (Gurib-Fakim and Gueho 1996). In Réunion Island and Mauritius, decoction of the leaves is consumed as astringent and purgative and is used to eliminate intestinal worms. A bath in the leaf decoction is taken to treat skin infections. The leaves, stems and roots have been reported to contain saponins, tannins, sterols, terpenes and traces of alkaloids (Gurib-Fakim and Gueho 1996; Seebaluck et al. 2015).

Leaf decoction, dichloromethane (DCM), ethyl acetate (EtOAc) and methanol (MeOH) extract were found active against ATCC bacterial strains of *Escherichia coli*, *Pseudomonas aeruginosa* and *Staphylococcus aureus*. MeOH extract also inhibited the growth of clinical isolates namely *Escherichia coli*, *Staphylococcus aureus*, *Acinetobacter baumannii*, *Klebsiella pneumonia*, *Streptococcus* group B, *Pseudomonas aeruginosa*, *Enterococcus faecalis*, *Proteus vulgaris* and *Candida albicans*. Hexane, DCM and EtOAc extracts showed antifungal activity against *Candida albicans*, ATCC 10231 and *Candida tropicalis*, ATCC 750 (Unpublished data).

EtOAc and MeOH extracts of the plant scavenged DPPH radical, ABTS cation radical and hypochlorus acid. MeOH extract showed potential antioxidant and reducing activity. Phytochemical screening of MeOH extract revealed the presence of tannins, polyphenol, terpenes, saponins, anthocyanins, coumarins, flavonoids and sugars (Unpublished data).

## Antidesma madagascariense *Lam. (Euphorbiaceae)*

*Antidesma madagascariense* is native to the Mascarene Islands where it occurs in Mauritius and Reunion Islands. Commonly known as 'Bois bigaignon bâtard' and 'Bois bigayon' in Mauritius, it is widely distributed in humid forests but can also be seen in Reunion Island in dense thickets of medium altitude (100–1600 m). Other familiar vernacular names include 'Bois de cabri', 'Bois de cabri blanc', 'Bois de gaulette blanc' and 'Bois d'oiseaux' in Reunion island (Centre de cooperation international en recherché agronomique pour le développement 2006). The genus consists of about 170 species growing in tropical and warm habitats. It is probably also present in Madagascar.

This plant is common in humid forests in Mauritius and shows ample variations in the texture and shape of the leaves which can be coriaceous to papery, narrowly elliptic to broadly oval and the domatia very visible or missing in some cases. This plant is characterised by the ample variations in the size, texture and shape of the leaves which can be coriaceous to papery, often oval to elliptic, 4–10 cm long and 3–5 cm wide, the margins entirely or slightly sinuous. The presence of domatia is very visible in the axils of primary nerves. This tree bears minute greenish or red flowers of different sexes (Hoffman 1999; Kuete 2013). Often abundant in clusters of small berries, the ovoid dark red fruits which turn into shiny purple black when ripe, are more or less flattened and 6–7 mm long (Hoffman 1999; Centre de cooperation international en recherche agronomique pour le développement 2006; Jolly et al. 2006).

*A. madagascariense* is very famous and widely used in folkloric medicine among the local population of the Mascarene Islands for the treatment and management of various ailments. In Mauritius, a decoction of the leaves and bark of *Antidesma madagascariensis* is used as a diuretic, against fever, diabetes and as an astringent. A decoction of the leaves is used to treat skin infections as well as rheumatism. Teas are used against dysentery and albumin in urine. To treat jaundice a mixture of *Aphloia theiformis*, *Toddalia asiatica* and *Antidesma madagascariensis* is used (Narod et al. 2004; Gurib-Fakim 2011). The leaves and bark decoctions are diuretic and are also used against fever, and for their astringent properties. The leaf tea is used to treat dysentery. A bath in the leaf decoction is used against skin infections. Along with other plants, it is reported as effective against jaundice. The decoction obtained after boiling 10 leaves of *A. madagascariense* in 1 liter of water can be consumed regularly for the treatment of albumin in urine (Kuete 2013). Furthermore, the leaves and barks have been reported to possess diuretic, astringent as well as febrifuge properties and also used diabetes management (Kuete 2013). A bath in the leaf decoction has been reported to alleviate skin infections, rheumatic and body aches. The leaves of *A. madagascariense* mixed with those of *Aphloia theiformis* (Flacourtiaceae) and *Toddalia asiatica* (Rutaceae) are used to treat jaundice (Hoffman 1999). Interestingly, *A. madagascariense* can be used to treat edema in pregnancy and can also be employed in the case of stroke depending on the type of decoction which is prepared (Kuete 2013).

Preliminary phytochemical screening of the leaves of *A. madagascariense* indicates the presence of phenols, tannins, alkaloids, flavonoids, cyanogenetic heterosides as well as leucoanthocyanins, sterols and saponins (Hoffman 1999; Narod et al. 2004; Picot et al. 2013). Interestingly, *A. madagascariense* has also been found to contain triterpenes and hydrolysable tannins, carpusin, and a dimer – antidesmin, a common constituent also characterised in other *Antidesma* species (Mahomoodally et al. 2006; Rangasamy et al. 2007). The bacteriostatic and bactericidal properties of *A. madagascariense*, validated from the local folk medicine, can be attributed to the presence of tannins in the plant. Furthermore, the aqueous and methanol extracts of the leaves demonstrated their molluscicidal properties against species of *Biomphalaria* and *Bulinus* and antifungal properties against *Cladosporium cucumerinum* (Hoffman 1999; Kuete 2013). Different fractions of the leaves and stems (water, methanol, chloroform and hexane) of *A. madagascariense* were previously reported to have significant inhibitory effects on Gram-positive bacteria *Staphylococcus aureus*, Gram-negative bacteria *Pseudomonas aeruginosa* and the fungus *Aspergillus niger* showing their potent antimicrobial activities (Tchinda et al. 2006; Gurib-Fakim et al. 2005). The ability of the methanol stem extracts of *A. madagascariense* to exhibit contractile properties on rat ileal smooth muscles coupled with the recent antioxidant and antimicrobial findings on *A. madagascariense* validate its ethnobotanical use in the effective treatment of dysentery (Hoffman 1999; Tchinda et al. 2006; Narod et al. 2004).

The different crude extracts of *A. madagascariense* exhibited potent antimicrobial activity which was found to increase with increasing polarity. The methanol leaves extracts of *A. madagascariense* had potent inhibitory effects against *Enterococcus faecalis* (minimum inhibitory concentration [MIC] = 60 μg/ml), *Staphyloccocus aureus* (MIC = 500 μg/ml), Methicillin-resistant *Staphylococcus aureus* (MRSA) (MIC = 250 μg/ml) and *Candida albicans* (MIC = 500 μg/ml) (Narod et al. 2004).

We previously studied the antioxidant propensity of different crude extracts and fractions of leaves of *A. madagascariense* using standard *in vitro* assays. The $IC_{50}$ values ranged from 3.94 to 87.05 μg/ml, 3.18 to 13.26 μg/ml and 6.29 to 25.24 μg/ml for the 2,2-diphenyl-1-picrylhydrazine (DPPH), superoxide (SO) and nitric oxide (NO) radical scavenging assays respectively. On the other hand, the *n*-butanol extracts of *A. madagascariense* proved to be the most potent antioxidant against for DPPH (% radical scavenging potential [RSP] = 93.68 ± 8.69%) and NO (% RSP = 65.56 ± 7.56%) assays, ethyl acetate extracts had a high percentage RSP (99.53 ± 7.53%) in SO assay (Mahomoodally et al. 2015). Likewise, a concentration of 0.5 mg/ml of methanol *A. madagascariense* leaves extracts was able to scavenge 70.6 ± 2.2% and 64.5 ± 1.8% of hydroxyl and hypochlorous acid radicals correspondingly, thus validating the antioxidant properties of *A. madagascariens* (Narod et al. 2004).

Antiglycation activities were confirmed in the ethylacetate (96.63 ± 10.36%), methanol (86.35 ± 5.65) and *n*-butanol (84.65 ± 6.35%) fractions of *A. madagascariense* and this was comparable to the anti-glycation drug aminoguanidine. Nevertheless, *A. madagascariense* extracts were found to have no activity against mitochondrial respiration in a MTT cytotoxicity assay ($P > 0.05$) (Mahomoodally et al. 2015). The efficacy of *A. madagascariense* extracts in managing diabetes was assessed through the inhibition of key carbohydrate hydrolysing enzymes. All the extracts exhibited variable inhibitory effects on α-amylase activity with ethylacetate fraction having the best inhibitory effect ($IC_{50}$ = 61.52 ± 11.09 μg/ml) which was lower than acarbose ($IC_{50}$ = 75.86 ± 8.16 μg/ml). Moreover, active fractions of *A. madagascariense* were found to inhibit significantly amylase activity in mouse plasma from 7.80% to 49.37%. α-glucosidase activity was significantly inhibited by *A. madagascariense* extracts with $IC_{50}$ values ranging from 19.70 ± 2.87 μg/ml to 44.92 ± 5.67 μg/ml which was comparable to the drug 1-deoxynojirimycin.

The repressive capacity of *A. madagascariense* extracts to increase blood glucose concentration in mice was investigated by *in vivo* studies in glycogen-loaded mice. Ethylacetate fraction was found to be more potent with glucose-lowering properties (−59.4%) comparable to acarbose (−55.1%) (Picot et al. 2013). Moreover, *in vivo* studies of *A. madagascariense* on rat everted intestinal sacs indicated that aqueous extract *A. madagascariense* significantly enhanced the uptake of D-glucose and fluid transport. It was also noted that a concentration of above 0.375 mg/ml of the extract was needed to enhance mucosal disappearance, gut wall content and serosal appearance of fluid. However, L-tyrosine and $K^+$ transport was not significantly enhanced to the contrary of $Na^+$. Therefore, the ability of *A. madagascariense* extracts to promote the transport of glucose, fluid and $Na^+$ across rat everted intestinal sacs might be attributed to the presence of bioactive phytochemicals, for instance, flavonoids, alkaloids, leucoanthocyanins, phenols and saponins, in *A. madagascariense* leaves which have possibly interact with the $Na^+$/glucose co-transporter in the enterocytes (Mahomoodally et al. 2012a, b, c).

The *in vitro* immunomodulatory property of *A. madagascariense* showed that extracts were able to modulate significantly the immune response of phagocytes and monocytes at different steps. At a concentration of 100 μg/ml, the inhibitory activity of crude methanol *A. madagascariense* extracts on whole blood phagocytes for reactive oxygen species (ROS) production was 94.2%. It was also suggested that *A. madagascariense* directly inhibited a final common biochemical target such as NADPH oxidase enzyme or scavenge ROS since it did not affect a specific transductional pathway.

## Aphloia theiformis *Benn. (Flacourtiaceae)*

*Aphloia theiformis* is indigenous to Madagascar, the Comoros, the Mascarenes, the Seychelles and tropical Africa, the species is still common in Mauritius and Reunion Island. It has become extremely rare or is already extinct on Rodrigues. Common in remnants of the native forest in the humid upland zone of Mauritius. In Comoros, *A. theiformis* is known locally as the "Mfandrabo", the leaves of which are widely used alone or in combination for the preparation of tea. Previously included in the Flacourtiaceae family, *Aphloia theiformis* is currently the only representative of the Aphloiaceae family (Danthu et al. 2010). *Aphloia theiformis*is known to have high morphological variations due to ecological factors.The synonyms are; *Aphloia mauritiana* Baker, *A. integrifolia* (Vahl.) Benn., *Lightfootia theiformis* Vahl., *Ludia heterophylla* auct. Non Lam.

In Comoros, the *A. theiformis* leaves areused for treating dizziness and leucorrhoea. In Mauritius, they are also used as a febrifuge. In Madagascar, the bark of this species is used as emetic; its young leavesare demonstrated effective against haematuria and thc oldones against rheumatisms. In terms of chemical composition, the literature reports the presence of saponins andxanthone in A. theiformis leaves (Hsoidrou et al. 2014).

A tea made from the leaves of *A. theiformis* withthe bark of *Terminalia bentzoe* has been reported to treat dysentery. The leaf decoction is used for fever (Adjanohoun et al. 1983) and a diuretic (Sussman 1980; Wong Ting Fook 1980). It also helps against rheumatism, ulcers, jaundice and gastrointestinal infections. Daruty (1886) reports on the use of this plant as an emetic. The leaves, mixed with the bark of *Erythroxylum laurifolium* and the roots of *Cocos nucifera* help against fever. In the Comoros, a leaf decoction is used for stomach pains; in the Seychelles, a root and leaf decoction helps against impetigo. Mixed withthose of *Cinnamomum camphora* or those of *Piper nigrum*, a refreshing drink is obtained. A decoction of the leaves mixed with those of *Dracaena reflexa* and those of *Centella asiatica*is given to teething children. The root decoction is used to wash skin infections, while a leaf poultice is applied to nail infection. A decoction of the root is reported to be a good purgative. In Madagascar, Boiteau-Allorge (1999) reports that an infusion of the leaves is diuretic, used against malarial fevers, haematuria, and seems to protect the red blood corpuscles against lysis *in vitro*. In has also been reported that bile-related problems are treated and/or managed by taking an oral infusion prepared from the following mixture of a panoply of plants: leaves of *A. theiformis*, maize stigma, pumpkin leaves, leaves of *Cyperus* sp., and plants of *Euphorbia hirta*.

The main constituent of *A. theiformis*is mangiferin, also known as aphloiol and whose structure has been established as being C-glucoside of 1,3,6,7-tetrahydroxyxanthone (Paris and Etchepare 1964). Preliminary screenings of the leaf extract showed the presenceof saponins and tannins (Smadja and Vera 1991). Lavergne (1990) reported on the composition of the leaves and stems: they were found to contain phenols, saponins, terpenes, triterpenes, flavones, flavonoids and traces of tannins. The plant growing in Madagascar is reported to be rich in leucoanthocyanins, flavonoids, tannins, steroids, 2-desoxy-2 sugars, saponins and traces of polyphenols (Gurib-Fakim and Guého 2000). The methanolic extract of the leaves from Mauritius is reported to contain three triterpenes, namely tormenticacid glucoside ester, 23-hydroxytormenticacid glucoside ester and 6β-hydroxytormentic acid (Gopalsamy et al. 1988). Previous works have reported the presence of aphloiol, a tetrahydroxyflavone (Pernet 1957).

The anti-infective properties of *A. theiformis* were studied against different strains of pathogenic bacteria and fungi, namely *Salmonella enteritidis*, *Enterobacter cloacae*, and *Bacillus subtilis*, which were all found to be susceptible to the crude methanolic extract (Rangasamy et al. 2007). The methanolic extracts of the leaves kill the snail *Biomphalaria glabra* at a concentration of 400 ppm. This snail is a vector for *Schistosoma mansoni*, the latter being responsible for the transmission of the disease bilharzia (Gopalsamy et al. 1988).

Recently, Hsoidrou et al. 2014 investigated the immunomodulatory potential of the phenolic fraction of *A. theiformis* leaves and their anti-inflammatory potent. Phenolic fraction from the leaves of A. theiformis was obtained by methanol extract, and recovered by an aqueous solution of 5% HCl, then filtered. We investigated the potential immunomodulatory effect in rat, by flow cytometer. Anti-inflammatory activity was also investigated in carrageenan-induced paw oedema rat model. Results demonstrated an immunosuppressive effect on lymphocyte for all doses administered. However, we noted an interesting immunostimulant effect on monocytes and granulocytes number at high doses. Then, it was noted an important anti-inflammatory-like effect for these doses *in vivo*. This work suggests the eventual use of this plant in the treatment of some inflammatory symptoms and immune dysfunctions.

## Elaeodendron orientale *Jacq. (Celastraceae)*

*Elaeodendron orientale* Jacq. also known as *Cassine orientalis* Jacq. Kuntze. (Patrimoine Réunion 2015) is an endemic plant of Madagascar and the Mascarene Islands (Osorio et al. 2014). *E.orientale* belongs to the Celastraceae family and is commonly known as 'Bois d'olive' and 'Bois rouge' in Mauritius and Réunion island respectively (Patrimoine Réunion 2015). The bark and leaves of *E. orientale* have been used in traditional medicine to treat various ailments namely, skin infections, veneral disease, scorpion fish poisoning (Osorio et al. 2014; RIS 2011), urinary infections, 'tambave' (RIS 2011), hypertension (Neergheen et al. 2005) and fish allergies (Gurib-Fakim et al. 2005). Traditionally, the bark is used for treating venereal diseases, skin infections, and scorpion fish poisoning. The leaves have astringent and emetic properties and are used to treat hypertension, and when mixed with those of *Kalanchoe pinnata* (Crassulaceae), which produce bufadienolides, can be used against fish allergies. Additionally, the leaves of *E. orientale* were reported to possess astringent and emetic properties (Osorio et al. 2014).

Recently, it was reported that cardenolide glycosides isolated from the leaves and fruits of *E. orientale* showed growth inhibition activity comparable to or higher than digoxin and digitoxigenin against cancer cell lines namely, carcinoma of the cervix, lung carcinoma and breast adenocarcinoma (Osorio et al. 2014). A new series of cardenolides isolated from *E. orientale* have been evaluated as anticancer agents. All the assayed cardenolide glycosides showed a potent growth inhibition activity against human tumour and leukaemia cell lines, and four of showed an effect comparable to the reference agents, digoxin and digitoxigenin. In addition, and taking into account the narrow security range of the GCs used in clinic, the authors proposed that these new compounds show similar SI to reference drugs on various cancer cell lines, thus increasing their potential for future study (Osorio et al. 2014).

Furthermore, the crude methanolic extract of *E. orientale* leaves was reported to possess α-amylase ($IC_{50}$ value 1745.58 ± 31.66 μg/ml compared to acarbose 1100.06 ± 0.03 μg/ml) and α-glucosidase (1.75 ± 0.26 μg/ml compared to acarbose 5115.73 ± 3.91 μg/ml) inhibitory properties. The extract exhibited an uncompetitive type of inhibition against both enzymes (Picot et al. 2014). Catechin, procyanidin dimers, kaempferol, myricetin and quercetin were found to be present in *E. orientale* (Neergheen et al. 2005). Further scientific investigation of the prophylaxis properties of *E. orientale* is crucial for the validation of ethnopharmacological data.

## Erythroxylum laurifolium *Lam. (Erythroxylaceae)*

*Erythroxylum laurifolium* Lam. also known as *Erythroxylum sideroxyloides* Lam. is an endemic plant of Mauritius and Réunion Island (Tisanes Indigènes 2014). *E.laurifolium* belongs to the Erythroxylaceae family and is commonly known as 'Bois de ronde' and 'Bois de rongue' in Mauritius and Réunion Island respectively

(Ministry of Agriculture and Natural resources 1996; Tisanes Indigènes 2014). *E.laurifolium* has been used in the Mauritian traditional medicinal system to treat diabetes (Mootoosamy and Mahomoodally 2014; Tisanes Indigènes 2014), renal stones (Richards 2012), nephritic spasm (Ministry of Agriculture and Natural Resources 1996), fever, throat infections and as diuretic (Neergheen et al. 2005).

According to Lavergne (1990) the old medicinal use of *E. laurifolium* was as a diuretic, an astringent and a remedy to treat renal colic, gravel, croup and inflammation of the throat. Most herbalists today use it as a remedy to treat renal colic, against kidney disorders, kidney and urinary stones and as a diuretic (Lavergne 1990). *E. laurifolium* was reported to be an herbal ingredient of an herbal tea used for the treatment of dysentery and kidney problems and composed of 15 cm of the bark of *E. laurifolium* Lam., 10 leaves of *Antidesma madagascariense* Lam., a handful of roots of *Coix lacryma-jobi*, 15 leaves of *Aphloia theiformis*, 0.5 cm of the root of *Rhizophora mucronata*, 3 whole plants of *Bidens pilosa* and 10 leaves of *Piper* species. The mixture was boiled in 1.5 l of water and was allowed to reduce to 1 l (Narod et al. 2004).

Ethanolic extract (99%) of *E. laurifolium* leaves showed angiotensin converting enzyme inhibitory activity. The activity was found to be due to proanthocyanidins or condensed tannins and the flavonoids quercetin-3-O-rhamnoside (quercitrin) ($IC_{50} = 0.67$ mM) and kaempferol-3-O-rhamnoside (afzelin) ($IC_{50} = 2.8$ mM). The combination of quercitrin and afzelin caused synergistic inhibitory activity on angiotensin converting enzyme (Hansen et al. 1996). Hansen et al. (1996) concluded that this laboratory study tend support the traditional use of *E. laurifolium* as a diuretic (and therefore an anti-hypertensive remedy) by the herbalists at Reunion Island and Mauritius. The presence of phenols, flavonoids, saponins, tannins, kaempferol, myricetin and quercetin were also reported (Neergheen et al. 2005).

Anti-microbial screening of *E. laurifolium* methanolic extracts showed inhibitory activity on *Salmonella enteritidis*, *Pseudomonas aeruginosa*, *Enterobacter cloacae*, *Bacillus subtilis*, *Scleorotinia sclerotium* and *Candida albicans* using the disk diffusion method (Rangasamy et al. 2007). The minimum inhibitory concentrations of *E. laurifolium* methanolic extracts on *Bacillus subtilis*, *Candida albicans* and *Salmonella enteritidis* were 1000, 125 and 500 μg/ml respectively (Rangasamy et al. 2007). The crude methanolic extract of *E. orientale* leaves was reported to possess α-amylase ($IC_{50}$ value 7472.92 ± 5.99 μg/ml compared to acarbose 1100.06 ± 0.03 μg/ml) and α-glucosidase (1.02 ± 0.02 μg/ml compared to acarbose 5115.73 ± 3.91 μg/ml) inhibitory properties. The extract exhibited an uncompetitive and mixed type of inhibition on α-glucosidase and α-amylase respectively (Picot et al. 2014). The capacity of *E. laurifolium* to inhibit lipid peroxidation was demonstrated using 2 systems namely, $Fe^{3+}$/ascorbate system ($IC_{50}$ 0.0435 ± 0.001 mg FW/ml) and 2, 2′-azobis (2-amidinopropane) dihydrochloride system ($IC_{50}$ 0.05 ± 0.002 mg FW/ml) (Soobrattee et al. 2008).

## Faujasiopsis flexuosa *(Lam.) C. Jeffrey (Asteraceae)*

*Faujasiopsis flexuosa* is an endemic plant of Mauritius traditionally used for the treatment and management of dysentery and diabetes. The use of medicinal plants is well anchored in the Mauritian culture. In the quest for therapeutic agents the study of local folk plants has gained much momentum. Based on ethnopharmacological data of the traditional use of *F. flexuosa*, scientists have attempted to validate its hypoglycemic and antimicrobial properties, hereby justifying its medicinal virtues. It was suggested that the possible presence of polyines and alkaloids might justify *F. flexuosa* antimicrobial and hypoglycemic properties. This monograph appraises the pharmacological studies undertaken in order to rationalize the use of *F. flexuosa* in traditional medicine and provides an account of its botanical description and ethnopharmacological uses.

*Faujasiopsis flexuosa* originally known as *Eupatorium flexuosum* Lam. (GBIF 2015) is a tropical plant of the Asteraceae family endemic to Mauritius (Mootoosamy and Mahomoodally 2014). The genus *Faujasiopsis* consists of 3 species which belong to the Mascarene has Archipelago (Gurib-Fakim and Brendler 2004a, b). The species *flexuosa* and *erecta* are commonly found in Mauritius and *bourbonensis* originates from Réunion Island (Gurib-Fakim and Brendler 2004a, b). Commonly known as 'Bois cassant' in Mauritius (Mootoosamy and Mahomoodally 2014; Gurib Fakim and Brendler 2004a, b). *F. flexuosa* is found in humid forests at medium altitudes (600–700 m) (Gurib-Fakim and Brendler 2004a, b). Other vernacular names commonly used in Réunion island include 'zigzag' and 'petit café' (CIRAD 2008). The plant was named after Faujas de Saint Fond the senior curator of the Natural History Museum in Paris and the epithet 'flexuosa' came from the latin word 'flexus' meaning 'full of bends' (Gurib-Fakim and Brendler 2004a).

*F. flexuosa* is presented as an erected or lianescent shrub of 1–2 m tall (Gurib-Fakim and Brendler 2004a, b; CIRAD 2008). The slender spreading branches are easily breakable (Gurib-Fakim and Brendler 2004a, b). The petioled leaves measuring between 6 to 17 cm long and 2 to 5 cm wide are ovale to lanceolate and have serrated to denticulated margins (Gurib-Fakim and Brendler 2004a, b) which turn slightly purple when mature (CIRAD 2008). Flower clusters are numerous and appear at branches ends (Gurib-Fakim and Brendler 2004a, b). Flower heads are of disc shape and the white corollas measure between 3 and 3.6 mm long.

The leaves and the whole plant of *F. flexuosa* were used by folk Mauritians for the treatment of diabetes, asthma and dysentery (Mootoosamy and Mahomoodally 2014; Gurib-Fakim and Brendler 2004a, b). Several preparations have been used for the treatment of diabetes.

Leaves of *F. flexuosa* are known to contain alkaloids, phenols, flavonoids, anthocyanins, anthraquinones, tannins, sterols, terpenes, triterpenes and traces of saponins (Gurib-Fakim and Brendler 2004a, b; Mahomoodally et al. 2012a; Picot et al. 2014). It was also reported that polyines present in several species of the Asteraceae family might be responsible for antibacterial activity against gram positive and gram negative bacteria and also viruses (Gurib-Fakim and Brendler 2004a, b). The hypoglycemic and antiasthmatic properties of *F. flexuosa* might be attributed to the presence of alkaloids in the plant (Gurib-Fakim and Brendler 2004a, b).

The crude water fraction of stem and leaves of *F. flexuosa* showed potent inhibitory effects on Gram negative bacteria namely *Escherichia coli* (minimum inhibitory concentration [MIC] = 4 mg/ml), *Pseudomonas aeruginosa* (MIC = 8 mg/ml), *Salmonella typhimurium* (MIC = 4 mg/ml) and Gram positive bacterium *Staphylococcus aureus* (MIC = 4 mg/ml). Stem methanol: chloroform (1:1) fraction, leaf hexane fraction, leaf methanol: chloroform (1:1) fraction and leaf methanol fraction inhibited fungi *Candida albicans* (MIC = 16 mg/ml) (Narod et al. 2004).

The antioxidant potential of crude extracts and fractions of *F.flexuosa* leaves was validated using standard *in vitro* assays. Crude water/methanolic extracts, dichloromethane/ethyl acetate/n-butanol and water fractions of *F. flexuosa* leaves showed $IC_{50}$ values ranging between 3.39 to 4.87 μg/ml, 4.95 to 37.01 μg/ml, 6.06 to 10.30 μg/ml and 3.89 to 6.13 μg/ml on 2,2-diphenyl-1-picrylhydrazyl, superoxide radical, nitric oxide and lipoxygenase respectively. $IC_{50}$ values of *F. flexuosa* for DPPH, SO and NO were comparable to the control, ascorbic acid. *F. flexuosa* crude extracts and fractions showed $IC_{50}$ values lower than the control quercetin on lipoxygenase (Mahomoodally et al. 2012b).

The immunomodulating capabilities of *F. flexuosa* extracts were assessed by evaluating the immune response of phagocytes and monocytes at different steps. The inhibitory activity of *F. flexuosa* crude methanolic extract on whole blood phagocytes for reactive oxygen species production was 88.7%, 72.4% and 60.7% at 100, 50 and 25 μg/ml respectively. It was suggested that *F.flexuosa* inhibited biochemical targets such as NADPH oxidase enzyme or scavenge reactive oxygen species rather than affecting a specific transductional pathway (Mahomoodally et al. 2012c).

The inhibitory potential of the extracts *F. flexuosa* leaves on α-amylase was validated using *in vitro* assays. Crude water/methanolic extracts, dichloromethane/ethylacetate/*n*-butanol and water fractions gave $IC_{50}$ values ranging from 27.36 ± 4.17 to 384.72 ± 39.9 μg/ml. Ethyl acetate fraction ($IC_{50}$ value 27.36 ± 4.17 μg/ml) showed best inhibitory activity on α-amylase with an $IC_{50}$ value significantly ($P < 0.05$) lower than the known antidiabetic drug acarbose ($IC_{50}$ value 75.86 ± 8.16 μg/ml). Further investigation on the activity of *F. flexuosa* ethylacetate fraction on amylase activity in mouse plasma showed that *F. flexuosa* inhibited amylase activity significantly ($P < 0.05$) from 7.54 to 87.39% with increasing extract concentration (100–800 *μg*/ml) (Mahomoodally et al. 2012a). No significant inhibition on α-glucosidase was reported (Mahomoodally et al. 2012a; Picot et al. 2014).

The possible glucose diffusion retardation properties of crude methanolic and aqueous extracts of *F. flexuosa* leaves were also investigated using a simple *in vitro* dialysis model. It was reported that both extracts (concentration of glucose in external solution after 4 hr: 0.738 ± 0.020 mM and 0.732 ± 0.009 mM for methanolic and aqueous extract respectively) retarded glucose diffusion significantly ($P < 0.05$) compared to the negative control experiment (concentration of glucose in external solution after 4 hr: 0.906 ± 0.015 mM) (Picot et al. 2014). These findings tend to suggest that *F.flexuosa* exerts hypoglycemic activity by inhibiting α-amylase, a key carbohydrate hydrolyzing enzyme and impeding glucose diffusion.

Additionally, it was reported that *F.flexuosa* crude extracts and fractions possessed antiglycation activities as compared to the negative control experiment and comparable ($P < 0.05$) to the known antiglycation drug aminoguanidine. $IC_{50}$ values ranged from 25.31 ± 1.98 to 69.54 ± 2.32 μg/ml (Mahomoodally et al. 2012b).

## Toddalia asiatica *Lam. (Rutaceae)*

*Toddalia asiatica* is indigenous to the Mascarene region and is also found in Malawi (Morris 1996) and Mozambique. It is fairly common and is found in numerous locations from sea level to high altitude forests. *T asiatica* is a scandent shrub or woody liana, bearing short, retrorse, dark brown prickles, usually pubescent to shortly hispid on younger parts, sometimes glabrous. The leaves are mainly consumed as remedies in traditional medicines. Two teaspoons of the root decoction, along with the leaves and root bark of *Rhus vulgaris*, the root bark of *Symphonia globulifera*, are drunkonce daily against colds (Neuwinger 1996). A bark maceration is used against blenorrhagia (gonorrhoea). The leaves and twigs are boiled and the vapours inhaled against bronchial diseases and malaria. The fruit or fruit decoction is used against cough, colds, and stomach pain. A decoction of the rootsap is taken as an emetic and against paralysis resulting from snake bites. The root sap is also applied to scarification resulting from snake bites. The root decoction is also drunk against cardiac and back pain, and also when there are digestive troubles resulting from over consumption of meat. The root decoction is also taken in the event of rheumatism and sterility. A poultice made from the root is applied to swollen limbs. The leaves are boiled with butter and drunk against pneumonia and rheumatism. A glassful of the leaf decoction is drunk twice daily against rheumatic pains. This decoction is also gargled against toothache. The leaves are also chewed against toothache, and a decoction of the aerial parts of the plant is drunk against malaria. A poultice made from the ground leaves is applied on scabies. Boiled in water, the decoction made from the ground leaves is drunk against snake bite. The plant resin is drunk by pregnant women as an oxytocicagent (Kokwaro 1976; Neuwinger 1996).

The phytochemicals compounds isolated are; Coumarins (toddaculin, coumurrayin, toddanone, 8-(3,3-dimethylallyl)-6,7-dimethoxycoumarin, isopimpinellin,6-(3-chloro-2-hydroxy-3-methylbutyl)-5,7-dimethoxycoumarin,6-formyllimettin,5,7,8-trimethoxycoumarin, toddasin,(+)-toddanol, 6-(2-hydroxy-3-methoxy-3-methylbutyl)-5,7-dimethoxycoumarin,and toddalolactone); five new coumarins (toddalenol, toddalosin, 5-methoxysuberenon, toddalenone, and 8-formyllimettin); alkaloids (benzo-phenanthridine, alkaloids (des-*N*-methylchelerythrine, oxychelerythrine, arnottianamide, oxyavicine, avicine, chelerythrine and chelerythrine-psicyanide); four known quinoline alkaloids (*N*-methylflindersine,4-methoxy-1-methyl-2-quinolone, skimmianine, integriquinolone); and one known triterpenoid (amyrin). In vitro assay of the root bark of *T. asiatica* has shown significant activity against the malarial vector *Plasmodium falciparum*. Bioguided fractionations of the bioactive extracts has revealed that the alkaloid nitidine, as well as other coumarins present in the plant (Gakunju et al. 1995; Oketch-Rabah et al. 2000), is largely responsible for this activity. Nitidine has previously received considerable attention as it is a well-known cytotoxic agent, particularly against leukaemia (Gakunju et al. 1995). Dihydronitidine, on the other hand, has also been reported to exhibit tumour-specific cytotoxicity *in vitro* (Iwasaki et al. 2006).

The aerial parts of *T. asiatica* var. *floribunda* have been tested for spasmolytic *activity in vitro* on guinea-pig ileum. Although the activity was significant, it was not attributed to the presence of the coumarins toddalolactone and toddanone, as previously reported. Attempts have been made to assign this activity to the presence of chlorocoumarin *in vitro* (Rastogi and Mehrotra 1998). However, these observations would justify the use of this plant against colics (Lakshmi et al. 2002).

Chinese scientists have tested the plant extracts on the cardiovascular system and haveisolated an active constituent, which has been identified to be isopimpinellin (Guo et al. 1998). The wood extract yielded seven compounds having strong anti-platelet aggregation activity in vitro (Tsai et al. 1998). The aqueous leaf extracts have also been tested for their antifeedant activity, and a 1% concentration showed 86.1% mortality (Sundararajan and Kumuthakalavalli 2001).

The essential oil extracted from the leaves (Veeramuthu et al. 2006) and the leaf extracts (Duraipandiyan et al. 2006) inhibited the growth of several bacteria in vitro, namely *Bacillus subtilis, Staphylococcus aureus* and *S. epidermidis*, among others *T. asiatica* has also been shown to have anti-inflammatory and analgesic effects in rats, and it appears that long-term administration does not cause injury to the liver (Hao et al. 2004).

Recent *in vivo* tests on mice using extracts of the root bark of *T. asiatica* has shown that the latter shows remarkable suppression of parasitaemia (38–66%) either alone or in combination with chloroquine. The survival rate of the mice was found to be longer than that of controls (Muregi et al. 2007a, b).

# 3 Conclusion

The tropical island of Mauritius is considered as a biodiversity hotspot in the Indian Ocean. It is currently the habitat of around 700 species of indigenous and endemic plants. Interestingly, the local inhabitants of Mauritius hold an inherent use of medicinal plants emanating from their ancestors for decades. Nonetheless, there is still a dearth of scientific validation and compilation of some indigenous plants in common use. Hence, such practice need proper documentation, conservation and scientific validation.

## References

Adjanohoun E, Ahyi MRA, Eymê J, Gassita JN, Goudotê E, Guêho J, Ip FSL, Jackaria D, Kalachand SKK, Keita A, Koudogbo B, Landreau B, Owadally AW, Soopramanien A (1983) Contributions aux etudes ethnobotaniques et floristiques à Maurice (Ile Maurice et Rodrigues). A.C.C.T, Paris

Boiteau-Allorge L (1999) CD-ROM-Plantes Medicinales de Madagascar. Editions Lune Rouge, Paris

Centre de cooperation international en recherché agronomique pourle développement (CIRAD) (2008) Faujasiopsis flexuosa (Lam.) C. Jeffrey. http://arbres-reunion.cirad.fr/especes/asteraceae/faujasiopsis_flexuosa_lam_c_jeffrey. Accessed 7 June 2016

Centre de cooperation international en recherché agronomique pour le développement. Antidesma madagascariense Lam (2006) http://arbres-reunion.cirad.fr/especes/euphorbiaceae/antidesma_madagascariense_lam. Accessed 7 Jun 2016

CSO (2014) Central Statistical Office, Mauritius

Danthu P, Lubrano C, Flavet L (2010) Biological factors influencing production of Xanthones in *Aphloia theiformis*. Chem Biodivers 7:140–150

Daruty DC (1886) Plantes Médicinales de Maurice. Port Louis, Maurice

Duraipandiyan V, Ayyanar M, Ignacimuthu S (2006) Antimicrobial activity of some ethnomedicinal plants used by Paliyar tribe from Tamil Nadu, India. BMC Complement Altern Med 6:35

Gakunju DMN, Mberu EK, Dossaji SF, Gray AI, Waigh RD, Waterman PG, Watkins WN (1995) Potent antimalarial activity of the alkaloid nitidine isolated from a Kenyan herbal remedy. Antimicrob Agents Chemother 39(12):2606–2609

Global Biodiversity Information Facility (GBIF) (2015) *Faujasiopsis flexuosa* (Lam.) C. Jeffrey. http://www.gbif.org/species/107958745. Accessed 7 Jun 2016

Gopalsamy N, Vargas D, Guého J, Ricaud C, Hostettmann K (1988) Saponins from the leaves of *A. theiformis*. Phytochemistry 27:593–3595

Guo S, Li S, Peng Z, Ren X (1998) Isolation and identification of active constituent of *Toddalia asiatica* in cardiovascular system. Zhong Yao Cai 21:515–516

Gurib-Fakim A (2011) Medicinal plants of Mauritius and of the world. Graphic Press Ltd, Baie du Tombeau

Gurib-Fakim A, Brendler T (2004a) Medicinal and aromatic plants of Indian Ocean Islands: Madagascar, Comoros, Seychelles and Mascarenes. MedpharmSci Publication, Stuttgart

Gurib-Fakim A, Brendler T (2004b) Medicinal plants of Indian Ocean Islands. MedPharm Scientific Publishers, Stuttgart

Gurib-Fakim A, Gueho J (1996) Plantes Médicinal de Maurice, TOME2. Edition de L'Ocean Indien, Rose-Hill, pp 71–77

Gurib-Fakim A, Guého J (2000) Inventaire et Etudes des Plantes Aromatiques et Médicinales des Etats du Sud-Ouest de l'Océan Indien. CD-ROM, COI/UE, Nouveaux Territoires, St. Denis, La Réunion (France)

Gurib-Fakim A, Subratty H, Narod F, Govinden-Soulange J, Mahomoodally F (2005) Biological activity from indigenous medicinal plants of Mauritius. Pure Appl Chem 77(1):41–51

Hansen K, Adsersen A, Smitt UW, Nyman U, Christensen SB, Schwartner C, Wagner H (1996) Angiotensin converting enzyme (ACE) inhibitory flavonoids from *Erythroxylum laurifolium*. Phytomedicine 2(4):313–317

Hao XY, Peng L, Ye L, Huang NH, Shen YM (2004) A study on anti-inflammatory and analgesic effects of alkaloids of *Toddalia asiatica*. Zhong Xi Yi Jie He Xue Bao 6:450–452

Hoffman P (1999) Kew bulletin. Royal Botanic Gardens, Kew

Hollup O (1994) The disintegration of caste and changing concepts of Indian ethnic identity in Mauritius. Ethnology 33:297–316

Hollup O (2012) Nurses in Mauritius motivated by extrinsic rewards: a qualitative study of factors determining recruitment and career choices. Int J Nurs Stud 49:1291–1298

Hsoidrou S, Bousta D, Lachkar M, Said Hassane SO, El Youbi-Hamsas A, Mansouri LEL, Benjilali J, El-Hajaji H, Farah A (2014) Phytochemical screening, immunomodulatory and anti-inflammatory activities of the phenolic fraction of *Aphloia theiformis* (Vahl) Benn. Plant used in comorian traditional medicine. Phytothérapie 21:354–359. doi:10.1007/s10298-014-0875-1

Iwasaki H, Oku H, Takara R, Miyahira H, Hanashiro K, Yoshida Y, Kamada Y, Toyokawa T, Takara K, Inafuku M (2006) The tumor specific cytotoxicity of dihydronitidine from *Toddalia asiatica* Lam. Cancer Chemother Pharmacol 58:451–459

Jolly A, Sussman RW, Koyama N, Rasamimanana H (2006) Ring tailed lemur biology: *Lemur catta* in Madagascar. Springer, Spring Street, New York. http://books.google.mu/books?id=WQ5yWlEJVVYC&pg=PA146&dq=antidesma+madagascariensis&hl=en&sa=X&ei=HSniUuGnKu6e7AaTkYHwBg&ved=0CCsQ6AEwAA#v=onepage&q=antidesma%20madagascariensis&f=false. Accessed 7 Jun 2016

Joseph A, Troester B (2012) Can the Mauritian miracle continue? The role of financial and ICT services as prospective growth drivers. Working Paper No. 01/2013, Berlin

Kokwaro JO (1976) Medicinal plants of East Africa. University of Nairobi Press, Nairobi

Kuete V (2013) Medicinal plant research in Africa. Elsevier, London

Lakshmi V, Kapoor S, Pandey K, Patnaik GK (2002) Spasmolytic activity of *Toddalia asiatica* var. floribunda. Phytother Res 16:281–282

Lavergne R (1990) Tisaneurs et Plantes Médicinales Indigènes de l'Ile de La Réunion. Editions Orphie, Chevagny sur Guye

Mahomoodally F, Gurib-Fakim A, Subratty AH (2006) Stimulatory effects of *Antidesma madagascariense* on D-glucose, L-tyrosine, fluid and electrolyte transport across rat everted intestine, comparable to insulin action *in vitro*. Br J Biomed Sci 63(1):12–17

Mahomoodally F, Mesaik A, Choudhary MI, Subratty AH, Gurib-Fakim A (2012a) In vitro modulation of oxidative burst via release of reactive oxygen species from immune cells by extracts of selected tropical medicinal herbs and food plants. Asian Pac J Trop Med 5:440–447

Mahomoodally FM, Subratty AH, Gurib-Fakim A, Choudhary MI (2012b) Antioxidant, antiglycation and cytotoxicity evaluation of selected medicinal plants of the Mascarene Islands. BMC Complement Altern Med 165:1–12

Mahomoodally MF, Subratty AH, Gurib-Fakim A, Choudhary MI, Khan SN (2012c) Traditional medicinal herbs and food plants have the potential to inhibit key carbohydrate hydrolyzing enzymes in vitro and reduce postprandial blood glucose peaks *in vivo*. Sci World J 12:1–9

Mahomoodally MF, Korumtollee HN, Chady ZZBK (2015) Ethnopharmacological uses of *Antidesma madagascariense* Lam. (Euphorbiaceae). J Intercult Ethnopharmacol 4(1):86–89

Ministry of Agriculture and Natural resources (1996) Mauritius: Country report to the FAO international technical conference on plant genetic resources. http://www.fao.org/fileadmin/templates/agphome/documents/PGR/SoW1/asia/MAURITIU.PDF. Accessed 7 Jun 2016

Ministry of Agro Industry and Food Security (MOAIFS) (2012) Forestry-indigenous and endemic plants. Republic of Mauritius. http://agriculture.gov.mu/English/Pages/Forestry/Indigenous-and-Endemic-Plants.aspx. Accessed 7 Jun 2016

Mootoosamy A, Mahomoodally MF (2014) Ethnomedicinal application of native remedies used against diabetes and related complications in Mauritius. J Ethnopharmacol 10:413–444

Morris B (1996) Chewa medical botany: a study of herbalism in Malawi, Monographs from the International African Institute. LIT Verlag, Hamburg, pp 227–228

Muregi FW, Ishih A, Suzuki T, Kino H, Amano T, Mkoji GM, Miyase T, Terada M (2007a) *In vivo* antimalarial activity of aqueous extracts from Kenyan medicinal plants and their chloroquine (CQ) potentiation effects against a blood-induced CQ resistant rodent parasite in mice. Phytother Res 21:337–343

Muregi FW, Ishih A, Miyase T, Suzuki T, Kino H, Amano T, Mkoji GM, Terada M (2007b) Antimalarial activity of methanolic extracts from plants used in Kenyan ethnomedicine and their interactions with chloroquine (CQ) against a CQ-tolerant rodent parasite in mice. J Ethnopharmacol 111:190–195

Narod FB, Gurib-Fakim A, Subratty AH (2004) Biological investigations into *Antidesma madagascariense* Lam. (Euphorbiaceae), *Faujasiopsis flexuosa* (Lam.) C. Jeffrey (Asteraceae), *Toddalia asiatica* (L.) Lam. and *Vepris lanceolata* (Lam.) G. Don (Rutaceae). J Cell Mol Bio 3:15–21

Neergheen VS, Bahorun T, Soobrattee MA, Aruoma OI (2005) A survey of the Mauritian endemic flora for potential prophylaxis. Food and Agricultural Research Council, Réduit, Mauritius. http://farc.govmu.org/English/Documents/Policy%20and%20promotion/information%20outreach/MAS20051stDraft.pdf. Accessed 7 Jun 2016

Neuwinger HD (1996) African Ethnobotany: poisons and drugs. CRC Press, Boca Raton

Oketch-Rabah HA, Mwangi JW, Lisgarten J, Mberu EK (2000) A new antiplasmodial coumarin from *Toddalia asiatica* roots. Fitoterapia 71:636–640

Osorio AA, Lope MR, Jimenez IA, Moujir LM, Rodriguez M, Bazzocchi IL (2014) *Elaeodendron orientale* as a source of cytotoxic cardenolides. Phytochemistry 105:60–67

Paris RR, Etchepare S (1964) A propos de la structure de l'aphloiol, parenté de cettexanthone avec la Mangiférine. Comptes Rendus de Académie des Sciences 258(21):5277–5279

Patrimoine Réunion (2015) La flore de La Réunion, *Elaeodendron orientale* Jacq. http://patrimoine-reunion.org/patrimoine-mondial/notre-flore/details/1/37. Accessed 7 Jun 2016

Pernet R (1957) Les Plantes Médicinales Malgaches. Catalogues de nos connaissances chimiques et pharmacologiques. Mémoires de l'Institut Scientifiques de Madagascar, Série B-Tome VIII

Picot CMN, Subratty AH, Mahomoodally FM (2013) Phytochemical profile and antioxidant properties of six medicinal plants traditionally used in the management of diabetes in Mauritius. Pharmacologia 5:42–49

Picot CMN, Subratty AH, Mahomoodally MF (2014) Inhibitory potential of five traditionally used native antidiabetic medicinal plants on α-amylase, α-glucosidase, glucose entrapment and amylolysis kinetics *in vitro*. Adv Pharmacol Sci 2014:1–7

Rangasamy O, Raoelison G, Rakotoniriana FE, Cheuk K, Urverg-Ratsimamanga S, Quetin-Leclercq J, Gurib-Fakim A, Subratty AH (2007) Screening for anti-infective properties of several medicinal plants of the Mauritians flora. J Ethnopharmacol 109:331–337

Rastogi RP, Mehrotra BN (1998) Compendium of Indian medicinal plants. Central Drug Research Institute, Lucknow

Research and Information System for Developing Countries (2011) Asian biotechnology and development review http://www.ris.org.in/images/RIS_images/pdf/abdr-nov-mail.pdf. Accessed 7 Jun 2016

Richards A (2012) Mauritius- Réunion-Rodrigues. Globe Pequot Press Inc, USA. https://books.google.mu/books?id=JnIb2OCzDvEC&pg=PA8&lpg=PA8&dq=erythroxylum+laurifolium+%2B+medicinal+properies&source=bl&ots=QbR4NpESR4&sig=nUJevEf7hos5CafiBoxhG2sPi8M&hl=en&sa=X&ei=u8f2VOP3BsP1Uu2CguAC&redir_esc=y#v=onepage&q=erythroxylum%20laurifolium%20%2B%20medicinal%20properties&f=false. Accessed 7 Jun 2016

Seebaluck R, Gurib-Fakim A, Mahomoodally F (2015) Medicinal plants from the genus *Acalypha* (Euphorbiaceae)–a review of their ethnopharmacology and phytochemistry. J Ethnopharmacol 159:137–157

Smadja J, Vera R (1991) Essais préliminaires sur quelques plantes de la Reunion. Plantes Médicinales et Phytothérapie 25(4):212–241

Soobrattee MA, Bahorun T, Neergheen VS, Googoolye K, Aruoma OI (2008) Assessment of the content of phenolics and antioxidant actions of the Rubiaceae, Ebenaceae, Celastraceae, Erythroxylaceae and Sterculaceae families of Mauritian endemic plants. Toxicol In Vitro 22:45–56

Sreekeeson DP, Mahomoodally MF (2014) Ethnopharmacological analysis of medicinal plants and animals used in the treatment and management of pain in Mauritius. J Ethnopharmacol 157:181–200

Sundararajan G, Kumuthakalavalli R (2001) Antifeedant activity of aqueous extract of *Gnidia glauca* Gilg. and *Toddalia asiatica* Lam. on the gram pod borer, *Helicoverpa armigera* (Hbn). J Environ Biol 22:11–14

Sussman LK (1980) Herbal medicine in Mauritius. J Ethnopharmacol 2:259–269

Tchinda AT, Teshome A, Dagne E, Arnold N, Wessjohann LA (2006) Squalene and amentoflavone from *Antidesma laciniatum*. Chem Soc Ethiop 20(2):325–328

Tisanes Indigènes (2014) *Erythroxylum* laurifolium Lam., Erythroxylaceae, Endémique. http://www.tisanes-indigenes.re/wp-content/uploads/2014/08/Erythroxylum-laurifolium.pdf. Accessed 7 Jun 2016

Tsai IL, Wun MF, Teng CM, Ishikawa T, Chen IS (1998) Anti-platelet aggregation constituents from Formosan *Toddalia asiatica*. Phytochemistry 48:1377–1382

Veeramuthu D, Muniappan A, Savarimuthu I (2006) Antimicrobial activity of ethnomedicinal plants used by Paliya tribe from Tamil Nadu, India. BMC Complement Altern Med 6:35

Wong Ting Fook WTH (1980) The medicinal plants of Mauritius. ENDA, Dakar

World Health Rankings (2014) http://www.worldlifeexpectancy.com/mauritius-life-expectancy. Accessed 8 June 2016

# Some Wild Herbaceous Plants of Nigeria: A Biological, Pharmacognostic and Phytochemical Review

**Sunday E. Atawodi, Mubarak L. Liman, Olufunsho D. Olowoniyi, and Godwin O. Adejo**

**Abstract** In the twentieth century, science makes substantial breakthrough in management and cure of diseases by chemotherapy due to discovery of antibiotics and other chemotherapeutic agents. In the twenty first century, man and other animals are still, however being challenged by emergence of new infectious and non-infectious diseases that have proven to be resistant to the available novel therapeutic drugs. The available orthodox medications have either been ineffective against some causative agents of these diseases; too costly or basically unavailable to the average citizen of developing countries like Nigeria. The use of herbal remedies for prevention, management and cure of diseases is as old as antiquity, yet still common practice among the African population. A number of plants used in traditional herbal medicine have been evaluated by different researchers. This book chapter provides a concise review of traditional use, phytochemical contents, pharmacognostic and biological activities of some wild herbaceous plants that of medicinal importance in Nigeria. Plants including; *Cuminium cyminum* Linn, *Rauwolfia vomitoria* Afzel, *Cassia sieberiana* D.C, *Piliostigma thonningii* (Schumach.) Milne-Redh, *Guiera senegalensis* J.F.Gmel, *Acalypha indica* Linn, *Euphorbia hirta* Linn, *Euphobia unispina* N.E.Br, *Phyllantus muellerianus* (Kuntze) Exell, *Senna occidenatlis* Linn and *Grewia mollis* Juss are comprehensively reviewed. Phytochemicals like glycosides, alkaloids, polyphenols, flavonoids, phytosterols, terpenoids, saponins etc are widely reported to be present in various parts of these plants. The traditional herbal use of the plants in management and cure of ailments, as well as reported biological and pharmacological activities were linked to the presence of these phytochemicals. Based on the information reviewed, it is concluded that these wild herbaceous plants are potential sources of new natural bioactive substances that could be explored and exploited for their therapeutic and industrial applications.

Dedicated to Our Teachers and Mentors for Enduring Knowledge Impacted

S.E. Atawodi (✉) • M.L. Liman • O.D. Olowoniyi • G.O. Adejo
Biochemistry Department, Ahmadu Bello University, Zaria, Nigeria
e-mail: atawodi_se@yahoo.com





M. Neffati et al. (eds.), *Medicinal and Aromatic Plants of the World - Africa Volume 3*, Medicinal and Aromatic Plants of the World,
DOI 10.1007/978-94-024-1120-1_10

**Keywords** Herbaceous plants • Nigeria • Wild plants • Pharmacognostic effects • Phytochemicals • Traditional medicine • Herbal medicine

## 1 Introduction

For time *in-memoriem* plants have been a source of food and therapeutic remedy for humans and his animals. Traditional societies in Africa and elsewhere have always used herbs to promote healing (Atawodi et al. 2009, 2011; Ene et al. 2009; Focho et al. 2009). Even presently, traditional medicine is still the predominant means of health care delivery in many developing countries where larger percentage of their total population depends on it for their well being (Atawodi et al. 2014a, b, c). In addition, plants are important source of food as well as industrial raw materials for various products, and are also the basis for the development of many orthodox drugs. The medicinal values of plants lie in their component phytochemicals such as different classes of alkaloids, tannins, flavonoids and other phenolic compounds, which produce definite physiological actions on animal and human body (Ullah and Khan 2008; Atawodi and Ogunbusola 2009; Ogbe et al. 2009; Asmah et al. 2006).

Nigeria is situated in the Western part of Africa, with coastal boundary delimited by the Gulf of Guinea in the south, and shared land boundary with Cameroon and Chad in the east, Niger Republic in the north and Benin in the west. Nigeria covers a total area of 923.758 $km^2$ with a coastline of 853 km and land boundry of 4047 km. The latitudinal and longitudinal extent of the country is 4° to 14° N and 2° to 15° E making it one of the few countries in the world to have both tropical and sub-Saharan vegetation. As a tropical country, Nigeria is blessed with an enormous biodirversity of natural resources but unfortunately receiving an unfair share of the sun's ultraviolet radiation in addition to the tropical ecosystem's myriad of pathogenic microbes, as well as man-made disease causing agents like tobacco (Atawodi et al. 1995). African plants which were able to withstand and survive in these stressful conditions due to their abilities to accumulate biologically active phytochemicals (Atawodi et al. 2014a), should also be able to produce secondary metabolites that will protect the population.

From the mangrove rain forest of the south to the sunny, seasonal rainy savannah in the north, Nigeria is blessed with an enormous plant biodiversity. The presence of these numerous floras with ancient ethnomedicinal uses in herbal medicine inspired several studies by various researchers on the biological activities and chemical constituents inherent in these plants (Ashaf and Olunu 2011; Adejo et al. 2015). However, many of such plants which include large numbers of wild plants whose phytochemical constituents have been identified and potential biological activities confirmed are yet to be exploited pharmaceutically or industrially. This chapter attempts to concisely provide an update on the traditional use, phytochemical contents, pharmacognostic and biological activities of some Nigerian plants with particular reference to wild herbaceous plants.

**Pictures 1 and 2** *Cuminia cyminum* J.F. Gmel (Source: http://magicgardenseeds.com)

## *Apiaceae*

***Cuminium cyminum* Linn.** Synonyms: *Cuminia cyminum* J.F. Gmel., *Cuminum aegyptiacum* Merax ex DC., *Cuminum hispanicum* Bunge, *Cuminum odorum* Salisb., *Cuminum officinale* Garsault (Inval.), *Cuminum sativum* J. Sm., *Cyminon longeinvolucellatum* St.-Lag., *Linguisticum cuminum* (L.) Crantz, *Selinum cuminum* E.H. Crantz.

**Common Names**
Commonly called Cumin (Johri 2011); Cummin, Roman Caraway, Jeera (India) (Parashar et al. 2014) (Pictures 1 and 2).

**Morphological Description**
*Cuminum cyminum* is an annual herbaceous plant with a slender branched stem. It grows up to 30–60 cm tall. The leaf is alternate, simple or compound and have a sheathing base below. The flowers are small, pink and characteristically borne in umbrella shaped clusters. The flowers have both male and female structures together and an inferior ovary that develops into a very characteristic fruit called a cremocarp, which is a capsule that upon maturity breaks into two one-seeded bits (Parashar et al. 2014).

**Geographical Distribution**
*Cuminium cyminum* grows in many countries of Africa, as well as Asia and Europe (Johri 2011)

**Ecological Requirements**
Cumin needs fairly cool, less humid climate and temperatures of between 25 and 30 °C to grow. It is very sensitive to rainfall, such that rainfalls at times of harvest can drastically lower yield and crop quality. The plant is also susceptible to frost damage during flowering and initial stages of seed formation. It grows best on well drained sandy-loam soil and optimally alkaline pH range of 6.8–8.3 (Anandaraj and Sastry 2015). *Cuminium cyminum* has however, been experimented to thrive well in relatively low temperature zones as higher temperatures lower yields and affects both seedling weight and height. A temperature range of 15–25 °C has been found to be the optimum temperature for cumin seed germination rate, optimum seedling growth, seedling weight and height. With respect to timing, the last week of March – first week of April was established to be the best planting time, as this period offers significantly shorter emergence time, highest plant height and yield (Khosh-Khui and Bonyanpour 2006).

**Major Chemical Constituents and Bioactive Compounds**
The major phytochemicals occurring in cumin are cuminaldelyde, limonene, $\alpha$- and $\beta$- pinene, 1,8–cineole, *o*- and *p*-cymene, $\alpha$- and $\gamma$-terpinene, sufranal and linalool (Johri 2011). Cumin seeds contain up to 5% of a volatile oil composed primarily of aldehydes. In addition, the seeds yield 22% fats, numerous free amino acids and a variety of flavonoid glycosides, including the derivatives of apigenin and luteolin. The cuminaldelyde content varies considerably, depending on the source of the oil (Parashar et al. 2014). Ahmad and Saeidnia (2011) also reported monoterpene hydrocarbons as major components of the plant, while sesquiterpenes are minor components. The analysis of the volatile oil from the fruit of cumin showed the presence of trans-dihydrocarvone, $\gamma$-terpinene, *p*-cymene, $\alpha$-phellandrene and *p*-menth-2-en-7-ol (Chaudhary et al. 2014).

**Traditional Uses, Part(s) Used and Common Knowledge/Uses in Traditional Medicine**
In Tunisian traditional medicine, cumin is considered abortive, galactagogue, antiseptic and antihypertensive (Leporatti and Ghedira 2009). In Indian traditional medical practices, cumin seeds are considered carminative, eupeptic, antispasmodic, and astringent. It is used in Nigeria for the treatment of mild digestive disorders, diarrhea, dyspepsia, flatulence, morning sickness, colic, dyspeptic headache and bloating. It promotes the assimilation of other herbs and improves liver function (Johri 2011). Cumin is also used in bronchio-pulmonary disorders and as a cough remedy.

**Biological Activities**
Several studies have verified the antibacterial actions of cumin against a range of useful and pathogenic gram-positive and gram-negative bacterial strains (DeMartino et al. 2009). Specifically, Chaudhary et al. (2014) reported the potency of the essential oil of cumin against *Staphylococcus epidermidis*, *Staphylococcus aureous*, *Staphylococcus haemolyticus*, *Propionibacterium acnes*, *Corynebacterium diphtheriae*, *Erysipelothrix rhusiopathiae*, *Bacillus cereus*, *Clostridium tetani*,

*Clostridium difficile, Escherichia coli, Salmonella typhi, Klebsiella pneumoniae, Vibrio cholera, Aeromonas hydrophila, Mycobacterium tuberculosis* and *Neisseria gonorrhoeae*. Bokaeian et al. (2014) have also demonstrated that essential oil from cumin possess potent antimicrobial activity against multi-drug resistant *Escherichia coli*

Antifungal activity of cumin has also been reported against soil, food, animal and human pathogens including, dermatophytes, *Vibrio* sp., yeast, aflatoxins and other mycotoxin-producing fungi (DeMartino et al. 2009; Boyraz and Ozcan 2005). Using agar-well diffusion method El-Said and Goder (2014) reported that essential oils of *Cuminum cyminum* inhibited the growth of twenty six fungal isolates, while Chaudhary et al. (2014) confirmed the antifungal activity of cumin oil against *Aspergillus niger, Saccharomyces cererisae* and *Colletrotrichum gloeosporiodes.* Other biological activities reported for cumin include antioxidant, anticarcinogenic, antimutagenic, antidiabetic, diuretic, immunomodulatory, and estrogenic (Johri 2011).

## *Apocynacea*

*Rauwolfia vomitoria* **Afzel** (Synonyms: *Rauvolfia stuhlmannii* K.Schum) (Pictures 3 and 4)

### Botanical Name and Family

*Rauwolfia vomitoria* belongs to the family of Apocynaceae, Its common names are swizzle stick, devil's pepper (English) and Asofeyeje (Fapojuwomi and Asinwa 2012).

### Morphological Description

*Rauwolfia vomitoria* is a shrub or small tree that grows up to 8m, with parts, except old ones containing latex. The branches are usually whorled, while the nodes are enlarged and leaves, lumpy in threes, elliptic-acuminated to broadly lanceolate. Usually, flowers are minute, sweet-scented with branches of inflorescences that are distinctly superfluous with hardly ant-free corolla lubes, while the fruits are fleshly and red in colour (Fapojuwomi and Asinwa 2012).

### Geographical Distribution

*Rauwolfia vomitoria* is native to tropical Africa from Senegal east to Sudan and Tanzania, south to Angola; and naturalized in China, Bangladesh, and Puerto Rico. *Rauvolfia vomitoria* occurs naturally in forest, but is mostly found in forest regrowth where fallow periods are prolonged.

### Ecological Requirements

*Rauwolfia vomitoria* occurs in bush vegetation, gallery forest, secondary vegetation where fallow periods are long, and along roadsides, from sea-level up to 1600 m altitude. *Rauwolfia vomitoria* can be found flowering and fruiting almost throughout the year, but sometimes not and usually less abundantly during the rainy season.

**Pictures 3 and 4** *Rauwolfia stuhlmannii* (Sources: Authors and prota4u.org)

The flowers are pollinated by insects such as small bees and flies, and the fruits are dispersed by birds.

**Major Chemical Constituents and Bioactive Compounds**
*Rauwolfia vomitoria* contains a large number of indole alkaloids, between 40 and 80. Most occur in very small amounts and several are disputed. Most alkaloids occur in an unstable complex, and seasonal variation is present as well. Leaves contain 0.03–0.8% total alkaloids, stem bark about 0.6%, roots 0.15–0.2% and root bark 1.5–2.0%.The alkaloids of *Rauwolfia vomitoria* can be grouped into 5 main types: (1) yohimbine and derivatives, including reserpine and deserpidine (11-demethoxyreserpine); (2) the heteroyohimbine type, including ajmalicine (raubasine), reserpinine (rescinnamine) and reserpiline; (3) sarpagane derivatives, including sarpagine (raupine); (4) the dihydro-indole type, including ajmaline; (5) the anhydronium bases, including alstonine, serpentine and serpenticine. Other groups include the oxindoles and pseudoindoxyls. Serpentinine is the only dimeric yohimbin-related alkaloid isolated so far.

In the root bark reserpiline is the major component, followed by reserpine, reserpinine and ajmaline. In the stem bark reserpiline is also the major component, with small amounts of isoreserpiline and yohimbine. The leaves were found to contain mainly geissoschizol, but no reserpine, reserpinine or ajmaline. The alkaloids in the leaves comprised about 41% heteroyohimbines and 52% oxindoles. The unripe fruit contains several alkaloids, amongst which is 2,6-Dimethoxybenzoquinone, a benzoquinone that is a toxic chemical compound (Van-Dilst and Leeuwenberg 1991).

**Traditional Uses, Part(s) Used and Common Knowledge/Uses in Traditional Medicine**
In the entire distribution area of *Rauwolfia vomitoria,* a root decoction, root macerate or powdered root in water is taken to treat diarrhoea, rheumatism, jaundice, venereal diseases and snakebites. Root products are also widely taken to treat hypertension, and as a sedative to calm people with epilepsy, and people who are

psychotic or mentally ill; they are also used to wash children with colic or fever. Externally, macerated or powdered root or sometimes pulped fruit are applied to a range of skin problems, such as rash, pimples, chicken pox, wounds, scabies, psoriasis, leprosy, hemorrhoids, head lice and parasitic skin diseases. A root decoction is used as a mouth wash against gingivitis or thrush. The stem bark or leaves are also used for these purposes, but to a lesser extent. The stem bark, leaf decoction and latex of young twigs are widely used as purgative or emetic (Burkill 2000).

In Guinea the root maceration is applied to tumors. In Liberia a bark infusion is taken to cure fever. A leaf infusion is rubbed in against yaws. Dried or fresh pulverized roots in palm wine or oil are taken to treat female sterility. A root decoction is used in massages and baths to treat rheumatism, tiredness and rachitis. In Togo pulverized root bark in brandy is taken to treat tuberculosis. In Cameroon a decoction of powdered roots is taken to treat diabetes and malaria. In the Central African Republic a root decoction is taken to treat hernia. In Côte d'Ivoire leaf sap is rubbed between the toes to treat infections caused by humidity. A leaf maceration is used for bathing children with fever. In Gabon chopped and boiled leaves mixed with fat are applied to the skin to cure rheumatism and sprains. A mixture of pulverized root or leaf sap with plant oil or lemon juice is applied to the hair to stop hair loss. In Nigeria the root and leaves in decoction are taken to treat indigestion, as a tonic, and as an abortifacient. In Equatorial Guinea the latex is used for cicatrization of wounds (Terashima and Ichikawa 2003).

In the Central African Republic the roots of *Rauwolfia vomitoria*, alone or together with seeds of *Strophanthus gratus* (Wall & Hook) Franch. are pounded to a paste, which serves as arrow poison. In DR Congo, the roots are a common additive to *Periploca nigrescens* Afzel. hunting poison, while in Equatorial Guinea, the root scrapings are mixed with cassava meal and used as a rat poison. In Nigeria and other West African countries, the root is considered aphrodisiac when taken in palm wine (Latham 2004).

*Rauvolfia vomitoria* is used to treat leprosy in the Democratic Republic of Congo. The plant is very important and useful in the treatment of lunatic patients; the root is added to gin and given to mentally ill persons. It can also be ground into powder and taken with pap, and can be taken in form of decoction used for rheumatic pains. An infusion of the root bark is used to treat jaundice and gastro-intestinal disturbance (Fapojuwomi and Asinwa 2012).

**Biological Activities**

Of the *Rauwolfia* alkaloids, five are used in medicine: reserpine, reserpinine, deserpidine, ajmalicine and ajmaline. There are several patented methods for the extraction of the main component reserpine. Furthermore, several simple and accurate methods have been developed to identify *Rauvolfia* alkaloids, e.g. reserpine, serpentine and ajmaline.

Reserpine is a well-known antihypertensive, antipsychotic and sedative. It is a sympatholytic agent acting indirectly on the peripheral and central nerve terminals (Bedu-Addo 1993). It impairs the storage of biogenic amines resulting in depletion of norepinephrine, dopamine, and serotonin. Depletion of norepinephrine induces a lasting drop in blood pressure. Contra-indications for using reserpine are depression, peptic ulcer, and hypersensitivity to the alkaloid, while the side effects of the medication include drowsiness, nasal congestion, salivary and gastric hypersecretion, paradoxical anxiety, depression and retention of water and $Na^+$. Overdose may also cause respiratory depression, slowed heartbeat, hypotension, confusion, tremors, convulsions and gastro-intestinal distress. Reserpine has been shown to enhance the hypoglycaemic effect of insulin and the hyperglycaemic effect of adrenalin, and has inhibited the physiological hyperglycaemic response in diabetic patients (Nwodo et al. 2003). Because of the necessary high doses and the resulting dangerous side effects, reserpine lost its importance as a medicine. It is only used in low doses for mild to moderately severe high blood pressure, often together with ajmalicine. Reserpinine and deserpidine are reserpine analogues. Both alkaloids have the same effects as reserpine, and can be used to treat the same conditions, while their side effects are reported to be less pronounced. Reserpiline is marked sympatholytic and hypotensive with no noticeable depressant effects on the central nervous system and no sedative properties, and also lacks most of the side effects of reserpine and its analogues.

Ajmalicine is an α-adrenergic blocking spasmolytic which, at high doses, moderates the activity of the vasomotor centres, especially in the brain stem causing an increase of the blood flow to the brain. It is mainly used in products that treat the psychological and behavioural problems associated with senility, stroke and head injuries. Ajmaline is an anti-arrhythmic, which substantially decreases the rate of depolarization of artrial and ventricular cells. Its toxicity has limited its uses and it is mainly prescribed against rapid irregular cardiac beat. Because of its toxicity, it is no longer marketed in several countries. Several other *Rauwolfia* alkaloids have hypotensive or sedative activities, but most are less effective (Nwodo et al. 2003).

An ethanolic leaf extract of *Rauwolfia vomitoria* caused a reduction in blood sugar levels of normal and alloxan-induced diabetic rabbits, comparable to that of tolbutamide, while decoction of the root has no adverse effect on the oestrous cycle, fertilization or implantation, and also has no effect on the foetus or hormone-induced infertility in rats. The root bark extract has been reported to show antibacterial, antiprotozaol activity in vitro against several human pathogens (Fapojuwomi and Asinwa 2012).

## *Caesalpinacea*

***Cassia sieberiana* D.C** (Synonym: *Cassia kotschyana* Oliv.) (Picture 5)

*Cassia sieberiana* DC, (family: Caesalpiniaceae) is distributed widely from Senegal, through Nigeria, and Gambia eastward to DR Congo and Uganda (Van der Maesen 2007).

**Picture 5** *Cassia sieberiana* (Source: http://www.zimbabweflora.co.zw)

**Common Names**
African laburnum, West African laburnum (English), Casse du Senegal, Casse de sieber, Casser agrappes, Casse-flute (French), Mzangaya, Mzangaye (Swahili) (Van der Maesen 2007).

**Morphological Description**
*Cassia sieberiana* is an annual plant that grows in the world tropical zones. It is an upright growing plant that can reach a height of 4 ft with large green leaves and very bright yellow flowers. The seed range in colour from greenish brown to dark brown with smooth surfaces and may have small bright colored bands on the outer surface (Olapade et al. 2014).

**Geographical Distribution**
*Cassia sieberiana* is distributed widely from Senegal, through Nigeria and Gambia east to DR Congo and Uganda (Van der Maesen 2007).

**Ecological Requirements**
It occurs in tree or shrub savanna with less than 800 mm annual rainfall. Acid sandy soil is preferred (Van der Maesen 2007).

**Major Chemical Constituents and Bioactive Compounds**
*Cassia sieberiana* leaves contain flavones (quercitrin, isoquercetrin), an anthraquinone (rhein) and tannins. Anthraquinones and sterols are also present in the root (Van der Maesen 2007). Polyhydroxy and phenolic substances were identified in the root extract (Nartey et al. 2012) while the seeds contain tannins, alkaloids, phenols, oxalate, cardiac glycosides and flavonoids (Olapade et al. 2014).

*Cassia sieberiana* is included in Nigerian Herbal Pharmacopoeia (NHP) and West African Herbal Pharmacopoeia (WAHP) (Ajayi et al. 2014).

**Traditional Uses, Part(s) Used and Common Knowledge**

The entire plant is purgative and diuretic. In Senegal, an infusion of the entire plant is given against all children's diseases. Different parts of the plant are applied to teeth to cure toothache. Across Africa, it is also used to treat skin diseases, malaria, stomachache, ulcer, diarrhea, gonorrhea and sleeping sickness. The decoction of the root is used to treat haemorrhoids, bilharzia, dropsy, bloody dysentery and intestinal worms. The root decoction is also considered an aphrodisiac (Van der Maesen 2007).

In Senegal, the aqueous root extract of *C. Sieberiana* was used in traditional medicine to treat pain and inflammation (Kerharo-Adam 1974). The roots, boiled in water, are used to treat haemorrhoids, bilharzia, leprosy, dropsy and bloody dysentery. In Côte d'Ivoire the decoction is taken in large doses to treat intestinal worms including tapeworms, although this is risky. An infusion of the root bark is employed against venereal diseases, sterility and dysmenorrhoea. After soaking the roots in water, the liquid is used for a bath against tiredness and for body massage. A decoction of the roots is considered an aphrodisiac. In Burkina Faso a pinch of powdered dried decorticated roots taken at the end of each meal is said to prevent malaria. Crushed roots are rubbed on the temples to treat headache. Debarked roots are boiled with bark of *Terminalia macroptera* Guill. & Perr. to combat eczema (Van der Maesen 2007).

**Biological Activities**

Leaf extracts of *Cassia sieberiana* exhibit antibacterial action against *Staphylococcus lutea*, *Mycobacterium phlei, Bacillus subtillis* and *Proteus* sp (Van der Maesen 2007). Several compounds exhibiting a broad spectrum antimicrobial activity have been isolated from *Cassia sieberiana*. A crude extract of the leaves showed antiplasmodial activity against *Plasmodium* including activity against chloroquine-resistant strains of *Plasmodium falciparum*, and flavonoids and isoflavonoids have been isolated as active ingredients. The plant has also been reported to have antioxidant, gastroprotective, laxative, anti-diarrheal anti-inflammatory and analgesic activities (Asase et al. 2005; Nartey et al. 2012)

*Piliostigma thonningii* **(Schumach.) Milne-Redh** (Synonym: *Bauhinia thonningii* Schumach. & Thonn.)

**Botanical Name and Family**

*Piliostigma thonningii* (Schumach.) Milne-Redh is a leguminous plant belonging to the Caesalpiniaceae family (Jimoh and Oladiji 2005).

**Common Names**

Camel's foot, Monkey bread (English), Kalgo (Hausa), Abefe (Yoruba), Okpoatu (Ibo) (Jimoh and Oladiji 2005) (Picture 6).

**Morphological Description**

*Piliostigma thonningii* is a deciduous dioecious tree that grows up to 10–40 m tall, but the bole can be branchless for 2–3 m. The outer bark is rough, longitudinally fissured and dark brown to grey or black. The leaves are alternate, conspicuously

**Picture 6** *Piliostigma thonningii* (Source: Authors)

bi-lobed for one eighth to one third of the length. The stipules are 3–6 mm long, the petiole is 2–7 cm long and the blade grows to between 17 and 21 cm, with base, usually a strong cordate. The apex of lobes are rounded to acute, leathery, upper surface glabrous. The lower surface has rusty brown crisped hairs and conspicuous reticulate venation, palmately veined with 11–15 basal veins. Inflorescence is a panicle, usually alternately leaf-opposed and axillary along branches, male inflorescence very narrowly pyramidal, up to 25 cm × 5.5 cm, female inflorescence up to 7 cm long, few flowered. The many-seeded fruit is an oblong to linear-oblong pod, 12–37 cm × 3–7 cm, woody, brown-pubescent when young but later glabrescent, persisting on the tree but finally decaying on the ground. The seeds are obovoid to ellipsoid, 4–9 mm × 2–7 mm × 3–4 mm dark brown to blackish, compressed. Seedling is with epigeal germination (Lemessa 2010).

**Geographical Distribution**
Commonly called camel's foot tree, *Piliostigma thonningii* is native to tropical Africa, widespread in the Sudano-Guinean region from Senegal eastward to Eritrea and occurs southward to Namibia, Botswana, Mozambique and South Africa. It also occurs in Yemen (Lemessa 2010).

**Ecological Requirements**
*Piliostigma thonningii* occurs from sea level up to 2200 m altitude in areas with annual temperature of 20–2 °C, an average rainfall of 400–1500 mm and a dry season of 6–11 months. It is sensitive to frost, grows on all soil types but heavy clay soils or medium loamy soils are preferred. *Piliostigma thonningii* is common in secondary and gallery forest, woodland, wooded grassland, bush land and also river valleys (Lemessa 2010).

**Major Chemical Constituents and Bioactive Compounds**

The phytochemicals contained in *Piliostigma thonningii* seed include saponins, flavonoids, phenolics, glycosides, anthraquinones and cardiac glycosides. It is also a good source of antioxidant micronutrients such as iron, calcium, selenium, zinc and manganese (Jimoh and Oladiji 2005; Bello et al. 2013) Epicatachin and inositol have been identified in the fruit while saponins, flavonoids, phenolics, anthraquinones and glycosides are in the seeds. The bark contains tannins whereas the leaf contains 2-phenoxychromone, 6,8-di-*C*-methylquercetin-3-methyl ether, 6-*C*-methylquercetin-3,7-dimethyl ether and 6,8-di-*C*-methylquercetin-3,7-dimethyl ether. Quercetin and quercitrin are also present in the leaf (Lemessa 2010).

**Traditional Uses, Part(s) Used and Common Knowledge**

The bark of *Piliostigma thonningii* is commonly used in tying hut, fence and bridge building. Fibre extracted from the bark and root is widely used for making string, rope, and cloth. The fruit is eaten as a snack or as an emergency food, especially by children and herdsmen. An infusion of the bark, leaf or pods is used to coagulate Funtumia latex used in making rubber. The leaf extract is effective in treating inflammations, bacterial infections, worm infestation and to arrest bleeding (Ozolua et al. 2009).

**Biological Activities**

The root, bark and leaf extracts possess antiviral activity against Herpes simplex virus type 1 and 2, HIV and different strains of influenza and syncytial viruses. The root bark also possesses antitussive, anti-inflammatory and analgesic properties. The stem bark has antibacterial activity against *Bacillus subtilis*, *Corynebacterium pyogenes*, *Escherichia coli*, *Proteus vulgaris*, *Shigella dysenteriae* and *Staphylococcus aureus*. The bark extract also have larvicidal and antihelminthic properties. Butanol and ethylacetate extracts have antioxidant activity (Lemessa 2010).

**Uses in Traditional Medicine**

A root preparation is applied on wounds and ulcers as a haemostatic and to promote healing. It is also used as a diuretic and for the treatment of diarrhea, dysentery, worms and other intestinal problems. Root preparations are also used as remedy for cough and snake bite (Jimoh and Oladiji 2005). Preparations of the bark are used to treat pneumonia, skin infections, pains, rheumatism and inflammation. Preparations of the leaf are used for respiratory problems, as antiseptic and to promote wound healing. They are also administered to treat epilepsy and heavy menstrual flow (Lemessa 2010; Bello et al. 2013). The leaf extract is effective in treating inflammations, bacterial infections, worm infestation and to arrest bleeding (Ozolua et al. 2009).

**Picture 7** *Guiera senegalensis* (Source: Authors)

## *Combretaceae*

### *Guiera senegalensis* J.F Gmel. (Picture 7)

**Botanical Name and Family**
*Guiera senegalensis* (Family: Combretaceae) is one of the most important West African medicinal plants often used to treat a variety of microbial infections and the most frequently used part is the leaf (Silva et al. 2008).

**Common Names**
Guier du Senegal (French). Sabara (Hausa), Shafi pitu (Marghi), Tadar (Tangale), Kashishi (Kanuri)

**Morphological Description**
*Guiera senegalensis* is a shrub that can grow to a height of 3–5 m according to habitat. Its stem presents numerous knots that send out branches. The ash-grey stem and branches have fibrous and pubescent bark and bear opposing short petiolated oval leaves, sometimes mucronated, sometimes even cordate, about 2 to 4 cm long by 1 to 2 cm wide. The grey-green leaves, darker on their upper surface display black spots on their lower surface and are slightly downy on both sides (Somboro et al. 2011).

**Geographical Distribution**
*Guiera senegalensis* is a shrub found abundantly in the savannah region of West Africa (Oshobu and Geidam 2014). It is widely distributed in the savannah region of West and Central Africa – Nigeria, Senegal, Gambia, Mali, Niger, Burkina Faso and Ghana (Salihu and Usman 2015).

**Ecological Requirements**

*Guiera senegalensis is* is found in shrub savanna, tree savanna and fallow land from sea level up to 1000 m altitude. It grows in areas with 200–800 mm annual rainfall and can thrive on all types of soil, but mainly on dry sandy or degraded soils, sometimes in areas which are temporarily flooded. It does not tolerate heavy shading. *Guiera senegalensis* is considered an indicator of overgrazing and is very drought resistant (Sanogo 2012).

**Major Chemical Constituents and Bioactive Compounds**

Among the phytochemicals identified in different parts of *Guiera senegalensis* such as leaves, fruits, root and stem bark are resins, alkaloids, tannins, saponins, glycosides and terpenes (Onwuliri et al. 2009), terpenoids, anthraquinones, coumarins, flavonoids, cardiotonic and cyanogenic heterosides (Somboro et al. 2011). The gall extract is particularly rich in flavonoids and polyphenols (Sombie et al. 2011).

**Traditional Uses, Part(s) Used and Common Knowledge**

*Guiera senegalensis* is active against cough, respiratory congestion and fever. It is prescribed as an antitussive, to ease breathing and to treat lung and bronchial disorders and malaria. Various parts are prescribed for stomach pains, dysenteric diarrhea, syphilis, beriberi, leprosy and impotence. A decoction of the leaf is applied against eczema, cold and chest conditions. Veterinary uses of the plant are in diets designed to increase body weight, reproductive capacity and milk secretion in animals (Somboro et al. 2011). It is one of the most important West African medicinal plants often used to treat a variety of microbial infections, including trypanosomiasis (Atawodi et al. 1995), and the most frequently used part is the leaf (Silva et al. 2008).

**Biological Activities**

Aqueous (cold and hot) and methanolic extracts of *Guiera senegalensis* were tested for antimicrobial efficacy against *Staphylococcus aereus, Salmonella typhi, Escherichia coli*, and *Streptococcus pyrogenes* by determining the minimum inhibitory concentration (MIC) and minimum bactericidal concentration (MBC) of the extracts using standard methods. Results showed that all the extracts inhibited the growth of *S. aereus, S. typhi* and *E. coli* at all concentrations while *S. pyrogenes* was only inhibited at 250 and 200 mg/ml on some of the extracts (Onwuliri et al. 2009). Other biological activities identified in *Guiera senegalensis* include antioxidant (Atawodi and Onaolapo 2010), trypanocidal (Atawodi 2005), effect on central nervous system, on cancer cells and as a snake venom detoxicant (Somboro et al. 2011).

**Uses in Traditional Medicine**

In certain climes, people mix galls from *G. senegalensis* (gall nuts are frequently formed on the above-ground parts of the plant) with charcoal to make a highly diuretic powder prescribed in serious cases of oligouria and even anuria, and in particular for cerebral malaria. Other people reduce the galls to powder with the pith from *Combretum aculeatum* stems and salt. This powder is diluted in water immediately before use and is prescribed for painful stomach cramps with mucous stools

**Picture 8** *Acalypha indica* (Source: http://pixgood.com)

and vomiting. A tea made from leaves is prescribed by the oral route (one litre daily) to treat eczema and is also used against attacks of fever, and to cure chest conditions and colds. Fresh mashed, chewed or cut leaves when placed on a wound staunch bleeding. Powdered dried leaves associated with *Melanther ascandens* are administered by the nasal route to treat headaches and sinusitis. The leaves are also used as a poultice on tumors and against Guinea worm (Somboro et al. 2011).

## *Euphorbiaceae*

***Acalypha indica* Linn** (Synonyms: *Acalypha ciliata* Wall., *Acalypha canescens* Wall., *Acalypha spicata* Forsk.)

**Common Names** Three-seeded mercury, Acalypha (Indian), Ricinelle des Indes, orielle de chatte, herbe chatte (French) (Picture 8)

**Botanical Name and Family**
*Acalypha indica* belongs to the family Euphorbiaceae and is commonly called 'Indian acalypha', or Indian copperleaf (Jagatheeswari et al. 2013).

**Morphological Description**
*Acalypha indica* is a monecious, annual to sometimes short-lived perennial herb up to 1.5 to 2.5 m tall. Stems are sparingly to densely hairy and leaves are simple and arranged spirally.Stipules are linear, about 2mm long and the petiole up to 12 cm long. The flowers are unisexual. The male flowers have four lobed, minute, granular,

dotted, greenish calyx, they have eight stamens. The female flowers have three triangular-ovate, ciliate sepal and superior ovary. The fruit is a 3-lobbed capsule splitting into three cocci, each 2-valved and 1-seeded (Schmelzer and Gurib-Fakim 2008).

**Geographical Distribution**

*Acalypha indica* occurs in Nigeria and from Sudan eastward to Somalia and Southwards through Democratic Republic of Congo and East Africa and up to South Africa. Also found in Indian Ocean Islands, India, South East Asia and Oceania (Schmelzer and Gurib-Fakim 2008).

**Ecological Requirements**

*Acalypha indica* thrives on sandy margins of rivers and seasonal water courses, usually in the shade of thickets. It also grows on rocky hillside and as a weed on fields from sea level up to 135 m altitude. *Acalypha indica* flowers throughout the year in regions without a pronounced dry season (Schmelzer and Gurib-Fakim 2008).

**Major Chemical Constituents and Bioactive Compounds**

*Acalypha indica* is reported to contain alkaloids, tannins, saponins, steroids and proteins (Rajaselvam et al. 2012). The dried aerial parts contain a cyanogenic glycoside, acalyphin, which is a 3-cyanopyridone derivative. Flavonoids, such as the kaempferol glycosides mauritianin, clitorin, nicotiflorin, and biorobin have been isolated from the flowers and the leaves. The plant also contains tannins, $\beta$-sitosterol, acalyphamide, aurantiamide, succinimide and the pyraquinolinone alkaloid flindersin.

**Traditional Uses, Part(s) Used and Common Knowledge**

*Acalypha indica* is used as a diuretic, antihelmintic and for respiratory problems such as bronchitis, asthma and pneumonia (Rajaselvam et al. 2012).

The main parts that are usually exploited in medicine and the roots, leaf, stalk and flowers they are used as emetic, expectorant, laxative and diuretic. The plant is useful in bronchitis pneumonia and pulmonary tuberculosis. It also has anti-diabetic properties (Ishak et al. 2013).

In East Africa, the leaf sap is used to treat eye infections, while the leave powder is applied to maggot infested wounds. It is used in Comoros to treat joint pains and a root decoction is used in the Seychelles to treat stomach ache and intestinal wounds. In Madagascar, the crushed aerial parts are applied to skin parasite and an infusion is taken as a purgative and vermifuge, while in Mauritius, the juice of the crushed leaves mixed with salt or a decoction of the aerial parts is applied to scabies and other skin problems (Schmelzer and Gurib-Fakim 2008).

**Biological Activities**

Antimicrobial studies of the aqueous and acetone extracts of *Acalypha indica* shared that the extracts inhibited the growth of *Escherichia coli, Klepsiella* sp, *Staphylococcus aureous* and *Bacillus Subtilis* (Rajaselvam et al. 2012). Ishak et al. (2013) also reported that the petroleum ether, chloroform and methanol extracts of fresh parts of *Acalypha indica* exhibited antibacterial effect on *Staphylococcus aureous* and antifungal effect on *Candida albicans*.

Cytotoxic activity against Hela cell lines has also been reported while and ethanol leaf extract of the plant showed significant inhibition to *Viper russelli* venom-induced lethality, hemmorhage, necrotizing and mast cell degranulation in rats. It also has cardiotoxic and neurotoxic effects on isolated frog tissue (Schmelzer and Gurib-Fakim 2008).

*Acalypha indica* was formerly listed in the British Pharmacopoeia, and because of its numerical uses in India, it is listed in the pharmacopoeia of India (Schmelzer and Gurib-Fakim 2008).

**Uses in Traditional Medicine**

It has been reported to be useful in treating pneumoniae, asthma, rheumatism and several other ailments. The dried leaves of *Acalypha indica* was made into a poultice to treat bedsores and wounds and the juice of *Acalypha indica* is added to oil or lime and used to treat a variety of skin disorders (Jagatheeswari et al. 2013). It is also used as a purgative, a very good remedy in the treatment of piles, and the root is used as a dewormer in children when administered in the morning empty stomach. When the leaves are used with turmeric, it is useful in relief from acne and pimples (Kucinggalak 2015).

***Euphorbia hirta* Linn** (Synonyms: *Chaemescyle hirta*)

**Common Names** Asthma herb (English), Odane Inenemili (Ibo), Kinkerechintara, Lupka (Nupe), Nonon kurciya, Janyaro (Hausa), Emile/Egele (Yoruba) (Picture 9)

**Morphological Description**

*Euphorbia hirta* is a plant belonging to the family: 'Euphorbiaceae'. It is a branched annual herb, prostrate to ascending, with branches up to 50 cm long, with latex; all parts short-hairy and with sparse yellow hairs c. 1.5 mm long. Leaves opposite, distichous, simple; stipules linear, up to 2.5 mm long; petiole up to 3.5 mm long; blade ovate, 1–4 cm × 0.5–2 cm, base very unequal, one side cuneate, the other side rounded, apex almost acute, margin finely toothed, often with a purple blotch near the midvein. Inflorescence a terminal or axillary cluster of flowers, called a 'cyathium', with several cyathia densely clustered into a cyme c. 15 mm in diameter; peduncle up to 15(–20) mm long; cyathia with a cup-shaped involucre c. 1 mm in diameter, tinged purple, lobes triangular, fringed, glands 4, tiny, elliptical, green or purplish, with minute white to pink appendages, each involucre containing 1 female flower surrounded by many male flowers. Flowers unisexual; male flowers sessile, bracteoles linear, fringed, perianth absent, stamen 1, c. 1 mm long; female flowers with short pedicel, perianth a rim, ovary superior, short-hairy, 3-celled, styles 3, minute, apex 2-fid. Fruit a just exserted, acutely 3-lobed capsule c. 1 mm in diameter, base truncate, short-hairy, 3-seeded. Seeds oblong-conical, c. 1 mm long, slightly wrinkled, pinkish brown, without caruncle (Johnson et al. 1999).

**Picture 9** *Euphorbia hirta* (Source: Authors)

## Geographical Distribution

*Euphorbia hirta* is said to be native to Central America, but a very common weed of the tropics and subtropics; it occurs throughout tropical Africa and also in South Africa.

## Ecological Requirements

*Euphorbia hirta* grows in cultivated fields, gardens, roadsides and waste places, from sea-level up to 2000 m altitude.*Euphorbia hirta* is considered a weed, and can be a nuisance in crops due to the large number of seedlings.

## Major Chemical Constituents and Bioactive Compounds

Important constituents of the aerial parts are terpenoids, including triterpenes: $\alpha$-amyrin, $\beta$-amyrin, friedelin, taraxerol, and esters of it: taraxerone, 11α, 12α-oxidotaraxerol, cycloartenol, 24-methylene-cycloartenol, and euphorbol hexacosoate. The aerial parts and roots also contain diterpene esters of the phorbol type and ingenol type, including 12-deoxyphorbol-13-dodecanoate-20-acetate, 12-deoxyphorbol-13-phenylacetate-20-acetate, ingenol triacetate, as well as the highly toxic tinyatoxin, a resiniferonol derivative. Other terpenoids isolated are sterols including β-sitosterol, campesterol, cholesterol and stigmasterol (Abu-Sayeed et al. 2005).

Tannins isolated from the plant include the dimeric hydrolysable dehydroellagitannins euphorbins A, B, C, E and terchebin, the monomeric hydrolysable tannins geraniin, 2,4,6-tri-*O*-galloyl-$\beta$-*D*-glucose and 1,2,3,4,6-penta-*O*-galloyl-$\beta$-*D*-glucose and the esters 5-*O*-caffeoylquinic acid (neochlorogenic acid) and 3,4-di-*O*-galloylquinic acid, and benzyl gallate (Adedapo et al. 2005)

Acids isolated include ellagic acid, gallic acid, tannic acid, maleic acid and tartaric acid (Adedapo et al. 2005). Flavonoids isolated include quercetin, quercitrin, quercitol and derivatives containing rhamnose, quercetin-rhamoside, a chlorophenolic acid, rutin, leucocyanidin, leucocyanidol, myricitrin, cyanidin 3,5-diglucoside, pelargonium 3,5-diglucoside and camphol. The flavonol glycoside xanthorhamnin was also isolated. The stems contain the hydrocarbon hentriacontane and myricyl alcohol. The latex contains inositol, taraxerol, friedelin, β-sitosterol, ellagic acid, kaempferol, quercitol and quercitrin (Abu-Sayeed et al. 2005).

The mineral content of a sample of the dried leaves was: Ca 1.1%, P 0.3%, Fe 0.03%, Mg 0.5%, Mn 0.01%, Zn 0.01% and Cu 0.002%. Fresh leaves from *Euphorbia hirta* plants of Nigerian origin were found to contain high levels of Mn (189 ppm), Cu (30.5 ppm), Zn (152 ppm), and $NO_3$ (4600 ppm). Varying proportions of Fe, Mg, K, Ca and Na were found (Wallace et al. 1990).

**Traditional Uses, Part(s) Used and Common Knowledge/Uses in Traditional Medicine**

*Euphorbia hirta* is an important medicinal herb used throughout its distribution area, including tropical Africa. It is held in high esteem, as a decoction or infusion, to treat gastrointestinal disorders, including intestinal parasites, diarrhoea, peptic ulcers, heartburn, vomiting and amoebic dysentery (Oyewale et al. 2002).

It is also regarded as an outstanding medication to treat respiratory system disorders, including asthma, bronchitis, hay fever, laryngeal spasms, emphysema, coughs and colds. The leaves are mixed with those of *Datura metel* L. in preparing 'asthma cigarettes (Hienmann and Bucar 1994).

Other principal uses are as a diuretic to treat uro-genital diseases, such as kidney stones, menstrual problems, sterility and venereal diseases. The plant is also used to treat affections of the skin and mucous membranes, including warts, scabies, tinea, thrush, aphthae, fungal afflictions, measles, guinea-worm and as an antiseptic to treat wounds, sores and conjunctivitis (Edwin et al. 2007).

The plant has a reputation as an analgesic to treat severe headache, toothache, rheumatism, colic and pains during pregnancy. It is used as an antidote and pain relief of scorpion stings and snakebites. It is antipyretic and anti-inflammatory (Ogbulie et al. 2007).

It is also used in the treatment of jaundice, hypertension, oedema, anaemia and malaria, as an aphrodisiac, and to facilitate childbirth. In West Africa the plants are widely used as a galactagogue, and in Nigeria they are marketed for this purpose. In Uganda whole plants are chewed to induce labour during childbirth (Hienmann and Bucar 1994).

**Biological Activities**

Several of the traditional medicinal uses of *Euphorbia hirta* have been supported by in-vitro studies. Ethanol, petroleum ether and dichloromethane extracts of whole plants showed significant in-vitro antiplasmodial activity ($IC_{50}$ = 3 μg/ml) and decreased growth of *Plasmodium falciparum* by 89–100% at a test concentration of 6 μg/ml (Tona et al. 2004). *In vivo*, the extracts reduced parasitaemia in mice infected with *Plasmodium berghei berghei* at oral doses of 100–400 mg/kg per day.

From a methanolic extract of the aerial parts the flavonol glycosides afzelin, quercitrin and myricitrin were isolated, which showed proliferation inhibition of *Plasmodium falciparum*, with $IC_{50}$ values of 1.1, 4.1, 5.4 μg/ml respectively, while they exhibited little cytotoxic effect against human epidermoid carcinoma KB 3–1 cells.

An ethanolic extract was active in selectively inhibiting Herpes simplex virus type-1 (0.001–0.1 mg/ml) (Somchit et al. 2001). Lyophilized aqueous extract of the aerial parts has been evaluated for analgesic, antipyretic and anti-inflammatory properties in mice and rats. The extract exerted central analgesic properties at doses of 20 and 25 mg/kg, and antipyretic activity at doses of 100 and 400 mg/kg, whereas anti-inflammatory effects against carrageenan-induced oedema in rats were observed at a dose of 100 mg/kg (Ogbulie et al. 2007).

The aqueous extract of the aerial parts has been found to strongly reduce the release of prostaglandins, and thus depress inflammation (Hienmann and Bucar 1994). An ethanolic extract of the aerial parts was found to possess a prominent anti-anaphylactic activity and also showed significant antihistaminic, anti-inflammatory and immunosuppressive properties in various animal models. Water and ethanolic leaf extracts produced a time-dependent increase in urine output in rats. A methanol extract of leaves and stems inhibited the activity of angiotensin-converting enzyme by 90% at 500 μg and 50% at 160 μg. The extract (10 mg/100 g, intraperitoneally) significantly decreased the amount of water consumed by rats. An ethanolic extract of the whole plant showed a dose-dependent ulcer protective effect in rats (Rao et al. 2003). The active compound was found to be quercetin, which had an anti-ulcer activity ranging from 48–64% comparable to 61–80% of the standard drug ranitidine. An ethanolic extract of the aerial parts showed significant hepatoprotective activity in rats. Extracts of whole plant material have oestrogenic activity in female guinea pigs, when given orally. In organ bath tests with ileum preparations, shikimic acid and choline extracted from the aerial parts had relaxing and contracting properties, respectively (Lanhers et al. 1990). Shikimic acid also has acute toxicity, mutagenicity and carcinogenicity. The aqueous crude extract significantly reduced the faecal egg count of helminths in dogs (Johnson et al. 1999).

Several of the extracts of *Euphorbia hirta* showed potential for controlling plant diseases and pests (Hienmann and Bucar 1994). For example, a whole plant extract inhibited growth of vascular wilt (*Fusarium oxysporum*) and the causal agent of sheath rot of rice, *Sarocladium oryzae*; aqueous extracts of the aerial parts inhibited aflatoxin production by *Aspergillus parasiticus* on agricultural crops, including rice, wheat, maize and groundnuts. Leaf extracts completely inhibited soft rot infection caused by the bacteria *Erwinia carotovora* pv. *carotovora*. The infectivity of tobacco mosaic virus on *Nicotiana glutinosa* L. was strongly inhibited (>80%) by tannins extracted from the aerial parts. The latex inhibited sugarcane mosaic virus-A by 78.5% and sugarcane mosaic virus-F by 80%. Root and leaf extracts showed nematicidal activity against *Meloidogyne incognita*; a whole plant extract effectively reduced hatching in the nematode *Heterodera avenae*. A 10% ethanol crude extract showed significant larvicidal action against the larvae of the tick *Boophilus microplus*. Aqueous stem, latex and leaf extracts have potent molluscicidal activity against the freshwater snails *Lymnaea acuminata* and *Indoplanorbis exustus*, both intermediate hosts of *Fasciola hepatica* and *Fasciola gigantica*, which cause endemic fascioliasis in cattle and livestock (Singh and Khandelwal 2010). Toxicity of the extracts was time dependent and dose dependent against both snails. The doses that can be used for killing 90% of the *Lymnaea acuminata* populations are safe for the fish *Channa punctatus*.

An aqueous extract of the whole plant acts as an antidiarrhoeic agent by anti-amoebic, antibacterial and antisplasmodic activities. The antidiarrhoeal activity is attributed to quercitrin through the release of the aglycone quercetin in the intestine (Singh and Khandelwal 2010). Quercitrin showed antidiarrhoeic activity at doses of 50 mg/kg in mice. A crude plant extract and an ethanolic extract had significant anti-amoebic activity against *Entamoeba histolytica* in vitro at 35 mg/ml. An aqueous lyophilysate of the whole plant showed higher activity against *Entamoeba histolytica* than either the ethyl acetate or methanol extracts, at 30 mg/ml. An aqueous plant extract showed concentration-related activity against non-pathogenic amoebae of the *Amoeba proteus* type. Different extracts from the aerial parts showed antibacterial activity against a wide spectrum of both gram-positive and gram-negative bacteria. Extracts of the aerial parts showed strong antibacterial activity against *Shigella dysenteriae*, a causal agent for dysentery in humans. The active compound was found to be ethyl gallate, which has broad spectrum antibiotic activity at non-toxic doses. A crude ethanol extract of the whole plant showed dose-dependent activity against *Candida albicans*, but not against several other pathogenic fungi. Some of the isolated antibacterial compounds were taraxerone and 11α, 12α-oxidotaraxerol, which showed low cytotoxicity.

The levels of chemicals in *Euphorbia hirta* are high enough to constitute a source of toxicosis to animals consuming the plants and should also be a source of concern in medicinal use. Few toxic effects have been documented for *Euphorbia hirta*. An ether extract was found to be toxic in a brine shrimp lethality test, whereas ethyl acetate and aqueous extracts were within safe limits (Ogbulie et al. 2007). In another test, however, an aqueous crude extract was found to cause testicular degeneration in sexually mature male rats as well as a reduction in the mean seminiferous tubular diameter. Several other extracts given orally to rats caused dullness and anorexia and induced a 20% mortality rate. Some fractions from the ethanolic extract showed potentially deleterious effects on the blood serum chemistry of rats (Adedapo et al. 2005). In feeding experiments with rats however, no difference in the blood serum was found after a prolonged period of adding *Euphorbia hirta* to the diet. It was also found that drying *Euphorbia hirta* prior to extraction considerably reduces the cytotoxic activity of certain of its extracts.

### *Euphorbia unispina* N.E. Br

**Common and Local Names** Candle plant (English) (Newton 2008) and Oro Adete in Yoruba language (Picture 10).

**Picture 10** *Euphorbia unispina* (Source: Authors)

**Morphological Description**

*Euphorbia unispina* is a monoecious, sparsely branching shrub that can attain 3.5 m tall. The branches are cylindrical, up to 2.5 cm in diameter. It is silvery-grey in colour and covered with shallow tubercles and horny spine. The leaves are simple and arranged spirally at the stem apex in 4–5 ranks, the stipules are modified into two stout spines, 6–10 mm long. The petioles are short and thick, and blades are oblong to spoon-shaped, 5–12 cm × 1.5–5 cm. The base is long-cuneate, apex notched and fringed, acute or rounded, almost entire, fleshy, glabrous, pinnately veined. The inflorescence is an axillary cyme at the ends of branches, consisting of clusters of flowers, each cluster called a cyathium. The flowers are red, unisexual, male flowers sessile, perianth absent, stamen shortly exserted. Female flowers are with curved pedicel, 4–8 mm long in fruit, perianth 5-lobed, ovary superior, 3-celled, glabrous, styles 3, up to 2 mm long, slender, fused at base, bifid at apex. The fruit is an obtusely 3-lobed capsule, about 6 mm in diameter, glabrous and three-seeded. The seeds are ovoid in shape (Newton 2008).

**Geographical Distribution**

*Euphorbia unispina* is found in Africa from Guinea and Mali east to southern Sudan.

**Ecological Requirements**

*Euphorbia unispina* grows on rocky hills and slopes in savanna (Newton 2008)

**Chemical Constituents**

The latex of *Euphorbia unispina* contains esters of diterpene alcohols of the tigliane type, 12-deoxyphorbol and 12-deoxy-16-hydroxyphorbol, and the daphnane type, resiniferonol, as well as several macrocyclic esters of the diterpene alcohol 18-hydroxyingol (Rao et al. 2003).

**Uses**

The latex of *Euphorbia unispina* is applied to the neck to cure sleeping sickness in the West African countries of Guinea, Mali and Cote d'Ivoire because it is believed that the disease is caused by ganglia in the neck. In Cote d'Ivoire and Nigeria, the latex is applied to leprosy sores. An inhalation of the stem ash is used to treat asthma in Benin, while a mixture of palm oil with the latex is taken to treat constipation and colic. Skin diseases and hemmorroids are treated by a macerate of the stem in water while in Northern Nigeria, the latex is rubbed onto the body to treat mental illness. The latex is also use in dental care by dropping it on carious tooth to relieve toothache or to help loosen the tooth and render extraction easier. Smoking of the dried leaves in a pipe is used to treat bronchitis. In a mixture with other ingredients such as *Strophanthus* species, the latex is used in the preparation of arrow poison (Newton 2008). The root of *Euphorbia unispina* is also used to treat cancer (Soladoye et al. 2010).

**Biological Activities**

*Euphorbia unispina* latex is very caustic and toxic. It is very irritating to the skin and mucous membranes. It can cause blindness when in contact with the eyes. In Northern Nigeria, the latex is used as a poison to commit murder and suicide (Newton 2008).

## *Phyllanthus muellerianus* (Kuntze) Exell (Pictures 11 and 12)

**Botanical Name and Family**

*Phyllantus muellerianus* belong to the family Euphorbiaceae.

**Morphology**

It is a monoecious, glabrous, straggling or climbing shrub or small tree up to 12 m tall; branches spreading or pendulous, main branches stout, angular, reddish tinged, branchlets 15–20(–25) cm long, with several short axillary shoots; branch basis transformed into a pair of spines c. 4 mm long, purplish brown. Leaves alternate, distichous along lateral twigs, simple, glabrous; stipules lanceolate, *ca*. 2 mm long,

**Pictures 11 and 12** *Phyllantus muellerianus* (Source: faunaandfloraofvietnam.blogspot.com, www.prota4u.org)

acuminate; petiole 3–5 mm long; blade ovate, elliptical-ovate to ovate-lanceolate, 3–9 cm × 2–4.5 cm, base cuneate to rounded, apex acute to obtuse, with 10–14 pairs of lateral veins. Inflorescence a false raceme on short axillary shoots, 2–6 cm long, solitary or several together, with flowers in clusters having 2–3 male flowers and 1 female flower in each cluster. Flowers unisexual; perianth lobes 5, elliptical, c. 1 mm long, rounded, greenish white or greenish yellow; male flowers with pedicel c. 1.5 mm long, disk glands 5, free, minutely warted, fleshy, stamens 5, free, unequal, anthers very small; female flowers with stout pedicel c. 1 mm long, disk glands 5, free or fused, knobbly, fleshy, ovary superior, ellipsoid, warty, 4–5-celled, styles 4–5, free, c. 0.5 mm long, 2-fid at apex. Fruit a fleshy, nearly globose capsule 3–4 mm in diameter, usually smooth, green, becoming red, later black, 6-seeded. Seeds angular, c. 1 mm long, with faint ridges, bright reddish brown or yellowish brown.

**Geographical Distribution**

*Phyllanthus muellerianus* occurs from Senegal and Guinea Bissau east to Sudan and Kenya and south to northern Angola and northern Mozambique. It is a large genus comprising about 750 species in tropical and subtropical regions, with about 150 species in mainland tropical Africa and about 60 in Madagascar and the Indian Ocean islands (Burkill 1994).

**Ecological Requirements**

*Phyllanthus muellerianus* occurs in riverine forest and wooded grassland, on deep and well-drained soils, from sea-level up to 1600 m altitude. In Nigeria, *Phyllanthus muellerianus* is reported as a weed of rice fields. *Phyllanthus muellerianus* can be propagated through seeds and stem cuttings (Arbonnier 2004).

**Major Chemical Constituents and Bioactive Compounds**

From the extracted essential oils of the leves was found to be (*E*)-isoelemicin (Brusto et al. 2012). Preliminary phytochemical screening of the leaves and stem methnolic extracts showed that both the extracts contained tannins, flavonoids, saponins, alkaloids, and anthraquinones. From the stem bark the triterpenoids 22-$\beta$-hydroxyfriedel-ene and 1$\beta$,22$\beta$-dihydroxyfriedelin were isolated.

**Traditional Uses, Part(s) Used and Common Knowledge/Uses in Traditional Medicine**

*Phyllanthus muellerianus* is widely used to treat intestinal troubles. An infusion of the young shoots is taken to treat severe dysentery. In Sierra Leone a leaf decoction is taken to treat constipation. In Ghana and Nigeria cooked roots, sometimes with maize meal or other plants, are taken to treat severe dysentery. In Congo powdered roasted roots with palm oil are taken to treat stomach problems and as an anti-emetic. In Tanzania roots are pounded in water and the liquid is drunk to treat diarrhoea. Boiled roots are also applied as enema to treat stomach-ache.

In West Africa leaf sap or sap from the thick hollow stem is applied as eye drops to treat pain in the eyes, eye infections or to remove a foreign body. In Côte d'Ivoire and Burkina Faso twigs are sucked to prevent toothache. Powdered roots are used as a snuff and a bark decoction is taken to treat a sore throat, cough, pneumonia and enlarged glands. Pulped leafy twigs are rubbed on the body to treat paralysis. In Nigeria a root bark decoction is taken as an alternative to treat fever. A twig and root decoction is taken to treat jaundice and urethral discharges. In the Central African Republic fresh root bark is crushed and macerated in water or palm wine, and the liquid drunk as an aphrodisiac. In Gabon roasted powdered twigs are eaten with plant ash to treat dysmenorrhoea. In DR Congo dried bark powder is sniffed to treat colds and sinusitis. A root bark decoction is applied to swellings and is drunk to treat gonorrhea. Stem ash is applied to scarifications to treat rheumatism and intercostal pain. In Tanzania a root decoction is taken to treat hard abscesses; powdered dried roots and stem bark are sprinkled on wounds as a dressing.

Throughout West Africa pounded leaves are applied as wound dressing. In Côte d'Ivoire the leaves are eaten, together with young leaves of *Funtumia elastica* (Preuss) Stapf, to improve male fertility. In Ghana and Nigeria leaves boiled with palm fruit are given to women after delivery as a general tonic. In Cameroon a maceration of the leaves and roots is used to wash the body to treat rash with fever in children. In DR Congo a leaf decoction is taken to treat anaemia and also used as a mouthwash to treat toothache. A leaf extract is used as a bath and a vapour bath to treat venereal diseases. Cooked leaves are applied to the gums to treat toothache. A flower infusion is cooling and gently aperient. The fruits are edible and slightly acidic. In Sierra Leone and Nigeria the sap from the hollow branches is considered potable.

In Cameroon the bark is sometimes added to palm wine to render it strongly intoxicating. In Kenya the stems are considered excellent firewood; branches thicker than 15 cm become hollow and are less used. In East Africa the brown dye from the bark is used to dye mats and fishing lines. From the whole plant a black dye is obtained used to colour fibres. In Zambia the wood is used for rafters and other construction work. It is also used to make fish traps and basketry. The leaves are used as fodder. In Sierra Leone and Nigeria leaves are sometimes cooked with food or in soup as a seasoning. In Nigeria twigs are used as chew-sticks after removal of the spines. Fruit pulp is used as a hair fixative. In Gabon *Phyllanthus muellerianus* is used in magic to lift taboos.

**Biological Activities**

Phytochemical screening of the leaves and stem bark showed the presence of tannins, flavonoids, saponins, alkaloids and anthraquinones. A leaf extract showed moderate antiplasmodial activity ($IC_{50} = 9.4$ µg/ml) and low cytotoxicity on mammalian cell lines. Both the aqueous and methanol extracts of the leaves and stem bark showed high antibacterial activity against *Staphylococcus aureus* and *Pseudomonas aeruginosa*. The activity of the extracts was relatively stable at high temperatures and was enhanced at low pH. In another test a chloroform extract showed high antifungal activity against *Candida albicans* and antibacterial activity against *Escherichia coli*. A crude aqueous extract orally administered to rats caused significant changes in haematological and biochemical parameters, which are used as indices of toxicity (Anuka et al. 2005).

## *Fabaceae*

***Senna occidentalis* Linn** (Synonyms: *Cassia occidentalis* L., *Ditremexa occidentalis* (L.) Britt. & Rose. *Senna occidentalis* (L.) Roxb.; *Senna occidentalis* var. sophera (L.) (Pictures 13 and 14)

**Botanical Name and Family**
*Senna occidentalisis* of the family- Fabaceae (Leguminosae) and sub-family Caesalpinioideae, is commonly called stink weed (Saidu et al. 2011).

**Common Names**
Coffee senna, ant bush, antbush, arsenic bush, negro coffee, Nigerian senna, septicweed, sickle pod, stink weed, stinking pea, stinking weed, stinkingweed, stinkweed, styptic weed.

**Morphological Description**
*Senna occidentalis* is an erect herb commonly found by the road sides in ditches and waste dumping sites (Saidu et al. 2011). The leaves are alternate, compound and pinnate, consisting of four or five pairs of leaflets widely spaced along a common stalk. The leaflets are pointed at the tips and the flowers are yellow and produced in loose clusters in the terminal leaf axils. The fruits are in form of thin pods, 3–4 in long and pale green when tender, thick and dark green when mature. It brings out pods from September to November and the pods are slightly curved, with paler longitudinal stripes along the edges. Each pod contains 50–60 small seeds together weighing 1.90 to 2.25 g. From December onwards, the pods start drying up, turn brown and the seeds turn dark brown (Vashishtha et al. 2009).

**Geographical Distribution**
*Senna occidentalis* is found across West Africa (Saidu et al. 2011).

**Ecological Requirements**
*Senna occidentalis* is a weed found in degraded pastures and plantations, which is poisonous to cattle and other domestic animals (Modena et al. 2012). It was shown experimentally that *Senna occidentalis* can survive under widely variant

**Pictures 13 and 14** *Senna occidentalisis* (Source: Anand Kumar Reddy, http://pinpicsnow.com)

light intensities. Although it occurs at optimal temperature of 25.0 °C, it also survives at its extreme temperatures of 10.0–12.5 °C (low) and 45 °C (high). Optimal pH is 6 although it relatively prefers more acidic condition than basic. The sandy loam soil is better preferred by the plant.

**Major Chemical Constituents and Bioactive Compounds**
Phytochemical screening of aqueous extracts of *Senna occidentalis* revealed the presence of flavonoids, cardenolides, saponins, anthraquinones and alkaloids (Saidu et al. 2011). The hexane, ethyl acetate and methanol extracts revealed the presence of tannins, alkaloids, reducing sugar, phenols, anthraquinones, resins, saponins and glycosides (Odeja et al. 2014). Senna occidentalis has also been reported to contain carbohydrates, saponins, sterols, flavonoids, resins, alkaloids, terpenes, anthraquinones, glycosides and balsam, which are indicators to its medicinal attributes. Flavonoids and resins might be responsible for its anti-inflammatory properties; alkaloids known for decreasing blood pressure, balancing the nervous system in case of mental illness and antimalarial properties. Tannins help in wound healing and anti-parasitic properties. Tannins possess antitumor and antviral properties and eudesmane sesquiterpenes possess antibacterial properties (Vijayalakshmi et al. 2013). Other compounds reported include 1,8-dihydroxyl-2-methyl anthraquinone, 1,4,5-trihydroxyl-3-methyl-7-methoxy anthraquinone, cassiaoccidentalin A,B and C which are C-glycosides, achrosine, anthrones, apigenin, aurantiobtusin, campesterol, cassiollin, chrysotusin, chrysophanic acid, chrysarobin, chrysoeviol, essential oils, funicolosin, galactopyranosyl, helminthosporin, islandicin, kaempferol, lignoceric acid, linoleic acid, linolenic acid, mannitol, mannopyranosyl, matteucinol, obtusifolin, oleic acid, physcion, quercetin, rhamnosides, rhein, rubrofusarin, sitosterols and xanthorin (Vijayalakshmi et al. 2013).

**Traditional Uses, Part(s) Used and Common Knowledge**
*Senna occidentalis* is effective in the treatment of dysentery and diarrhea which are mainly caused by microorganisms. An infusion of the leaf is also used to treat typhoid fever (Saidu et al. 2011). The leaves and roots are used as remedy for bacterial and fungal infections and can also boost immune function. The leaves and roots are ingredients of tonics used to treat liver disorders, abscesses, insect bites, scorpion sting, constipation, diabetes, oedema, fever, inflammation, itch, rheumatism, ringworm, scabies, skin diseases, snakebite and wounds (Vashishtha et al. 2009)

The pods and seeds are considered toxic to grazing animals that feed on them as they become seriously ill and may die (Vashishtha et al. 2009)

**Biological Activities**
The hexane, ethylacetate and methanol extracts of *Senna occidentalis* at varying concentrations inhibited the growth of *Staphilococcus aureus, Escherichia coli, Bacillus subtilis, Pseudomonas aeruginosa, Salmonella typhi, Klebsiella pneumonia, Candida albicans, Aspergillus niger, Penicillium notatum* and *Rhizopus stolonifer* to varying extent (Odeja et al. 2014). The methanol, aqueous, benzene, petroleum ether and chloroform extract of *Cassia occidentalis* leaf were screened for antibacterial activity. The methanol extract inhibited the growth of *P. aeruginosa*,

*K. pneumonia, P. mirabilis, E. coli, S. aureus* and *S. epidermidis*. The aqueous extract was effective against *P. vulgaris, K pneumonia*, and *P. aeruginosa* while the benzene and petroleum ether extracts were effective against *P. mirabilis* and *E. coli.* The chloroform extract was ineffective against all the test organisms (Arya et al. 2010). The extract of *Senna occidentalis* flower showed maximum growth inhibition against *Klebsiella pneumonia* but no activity against *Staphylococcus aureus, Streptococcus pneumonia* and *Pseudomonas aeruginosa* (Daniyan et al. 2011). The crude leaf extract also inhibited the growth of *Pseudomonas aeruginosa, Staphylococcus aureus, Escherichia coli, Bacillus subtillis* and *Candida albicans*.

*Senna occidentalis* also possess antioxidant, hepatoprotective, anti-inflammatory, larvicidal, antimalarial, antidiabetic, analgesic, antipyretic and antidepressant properties (Vijayalakshmi et al. 2013).

**Uses in Traditional Medicine**

*Senna occidentalis* is used as a broad spectrum internal and external antimicrobial to treat bacterial and fungal infections. Also used for liver disorders (jaundice, lepatitis, cirrhosis, detoxification, injury and failure, bile stimulant, etc) and as treatment for intestinal worms, internal parasites, skin parasites. Moreso, used as a cellular protector and preventative to cell damage, viz: immune, liver, kidney and cancer (Sadiq et al. 2012).

## *Malvaceae*

*Grewia mollis* **Juss** (Synonyms: *Grewia venusta* Fresen and *Grewia pubescens* P.Beauv) (Picture 15)

**Picture 15** *Grewia mollis* (Source: Author)

**Morphological Description**

*Grewia mollis* is a shrub or small tree up to 20 ft tall, of Malvaceae family (previously belonging toTiliacaea family) widely distributed within the Northern, middle belt of Nigeria and some African countries. *Grewia mollis* is often gregarious, with branch-suckering leading to the formation of thickets various parts of the plant are used in food and medicine (Asuku et al. 2012). It is a shrub or small tree up to 10.5 m tall, often multi-stemmed; stem diameter up to 30 cm; young branches densely stellate-pubescent, turning dark grey to purple with age; outer bark black, thick, rough, flaking and deeply fissured, inner bark yellowish to brown, fibrous. Leaves alternate, simple; stipules lanceolate, 5–10 mm long, slightly hairy, caducous; petiole 4–13 mm long, greyish to reddish brown pubescent; blade elliptical to elliptical-oblong. Inflorescence is a cyme. Flowers bisexual, regular, 5-merous, slightly scented; pedicel 3–11 mm long; sepals linear-oblong, 6–11 mm long, greyish hairy outside; petals obovate to oblong, 4–6(–8) mm × c. 2 mm, sometimes notched at the apex, bright yellow; apex densely hairy; stamens numerous, 3–6 mm long; ovary superior, 1.5–2 mm long, densely hairy. Fruit a globose drupe 4–8 mm × 5–8 mm, finely whitish hairy, yellow turning black; endocarp hard and woody.

**Geographical Distribution**

*Grewia mollis* occurs widely in tropical Africa, from Senegal and Nigeria eastward to Somalia and southward to Zambia and Zimbabwe.

**Habitat and Ecology**

*Grewia mollis* occurs in tropical climates with annual average rainfall of 600–1400 mm. It is usually distributed in lowlands of West Africa up to high altitude zones in East Africa and thrives in forest, open woodland, riverine and seasonally flooded grassland. It is known to be highly resistant to fire.

**Collection Practices**

*Grewia mollis* is distributed in the tropical and subtropical parts of Africa and Asia .Germination of *Grewia mollis* often occurs after a bush fire followed by rains but growth is slow. Flowering is around the end of March and October depending on locale. The parts of *Grewia mollis* are normally collected from the wild but it can be propagated by seed or seedlings. Seeds are usually collected from dried fruits fallen on the ground.

**Chemical Composition**

Hanan and co-workers (2012) have investigated extensively the plant *Grewa mollis* and isolated the following compounds, 7-(1-*O*-*β*-*D*-galacto turoniole-4-(1-*O*-*β*-glucopyranesyl-3,4,5,7 tetrahydroxyflavone, lutelion, 7-*β*-hydroxy-2,3-en-deoxojessic acid, 7-*β*- hydroxy-2,3-deoxojessic acid sitesterol and *β*-sitosterol-3-*O*-glucoside,from the aerial part of the plant. Further phytochemical investigation on biologically active methanolic extract resulted in the isolation of luteloin, 7-(1-*O*-*β*-*D*-galacturonide)-4′-(1-*O*-*β*-glucopyranosyl)-3′4′,5,7-tetrahydroxyflavone, 7-*β*-hydroxy-23 enedeoxojessic acid , 7-*β*-hydroxy-23- deoxojessic acid , *β*-sitosterol and *β*-sitosterol-3-*O*-glucoside. Two other compounds; triterpene lup-20-en-2-ol and 1,3-hexyloxacyclotridec- 10-en-2-one, have been isolated from the roots of the plant (Efiom and Oku 2012).

**Uses in Traditional Medicine**
In Nigeria the fruit of *Grewia mollis* is used as a febrifuge. The stem bark, branch, trunk, decoction is used to cure constipation. The mucilaginous bark and leaves are applied to ulcers, cuts, sores and snakebies and acclaimed to have laxative property. Bark and root preparations are taken to treat cough. Extracts of bark and leaves are reportedly drunk to treat fever. In Togo a decoction of the stem bark is drunk to treat diarrhoea, and a macerate is taken to ease childbirth. An infusion of the bark is used to treat colic dysentery (Adamu et al. 2005). In East Africa leaves pounded and mixed with water are taken against stomach problems and also given to constipated domestic animals. In Côte d'Ivoire a decoction of the leaves is used in baths and drinks to cure rickets in children. A decoction of the roots is drunk in Senegal in case of palpitation. In Central Africa sap from root-shavings is placed under the eyelid to treat sore eyes, whereas a liquid obtained by kneading the root bark in water is drunk to treat stomach-ache, colic and poisoning by certain plants. In Ghana a paste of ground roots is applied to rheumatic swellings and inflammation (Bekalo et al. 2009; Asuku et al. 2012).

**Biological and Cytotoxic Activities**
The *invitro* antibacterial activity of the aerial part of *Grewia mollis* was investigated by Hanan and co-workers (2012) against various strains of bacteria such as *Staphylococcus epidermidis, Bacillus subtilis, Staphylococcus aureus, Klebsiella pneumoniae, Pseudomonas aeruginosa* and *Escherichia coli*. The methanolic extract showed significant activity against *Staphylococcus epidermidis, Bacillus subtilis and Staphylococcus aureus* when compared with the standard antibiotics. Methanolic extract of *G. mollis* aerial part showed inhibitory activity against oral cavity pathogen with extract exhibiting a concentration-dependent killing of *Staphylococcus aureus* and also preventing the formation of water-insoluble glucan. This researchers further demonstrated the antiinflammatory and antihypertensive activities of the aerial part of *G.mollis*. The antioxidant an activity of the leaves extract of the plant has also been reported (Asuku et al. 2012). The researchers further demonstrated the protective and ameliorative potentials of this plant against experimentally induced liver injuries in rats.

## 2 Conclusion

The properties, uses and application of some wild herbaceous plants distributed throughout Nigeria have been reviewed. The active compounds variously isolated as well as their pharmacognostic and biological properties of the plants reviewed have been highlighted. Since many of the biological activities of these plants have been validated experimentally, further studies on the potential industrial use of these plants in developing countries like Nigeria should be seriously explored, such as in drugs development, as well as production of natural supplements, nutraceuticals, natural insecticide, phytochemicals for industrial usage etc. The cultivation of wild but useful plants could be an essential step that can bring about the much needed turn-around in the nations' agricultural system, economy and conservation of plant biodiversity in Nigeria.

**Acknowledgments** Professor S.E. Atawodi's research work on antioxidants was partly supported by Alexander von Humboldt Foundation of Germany and the German Cancer Research Centre (DKFZ), Heidelberg.

## References

Abu-Sayeed M, Ali MA, Bhattacharjee PK, Islam A, Astaq GRM, Khan M, Yeasmin S (2005) Biological evaluation of extracts and triterpenoids of Euphorbia hirta. Pakistan J Sci Ind Res 48(2):122–125

Adamu HM, Abayeh OJ, Agho MO, Abdullahi AL, Uba A, Dukku HU, Wufem BM (2005) An ethnobotanical survey of Bauchi State herbal plants and their antimicrobial activity. J Ethnopharmacol 97:421–427

Adedapo AA, Abatan MO, Idowu SO, Olorunsogo OO (2005) Effects of chromatographic fractions of *Euphorbia hirta* on the rat serum biochemistry. Afri J Biomed Res 8(3):185–189

Adejo GO, Eseyin JS, Akintayo CO, Obinna M (2015) *Morinda lucida* leaf and fruit extracts may attenuatediarrhea arising from ebola. Basic Res J Med Clin Sci 4(1):95–100

Ahmad RG, Saeidnia S (2011) Photochemistry of *Cuminum cyminum* seeds and its standards from field to market. Pharma J 3(25):1–5

Ajayi CO, Elujoba AA, Bejide RA, Akinloye JA, Omonisi AE (2014) Toxicity and pharmacognostic standards for laxative properties of Nigerian *Cassia sieberiana* and *Senna obtusifolia* roots. Eur J Med Plants 6(2):110–123

Anandaraj M, Sastry EVD (2015) Soils, Plant growth and Crop production – Cumin, Fenneland Fenugreek. http://www.eolss.net/Eolss-sampleAllChapter.aspx. Accessed on 12 Apr 2015

Anuka JA, Yaro AH, Wannang NN, Ezenwanne EB, Yakasai IA (2005) Some in vivo and in vitro studies of the aqueous leaf extract of *Phyllanthus muellerianus* (Euphorbiaceae) in laboratory animals. J Pharm Biores 2(2):93–99

Arbonnier M (2004) Trees, shrubs and lianas of West African dry zones. CIRAD Margraf Publishers Gmbh, MNHN, Paris, France, p 573

Arya V, Yadav S, Kumar S, Yadav JP (2010) Antimicrobial activity of *Cassia occidentalis* L (leaf) against various human pathogenic microbes. Inter J Life Sci Med Res 9:1–11

Asase A, Oteng-Yeboah AA, Odamtten GT, Simmonds MSJ (2005) Ethnobotanical study of some Ghanaian anti-malarial plants. J Ethnopharmacol 99:273–279

Ashaf AT, Olunu OO (2011) Toxicological evaluation of ethanolic root extract of *Morinda lucida* (L.) Benth. (Rubiaceae) in male Wistar rats. J Nat Pharm 2:108–114

Asmah R, Susi E, Abdah MA, Patimah I, Taufiq YH, Mohd FAB (2006) Anticarcinogenic properties of *Strobilanthes crispus* extracts and its compounds *in vitro*. Inter J Cancer Res 2(1):47–49

Asuku O, Atawodi SE, Onyike E (2012) Antioxidant, hepatoprotective, and ameliorative effects of methanolic extract of leaves of *Grewia mollis* Juss. on carbon tetrachloride–treated albino rats. J Med Food 15(1):83–88

Atawodi SE (2005) Antioxidant potentials of African medicinal plants. Afr J Biotechnol 4(2):128–133

Atawodi SE, Preussmann R, Spiegelhalder B (1995) Tobacco-specific nitrosamines in some Nigerian cigarettes. Cancer Lett 97:1–6

Atawodi SE, Atawodi JC, Idakwo P, Pfundstein B, Haubner R, Wurtele G, Spiegelhalder B, Bartsch H, Owen RW (2009) Evaluation of the polyphenol composition and antioxidant activity of african variety of *Dacryodes edulis* (G.Don) H.J Lam Fruit. J Med Food 12(6):1321–1325. doi:10.1089/jmf.2008.0215

Atawodi SE, Ogunbusola F (2009) Evaluation of anti-trypanosomal properties of four extracts of leaves, stem and root barks of prosopis africana in laboratory animal. Biokemistri 21(2):101–108

Atawodi SE, Onaolapo GS (2010) Comparative *in vitro* antioxidant potential of different parts of *Ipomoea asarifolia*, Roemer & Schultes, *Guiera senegalensis*, J. F. Gmel and *Anisopus mannii* N. E. Brown. Braz J Pharm Res 46(2):245–250

Atawodi SE, Adekunle OO, Bala I (2011) Antioxidant, organ protective and ameliorative properties of methanol extract of Anogeissus *leiocarpus* Stem Bark against carbon tetrachloride-induced liver injury. *Internl J Pharmaceut Sci Res* 2(6):748–753

Atawodi SE,Olowoniyi OD, Adejo GO, Liman ML (2014a) Review of the antioxidant potential of African medicinal plants and food plants. In: Dubey NK (ed) Plants as a source of natural antioxidants. CAB International, UK, 63 pp

Atawodi SE, Olowoniyi OD, Obari MA, Ogaba I (2014b) Ethnomedical survey of Adavi and Ajaokuta Local Government Areas of Ebiraland, Kogi State, Nigeria. Ann Res Rev Biol 4(24):4344–4360

Atawodi SE, Olowoniyi OD, Daikwo MA (2014c) Ethnobotanical survey of plants used for the management of hypertension in the Igala Speaking Area of Kogi State, Nigeria. Ann Res Rev Biol 4(24):4535–4543

Bedu-Addo K (1993) Effect of the crude extract of the root of *Rauwolfia vomitoria* on the reproduction of female albino rats (*Rattus norvegicus*). Kwame Nkrumah University of Science and Technology, Kumasi, Ghana, Thesis

Bekalo TH, Woodmatas SD, Woldemariam ZA (2009) An ethnobotanical study of medicinal plants used by local people in the lowlands of Konta Special Woreda, southern nations, nationalities and peoples regional state. Ethiopian J Ethnobiol Ethnomed 5:26

Bokaeian M, Shiri Y, Bazi S, Saeids S, Sabi Z (2014) Antibacterial activities of *Cuminum cyminum* Linn essential oil against multi-drug resistant *Escherichia coli*. Inter J of Infect 1(1)

Bello OO, Bello TK, Bankole SA (2013) Occurrence of antibioticresistant Staphylococcus aureus in some street-vended foods in Ogun State, Nigeria. J Adv Biol 1(1):21–28

Boyraz N, Ozcan M (2005) Antifungal effect of some spicehydrosols. Fitoter 76:661–665

Brusto G, Cesari I, Gilardoni G, Tosi S, Grisoli P, Picco AM, Caccialanza G (2012) Chemical composition and antimicrobial activity of *Phyllanthus muellerianus* (Kuntze) Excel essential o. J Ethnopharm 143(3):657–662

Burkill HM (1994) The useful plants of West Tropical Africa, 2nd edn, vol 2, Families E–I. Royal Botanic Gardens, Kew, Richmond, UK, 636 pp

Burkill HM (2000) The useful plants of West Tropical Africa, 2nd edn, vol 5, Families S–Z, Addenda. Royal Botanic Gardens, Kew, Richmond, UK, 686 pp

Chaudhary N, Husain SS, Ali M (2014) Chemical Composition and antimicrobial activity of volatile oil of the seeds of *Cumimum cyminum*. World J Pharm Pharma Ser 3(7):1428–1441

Daniyan SY, Oloruntimilehin JB, Ifeadi O (2011) Antibacterial activity of *Cassia occidentalis* flower vegetable extract on selected bacteria. Asian J Biomed Pharm Sci 1(1):23–27

DeMartino L, De Feo V, Fartianni F, Nazzaro F (2009) Chemistry, oxidant, antibacterial and antifungal activities of volatile oils and their components. Nat Prod Commun 4(12):1741–1750

Edwin JE, Dubey S, Gupta S, Gupta VB (2007) Hepatoprotective activity of *Euphorbia hirta* Linn. Geo Bios 34(1):81–83

Efiom OO, Oku E (2012) Isolation And Characterization of Triterpene Lup-20-EN-3-Ol And 1,3-Hexyloxacyclotridec-10-EN-2-One From the root of *Grewia Mollis*. N Y Sci J 5(11)

El-Said AHM, Goder EH (2014) Antifungal activities of *Cuminum cyminum* and *Pimpinella anisum* essential oils. Int J Curr Microbiol Appl Sci 3(3):937–944

Ene AC, Atawodi SE, Ndukwe I, Kwanashie HO (2009) Bioassay-guided fractionation and in vivo antiplasmodial effect of fractions of chloroform extract of *Artemisia maciverae* Linn. Acta Trop 112:288–294. doi:10.1016/j.actatropica. 2009.08.016

Fapojuwomi OA, Asinwa IO (2012) Assessment of medicinal values of *Rauvolfia vomitoria* (Afzel) in Ibadan Municipality. Greener J Med Sci 3(2):037–041

Focho DA, Ndam WT, Fonge BA (2009) Medicinal plants of *Aguambu Bamumbu* in the Lebialem highlands, southwest province of Cameroon. Afri J Pharm Pharmacol 3(1):001–013

Hanan MA, Musarat A, Azza ME (2012) Biological evaluation of constituents from Grewia mollis. J Chem Pharm Res 4(1):50 8–518

Hiermann A, Bucar F (1994) Influence of some traditional medicinal plants of Senegal on prostaglandin biosynthesis. J Ethnopharmacol 42(2):111–116

Ishak FD, So'ad SZM, Jauhari AHA, Mashid NN, Hassan NM (2013) *In vitro* study of antimicrobial activity of *Acalypha indica* Linn extract. Open Conf Proceed J 4(2):57–60

Jagatheeswari D, Deepa J, Ali HSJ, Ranganathan P (2013) *Acalypha indica L-* An important medicinal plant: a review of its traditional uses, and pharmacological properties. Int J Res Bot 3(1):19–22

Jimoh FO, Oladiji AT (2005) Preliminary studies on *Piliostigma thonningii* seeds: proximate analysis, mineral composition and phytochemical screening. Afri J Biotechnol 4(12):1439–1442

Johnson PB, Abdurahman EM, Tiam EA, Abdu AI, Hussaini IM (1999) *Euphorbia hirta* leaf extracts increase urine output and electrolytes in rats. J Ethnopharmacol 65(1):63–69

Johri RK (2011) *Cuminum cyminum* and *Carum carvi*: an update. Pharmacogn Rev 5(9):63–72

Kerharo-Adam JG (1974) Pharmacopée sénégalaise traditionnelle,plantes médicinales et toxiques. Vigot et Frères, Paris p 1011

Khosh-Khui M, Bonyanpour AR (2006) Effects of some variables on seed germination and seedling growth of Cumin (*Cuminum cyminum* L.) Int J Agric Res 1:20–24

Kucinggalak KG (2015) http://www.forestry.gov.my/herba/Kucinggalak_en.pdf. Accessed on the 8 May 2015

Latham P (2004) Useful plants of Bas-Congo province. Democratic Republic of the Congo. DFID, London, UK. 320 pp. Int J Pharm 34(5):355–358

Lanhers MC, Fleurentin J, Cabalion P, Rolland A, Dorfman P, Misslin R, Pelt JM (1990) Behavioral effects of *Euphorbia hirta* sedative and anxiolytic properties. *J Ethnopharmacol* 29(2):189–198

Lemessa D (2010) *Piliostigma thonningii* (Scumach) Milne-Redh. Internet: Brink M, Achigan-Dako EG (Ed). Prota 16: Fibres/Plantas a fibres. PROTA. Wageningen, Netherlands http://www.prota4u.org. Assessed 25 May 2015

Leporatti ML, Ghedira K (2009) Comparative analysis of medicinal plants used in traditional medicine in Italy and Tunisia. *J Ethnobiol Ethnomed* 5:31

Modena ÉDS, Pires ACV, Barônio GJ, Inforzato I, Demczuk SDB (2012) Do fruit traits of the *Senna occidentalis* weed influence seed predation by Bruchinae? Braz J Biosci 10(3):293–297

Nartey ET, Ofosuhene M, Kudzi W, Agbale CM (2012) Antioxidant and gastric cytoprotective prostaglandin properties of *Cassia sieberiana* root bark extract as an anti-ulcerogenic agent. BMC Compl Altern Med 12:65

Newton LE (2008) *Euphorbia unispina* N.E.Br. Internet: Schmelzer GH, Gurib-Fakim A (Ed) Plant Resources of Tropical Africa / Ressources vegetales de l'Afrique tropicale, Wageningen, Netherlands. http://www.prota4u.org/search.asp. Assessed 24 May 2015

Nwodo NJ, Okide GB, Okonta JM, Ebebe IM (2003) Antidiabetic effect of *Rauwolfia vomitoria* ethanolic leaf extract in rabbits. J Trop Med Plants 4(1):71–74

Odeja OO, Obi G, Ogwuche CE, Elemike EE, Oderinlo OO (2014) Phytochemical screening, antioxidant and antimicrobial activities of *Senna occidentalis* (L) leaves. Int J Herb Med 2(4):26–30

Ogbe AO, Ditse U, Echeonwu I, Ajodoh K, Atawodi SE, Abdu PA (2009) Potential of a wild medicinal mushroom, *Ganoderma* sp., as feed supplement in chicken diet: effect on performance and health of pullets. Int J Poult Sci 8(11):1052–1057

Ogbulie JN, Ogueke CC, Okoli IC, Anyanwu BN (2007) Antibacterial activities and toxicological potentials of crude ethanolic extracts of *Euphorbia hirta*. Afr J Biotechnol 6(13):1544–1548

Olapade AA, Ajayi OA, Ajayi IA (2014) Physical and chemical properties of *Cassia sieberiana* seeds. Int Food Res J 21(2):67–772

Onwuliri F, Ndako J, Olabode AO, Echeonwu GON, Onwuliri EA, Nwankiti PP, Ohaeri MC, Amona GE (2009) Phytochemical screening and antimicrobial activity of *Guiera senegalensis* leaves. Int J Trop Agric Food Sys 3(2)

Oshobu MA, Geidam MA (2014) Effects of aqueous ethanolic extract of the leaves of *Guiera senegalensis* J.F.Gmel (Combretaceae) on liver function in Wister strain albino rats. Int J Sci Res Manag 2(3):611–615

Oyewale AO, Mika A, Peters FA (2002) Phytochemical, cytotoxicity and microbial screening of *Euphorbia hirta* Linn. Glob J Pure Appl Sci 8(1):49–55

Ozolua RI, Alonge P, Igbe I (2009) Effects of leaf extracts of *Piliostigma thonningii* Schum on aortic ring contractility and bleeding time in rats. J Herb Spices Med Plants 15:326–333

Parashar M, Jakhar ML, Mabik CP (2014) A review on biotechnology, genetic diversity in cumin (*Cuminum cyminum*). Int J Life Sci Pharmacogn Rev 5(9):63–73

Rajaselvam J, Benila SJM, Meena R (2012) A study of antimicrobial activity of *Acalypha indica* against selected microbial species. Int J Pharma Sci Res (IJPSR) 3(9):473–476

Rao CV, Rao GMM, Kartik R, Sudhakar M, Mehrotra S, Goel RK (2003) Ulcer protective effect of *Euphorbia hirta* extract in rats. J Trop Med Plants 4(2):199–205

Sadiq IS, Shuaibu M, Bello AB, Tureta SG, Isah A, Izuagie T, Nasiru S, Kamaru MB (2012) Phytochemistry and antimicrobial Activities of *Cassia Occidentalis* used for herbal remedies. J Chem Eng 1(1):38–41

Saidu AN, Aina EO, Mann A, Leje UI (2011) The effect of aqueous extract of leaves on rat infected with *Salmonella typhi*. Aus J Basic Appl Sci 5(12):1863

Salihu SO, Usman AA (2015) Antimicrobial and phytochemical study of the bioactive fractions of *Guiera senegalensis* from Alasan Tambuwal, Nigeria. J Pharm Phytochem 3(6):106–111

Sanogo R (2012) *Guiera senegalensis* J.F. Gmel. Internet: Schmelzer GH, Gurib-Fakim A (ed). Prota 11(2): Medicinal plants/plantes medicinales 2. PROTA, Wageningen, Netherlands. http://www.prota4u.org. Accessed 4 May 2015

Schmelzer GH, Gurib-Fakim A (2008) Plant resources of tropical Africa medicinal plants 1 *PROTA* Foundation, Wageningen, Netherlands/Backhiys Publishers, Leiden. Netherlands CTA 11(1)

Silva O, Serrano R, Gomes ET (2008) Botanical characterization of *Guiera senegalensis* leaves. Micro Microanal 14(5):398–404

Singh P, Khandelwal P (2010) Cetyltriacontanoate and other constituents from *Acacia jacquemontii* and *Kigelia pinnata*. J Indian Chem Soc 87:1403–1407

Soladoye MO, Amusa NA, Raji-Esan SO, Chukwuma EC, Taiwo AA (2010) Ethnobotanical survey of anticancer plants in Ogun State, Nigeria. Annals Biol Res 1(4):261–273

Sombie PAED, Hilou A, Mounier C, Coulibaly AY, Kiendrebeogo M, Millogo JF, Nacoulma OG (2011) Res. J Med Plants 5(4):448–461

Somboro AA, PatelK DD, Sidibe L, Chalchat JC, Figuredo G, Ducki S, Troin Y, Chalard P (2011) An ethnobotanical and phytochemical study of the African medicinal plant *Guiera senegalensis* J.F.Gmel. J Med Plants 5(9):1639–1651

Somchit MN, Mutalib AR, Hasmawie MR, Murni A (2001) *In vitro* antifungal and antibacterial properties of *Euphorbia hirta*. J Trop Med Plants 2(2):179–182

Terashima H, Ichikawa M (2003) A comparative ethnobotany of the Mbuti and Efe hunter-gatherers in the Ituri forest, Democratic Republic of Congo. Afr Study Monogr 24(1–2):1–168

Tona L, Cimanga RK, Mesia K, Musuamba CT, De Bruyne T, Apers S, Hernans N, Van Miert S, Pieters L, Totté J, Vlietink AJ (2004) In vitro antiplasmodial activity of extracts and fractions from seven medicinal plants used in the Democratic Republic of Congo. J Ethnopharm 93:27–32

Ullah MF, Khan MW (2008) Food as medicine: potential therapeutic tendencies of plant derived polyphenolic compounds. Asian Pac J Cancer Prev 9:187–196

Van der Maesen LJG (2007). *Cassia sieberiana* DC. Internet: Schmelzer GH, Gurib-Fakim A (ed) Prota 11(1): Medicinal plants/plantes medicinales 1. [CD-Rom] *PROTA*, Wageningen, Netherlands. http://www.prota4u.org/search.asp. Assessed 24 May 2015

Van-Dilst FJH, Leeuwenberg AJM (1991) Rauvolfia L. in Africa and Madagascar. Series of revisions of Apocynaceae 33. Bulletin du Jardin Botanique Nationale

Vashishtha VM, John T, Kumar A (2009) Clinical and pathological features of acute toxicity due to *Cassia occidentalis* in vertebrates. Indian J Med Res 130:23–30

Vijayalakshmi S, Ranjitha J, Devi Rajeswari V, Bhagiyalakshmi M (2013) Pharmacological profile of *Cassia occidentalis* – a review. Int J Pharm Pharma Sci 5(3):29–33

Wallace AP, Marko KE, Wisdom AP, Dennis E (1990) Comparative clinical studies of three plant leaves as sources of protein using albino rats. J Ghana Sci Assoc 2(3):81–87

# Phytochemical, Pharmacological and Therapeutic Potentials of Some Wild Nigerian Medicinal Trees

**Sunday E. Atawodi, Olufunsho D. Olowoniyi, Godwin O. Adejo, and Mubarak L. Liman**

**Abstract** With its location in an extensive geographical area that spread from the Mangrove forest in the south to the Sahel savannah in the north, Nigeria has diverse tropical vegetation with varying economic and medicinal significance. The vegetation, among other uses is a source of herbal medicine on which a large proportion of its populace relies for primary health care. Trees have a large repository of phytochemicals, to which the various pharmacological and therapeutic attributes can be ascribed. This article is a concise review of nine trees with medicinal importance belonging to eight families, which are widely used in Nigerian traditional and indigenous medicine. For each plant, information on the taxonomy, morphology, geographical distribution, ecological requirements, major chemical constituents and bioactive compounds, traditional uses and medicinal uses based on biological activities, as well as uses supported by clinical data are provided.

**Keywords** Trees • Medicinal plants • Phytochemicals • Pharmacology • Therapeutic potential • Herbal medicine • Nigeria

## 1 Introduction

Nigeria, is the most populous country in Africa, having a landmass of over 923.768 $km^2$ enclosed within latitudes $4°16^1$ N and $13°52^1$ N of the equator and between longitudes $2°49^1$ E and $14°37^1$ E of Greenwich Meridian. Within the varied ecological zones to which the country belongs is an array of floral species. There are 7895 plant species from 338 families and 2215 genera that have been identified in Nigeria. Its moist forests are also rich in epiphytic ferns and orchids and contain

---

Dedicated to all those that have made life worth living

S.E. Atawodi (✉) O.D. Olowoniyi • G.O. Adejo • M.L. Liman
Biochemistry Department, Ahmadu Bello University, Zaria, Nigeria
e-mail: atawodi_se@yahoo.com



M. Neffati et al. (eds.), *Medicinal and Aromatic Plants of the World - Africa Volume 3*, Medicinal and Aromatic Plants of the World,
DOI 10.1007/978-94-024-1120-1_11

over 560 species of trees which attain heights of at least 12 m and girth of 60 cm (Borokini et al. 2010).

Apart from making the biomolecules that are universally required by living organisms, plants also synthesize a large number of complex chemicals that are unique to them. Phytochemicals are products of secondary metabolism defined variously as a metabolic *cul de sac* or peripheral metabolic pathways leading off from the universal (or primary) metabolism. Phytochemicals range from cell wall substances, through photosynthetic pigments, terpenes and terpenoids, the alkaloids, plant phenolics to plant hormones, non-protein amino acids and cyanogenic glycosides. Many phytochemicals exhibit bioactive activity in other living organisms and are useful as therapeutic agents, pesticides, food additives and other biologicals (Onwuliri et al. 2006).

Medicinal plants are plants whose one or more of their organs contain active ingredients which can be used for therapeutic purposes or contain compounds that can be used for the synthesis of useful drugs. The medicinal values of these plants lie in the bioactive phyto-constituents that produce definite physiological action on animals (Afolabi and Afolabi 2013). Herbal medicines have been widely utilized as effective remedies for the prevention and treatment of multiple health conditions by almost every known culture. Thus, for over 5000 years, the Chinese and Indians have used herbal medicine for their health care (Rivera et al. 2013), steming from the fact that herbal medicine is cheaper, easily accessible, easy to prepare and devoid of severe side effects (Fasola et al. 2010). In Nigeria, plants are used for the management of oxidation-associated diseases (Atawodi 2005; Atawodi et al. 2014a) and other ailments (Atawodi et al. 2014b; Atawodi et al. 2014c). To add to that body of knowledge in the most populous black nation, this article is a further compilation of notable Nigerian medicinal trees.

## 2 Annonaceae

***Annona muricata*** **Linn** (Synonyms: *Annona macrocarpa*, *A. bonplandiana*, *A. cearensis*, *Guanabanus muricatus)*

(Source: biogeodb.stri.si.edu)

**Taxonomy**
*Annona muricata* Linn commonly known as graviola, soursop, paw-paw, guanabana, adunu, sorasaka, Mullaatha, Thorny custard apple, Shul-ram-fal, or Hanuman fal. belongs to the family of Annonaceae (Sulaiman et al. 2012; Mishra et al. 2013).

**Morphological Description**
It is a flowering, evergreen tree with dark green leaves, which are lanceolate, glossy and broad, (Sulaiman et al. 2012; Mishra et al. 2013). Growing up 5–6 m in height, the young branchlets are rusty-hairy, the malodorous leaves, normally evergreen, are alternate, smooth, glossy, dark green on the upper surface, lighter beneath; oblong, elliptic or narrow-obovate, pointed at both ends, 6–20 cm long and 2–6 cm wide. The flowers are borne singly, may emerge anywhere on the trunk, branches or twigs. They are short stalked, 4–5 cm long, plump, and triangular-conical; the 3 fleshy, slightly spreading, outer petals yellow-green, with three close-set inner petals, pale-yellow. The fruit is more or less oval or heart-shaped, sometimes irregular, lopsided or curved, due to improper carpel development or insect injury. The size ranges from 10–30 cm long and up to 15 cm in width and the weight may be up to 4.5–6.8 kg. The fruit is compound and covered with reticulate, leathery-appearing but tender, inedible bitter skin from which protrude few or many stubby, or more elongated and curved soft, pliable "spines" (Adewole and Caxton-Martins 2006).

**Geographical Distribution**
It is a typical tropical tree with heart-shaped edible fruits and is widely distributed in most of tropical countries, (Adewole and Caxton-Martins 2006), including sub-Saharan African countries like Nigeria that lie within the tropics and sub-tropics (Adewole and Caxton-Martins 2006; Mishra et al. 2013; Sulaiman et al. 2012). However, it is native to Mexico, Cuba, Central America, the Caribbean and northern South American countries, including Colombia, Brazil, Peru and Venezuela (Mishra et al. 2013).

**Ecological Requirements**
It is adapted to areas of high humidity and relatively warm winters; usually, temperatures below 5 °C (41 °F) will cause damage to leaves and small branches, while temperatures below 3 °C (37 °F) can be fatal, as the fruit becomes dry and unfit for consumption (Mishra et al. 2013).

**Major Chemical Constituents and Bioactive Compounds**
Among the chemical constituents found in the leaf of *Annona muricata* are alkaloids, essential oils and annonaceous acetogenins (Atawodi 2011, Sulaiman et al. 2012).

**Traditional Use (Part(s) Used) and Common Knowledge**
Traditionally, the leaf has been used in the treatment of headaches, hypertension, cough, asthma and as an antispasmodic, sedative and nervine for heart condition (Sulaiman et al. 2012). It has also been reported to exhibit anti-inflammatory and analgesic effects (Sulaiman et al. 2012).

The fruit has been traditionally applied for cough, hypertension, rheumatism, tumors, cancer, asthma, childbirth, lactagogue, tranquilizer and in liver disorders, as well as arthritis, where extracts are applied externally. The leaf is also used for treating headaches, insomnia, cystisitis, liver problems, diabetes and as anti-inflammatory, anti-spasmodic and anti-dysenteric agents. The decoction of the leaf has parasitic, anti-rheumatic and neuralgic effects when used internally, while when cooked, the extract of leaf is applied topically against rheumatism and abscesses (Mishra et al. 2013).

**Modern Medicine Based on Traditional Medicine Uses (Biological Activities)**
The leaf and seed of the tree have long been used by native peoples for treating and managing an astounding variety of ailments, ranging from parasites (seed), to high blood pressure and cancer (Adewole and Caxton-Martins 2006; Mishra et al. 2013).

**Uses Supported by Clinical Data, Pharmacopoeias, Well Established Documents and in Traditional Medicine**
Annonaceous acetogenins from *Annona muricata* L have been found to possess anti-tumor and anticancer agents in numerous *in vitro* studies. The compound was also shown to be selectively toxic against various types of the cancerous cells without harming healthy cells (Sulaiman et al. 2012). Ethanolic leaf extract applied topically on mice skin papillomagenesis at graduated doses of 30, 100 and 300 mg/kg significantly reduced induced papillomagenesis at all doses in both pre-initiation and promotion protocols in a dose-dependent manner, and comparable to carcinogen–treated control (Sulaiman et al. 2012).

Moreover, at 100 and 300 mg/kg, the leaf extract completely inhibited the tumor development in all stages, thus suggesting the suppression of tumor initiation as well as tumor promotion even at lower dosage by *A. muricata* leaves extract (Sulaiman et al. 2012). Research has further validated the use of various parts of *A. muricata* for hypertensive, antispasmodic, vasodilatory and cardio depressive-conditions (Mishra et al. 2013).

## 3 Asclepiadaceae

### Parquetina nigrescens *(Afzel.) Bulock*

**Synonyms** *Periploca nigrescens*

**Common and Local Names**
Ewe Ogbo (Yoruba) (Odetola et al. 2006)

**Morphological Description**

*Parquetina nigrescens* has a woody base, up to 8 m long, glabrous, latex copious. The leaves are simple and entire, almost sessile, blade oblong, (7.5-)12.5–14(−16) cm × (3-)4.5–7.5(−11) cm, apex rounded, suddenly long-acuminate, base rounded to almost cordate, green and often glossy above, pale green beneath, (thin) leathery, with conspicuous lateral veins. Inflorescence an open axillary cyme, 15–30 flowered. Flowers bisexual, 5-merous, regular, pedicel 6–10 mm long; sepals very broadly ovate, about 2.5 mm long, apex obtuse; corolla with very short tube, lobes fleshy, spreading, elliptical to ovate, 10–12 mm × 4–5 mm, reflexed, apex rounded, above velvety, dark brown at apex; corona lobes bifid, filiform, 5–7 mm long, cross-shaped because of 2 lateral segments above middle, apical segments curled, greenish white to pale yellow; stamens with hairy filaments, pollen carriers spoon-shaped; ovaries 2, free, semi-inferior, almost globular, gynostegium excerted, styles fused, apex 5-angled. The fruit is a pair of follicles, horizontal, each follicle narrowly ovoid, 12–21 × 12–20 mm, 2-ridged, apex slightly curved, many-seeded. Seeds narrowly elliptical to elliptical, brown to dark brown, warty, tuft of hairs white (Alvarez Cruz 2012).

(Source: Authors)

**Geographical Distribution**

*Parquetina nigrescens* Afzel. occurs in a large part of Africa, from Senegal east to Sudan and south through Central and East Africa to Zambia, Angola and Eastern Zimbabwe (Alvarez Cruz 2012).

**Ecological Requirements**

*Parquetina nigrescens* grows in secondary forests, savanna, vegetation bordering roads and gallery forest, also commonly growing on ant-hills. It grows on various types of soil, including marshy areas (Alvarez Cruz 2012).

**Chemical Constituents**
*Parquetina nigrescens* leaf contain cardenolides, glycosides and alkaloids (Odetola et al. 2006) while saponins, alkaloids, tannins, anthraquinones, terpenoids, flavonoids, cardiac glycosides and ascorbic acid are reported by Akinyemi and Dada (2013). The rich cardiac glycosides (cardenolides) collectively called strophanthins are most abundant in the latex and are responsible for the activity in arrow poison.

**Traditional and Common Uses**
Different parts of *Parquetina nigrescens* (leaf, bark, latex and roots) are used for various disorders of the gastrointestinal tract disorders and as constituents of medications for treating rickets, diarrhea, skin lesion, menstrual disorders and gonorrhea. A decoction of the leaf is drunk as aphrodisiac (Odetola et al. 2006). The use of *Parquetina nigrescens* as herbal remedy for the management of sickle cell anemia has also been reported (Imaga 2013). A decoction of the leaf is taken as enema to treat kidney problems, constipation and to induce abortion. A decoction of the leaf or whole plant, sometimes with parts of other plant species is taken to treat measles, intestinal worms, diarrhea, dysentery, diabetes, menstrual disorders and venereal diseases. The latex and leaf sap cause a burning sensation on the skin and are externally applied to tumors, abscesses, sores and burns. They also blacken scars and are applied as a dressing for wounds. The latex is very toxic and widely used as an ingredient of arrow poison (Alvarez Cruz 2012).

**Biological Activities**
In a study to evaluate the antimicrobial activity of the aqueous leaf extract of *Parquetina nigrescens*, it showed antimicrobial effect against *Staphylococcus aureus, Salmonella typhi, Proteus mirabilis, Pseudomonas aeruginosa, Bacillus subtilis* and *Preteus vulgaris* while the ethanolic extract only had effect on *Pseudomonas aeruginosa* and *Salmonella typhi* (Odetola et al. 2006).

## 4 Bignoniaceae

### Kigelia africana *(Lam.) Benth*

**Synonym** *Kigelia pinnata, Kigelia aethiopica*

**Common and Local Names** Sausage tree (English); Rawuya (Hausa); Uturubein (Igbo); Pandoro, Iyan (Yoruba); Mwegea (Swahili); Umfongothi (Zulu) (Grace and Davis 2002).

(Source: Authors)

**Morphological Description**

The tree is a semi-deciduous to deciduous tree that grows up to 25 m tall. The bark is grey and smooth and flakes in older specimens. Leaves are crowded near the tips of branches, and young leaves are brownish-red. Flowers bloom in long, loose, pendulous spray of 5–12 flowers. Petals are a deep, velvety red with yellow veining on the outside. The cylindrical fruit is pendulous on a long fruit stalk. The fruit can grow up to 1.0 m long and 20 cm wide, and is grey and rounded at the apex. The tree flowers from August to October, and fruits from December to June.

Depending on the climate, the sausage tree is remarkably fast-growing and can mature in 4–5 years. Ripe fruits can weigh up to 12 kg and can cause considerable damage when they drop. With its fast growth rate, spreading canopy, and interesting flowers and fruit, *Kigelia africana* is a popular street tree. It can also be used successfully for bonsai, because the thick stem makes for an attractive feature (Watt and Brayer-Brandwijk 1962). The fruit is indehiscent, with woody wall and heavily marked with lenticels at the surface. It is grey-brown and many seeded when matured (Olatunji and Atolani 2009).

**Geographical Distribution**

*Kigelia africana* (Lam) occurs throughout tropical Africa (Saini et al. 2009) but does not occur in Mauritania, Sao Tome and Principe, or the Indian Ocean Islands. It has been introduced as an ornamental to Cape Verde and Madagascar, Pakistan, India, China, South East Asia, Australia, Hawaii and Central and South America (Grace and Davis 2002).

**Ecological Requirements**

*Kigelia africana* grows along watercourses, in riverine fringes, alluvial and open woodland, high rainfall savanna, shrub-land and in rain forest. It occurs on loamy red clay soils, sometimes, rocky, damp or peaty, from sea level up to zoom altitude (Olatunji and Atolani 2009).

**Chemical Constituents**

The root and bark of *Kigelia africana* have naphthoquinone, lapachol and the dihydroisocoumarin kigelin as major compounds. Several other compounds including the naphthaquinoids kigelinone, pinnatal and isopinnatal, and the sterols stigmasterol and β-sitosterol have been isolated from the bark. The flavonoids 6-hydroxyluteolin-7-α-glucoside and luteolin have been isolated from the fruits and the leaves, while the root has also yielded dihydroisocoumarins, lapachol and sterols, and the presence of iridoid glycosides also have been reported (Govindachari et al. 1971; Alamelu and Bhuwan 1974).

A study of the heartwood identified the presence of lapachol, dehydro-α-lapachone, tecomaquinone-I, D-sesamin, paulowin, kigeliol, kigelinone, β-sitosterol and stigmasterol (Singh and Khandelwal 2010). The major iridoids found in the root bark and stem bark of *Kigelia africana* are specioside, verminoside and minecoside (Houghton and Akunyili 1993). The iridoids possess anti-inflammatory properties (Gouda et al. 2003; Kupeli et al. 2005). Also, those same iridoids were isolated from the stem bark and they possess anti-amoebaic properties (Bharti et al. 2006). Further analysis of the fruits identified four new iridoids, namely, 7 – hydroxyviteoid II, 7-hydroxyeucommic acid, 7-hydroxyl-10-deoxyeucommiol and 10 – deoxyeucommiol. Seven known compounds, namely jiofuran, jioglutolide, 1-dehydroxy-3,4-dihydroaucubigenin, des-p-hydroxybenzoyl kisasagenol B, ajugol, verminoside and 6-transcaffeoyl ajugol were also identified (Gouda et al. 2003). A study of the antimicrobial properties of the aqueous stem back extract of *Kigelia africana* revealed the presence of two naphthaquinones, kigelinone and isopinnatal (Akunyili et al. 1991; Akunyili and Houghton 1993).

The hexane extract of the leaf of *Kigelia africana* has been reported to be rich in hydrocarbons and some volatile compounds. They include n–hentriacontane, 1-tricosene, 11-(2,2-dimethyl) propylheneicosane, 2,6,10–trimethyldodecane, penta fluoroheptadecyl ester, 2–ethylhexyloctadecyl sulfurous acid ester, heneicosane and hexyloctyl–sulfurous acid ester. Others are 4,4-dimethyl undecane, methyl-12-methyltetradecanoate, 1-iodohexadecane and 1- iododecane. Hentriacontane have been reported to have a possible anti–tumour activity while methyl-12-methyltetradecanoate has also been reported for its inhibition capacity on the development of coneal angiogenesis, which is responsible for blindness and other infections (Atolani and Olatunji 2010).

**Traditional and Common Uses**

The powdered fruit of *Kigelia africana* is applied as a dressing in the treatment of wounds, abscesses and ulcers. The green fruit is used as a poultice for syphilis and rheumatism, and a poultice made from the leaves is used as treatment for backache. An infusion is made from the ground bark and fruits to treat stomach problems in children, and an infusion from the root and bark is taken to treat pneumonia. A decoction made from the bark is gargled to relieve toothache (Jackson and Beckett 2012). *Kigelia africana* is also widely used for a variety of skin complaints. It is used in both traditional and orthodox medicines to treat malignant neoplasms such as skin melanoma, tumors and breast cancer. Traditional preparations include extracts, poultices and powders of the bark or fruit, topical creams containing extracts of the fruits are produced commercially (Grace and Davis 2002). Snakebite antidotes are made with an infusion of the parts, taken orally or rubbed onto the bite. A fruit decoction is also used to treat edema of legs (Grace and Davis 2002). Infectious diseases including leprosy and impetigo, dermal complaints and infections such as whitlows, eczema, cysts, acne, boils and psoriasis are treated with traditional medicines containing the fruits and less frequently the bark (Grace and Davis 2002).

The powdered fruit is applied as a dressing for ulcers, while unripe fruit is used as a dressing for wounds, hemorrhoids and rheumatism. Venereal diseases are commonly treated with the bark extracts, usually in palm wine as oral medication. The fruits and bark, ground and boiled in water are also taken orally or used as an enema in treating children's stomach ailments arising from infestation by worms (Grace and Davis 2002; Saini et al. 2009). Aqueous preparations of the root, fruit and flowers are administered orally or as a vaginal pessary to treat infections. The fruits and bark are used to promote breast development in young women, or in contrast, to reduce swelling and mastitis of the breast. The fruits are further employed as a galactogue and a decoction of the bark and leaves is administered as an abortifacient. Sexual complaints such as infertility, poor libido, sexual asthenia and impotence are treated with medicines containing the fruits, roots or leaves.

A small quantity of unripe fruit is chewed, while an aqueous preparation or preparations in local beer including palm wine is taken orally as a sexual stimulant and aphrodisiac (Grace and Davis 2002). Fungal infestation such as ringworm, mycosia and athletes' foot are washed with the water in which the bark has been macerated, and preparation containing the leaves and fruits applied locally (Grace and Davis 2002). Aqueous extract of the leaves is applied to treat diarrhea, while that of the stem bark is for epilepsy. Parts of the plant are also used to treat malaria, diabetes and postpartum hemorrhage. For ethno-veterinary medical application, the fruits and sometimes other parts of *Kigelia africana* are used to treat digestive system disorders, leg edemas, dermal irritations and infections, mastitis and retained placenta. Brucellosis and Newcastle disease are also treated with *Kigelia africana* (Grace and Davis 2002).

**Biological Activities**

Among the many uses of *Kigelia africana* in traditional medicine is in the treatment of bacterial and fungal infections. In a study to verify these properties, crude aqueous, ethanolic and ethyl acetate extracts of stem, bark and fruit were screened for antibacterial activity. The stem bark and fruit extracts showed similar antibacterial effects against Gram-negative and Gram-positive bacteria (Grace and Davis 2002). Owolabi et al. (2007) also reported the antibacterial and antifungal effects of the crude ethanolic extract of the stem bark, using agar diffusion technique against *Staphylococcus aureus* and *Candida albicans.* The effect of the extract was comparable to that of the standard amoxillin antibiotic.

In another study, the stem bark extract inhibited the growth of *Escherichia coli, Pseudomonas aeruginosa, Staphylococcus aureus* and *Candida albicans* (Akunyili et al. 1991). Evaluation of the antibacterial activities of ethanolic and aqueous extracts of *Kigelia africana* fruit against multi-drug resistant *Pseudomonas aeruginosa* showed that the ethanolic extract was more potent than the aqueous extract (Tyagi et al. 2011). Antiprotozoal, anti-inflammatory, analgesic, antidiarrheal, antidiabetic, antioxidant and anti-ulcer activities have been reported and scientifically evaluated for various parts of *Kigelia africana* (Atawodi and Olowoniyi 2015).

## 5 Chrysobalanceae

### Parinari curatellifolia *Planch. Ex Benth*

(Source: www.africamuseum.be, http://copperflora.org)

**Botanical Name and Family**
*Parinari curatellifolia* Planch. Ex Benth (Family: Chrysobalanaceae) is commonly called cork tree or hissing tree (Maharaj and Glen 2008). It is also called mbola plum, mobola plum, fever tree (Orwa et al. 2009).

**Common Names**
Hissing tree, Mbola plum, Mobola plum, fever tree (English), Pobeguin,mendoca (French), Mbura (Swahili), Grysappel (Afrikaans), (Orwa et al. 2009).

**Morphological Description**
*Parinari curatellifolia* is an evergreen, rough dark grey barked, spreading tree that can grow up to 20 m, and that is striking amongst surrounding vegetation because of its semi-circular, almost mushroom-shaped canopy depicting hues of blue-green and grey. The young shoots are densely covered with yellow wooly hairs. The leaves are alternate, single, elliptic to oblong, 3–8 × 2–4 cm, leathery dark green on top.

The leaves are finely velvety when young, but losing the hair later, densely hairy and grey to yellow underside. The apex is broadly tapering, often notched; base square, margin entire and has a short petiole. *Parinari curatellifolia* has small white and sweet scented flowers that exist in short branched heads or panicles, 4–6 cm in diameter in leaf axils.

The stalks and calyces are densely covered with yellowish wooly hairs, while the flowers are bisexual, have five sepals, five petals and seven or more stamens, joined at the base in a short ring inserted in the mouth of the receptacle. The flower has two-chambered ovary. The fruit is oval to round, up to 5 × 3.5 cm, russet-yellow to greyish, scaly and pitted, turns orange-yellow when ripe (Orwa et al. 2009).

**Geographical Distribution**

*Parinari curatellifolia* grows across Nigeria and other sub-Saharan Africa countries downwards up to South Africa, but it is exotic to Madagascar and Seychelles (Orwa et al. 2009).

**Ecological Requirement**

*Parinari curatellifolia* grows naturally in open deciduous woodland. It occurs in varying climatic regimes and is particularly common near rivers and in areas of poor drainage. It is sensitive to frost and cold wind. It grows in regions between 0–1900 m altitude, mean annual rainfall 0–2700 mm and mean annual temperature of 10–30 °C. A wide range of soil types is required for its growth- light-yellowish brown to reddish yellow, gritty, sandy clay loam and red to dark red friable clays with lateritic horizon (Orwa et al. 2009).

**Major Chemical Constituents and Bioactive Compounds**

Phytochemical screening of the stem bark extracts of *Parinari curatellifolia* revealed the presence of saponins, balsams, carbohydrates, alkaloids, tannins, cardiac glycosides, flavonoids, digitalis glycosides, phenol, terpenes and steroid (Peni et al. 2010).

**Traditional Uses, Part(s) Used and Common Knowledge**

*Parinari curatellifolia* is used as a remedy for dysentery, epilepsy, malaria, toothache and venereal diseases (Feitosa et al. 2012) and epiglottis (Wurochekke et al. 2013).

**Biological Activities**

The aqueous extract of stem bark of *Parinari curatellifolia*, by hole in plate bioassay procedure had inhibitory activity against *Staphylococcus aureus* and *Klebsiella* spp. The methanolic extract was effective against *Bacillus subtilis* and *Pseudomonas aeruginosa* (Peni et al. 2010), while the root bark is also reported to have strong antioxidant activity (Atawodi et al. 2013).

**Uses in Traditional Medicine**

*Parinari curatellifolia* is a valuable and cherished medicinal plant in which different parts of the plant are widely used by the traditional herbalist in the treatment of diabetes (Atawodi et al. 2013), and other conditions, like hypertension, liver-related diseases and epiglottitis. The medicinal uses of the fruit extract of *Parinari curatellifolia* has also been reported to include its use as cardiac tonic, dyspores and as diuretic. The leaves are used as expectorant, sedative, anti-inflammatory and in treating anemia. Traditionally the bark is used for washing clothes, for vaginal douches, itchy scalp, dandruff, cough and also for treating infections (Peni et al. 2010).

## 6 Combretaceae

### Anogeissus leiocarpus *(DC) Guill. & Perr.*

**Synonyms** *Anogeissus schimperi* Hochest. Ex Hutch & Dalz., *Concarpus leiocarpus* DC.

(Source: P. Poilecot)

**Taxonomy**
*Anogeissus leiocarpus* Guill. & Perr. (Combretaceae), Common name: Axle-wood tree (Victor et al. 2013) is an African indigenous tree commonly found in forest savannahs of Nigeria and other West African countries (Attiouaaf et al. 2011).

**Morphological Description**
*Anogeissus leiocarpus* is a woody deciduous tree that can grow up to a height of15–30 m tall and up to 1 m in diameter (Burkill 1985; Attiouaaf et al. 2011; Arbab 2014). The heights seem to vary as one moves from country to country from Senegal (shorter) to Nigeria (taller), but with generally sparse crown (Burkill 1985; Attiouaaf et al. 2011). Heart-wood is dark, dull brown, streaky, becoming sometimes almost ebony-black, and is very hard, dense and fine-textured, while the bark is generally grey and scaly (Burkill 1985). Branches are drooping and slender, with leaves arranged in alternate, ovate–lanceolate in shape, 2–8 cm long and 1.3–5 cm across (Arbab 2014). The leaves are acute at the apex and attenuate at the base, pubescent beneath. Inflorescence globose heads, 2 cm across, yellow; the flowers are bisexual, petals absent. Fruits are globose cone like heads (Arbab 2014).

The fruits of *A. leiocarpus* contain about 40 wind-dispersed seeds of 10 mg each. Each fruit is broadly winged, dark grey, 3 cm across. It can reproduce by seeds as well as vegetative propagation. The seeds ripen during the dry season and germinate mainly at the on-set of the rainy season (Hennenberg et al. 2005).

**Geographical Distribution**
In Africa, its occurrence extends from Senegal in West Africa to Sudan and Ethiopia in East Africa. Those growing in the driest area tend to have smaller leaves and hairier flowers than those growing under wetter conditions, but both differences are not sufficiently marked to create distinct varieties (Hennenberg et al. 2005; Atawodi 2011).

**Ecological Requirements**
*Anogeissus leiocarpus* thrives around the Guinea sub-humid dry forest stretching across a vast area that forms the 'Guineo-Congolian' rainforest, where rainfall ranges from 1200–1500 mm/year in the east to 1500–2000 mm/year in the west (Chidumayo and Gumbo 2010). Soils in the dry forest and woodland areas are relatively shallow with less than 1.5 m depth and high content of sand ranging from 30–90%, clay content of between 5–20% and silt making up between 5–45%. Silt-loamy soil is found in the eastern part of Africa to sandy-loam in the western part, falling within the areas where the *Anogeissus leiocarpus* can be found. The most favorable range of soil pH for the growth of the plant is 5.5–7.5 (Chidumayo and Gumbo 2010).

**Major Chemical Constituents and Bioactive Compounds**
Quantitative estimation of bioactive phytoconstituents showed that *A. leiocarpus* plant contains alkaloids, phenolics, flavonoids, alkaloids and flavonoids (Barku et al. 2013). It has been reported that there are 8 flavonoid types present in *A. leiocarpus,* namely, cathecin, 4H-1-Benzopyran-4-one, 7-[(6-deoxy-α-L-mannopyranosyl) oxy]-5-hydroxy-2-(4-hydroxy-3-methoxyphenyl), quercetin, isoquercetin, rutin, vitexin, kaempferol, and procyanidin B2 (Attiouaaf et al. 2011; Barku et al. 2013).

**Traditional Uses/(Part(s) Used) and Common Knowledge**
The leaf of *A. leiocarpus* is used externally as a decoction in the eastern part of Nigeria to treat skin diseases and psoriasis. Powders of the bark are applied on wounds, sores, boils, cysts and diabetic ulcers and the conditions are alleviated (Barku et al. 2013). Also in the Volta Region of Ghana, it was reported that the fresh leaves of *A. leiocarpus* have been used to heal wounds (Barku et al. 2013). The decoction and maceration of the stem bark are used against anorexia, constipation, jaundice, itching, wounds, carbuncles, boils and various forms of ulcers (Harouna et al. 2012).

**Modern Medicine Based on Traditional Medicinal Uses (Biological Activities)**
*Anogeissus leiocarpus* may serve as remedies for gonorrhea, general body pain, blood clotting agent, diabetes (Atawodi 2011), hypertension and other diseases with oxidative stress as major etiological factor (Atawodi et al. 2011).

**Uses Supported by Clinical Data, Pharmacopoeias, Well Established Documents and in Traditional Medicine**
It was reported that methanolic extract of the stem bark of *Anogeissus leiocarpus* possess antioxidant, hepatoprotective and ameliorative effect on hepatocellular injury via mechanisms believed to be closely related to its variously reported therapeutic properties (Atawodi et al. 2011).

## Terminalia catappa *Linn*

**Synonyms** *Terminalia procera*

(Source: Authors)

**Botanical Name and Family**
*Terminalia catappa* Linn (Family Combretaceae) is commonly called tropical almond.

**Common Names** Tropical almond, Umbrella tree.

**Morphological Description**
*Terminalia catappa* is a tree that can attain 25 m tall. The leaves are simple, alternate and clustered towards the end of branches (Jayaweera 1982). They are short petioled, 15–28 cm long, 10.5–16.5 cm broad, glabrous and shining above and tormentose below with two glandular depressions near the base of the midrib on

the underside. The leaves turn pink-red–yellow before falling. The flowers are regular polygamous in simple solitary, axillary, rusty-tormentose or pilose spikes, 6–18 cm long, upper flowers male and lower flowers bisexual, bracts minute, calyx lobes of five triangular valves, soon deciduous, petals absent, stamens ten inserted on the calyx lobes, biseriate, 5 lower stamens opposite the calyx teeth and the five higher ones alternating with them, ovary inferior, unilocular with 2 or 3 pendulous ovules, glabrous or hairy, ellipsoid, slightly compressed showing two edges. The fruit is an indehiscent drupe with a hard endocarp. The colour of the fruit is green, yellow or reddish (Jayaweera 1982).

**Geographical Distribution**
*Terminalia catappa* is native to countries of the tropical regions of Asia, Africa and Australia (Ajibade et al. 2014).

**Ecological Requirements**
A conspicuous semi-deciduous tree of coastal areas throughout the warm tropics, *Terminalia catappa* grows best in most tropical climates. It is well adapted to sandy and rocky coasts and flourishes on oolithic limestone. The natural habitat is in areas just inland from ocean beaches, near river mouths and on coastal plains, but also thrives when planted upland. It grows in regions on altitude 0–800 m, mean annual temperature 15–35 °C and mean annual rainfall 750–3000 mm. The favorite soil types include oolithic limestone, sands and loamy sands, silts, loam and clays. Soil pH is usually neutral to moderately alkaline, rich in bases. It also grows on acidic soil, but good drainage is required on clay soils (Orwa et al. 2009).

**Major Chemical Constituents and Bioactive Compounds**
Methanolic extract contains steroids, triterpenes, triterpenoidal saponins, alkaloids, flavonoids, tannins, glycosides and polyphenols (Babayi et al. 2004; Kankia 2012; Praveena 2014; Behera et al. 2014). The fruit extract mainly contains aldehydes, fatty acids, alkanols, ketones, esters, hydrocarbons and a heterocyclic compound. Specifically, it contains cis-9-hexadecenal, 9Z-9-tetradecenal, palmitic acid, cis-9-octadecenal, 6Z-6-octadecenoic acid, penta decenoic acid, arachidic acid, 1-(+)-ascorbic acid-2,6-dihexadecanoate (Ololade et al. 2014). Ikhwanuddin et al. (2014) also reported the presence of the following tannins: punicalagin, punicalin, terflavin A and B, tergallagin, tercatain, chebulagic acid, geranin, granatin B and corilagin; flavonoids- isovitexin, vitexin, isoorientin, rutin and triterpenoids- ursolic acid, 2a,3β,23-trihydroxy-urs-12-en-28-oic acid. Humic acid was also identified.

**Traditional Uses, Part(s) Used and Common Knowledge**
The tree is primarily used as an ornamental shade and salt-tolerant tree. The fruit is eaten like almonds with similar oil (Praveena 2014). It has a bitter, acrid and astringent taste and is an aphrodisiac. The leaf juice is used as ointment for scabies, leprosy and other cutaneous diseases (Nair and Chanda 2008). *Terminalia catappa* has therapeutic effect for liver-related diseases, dermatitis and hepatitis (Praveena 2014). The leaf is used to treat rheumatic joints, cough, asthma, dysentery, diarrhea, gonorrhea and skin ailments such as scabies (Behera et al. 2014). It is also a cardiac stimulant.

**Biological Activities**
Studies on the leaf and fruit have confirmed anticancer, antioxidant, anti-HIV reverse transcriptase, anti-inflammatory, antidiabetic and hepatoprotective activities (Praveena 2014). Anticancer, antioxidant, antifungal and anti-inflammatory properties have also been reported for different parts of the plant by Ikhwanuddin et al. (2014).

The methanolic extract of *Terminalia catappa* has been shown to inhibit the growth of two bacteria species, *Bacillus subtilis* and *Escherichia coli* at 200 mg/ml and two fungal species, *Aspergillus niger* and *Penicillium notatum* at 200 mg/ml (Praveena 2014). The organic extract also showed pronounced sensitivity against *Staphylococcus aereus* and *Klebsiella pneumoniae*. It has a high antioxidant potential (Ololade et al. 2014). The methanolic extract of the leaf is considerably more effective than the aqueous extract in inhibiting Gram positive, Gram negative bacteria and a fungal strain (Nair and Chanda 2008). The petroleum ether, water and ethanolic fractions of leaf extract also showed significant antibacterial activity using hole in plate bioassay procedure against *Pseudomonas aeruginosa, Streptococcus viridians* and *Staphylococcus aureus* (Kankia 2012).

The growth inhibitory activity of the methanolic extract of *Terminalia catappa* against *Bacillus subtilis* and *Staphylococcus aureus* has been reported by Babayi et al. (2004). Akharaiyi et al. (2011) demonstrated that the aqueous leaf extract of *Terminalia catappa* at different stages of maturity exhibited therapeutic effect on different microorganisms, with the young leaf showing more potency than the matured or red-pigmented leaf. The microorganisms studied include *Bacillus cerius, Staphylococcus aureus, Pseudomonas aeruginosa* and *Escherichia coli*. Others are *Proteus mirabilis, Salmonella typhi* and *Shigella dysenteriae*.

**Uses in Traditional Medicine**
*Terminalia catappa* has therapeutic effect against liver-related diseases, dermatitis and hepatitis (Praveena 2014). The leaf is used in treatment of rheumatic joints, cough, asthma, dysentery, diarrhea, gonorrhea and skin ailments such as scabies (Behera et al. 2014), as well as a cardiac stimulant. The tree is primarily used as an ornamental shade and salt-tolerant tree. The fruit is eaten like almonds with similar oil (Praveena 2014). It has a bitter, acrid and astringent taste and is an aphrodisiac. The leaf juice is used as ointment for scabies, leprosy and other cutaneous disease (Nair and Chanda 2008).

## 7 Fabaceae

### Detarium microcarpum *Guill. & Perr.*

**Synonyms** *Detarium senegalense* auct. non J.F.Gmel.

**Botanical Name and Family**
*Detarium microcarpum* Guill. (Fabacea) is an African leguminous medicinal plant.

**Common Names** Sweet Detar

(Source: www.prota4u.org, Ekpe/NSBP)

**Morphological Description**
When found in dry areas *Detarium microcarpum* appear as small trees of about 10 m tall with a dense rounded crown, while in wet areas it can reach a height of 25 m. The greyish bark breaks off into rectangular pieces to reveal a reddish inner surface. The twigs are covered with a smooth or peeling orange bark. The pinnate leaves are 8–12 cm long and consist of 3–6 pairs of alternate, almost opposite leaflets. The leaflets are 5–10 cm long and 3–5 cm wide, with a dull green upper surface and a greyish-green lower surface. The leaflets have a rounded, often notched apex and a rounded or subcordate base. The inflorescence is anaxillary raceme, of about 2–5 cm in length and congested. The creamy white fragrant flowers consist of 4 large sepals and 8–10 cream coloured prominent stamens, but petals absent (Vautier et al. 2007).

**Geographical Distribution**
*Detarium microcarpum* is distributed in the semi-arid sub-Saharan Africa, from Senegal to Cameroon, extending eastwards to the Sudan. It has an irregular distribution and can be locally very common. The species is often left when farmland is cleared. Typically, it is found in high rainfall savannah areas, dry forests and fallow land, on sandy or iron rich hard soils. It also occurs in open savannah of Northern Nigeria, parts of Chad and Niger republic as a more stunted tree with smaller fruits.

**Ecological Requirements**
*Detarium microcarpum* grows on dry soil in wooded savanna and open woodland, and is locally very common. It is most common in regions with an annual rainfall of 600–1000 mm. It is mainly found on shallow, stony and lateritic soils, and on hills. *Detarium microcarpum* regenerates well from shoots produced by the trunk or roots. Shoots from the trunk are much more vigorous than seedlings and can reach

a height of 1.5–2 m in 1–2 years. In Cameroon the average seedling height after 3 years is 0.6 m, and the seedling may reach 1.5 m in 4 years. *Detarium microcarpum* flowers during the rainy season, from July–September (–November), and bears fruit from September–January (–May). It sheds its leaves in November and produces new ones in March. The main flowering period of a tree is up to 8 days only, and flowers are pollinated by insects, especially in the mornings.

**Major Chemical Constituents and Bioactive Compounds**
The seed gum contains D-galactose as a major monosaccharide, as well as D-mannose and D-glucose. The fruit pulp contains 90% dry matter, of which 4–6 g/100 g protein and 3 mg/100 g ascorbic acid. The seeds yield 7.5% oil, with linoleic acid being the predominant fatty acid. The gum content (water-soluble polysaccharides) is high. The hulled seed flour contains per 100 g: water 3.5–6.5 g, crude fiber 3 g, crude fat 13–15 g, crude protein 13.5–27 g, carbohydrate 39 g, Ca 500 mg, Mg 500 mg, Fe 100 mg (Akpata and Miachi 2001).

**Traditional Uses, Part(s) Used and Common Knowledge/Uses in Traditional Medicine**
The roots, stems, bark, leaf and fruit are all used to treat ailments e.g. tuberculosis, meningitis, itching and diarrhoea. The fruits and leaves are used traditionally in the treatment of dysentery and syphilis. (Iwu 1993; Ikhiri and Ilagouma 1995) and the root water extract is used for leprosy (Collier and Chapman 2001). The fruit is edible and rich in vitamin C and the leaves and seeds are also used in cooking. The bark, leaves and roots of *Detarium microcarpum* are widely used throughout its distribution area because of their diuretic and astringent properties. They are prepared as infusions or decoctions to treat rheumatism, venereal diseases, urogenital infections, haemorrhoids, caries, biliousness, stomach-ache, intestinal worms and diarrhoea including dysentery (Burkill 1995). They are also used against malaria, leprosy and impotence. A decoction of the powdered bark is widely taken to alleviate pain, e.g. headache, sore throat, back pain and painful menstruation. The fresh bark or leaf is applied to wounds, to prevent and cure infections.

In Mali the bark is also used to treat measles, nocturia, hypertension, itch and tiredness, while a decoction of the leaves or roots is taken against paralysis, meningitis, tiredness, cramps and difficult delivery. The powdered seeds are applied to skin infections and inflammations, whereas the fruit is eaten to cure meningitis and malaria. In Burkina Faso the fruit pulp is used for treating skin infections. A preparation of the fruits is taken against dizziness in Niger and Togo (Keay et al. 1958), while in Senegal, a mixture of the leaf of *Detarium microcarpum*, *Sclerocarya birrea* (A. Rich.) Hochst. and *Acacia macrostachya* Rchb. ex DC. pounded in milk is considered very efficient for snakebites. In Benin a decoction of the leaf is taken to treat fainting and convulsions, while generally in West Africa, the root is part of a medico-magical treatment for mental conditions, and for protection against bad spirits.

In traditional veterinary medicine, the leaf and root are used to treat diarrhoea in cattle in southern Mali, and in Benin to treat constipation, while in Niger cattle are made to inhale the smoke of the leaves to treat fever (Abreu et al. 1998).

Incorporation of 0.5% gum in wheat flour increased the water absorption and the mixing tolerance index of the dough significantly. Oil-water emulsions stabilised by the seed flour or gum tolerate freezing and thawing better than commercial salad dressing, egg powder and gum tragacanth emulsions. Addition of the seed polysaccharide to fruit products (mango, orange, pineapple, tomato) improved their stability in storage (at 26 °C) for at least 2 months and was well acceptable to consumers.

**Biological Activities**

Olugbuyiro et al. (2009) reported on anti-viral activity and cytotoxicity of *D. microcarpium* column fractions in the Huh-7 Replicon assay. The active fraction MTH-1700 (6.155 g) which was eluted with EtOAc-MeOH (75:25) demonstrated good inhibitory and selective potency (83.87%) against Hepatitis C virus in a dose dependent manner compared with control (86.76%). The ethanolic extract of the bark showed antimicrobial action against *Pseudomonas aeruginosa*, *Citrobacter freundii*, *Klebsiella pneumoniae*, *Staphylococcus aureus*, *Streptococcus pyogenes* and *Listeria monocytogenes* (Abreu et al. 1998). The extract also showed moderate antitumour activity against breast cancer cells, while the flavanes present in the methanolic extract of *Detarium microcarpum* showed strong inhibitory effects on HIV-1 or HIV-2 infection (Abreu et al. 1999). The stem bark extract has been demonstrated to have significant molluscicidal activity against *Lymnaea natalensis*, probably because it contains tetranorditerpenes, the clerodane diterpenes catechine and cis-2-oxokolavenic acid (0.5%), the diterpene copalic acid (1.7%) and coumarin (1%). A methanol extract of the leaf exhibited strong feeding deterrent activity against the termite, *Reticulitermes speratus*.

## 8 Gramineae (Poaceae)

### Bambusa vulgaris *Schrad. ex J.C.Wendl.*

(Source: E. Boer)

**Botanical Name and Family**
*Bambusa vulgaris* belongs to the subfamily Bambusoideae of Poaceae (Koshy and Jee 2011).

**Morphological Description**
*Bambusa vulgaris* is the most easily recognized species of all known bamboos (Liese 2004). In a young stem, the primary branches are prominent; they are borne alternately along the stem, together forming a gigantic fan-like structure which is conspicuous from some distance. *Bambusa vulgaris* forms moderately loose clumps and has no thorns. It has lemon-yellow culms (stems) with green stripes and dark green leaves. Stems are not straight, not easy to split, inflexible, thick-walled, and initially strong. The densely tufted culms grow 10–20 m (30–70 ft) high and 4–10 cm (2–4 in) thick. Culms are basally straight or flexuose (bent alternately in different directions), drooping at the tips. Culm walls are slightly thick. Nodes are slightly inflated. Internodes are 20–45 cm (7.9–17.7 in). Several branches develop from mid-culm nodes and above. Culm leaves are deciduous with dense pubescence. Leaf blades are narrowly lanceolate.

**Geographical Distribution**
*Bambusa vulgaris* probably originated from tropical Asia, but grows throughout the tropics and subtropics, including African countries like Nigeria, Senegal, Ghana, Equatorial Guinea, Tanzania, and Cameroon.

**Ecological Requirements**
*Bambusa vulgaris* grows best in tropical climate at lower altitudes; above 1000 m altitude stems become smaller in length and diameter. It thrives under a wide range of moisture and soil conditions, growing in almost permanently humid conditions along rivers and lakes, but also in areas with a severe dry season, where the plants may become completely defoliated.

**Major Chemical Constituents and Bioactive Compounds**
Stem fibers from *Bambusa vulgaris* from tropical Africa (e.g. Côte d'Ivoire, Cameroon, Gabon and Congo) have an average chemical composition of: cellulose 41–44%, pentosans 21–23%, lignin 26–28%, ash 1.7–1.9%, silica 0.6–0.7%. The approximate chemical composition of young shoots per 100 g edible portion for green-stem and yellow-stem cultivars respectively is: water 90 g and 88 g, protein 2.6 g and 1.8 g, fat 4.1 g and 7.2 g, carbohydrates 0.4 g and 0 g, fiber 1.1 g and 1.2 g, Ca 22.8 mg and 28.6 mg, P 37 mg and 27.5 mg, Fe 1.1 mg and 1.4 mg, ascorbic acid 3.1 mg and 0 mg. Phytochemical screening of leaf extract of *Bambusa vulgaris* revealed the presence of carbohydrate, glycosides, alkaloids, flavonoids, phytosterols and triterpenoids (Kumar et al. 2011).

**Traditional Uses, Part(s) Used and Common Knowledge**
In Nigeria, a drink of macerated leaves is taken against venereal diseases, while in DR Congo, the leaves form part of preparations used for treatment of measles. Other ethno-medicinal usages include treatment of malaria fever, inflammations, ulcers and wound healing. The roots are diuretic, tonic depurative, laxative and cooling and

also used in burning sensation, arthralgia, and general debility and dysuria. The fruits are salutary in diabetes whereas, the seeds are useful in obesity to reduce fats.

**Biological and Cytotoxic Activities**
The antimalarial and antimicrobial potentials of the leaf and shoot have been demonstrated. A chloroform extract of the leaf is active against *Mycobacterium tuberculosis*. Stigmasterol has been isolated from *Arthrobacter globiformis* fermented shoot (Sarangthem et al. 1998). In a study (Ojo et al. 2009) comparing the antiviral potentials of Nigerian plant, *B. vulgaris* resulted in inhibition only on measles virus at MIC of 62.5 μg/ml, justifying the usage of the plant in traditional medicine to treat measles.

The shoots of *B vulgaris* contains taxiphyllin (a cyanogenic glycoside) that functions as an enzyme inhibitor in the human body when released, but degrades readily in boiling water. It is highly toxic, and the lethal dose for humans is about 50–60 mg. A dose of 25 mg cyanogenic glycoside fed to rats (100–120 g body weight) caused clinical signs of toxicity, including apnoea, ataxia and paresis (Satya et al. 2010).

Ingestion of large amounts of leaf has been recorded to cause neurological disorders in horses, but this was not related to poisoning by hydrocyanic acid (Barbosa et al. 2006).

**Uses in Traditional Medicine**
Ethno-medicinal usages include treatment of malaria fever, inflammations, ulcers and wound healing. The antimalarial and antimicrobial potential of the leaf and shoots have been reported.

## 9 Lauraceae

## Cinnamomum verum *J. S. Presl.*

**Synonyms** *Cinnamomum zeylanicum* Nees

(Source: Steven Foster)

**Botanical Name and Family**
*Cinnamomum verum* J.S. Presl. (family: Lauraceae) is commonly called the cinnamon tree.

**Morphology**
*Cinnamomum verum* J.S. Presl. is an evergreen tree that reaches a height of 8–17 m in the wild. In an un-harvested state, the trunk is stout, 30–60 cm in diameter, with a thick, grey bark and the branches set low down. Leaves are stiff, estipulate, opposite and somewhat variable in form and size.

Petiole is 1–2 cm long, grooved on the upper surface. The lamina is usually 5–18 × 3–10 cm, ovate or elliptic; base more or less rounded and the tip tends to be somewhat acuminate. There are 3, sometimes 5, conspicuous longitudinal veins found at the base of the lamina and running almost to the tip. The young leaves of the flush are reddish, later turning dark green above with paler veins and pale glaucous beneath. Flowers are borne in lax axillary and terminal panicles on the ends of twigs. Peduncles are creamy white, softly hairy, 5–7 cm long. Individual flowers are very small, about 3 mm in diameter, pale yellow, with a foetid smell, each subtended by a small, ovate, hairy bract. The calyx is campanulate and pubescent with 6 acutely pointed segments. Corolla is absent. Fruit is a fleshy ovoid drupe, black, 1.5–2 cm long when ripe, with the enlarged calyx at the base (Orwa et al. 2009).

**Geogaphical Distribution**
*Cinnamomum verum* J.S. Presl. is indigenous to Sri Lanka. It also grows in India, Brazil, Colombia, Comoros, Dominica, Fiji, Ghana, Haiti, Indonesia, Jamaica, Madagascar, Malaysia, Mauritius, Mexico, Nicaragua, Nigeria, Philippines, Puerto Rico, Seychelles, Sierra Leone, Tanzania and Uganda.

**Ecological Requirements**
*Cinnamomum verum* requires a warm and wet climate with no extremes of temperatures. Although there can be months in which there is less rain, no prolonged dry season should occur and rain received should be about 150 days per year. Rocky and stony ground is unsuitable, while waterlogged and marshy areas should be avoided, as they result in an undesirable, bitter product, which is much less aromatic. An altitude below 500 m, mean annual temperature of 27 °C, mean annual rainfall of over 2000 mm are required. It is tolerant to a wide range of soils.

**Major Chemical Constituents and Bioactive Compounds**
*Cinnamomum verum* contains a wide range of essential oils, such as *trans* cinnamaldehyde, cinnamyl acetate, eugenol, L-borneol, caryophyllene oxide, b-caryophyllene, L-bornyl acetate, E-nerolidol, -cubebene, -terpineol, terpinolene, and -thujene (Rao and Gan 2014). Flavonoids, saponins, terpenoids and cardiac glycosides are also present (Udaya Prakash et al. 2014).

**Traditional Uses, Part(s) Used and Common Knowledge**
*Cinnamon verum* bark is used as natural food preservative, preventing the decomposition of products due to its rich content of cinnamaldehyde which has been proven to be active against many pathogenic gram positive and gram negative bacteria (Asha et al. 2014). It is used as a flavouring agent in foods and chewing gum.

Cinnamon is also used to improve the health of the colon thereby reducing the risk of colon cancer. It is a coagulant and prevents bleeding (Rao and Gan 2014).

**Biological Activities**

Evaluation of essential oil of *Cinnamomum verum* and its major component cinnamaldehyde on growth, hyphal ultrastructure and virulence factors of *Aspergillus fumigatus* and *Trichophyton rubrum,* showed potent antifungal activity on both liquid and solid media. Further test also revealed inhibitory activity against germinated and un-germinated conidia in test fungi. Electron microscopic studies also revealed multiple sites of action of cinnamaldehyde mainly on cell membranes and endomembranous structures of the fungal cells. The essential oil and cinnamaldehyde also caused a 70% reduction in elastase activity in *Aspergillus fumigatus* (Mohammed and Igbal 2011).

At 500 mg/ml *Cinnamomum verum* aqueous extract completely inhibited *Aspergillus flavus*, confirming a fungicidal activity that can make it an alternative to synthetic fungicides (Lakshmeesha et al. 2014).

Mayaud et al. (2008) also tested the essential oils containing aldehydes from *Cinnamomum verum* bark and phenols from the leaves for antimicrobial activity against 55 bacterial strains, using two methods, agar dilution and time-killing curve methods. Results showed that the *Cinnamomum verum* bark had the highest antimicrobial activity, particularly against resistant strains. In a study to evaluate the role of *Cinnamomum verum* essential volatile oil extracts on human macrophages infected by *Brucella abortus* 544, the oil at a concentration of 1% had the highest antibacterial activity against *B. abortus* 544 inside human macrophages (Al-Mariri et al. 2012). Other biological activities documented for cinnamon include antioxidant, antidiabetic, anti-inflammatory, antiemetic, antimycotic and anticancer activities (Rao and Gan 2014).

**Uses in Traditional Medicine**

Cinnamon bark oil is employed in dental and pharmaceutical preparations. The drops are used as a tonic, a sedative in childbirth and a remedy for many common disorders. It is used for gastrointestinal and respiratory disorders and as an aphrodisiac. It is also used to relieve headache, speed up parturition, as a carminative and to treat catarrh. The bark extract is an intestinal astringent and the essence is used as a poultice for rheumatism and is taken orally for spasms and for stomach and intestinal gas (Orwa et al. 2009).

## 10 Conclusion

This review has shown that many wild Nigerian trees belonging to different families are widely used in traditional medicine, and have indeed been demonstrated to have several biological and therapeutic effects due the presence of a variety of phytochemicals. The need to establish methods of domesticating them by cultivation as a means of conservation and sustainability for the present and future generations cannot be over-emphasized.

**Acknowledgements** Some of Professor S.E. Atawodi's research work was partly supported by Alexander von Humboldt Foundation of Germany and the German Cancer Research Centre (DKFZ), Heidelberg.

## References

Abreu PM, Rosa VS, Araujo EM, Canda AB, Kayser O, Bindseil KU, Siems K, Seemann A (1998) Phytochemical analysis and antimicrobial evaluation of *Detarium microcarpium* bark extracts. Pharm Pharmacol Lett 8(3):107–109

Abreu PM, Martins ES, Kayser O, Bindseil KU, Siems K, Seemann A, Frevert J (1999) Antimicrobial, antitumor and antileischmania screening of medicinal plants from Guinea-Bissau. Phytomed 6(3):187–195

Adewole SO, Caxton-Martins EA (2006) Morphological changes and hypoglycemic effects of *Annona muricata* Linn. (Annonaceae) leaf aqueous extract on pancreatic β-cells of streptozotocin-treated diabetic rats. Afr J Biomed Res 9:173–187

Afolabi F, Afolabi OJ (2013) Phytochemical constituents of some plants in Southwest Nigeria. IOSR J Appl Chem 4(1):76–78

Ajibade VA, Ibikunle KSO, Owoseni DO (2014) Incidence, multidrug resistance and antimicrobial activity of extracts from *Terminalia catappa on Streptococcus pyogenes*. J Basic Appl Sci Res 4(4):6–10

Akharaiyi FC, Ilori RM, Adesida JA (2011) Antibacterial effect of *Terminalia catappa* on some selected pathogenic bacteria. Int J Pharm Biomed Res 2(2):64–67

Akinyemi OI, Dada EO (2013) Phytochemical screening and antityphoid properties of ethanolic leaf extract of *Parquetina nigrescens*. ARPN J Agric Biol Sci 8(11):732–739

Akpata MI, Miachi OE (2001) Proximate composition and selected functional properties of *Detarium microcarpum*. Plant Fd Hum Nutr 56(4):297–302

Akunyili DN, Houghton PJ (1993) Monoterpenoids and napthoquinones from *Kigelia pinnata* bark. Phytochemistry 32:1015–1018

Akunyili DN, Houghton PJ, Raman A (1991) Antimicrobial activities of the stem bark of *Kigelia africana*. J Ethnopharmacol 35:173–178

Alamelu I, Bhuwan CJ (1974) An iridoid glycoside from *Kigelia pinnata*. Herba Pol 20:319

Al-Mariri A, Saour G, Hamou R (2012) *In vitro* antibacterial effects of five volatile oil extracts against intra macrophage *Brucella abortus* 544. Iran J Med Sci 37(2):119–125

Alvarez Cruz NS (2012) *Parquetina nigrescens* (Afzel.) Bullock. In: Schmelzer GH, Gurrib-Fakim A (eds) Prota 11(2): medicinal plants/Plantes medicinales 2. PROTA, Wageningen, Netherlands. Accessed 6 May 2015

Arbab AH (2014) Review on *Anogeissus leiocarpus* a potent African traditional drug. Int J Res Pharm Chem 4(3):496–500

Asha S, Nithisa K, Bharath Kumar R, Ravi KV (2014) Deciphering the antimicrobial potential of *Cinnamon zealanicum* bark. J PharmTech Res 6(4):1226–1235

Atawodi SE (2005) Antioxidant potential of African medicinal plants – a review. Afr J Biotechnol 4(2):128–133

Atawodi SE (2011) Evaluation of the hypoglycemic, hypolipidemic and antioxidant effects of methanolic extract of "Ata-ofa" polyherbal tea (*A-Polyherbal*) in alloxan-induced diabetic rats. Drug Inv Today 3(11):270–276

Atawodi SE, Olowoniyi OD (2015) Pharmacological and therapeutic activities of *Kigelia africana* (Lam.) Benth. Ann Res Rev Biol 5(1):1–17

Atawodi SE, Adekunle OO, Bala I (2011) Antioxidant, organ protective and ameliorative properties of methanol extract of *Anogeissus leiocarpus* StemBark against carbon tetrachloride-induced liver injury. Int J Pharm Sci Res 2(6):748–753

Atawodi SE, Yakubu OE, Umar IA (2013) Effect of methanolic extract of root of *Parinari curatellifolia* Planch. Ex benth on acute and chronic oxidative stress and hepatotoxicity induced by carbon tetrachloride in rats. Int J Agric Biol 15:523–528

Atawodi SE, Olowoniyi OD, Adejo GO, Liman ML (2014a) Review of the antioxidant potential of African medicinal plants and food plants. In: Dubey NK (ed) Plants as a source of natural antioxidants. CAB International, Wallingford. 63pp

Atawodi SE, Olowoniyi OD, Obari MA, Ogaba I (2014b) Ethnomedical survey of Adavi and Ajaokuta local government areas of Ebiraland, Kogi State, Nigeria. Ann Res Rev Biol 4(24):4344–4360

Atawodi SE, Olowoniyi OD, Daikwo MA (2014c) Ethnobotanical survey of plants used for the management of Hypertension in the Igala speaking area of Kogi State, Nigeria. Ann Res Rev Biol 4(24):4535–4543

Atolani O, Olatunji AG (2010) Epicuticular wax and volatiles of *Kigelia pinnata* leaf extract. Ethnobotan Leaflets 14:79706

Attiouaaf B, Yeo LLD, Antheaume C, Kaiser M, Weniger B, Lobstein A, Vonthron-Sénécheau C (2011) *In vitro antiplasmodial* and *antileishmanial* activities of flavonoids from *Anogeissus leiocarpus* (Combretaceae). Int J Pharm Sci Rev Res 11(2), Article-001, pp 1–6

Babayi H, Kolo I, Okogun JI, Ijah UJJ (2004) The antimicrobial activities of methanolic extracts of *Eucalyptus camaldulensis* and *Terminalia catappa* against some pathogenic microorganisms. Biok 16(2)

Barbosa JD, de Oliveira CMC, Duarte MD, Riet-Correa G, Peixoto PV, Tokarnia CH (2006) Poisoning of horses by bamboo, *Bambusa vulgaris*. J Eq Vet Sci 26(9):393–398

Barku Y, Victor A, Abban G (2013) Phytochemical studies, *in-vitro* antibacterial activities and antioxidant properties of the methanolic and ethyl acetate extracts of the leaves of *Anogeissus leiocarpus*. Internl J Biochem Res Rev 3(2):137–145

Behera DR, Bhatnagar S, Mahapatra AK (2014) Cytotoxic and radical scavenging potential of Indian Almond (*Terminalia catappa*) leaf extracts. Br Biomed Bull 2(1):31–39

Bharti N, Singh S, Naqvi F, Azam A (2006) Isolation and *in vitro* antiamoebic activity of iridoids isolated from *Kigelia pinnata*. ARKIVOC 2006:69–76

Borokini IB, Okere AU, Giwa AO, Daramola BO, Odofin WT (2010) Biodiversity and Consevation of plant genetic resources in field Genebank of the National Center for Genetic Resources and Biotechnology, Ibadan, Nigeria. Int J Biodiv Conserv 2(3):037–050

Burkill HM (1985) The useful plants of West Tropical Africa, vol 1, 2nd edn. Royal Botanical Gardens, Kew

Burkill HM (1995) The useful plants of West Tropical Africa. 2nd edition. Volume 3, Families J–L. Royal Botanic Gardens, Kew/Richmond, 857 pp

Chidumayo EN, Gumbo DJ (2010) The dry forests and woodlands of Africa: managing for products and services. Earthscan Publisher, Quebec

Collier J, Chapman R (2001) Combination therapy with interferon-alpha and ribavirin for hepatitis C: practical treatment issues. BioDrugs 15(225):238

Fasola TR, Okeocha PC, Odetola A (2010) Screening for the hypoglycaemic potentials of the extract of *Vernonia amygdalina*. Ethnobotan Leaflets 14:759–765

Feitosa EA, Xavier HS, Randau KP (2012) Chrysobalanaceae: traditional uses, phytochemistry and pharmacology. Rev Bras Farm 22(5):1181–1186

Gouda Y, Abdel-baky A, Darwish F, Mohammed K, Kesai R, Yamasaki K (2003) Iridoids from *Kigelia pinnata*, DC. Phytochemistry 63:887–892

Govindachari TR, Patankar SJ, Vishvanthan N (1971) Isolation and structure of two new dihydroisocumarins from *Kigelia pinnata*. Phytochemistry 10:1603–1606

Grace OM Davis S (2002) *Kigelia africana* (Lam) Benth. Record from Protabase. In: Oyen LPA, Lemmens RHMJ (eds) PROTA (Plant Resources of Tropical Africa, Wageningen, Netherlands). http://database.prota.org/search.htm

Harouna S, Adama H, Djifaby SPAE, Moussa C, Roland M, Millogo J, Nacoulma OG (2012) Phytochemistry and biological activities of extracts from two Combretaceae found in BurkinaFaso: *Anogeissus Leiocarpus* (DC) Guill. and Perr. and *Combretum Glutinosum* Perr. Ex DC. Univers J Environ Res Technol 2(5):383–392

Hennenberg JF, Goetze D, Minden V, Traoré D, Porembski S (2005) Size-class distribution of *Anogeissus leiocarpus* (Combretaceae) along forest–savanna ecotones in northern Ivory Coast. J Trop Ecol 21:273–281

Houghton PJ, Akunyili DN (1993) Iridoids from *Kigelia pinnata* bark. Fitoterapia 65:473–474
Ikhiri K, Ilagouma AT (1995) Fitoterapia 46:274–274
Ikhwanuddin M, Moh JHZ, Hidayah M, Hidayati ABN, Aina-Lyana NMA, Juneta ASM (2014) Effect of Indian almond *Terminalia catappa* leaves water extract on the survival rate and growth performance of black tiger shrimp *Penaeus monodon* post larvae. Aquaculture, aquarium, conservation and legislation. Int J Bioflux Soc 7(2):85–93
Imaga NA (2013) Phytomedicines and nutraceuticals: alternative therapeutics for sickle cell anemia. Sci World J 2013:1–12
Iwu MM (1993) Handbook of African medicinal plants. CRC Press Inc, London
Jackson S, Beckett K (2012) Sausage tree *Kigelia pinnata*: an ethnobotanical and scientific review. HerbalGram 94:48–59
Jayaweera DMA (1982) Medicinal plants (indigenous and exotic) used in Ceylon, vol IV. The National Science Council of Sri Lanka, Colombo
Kankia HI (2012) Phytochemical screening and antibacterial activities of leaf extracts of *Terminalia catappa* (Umbrella tree). Int J Sci Res 3(12):2658–2661
Keay RWJ, Hoyle AC, Duvigneaud P (1958) Caesalpiniaceae. In: Keay RWJ (ed) Flora of West Tropical Africa. Volume 1, Part 2, 2nd edn. Crown Agents for Oversea Governments and Administrations, London, pp 439–484
Koshy KC, Jee G (2011) Studies on the absence of seed set in *Bambusa vulgaris*. Curr Sci 81(4):375–378
Kumar M, Krishnamurth S, Suseel S (2011) Phytochemical investigation and acute toxicity study of methanol extract of *Bambusa vulgaris* leaves. J Pharm Res 4(2):403
Kupeli E, Harput U, Varel M, Yesilada E, Saracoglu I (2005) Bioassay guided isolation of iridoid, glucosides with antinociceptive and anti-inflammatory activities from *Veronica anagallis-aquatic* L. J Ethnopharmacol 102:170–176
Lakshmeesha TR, Sateesh MK, Vedashree S, Sofi MS (2014) Antifungal activity of *Cinnamomum verum* on soybean seed-born *Aspergillus flavus*. Int J Adv Res 2(5):1169–1172
Liese W (2004) Preservation of bamboo structures. Gh J For 15–16:40–48
Maharaj V, Glen HF (2008) Parinari curatellifolia, KZN Herbarium, South African National Biodiversity Institution Plant Information Website
Mayaud L, Carricajo A, Zhiri A, Aubert G (2008) Comparison of bacteriostatic and bacteriocidal activity of 13 essential oils against strains with varying sensitivity to antibiotics. Lett Appl Microbiol 47(3):167–173
Mishra S, Ahmad S, Kumar N, Sharma BK (2013) *Annona muricata* (the cancer killer): a review. Glob J Pharmacet Res 2(1):1613–1618
Mohammed SAK, Igbal A (2011) *In vitro* antifungal, anti elastase and anti keratinase activity of essential oils of *Cinnamomum, Syzygium* and *Cymbopogon* species against *Aspergillus fumigatus* and *Trichophyton rubrum*. Phytomed 19(1):48–55
Nair R, Chanda S (2008) Antimicrobial activity of *Terminalia catappa, Manilkara zyota* and *Piper betel* leaf extracts. Indian J PharmaceutSci 70(3):390–393
Odetola AA, Oluwole FS, Adeniyi BA, Olatiregun AM, Ikupolowo OR, Labode O, Busari KO, Shorinola JA (2006) Antimicrobial and gastrointestinal protective properties of *Parquetina nigrescens* (Afzel.) Bullock. J Biol Sci 6(4):701–705
Ojo OO, Oluyege JO, Famurewa O (2009) Antiviral properties of two Nigerian plants. African J Plant Sci 3(7):157–159
Olatunji AG, Atolani O (2009) Comprehensive scientific demystification of *Kigelia africana*; a review. Afr J Pure Appl Chem 3(9):158–164
Ololade ZS, Olatunde OZ, Oyelese OJ, Olaoye OO, Odewande AA (2014) Total phenolic content, free radical inhibition, antioxidant and antibacterial potential of the medicinal organic compounds in the fruit of *Terminalia catappa* Linn. Nat Sci 12(2):46–50
Olugbuyiro JAO, Moody JO, Hamann MT (2009) Inhibitry effect of *Detarium microcarpum* extract against hepatitis C virus. Afr J Biomed Res 12:2

Onwuliri FC, Mawak ID, Wonang DL, Onwuliri EA (2006) Phytochemical, toxicological, toxicological and histopathological studies of some medicinal plants in Nigeria. Internl J Nat Appl Sci 2(3):225–229

Orwa C, Mutua A, Kindt R, Jamnadars R, Anthony S (2009) Agroforestree database: a tree reference and selection guide version 4.0. http://www.worldagroforestry.org/sites/treedbs/treedatabases.asp. Accessed on the 5 May 2015

Owolabi OJ, Omogbai EKI, Obasuyi O (2007) Antifungal and antibacterial activities of the ethanolic extract of *Kigelia africana* (Bignoniceae) stem bark. Afr J Biotechnol 6(14):1677–1680

Peni IJ, Elinge CM, Yusuf H, Itodo AU, Agaie BM, Mbongo AN, Chogo E (2010) Phytochemical screening and antibacterial activity of *Parinari curatellifolia* stem extract. J Med Plant Res 4(20):2099–2102

Praveena K (2014) Phytochemical, antimicrobial and *in vitro* antioxidant activity of *Terminalia catappa*. Int J Pharm Life Sci 5(2):3325–3329

Rao PV, Gan SH (2014) Cinnamon: a multi-faceted plant. Evid Based Complement Alternat Med Article ID 642942

Rivera JO, Loya AM, Cabalios R (2013) Use of herbal medicines and implications for conventional drug therapy medical sciences. Altern Integ Med 2:130. doi:10.4172/2327-5162.1000130

Saini S, Kaur H, Verma B, Ripudaman T, Singh S (2009) *Kigelia africana* (Lam.) Benth. An overview. Nat Prod Rad 8(2):190–197

Sarangthem K, Singh LJ, Srivastava RC (1998) Isolation of stigmasterol from fermented succulent bamboo shoots and its biotransformation into ADD using *Arthrobacter globiformis*. Indian J Plant Physiol 3(4):249–252

Satya S, Bal LM, Singhal P, Naik SN (2010) Bamboo shoot processing: food quality and safety aspect (a review). Trends Food 21:181–189

Singh P, Khandelwal P (2010) Cetyltriacontanoate and other constituents from *Acacia jacquemontii* and *Kigelia pinnata*. J Indian Chem Soc 87:1403–1407

Sulaiman H, Roslida AH, Fezah O, Tan KL, Tor YS, Tan CI (2012) Chemopreventive potential of *Annona muricata* L leaves on chemically-induced skin Papillomagenesis in mice. Asian Pac J Cancer Prev 13:2533–2539

Tyagi A, Singh V, Bharadwaj M, Kumar A, Thakur K (2011) Isolation and antibacterial susceptibility testing of multi-drug resistant *Pseudomonas aeruinosa* causing urinary tract infection. J Chem Pharm Res 3(4):342–347

Udaya Prakash NK, Bhuvaneswari S, Sripriya N, Arulmozhi K, Kavitha K, Aravitha R, Bharathiraja B (2014) Studies on phytochemistry, antioxidant, antibacterial, larvicidal and pesticidal activities of aromatic plants from Yelagiri Hills. Int J Pharm Pharm Sci 6(5):325–328

Vautier H, Sanon M, Sacande M (2007) *Detarium microcarpum* Guill. & Perr. Seed Leaflet 122

Victor Y, Barku A, Opoku-Boahen Y, Owusu-Ansah E, Dayie NTKD, Mensah FE (2013) *In-vitro* assessment of antioxidant and antimicrobial activities of methanol extracts of six wound healing medicinal plants. J Nat Sci Res 3(1):74–80

Watt JM, Breyer-Bradwijk MG (1962) The medicinal and poisonous plants of Southern and Eastern Africa, 2nd edn. Livingstone, London, p 52

Wurochekke AU, Eze HT, Nelson K (2013) Effect of methanolic leaf extract of *Parinari curatellifolia* on rat liver and kidney. Sky J Biochem Res 2(7):47–49

# Biological, Pharmacognostic and Phytochemical Review of Some Cultivated Medicinal Plants of Nigeria

**Sunday E. Atawodi, Godwin O. Adejo, Olufunsho D. Olowoniyi, and Mubarak L. Liman**

**Abstract** Medicinal plant cultivation is more of a rarity than common in Nigeria, similar to other areas in Africa, where these unquantifiable resources are rather shoddily harvested from the wild. This review identifies five cultivated medicinal plants species of Nigeria, i.e.: *Vernonia amygdalina* Del, *Cyperus esculentus, Cymbopogon citratus* Stapf, *Khaya senegalensis* (Desv.) A. Juss and *Moringa oleifera* Lam with purported biological and pharmacognostic effects and relates these effects to their phytochemical constituents, where possible. The findings are arranged according to general aspects, cultivation practices, post-harvest handling, production schedules and utilization are discussed. There appears to be no available information on *Khaya senegalensis* (Desv.) A. Juss. with respect to cultivars, grading, packing and storage. The low number and lack of comprehensive literature on cultivated medicinal plants, seems to be an indication of a near or total lack of attention or interest on the issue of medicinal plant conservation in Nigeria, and hence, this contribution could be regarded a wake-up call.

**Keywords** Cultivated • Medicinal plants • Phytochemicals • Pharmacognosy • Therapeutics • Potential • Herbal medicine • Nigeria

## 1 Introduction

The utilization of medicinal plants in preventing, managing and treating diseases play cardinal roles in most parts of the developing world, including Nigeria (Atawodi et al. 2009a, b; Ene et al. 2009). Of particular significance is the use of herbal plants or those regarded as medicinal food plants, which are vastly used as the primary

---

Dedicated to Our Students and Mentees for additional impetus towards continuous scientific contributions

S.E. Atawodi (✉) • G.O. Adejo • O.D. Olowoniyi • M.L. Liman
Biochemistry Department, Ahmadu Bello University, Zaria, Nigeria
e-mail: atawodi_se@yahoo.com



M. Neffati et al. (eds.), *Medicinal and Aromatic Plants of the World - Africa Volume 3*, Medicinal and Aromatic Plants of the World,
DOI 10.1007/978-94-024-1120-1_12

health care agents in rural areas where access to medical services are not readily available, accessible or affordable (Liman and Atawodi 2015). Surprisingly, despite the depletion of these forest resources due to factors ranging from over exploitation, bush burning, desert encroachment and drought, there exist great paucity in the inclination of the Nigerian farming population to cultivate these plants. It has indeed been reported elsewhere that the majority of plants with medicinal values are uncultivated (Mpondo and Dibong 2011) and simply collected from the wild.

A survey of cultivated and uncultivated medicinal plants carried out in Kwardu, Churkah Shigar and Tormik villages of the Himalayan mountain watershed showed that only 27%, 33% and 31% respectively, were cultivated (Raza and Shah 2011). Similarly, Getaneh and Girma (2014) have put the statistics at 75% for uncultivated and 25% for cultivated plants. A search through the literature showed that only a few plants with medicinal values are cultivated in Nigeria, the notable ones of which are here presented.

It is quite noticeable from the foregoing that cultivated medicinal plants are tremendously few, although possibility exists that larger numbers of these plants are cultivated in Nigeria, but are not documented. Therefore, this work was set to achieve two goals. First, to identify the cultivated medicinal plants in Nigeria and secondly, to glean through existing documented literature and build more on such plants' biological, pharmacognostic and phytochemical properties. Exposures arising from this piece could inform relevant stakeholders to take advantage of the huge gap that exist and provide opportunities for sustained harvests of useful medicinal plants, contribute to local needs on traditional medicines, highlight potential economic benefits, and contribute to the climatic change salvaging drive, as well as conservation efforts.

## *Asteraceae*

### *Vernonia amygdalina* Del.

*Vernonia amygdalina* Del. is a plant of medicinal importance which even though found in the wild, is also well cultivated for its useful culinary and medicinal values (Dansi et al. 2008). Because of its high requirement for water, it grows mainly during the rainy season and on irrigated soils. This mono-cropped or inter-row cropped plant, mostly at the backyard is propagated mainly by root cutting. The leaf is the most useful part in its fresh form, although mostly stored in the dried form. The leaf is highly used in food flavoring, while research has shown that *V. amygdalina* possess diverse pharmacological properties, including being an active hypoglycaemic, anticancer (Akah and Okafor 1992; Agbogidi and Akpomorine 2013) and immune fortifying agent.

*Vernonia amygdalina*, Bitter leaf (Photo by authors)

Part 1: General Aspects

**Morphological Description**

*Vernonia amygdalina* Del. occurs as shrub or small tree, about 10 m tall, branched; trunk up to 40 cm in diameter; bark grey to brown, smooth, becoming fissured; young branches densely pubescent. Leaves are alternate, simple; stipules absent; petiole 0.2–4 cm long; blade ovate-elliptical to lanceolate, 4–15(−28) cm × 1–4 (−15) cm, cuneate or rounded at base, shortly acuminate at apex, margin minutely toothed to coarsely serrate, finely pubescent but often glabrescent and pinnately veined. Inflorescence at head, arranged in terminal, compound, umbel-like cymes; stalk of head up to 1 cm long, pubescent; involucre cylindrical to broadly ellipsoid, 3–5 mm long, bracts 3–7-seriate, 1–4.5 mm long, appressed. Flowers bisexual, regular, 5-merous, strongly exerted from the involucre; pappus consisting of outer linear, caduceus scales up to 1.5 mm long and of inner creamy or brownish bristles 4–7 mm long; corolla tubular, 5–8 mm long, whitish, glandular, with erect lobes; stamens with anthers united into a tube, with appendages at apex; ovary inferior, 1-celled, pubescent and glandular, style hairy, 2-branched. Fruit occurs as a 10-ribbed achene 1.5–3.5 mm long, pubescent and glandular, brown to black, crowned by the much longer pappus bristles, while seedling occurs with epigeal germination (Ucheck 2004).

**Classification**
*Vernonia amygdalina* Del.family of *Asteraceae* [Compositae] (Ucheck 2004; Yedjou et al. 2008; Agbogidi and Akpomorine 2013).

**Origin and Distribution**
Widespread in East, Central and West African countries, including Nigeria (Yedjou et al. 2008).

**Soil Requirements**
They are usually planted in the morning, in loamy soil, at a depth of 2–5 cm deep (Agbogidi and Akpomorine 2013).

**Production Levels**
*Vernonia amygdalina* is a relatively short cycle crop that is harvested twice in a month, for up to 7 years. As the leaf is the material of interest, the leaves are usually harvested before onset of flowers as the yield of leaves drop sharply at this point up to time of seeds setting and seed maturation (Dansi et al. 2008). However, the highest yields are obtained during the rainy season, the peak being in May–August (Ucheck 2004).

**Major Production Areas in Africa**
It is grown in many countries in savannah zones and popular in most of West Africa countries including, Nigeria, Cameroon, Gabon and Congo Democratic Republic (Agbogidi and Akpomorine 2013), where they are also cultivated in the backyard of most households in the rural areas.

**Cultivars**
*Vernonia amygdalina* has two main cultivars which are the bitter and non-bitter. Chromosome numbers and karyotype in three species of the genus *Vernonia Schreber* in Southern Nigeria (Kemka-Evans and Okoli 2013).

## Part 2: Cultivation Practices

**Propagation**
Although it grows nearly everywhere, propagation is mainly by cutting. The stems of various length of *V. amygdalina* are cut in the afternoon, soothed and tied in bundles. The bundles are kept overnight, placed upright in a basin of water or covered with jute bag to avoid desiccation. They are now planted in the morning on loamy soil at a depth of 2–5 cm deep (Agbogidi and Akpomorine 2013).

**Soil Preparation**
Preparation of land for *V. amygdalina* planting is done with hand held hoes (Dansi et al. 2008). When propagated through the seed, planting is usually through broadcasting, and localized in portions of fields with debris from burnt grasses or wastes. Nurseries are also used to raise seedlings which are later transplanted on the fields. Although, the practice is rare among households (Dansi et al. 2008), it is most likely done among commercial vegetable farmers.

**Planting/Fertilization**
*Vernonia amygdalina* like many traditional leafy vegetables are planted mostly around or near homes, gardens, cattle enclosures, fertile portions of the cropland, and near riverbanks and swamps. Cultivation within close proximities to human abodes or easily accessible locations is mainly for protection from thieves, birds and animals, enriching soil fertility from dumping of refuse from kitchen to enhance yields (Dansi et al. 2008), to receive water from domestic use and to maintain freshness. The systems of cropping in the cultivation of *V. amygdalina* are mixed, inter-row, and mono-cropping. However, the most common cropping systems practiced are the mixed and inter-row cropping (Dansi et al. 2008).

**Irrigation**
Some farmers grow *V. amygdalina* near rivers and in swamps for easy irrigation especially during draught and mostly by those who produce in commercial quantities (Dansi et al. 2008).

**Weed Control**
Weed control among farmers of *V. amygdalina* is usually by tillage, which is manually carried out since the use of tractor and associated implements in tillage operations are beyond their reach (Nmor and Odeh 2013).

**Pest and Disease Control**
No major diseases affect the production of *V. amygdalina* apart from leaf curl virus. Pests do not cause major damage either, although many pest species have been recorded on bitter leaf in northern Nigeria like thrips, aphids, ants, white fly, *Empoasa* spp., *Sphearocoris annulus*, *Fabricius* spp., *Ptyelus grossus*, *Polyclaeis* spp. and *Xanthochelus vulneratus*. To control these, people traditionally sprinkle wood ash on the leaves to keep ants and aphids away. The bitter leaf weevil *Lixus camerunus* may damage stems and branches by making tunnels, causing branches to break (Ucheck 2004).

**Harvesting**
The young green leaves are plucked from the plant by hand or cut with a knife.

Part 3: Post-harvest Handling

**Part Collected and Collection Techniques Used**
The leaves are usually put to use in traditional medicine and as hops to make beer, where the tops of the shrubs are collected. In most other uses, the broad mature green leaves are collected. For human consumption, leaves are squeeze-washed or boiled to help remove the extreme bitterness and foam (Aregheore 1998).

**Grading**
The broad green leaves could be macerated and used as vegetables or condiments in cooking soup. In other instances, the water extract is collected and used as tonic for

prevention of certain ailments. The bitter leaves are washed in some climes and preserved by drying under the sun or by freezing (Aregheore et al. 1998).

**Packaging**

Stems of various lengths are cut in the afternoon then sorted and tied into bundles of roughly equal length. Depending on the season, stems brought to the market have a length of 30–50 cm, often longer during the peak period. Bundles of 15–20 stems weighing 1–2 kg are often made, but smaller ones are made during periods of scarcity. The bundles are kept overnight, placed upright in a basin of water and sometimes covered with jute bags to avoid desiccation. The bundles themselves are tied together into bigger bundles before they are carried to the market. During dry periods people pick only the leaves and leave the shoots intact (Ucheck 2004).

**Storage**

*Vernonia amygdalina* is rarely stored because it is grown only enough to meet household consumption needs. It also has low storage capacity being perishable in the fresh form. As such, even in commercial cultivation, they are sold soon after harvest in order to cut losses. However, some growers of *V. amygdalina*, dry the harvested leaves whole or chopped, under the sun. Recently, there is more of shade-drying in an airy environment and storage in traditional containers such as gourds.

Shade-drying reduces blanching, although blanching improves colour and carotene retention due to inactivation of enzymes. However, it results in ascorbic acid loss (Chweya and Eyzaguirre 1999). Reed mats or sacks are the common drying surfaces, which are placed on raised platforms or roof tops in order to protect vegetables from domestic animals and soil. However, the leaves are susceptible to infestation by insects and moulds. In some places, the leaves are spread on the ground and naturally occurring flat stone surfaces, where these are found (Dansi et al. 2008).

**Marketing**

According to (Dansi et al. 2008), the selling of vegetables, including *V. amygdalina* is a female-dominated enterprise, with participation by men only in the market gardens (or on the farm).

## Part 4: Production Schedules

Some *V. amygdalina* grown from seeds can thrive well in slightly acidic soil with low organic matter and high water holding capacity. Water is the key factor for the growth of its leaves. High yields can be obtained during rainy season (Kayode 2004). *V. amygdalina* is a short cycle crop which can be harvested twice per month for up to 7 years. Planting *V. amygdalina* can be easy because it is compatible with any type of crop and can be planted in a variety of arrangements (Biggelaar and Gold 1996).

Part 5: Utilization

**Cosmetic**
The stem and root divested of the bark are used as chewing-sticks in many West Africa countries like Cameroon, Ghana, and Nigeria (Yedjou et al. 2008).

**Pharmaceutical and Therapeutic**
*V. amygdalina* has been demonstrated to possess active anticancer (Izevbigie 2003; Agbogidi and Akpomorine 2013), anti-bacteria, antimalaria, and anti-parasitic (Tadesse et al. 2008) agents. The roots and leaves are used in ethnomedicine to treat fever, hiccups, kidney problems, stomach discomfort and leukemia (Burkill 1985; Hamowia and Saffaf 1994; Khalafalla et al. 2009). Pharmacological studies have also shown that the leaf extracts have both hypoglycaemic and hypolipidaemic properties in experimental animals and so could be used in managing diabetes mellitus (Akah and Okafor 1992; Agbogidi and Akpomorine 2013).

Polyherbal preparations with bitter leaf as the active ingredients strengthen the immune system through many cytokines and chemokines regulation (Agbogidi and Akpomorine 2013). The herb not only lowers the body sugar level sufficiently, it also plays a role in the repair of pancreas (Agbogidi and Akpomorine 2013). The beneficial use of *V. amygdalina* in animal nutrition in Nigeria has also been well documented (Onwuka et al. 1989; Aregheore et al. 1998).

Adding bitter leaf to your health diet may reduce the risk of chronic diseases like breast cancer and type-2 diabetes. Bitter leaf can reduce LDL and total cholesterol. It is an abundant source of antioxidant agents (Agbogidi and Akpomorine 2013).

**Food and Flavoring**
Bitter leaf is a traditional ingredient in African cuisine (Agbogidi and Akpomorine 2013). The leaves are one of the most widely leafy vegetables consumed by Nigerians, and even Cameroonians during special occasions such as marriages, baptisms, Christmas, and birthdays. It is an abundant source of the poly unsaturated fatty acids, linoleic and linolenic acids which have been found to be protective against cardiovascular disease (Agbogidi and Akpomorine 2013). But, the dietary use of the leaves of *V. amygdalina* has remained limited only to the cultures that use it as vegetable (after maceration to remove the bitter principles) in soup and porridge preparation, suggesting an under-utilization of the plants leaves (Ijeh and Ejike 2011).

**Industrial**
Bitter leaf can be put into use in soap making, ornaments, substituted for hops in beer brewing, restoration of stamina as well as contribution to carbon sequestration thereby reducing environmental degradation (Agbogidi and Akpomorine 2013). The aqueous leaf extract of *V. amygdalina* has been shown to be phytotoxic. Powders of dried leaves of *V. amygdalina* have been reported to have insecticidal potency against the larvae of *Calloso bruchus maculatus* and *Sitophilus zeamais* – insects responsible for heavy losses of stored cow pea and maize, respectively (Ijeh and Ejike 2011). It was also reported that the essential oil of *V. amygdalina* was effective in the control of *Sitophilus zeamais* (Ijeh and Ejike 2011).

**Others**

Unlike other plants that are native to certain areas, it can adapt to a variety of climates. It is grown in many countries, in savannah zones and cultivated fields (Agbogidi and Akpomorine 2013).

**Safety Data**

In an acute toxicity experiment, *V. amygdalina* was reported to be non-toxic (Ijeh and Ejike 2011). However, other earlier reports implicate the plant to be toxic in high doses (Ijeh and Akomas 2005; Ojiako and Nwanjo 2006). This could be inferred because of the presence of HCN above 4.0 mg/kg in the leaf, besides the appreciable saponin contents especially if not properly washed (Dike 2010).

## *Cyperaceae*

### *Cyperus esculentus* var. Sativa

*Cyperus esculentus* is a plant cultivated mainly for its food purpose. But it became subsequently known for its medicinal value and could safely be referred to a medicinal food plant. It thrives in moist sandy-loam soil though, largely insensitive to soil types (Bamishaiye and Bamishaiye 2011). It is propagated from the tubers or rhizomes, and cultivated anytime within the rainy season. It matures between 90 and 120 days with no requirement for manure or weeding except if necessary (FAO 1988). Being another women-dominated farming, harvest is manually carried out and labor intensive. It is sun-dried and stored unwashed after harvest. Used widely in cosmetics, it has been reported to prevent fibrosis, prostate and hernia cyst. It also has LDL-cholesterol lowering and HDL-cholesterol elevating properties.

*Cyperus esculentus*, Tiger nuts and plant

**Synonyms**
*Cyperus esculentus* var. *esculentus*, *Cyperus esculentus* var. *hermannii*, *Cyperus esculentus* var. *leptostachyus* and *Cyperus esculentus* var. *macrostachyus* (Oloyede et al. 2014).

**Common Names**
In Nigeria, it is called "Aya" (in Hausa), "imumu" (in Yoruba), "ofio" and "aki Hausa" (in Ibo), ópu-ópa (Igala); and elsewhere, Zulu nut, yellow nut grass, ground almond, chufa, edible rush, rush nut, yellow nut sedge or earth almond.

Part 1: General Aspects

**Morphological Description**
Tiger nut is a tough erect fibrous-rooted perennial plant, 1–3 ft. high, reproducing by seeds and by many deep, slender rhizomes, which form weak runners above the ground, and small tubers or nutlets at the tips of underground stems (Bamishaiye and Bamishaiye 2011). The leaves are long, narrow, shiny, light-green and arranged in three rows around the triangular stem. They are characterised by pointed tips separated from the rest of the leaves by a distinct shoulder. The flowers are yellowish or yellowish-brown with spikes which are usually about 1.0–1.5 cm long. The roots are in form of rhizomes which are yellow in colour and have single tubers seen as root swellings at the end. These tubers (or root swellings) are generally round, small and 5.0–20.0 mm in length, with a thin pale brown outer coat which gets darker with maturity. The tuber is a nicely flavoured sweet nut and crispy (FAO 1988).

**Classification**
Tiger nut (*Cyperus esculentus* var. *sativa*) is one members of the family *Cyperaceae* (Bamishaiye and Bamishaiye 2011).

**Origin and Distribution**
It is widely distributed in the temperature zones within South Europe as its probable origin, and has become naturalized in Ghana, Nigeria and Sierra Leone (Anon 1992).

**Soil Requirements**
Tiger nut is found inhabiting wet marshes and edges of streams and ponds where it grows in coarse tufts. According to Bamishaiye and Bamishaiye (2011), the plant grows best in moist sandy-loam soils but will grow even in the hardest clay and tolerates high soil moisture and is intolerant to shade.

**Production Levels**
It was observed that tiger nut production is predominantly the work of women farmers, as high as 70% (Bamishaiye and Bamishaiye 2011; Tetteh and Ofori 1998).

**Major Production Areas in Africa**
Ghana, Nigeria and Sierra Leone (Anon 1992).

**Cultivars**

There are mainly three varieties namely: black or dark brown ('Fanti'), brown and yellow ('Kwahu') (Tetteh and Ofori 1998; Okafor et al. 2003), and only yellow and brown are readily available in the Nigerian markets (Okafor et al. 2003). The yellow variety is preferred to all other varieties because of its inherent properties like its bigger fleshier size, and attractive colour. The yellow variety also yields more milk, contains lower fat, higher protein and less anti-nutritional factors especially polyphenols (Okafor et al. 2003).

Part 2: Cultivation Practices

**Propagation**

It is propagated from the tubers and their rhizomes (Bamishaiye and Bamishaiye 2011).

**Soil Preparation**

Experiments revealed that tuber production in tiger nut does not appear to be sensitive to soil type, but nutrient additions can increase tuber production. The increase in production is most likely tied to above ground vegetative biomass such that the larger the plant, the more tubers are produced. Therefore, from a production perspective, management techniques that enhance growing conditions are likely to be beneficial. Similarly, early sowing leads to a longer growing season which is correlated with higher production. Lastly, early emergence increases production because competing vegetation will not be as well established and will therefore have less impact on growing conditions (Dyer 2009).

**Planting**

It is usually sown in April and picked in November (Osagie et al. 1986). Before planting, the nuts are soaked in water for 3–4 days to facilitate germination. The water is changed every day to prevent rot. The soaked nuts may be planted directly, or kept in a basket until they begin to sprout between 7 and 12 days after soaking. The sprouted nuts are picked daily and planted. Pre-sprouting before planting ensures uniform emergence and a perfect stand. Although, tiger nut grows well when planted as a sole crop, most farmers in the study areas usually intercropped them with maize, cassava or cocoyam (Tetteh and Ofori 1998).

**Fertilization**

Fertilizer does not have a consistent effect on tiger nut tuber production (Dyer 2009), as the work of Tetteh and Ofori (1998) revealed that no form of manure (organic) or inorganic fertilizer needs to be applied to tiger nut. Most producers claim that when any form of manure is applied, the resulting nuts are less tasty and less attractive (Tetteh and Ofori 1998).

**Irrigation**
While in some parts with moderate rainfall, tiger nuts can thrive and produce good yields, in most parts, it requires copious amount of water and repeated watering especially during the dry season (FAO 1988).

**Weed Control**
According to Tetteh and Ofori (1998), weeding is done less often and mostly only once during the life of the crop. Weeding has been found to be unnecessary when pre-sprouted nuts were sown on newly made mounds, as they sprout early and grow faster to suppress weeds that germinate afterwards. Weeding becomes necessary only when inter-mound spacing is too wide. The weeds are hand-picked to avoid disturbing the developing nuts.

**Pest Control**
Termites and grass cutters have been reported as some challenges to tiger nuts. In order to reduce losses, harvest is done as soon as nuts mature (Tetteh and Ofori 1998). To avoid termites attack, plots where termite mounds are sited must be avoided.

**Disease Control**
Usually there are no common diseases that seriously affect tiger nuts. However, Negro bug known as *Corimelaena pulicaria* punctures the tubers while root knot nematodes (*Meloidogyne* spp.) also cause damages to the tubers before they mature (FAO 1988).

**Harvesting**
Signs of maturity include yellowing of leaves and cessation of new inflorescence. The nuts are harvested by hand-picking singly from the soil, a slow and labour – intensive process (Tetteh and Ofori 1998). However, most farmers harvest the nuts by gradually loosening the soil around the plants to expose them, and the plants are raised and the attached nuts are removed by hand. But the nuts left in the soil are slowly and tediously hand-picked singly (Tetteh and Ofori 1998).

Part 3: Post-harvest Handling

**Part Collected and Collection Techniques Used**
The tubers are the produce of interest in the tiger nut plant. These are considered to have reached maturity when the plants begin to wilt, and are found within the top 15–20 cm of the soil. They are harvested by hand-picking (Tetteh and Ofori 1998) or by running a small turning plough beneath them. Groundnut diggers have also been used. The tubers are then separated, soil removed, washed and allowed to dry for between 1 and 3 days before grading and storage in thin layers in sheds (FAO 1998).

**Grading**
They are generally dried out (mostly in the sun), a process that takes one or more months with occasional turning over to ensure uniform drying to preserve them and to eliminate prevailing rot or any microbial infection (Bamishaiye and Bamishaiye 2011).

**Packaging/Storage**
Usually, they are not washed after harvest, but are sun dried thoroughly and packed in jute bags or specially woven baskets called *kentenku* for up to 12 months (Tetteh and Ofori 1998). No preservatives are applied. Tiger nut can be processed into flower via drying under the sun or roasting to dehydrate the product. This is then introduced to a hopper for milling. The product is allowed to cool then packaged in air-tight containers and stored (Bamishaiye and Bamishaiye 2011). When processed into tiger nut oil, the purified cold extracted oil is collected in barrels and stored (Bamishaiye and Bamishaiye 2011).

**Marketing**
Tiger nuts are readily available in the Nigerian markets. In Northern Nigeria, the nuts can be bought in the market all year round (Bamishaiye and Bamishaiye 2011). According to Tetteh and Ofori (1998), units of sale include 50 kg minibags, 'size 34' bucket, 3 kg measure, and 450 g cans. Usually, farmers sell their nuts to middlemen who in turn sell to retailers in the urban centres. Most farmers sell their produce immediately after harvest because of lack of appropriate storage facilities, or the need for quick money (Tetteh and Ofori 1998).

## Part 4: Production Schedules

Tiger nuts usually take between 90 and 120 days to mature. On sandy soils the average yields are between 789 and 900 kg/ha. In large scale production aided by irrigation, yields could be as much as 8–14,000 kg/ha. The seed rates are reported to be between 17 and 45 kg/ha, with a spacing of 15 × 75 – 90 cm which requires 35–45 kg/ha of tubers (FAO 1988).

## Part 5: Utilization

**Cosmetic**
The oil is high oleic acid and low polyunsaturated fatty acid (linoleic acid and linolenic acid), and is suitable for salad making and for the skin. It also has higher oxidative stability than other oils, due to the presence of polyunsaturated fatty acids and $\gamma$-tocoferol.

**Pharmaceutical and Therapeutic**
Tiger nuts have been considered to have adequate properties to fight respiratory infections, stomach aches, promotes production of urine. Hence, it is considered

effective in the prevention of cyst, prostrate, hernia, rectum deformation and prolapse, endometriosis or fibrosis as well as fallopian tube blockage. The oil reduces low density lipoprotein-cholesterol (LDL-C) and increases high density lipoprotein-cholesterol (HDL-C), reduces level of triglycerides in blood and the risk of forming bloody clots, thereby preventing arteriosclerosis. It is also reported to stimulate the absorption of calcium in bones and production of new bony material, and is recommended for infants and the elderly because of its high vitamin E, and hence its antioxidant benefits in the cell membrane.

**Food and Flavoring**

Tiger nut has been cultivated as a livestock food and for human consumption. It can be eaten raw, roasted, grated, baked or used for ice cream and beverage making for its many useful benefits (Bamishaiye and Bamishaiye 2011). Tiger nut milk is a delicious and energizing drink that is usually served chilled, as it is, or packaged (Bamishaiye and Bamishaiye 2011).

**Industrial**

Tiger nut flour has a unique sweet taste, which is a good alternative to many other flours like wheat flour, used in the confectionery industry as it is gluten free and good for people who are averse to gluten in their diets (Belewu and Abodunrin 2006; FAO/WHO 1985). It is considered good flour or additive for the bakery industry (Anderson et al. 1994) as flavoring agent for ice creams and biscuits (Osagie and Eka 1998). In the textile industry, the oil is used to water proof fibres (Bamishaiye and Bamishaiye 2011).

**Safety Data**

The presence of anti-nutrients like tannins have been reported (Bamishaiye and Bamishaiye 2011), and may be responsible for the toxicity of the hexane extract in brine shrimp lethality assay (Tetteh and Ofori 1998). While these anti-nutrients can be eliminated by boiling in water (Bamishaiye and Bamishaiye 2011), the methanolic extract, is relatively non-toxic (Tetteh and Ofori 1998).

## *Gramineae (Poaceae)*

*Cymbopogon citratus* Stapf. *Cymbopogon citratus* Stapf is a plant cultivated mainly around homes. It performs well on sandy or clay-loam soils with good drainage. Being a plant considered on poor soils, application of fertilizer is usually a rare consideration. With adequate rainfall, irrigation is not required (Directorate 2012). Several diseases of *C. citratus* are mostly fungal. Thus, the application of appropriate fungicide before flowering in all cases is necessary. Harvest is done mechanically or manually. *C. citratus* possess anti-carcinogenic, anti-inflammatory, antioxidant, and cardioprotective properties. The issue of the toxicity or not of the plant is still in debate. However, *C. citratus* has been reported to be a good source of antioxidants such as flavonoids, and therefore, its tea remains a nutritionally acceptable and medicinally valuable beverage (Ekpenyong et al. 2014).

*Cymbopogon citratus* (Lemon grass)

## Part 1: General Aspects

### Description of Plant

Lemon grass is a tufted perennial grass growing to a height of 1 m with numerous stiff leafy stems arising from short rhizomatous roots (Karkala and Bhushan 2014). It has an economic lifespan of about 5 years (Carianne 2005). The leaf-blade is linear, tapered at both ends and grows to a length of 50 cm and width of 1.5 cm. The leaf-sheath is tubular in shape and acts as a pseudo-stem. The leaves are long, glaucous, green, linear, tapering upwards and along the margins. *C. citratus* plant produces flowers at matured stages of growth (Directorate 2012; Karkala and Bhushan 2014).

### Classification

*Cymbopogon citratus* Stapf is popularly known as citronella grass or lemon grass. This specie belongs to the Gramineae (Poaceae) family (Soares et al. 2013; Karkala and Bhushan 2014).

### Origin and Distribution

Lemon grass is widely cultivated in the tropics and subtropics (Directorate 2012).

### Soil Requirements

Lemon grass is widely adapted to a range of soils and performs well on sandy to clay loam soils with a pH range of 5.0–8.4 and good drainage. The lower the altitude and more alkaline the soil, the higher is the citral content of the oil. The variety with high citrates is in demand. Drier and loamier soils yield higher citral content. Lemon grass can be an option for consideration in poor soils, alkaline soils, steep slopes, and degraded forests and rehabilitation of non-forest mining and industrial waste lands (Directorate 2012).

**Production Levels**
The yield of oil is less during the first year of establishment and increases in the second year and reaches a maximum in the third and fourth years, after which it declines. For economy, the plantation is maintained only for 6 years. On average, 30–50 metric tons of fresh herbage is harvested per hectare per annum, which will yield 100–250 kg of oil at an oil yield ranging between 0.2 and 0.5%. Under irrigated conditions and better management with increased herbage, an oil yield of up to 500 kg/ha can be achieved (Directorate 2012).

**Major Production Areas in Africa**
Lemongrass is grown throughout Africa, in the Democratic Republic of the Congo (DRC), Angola, Madagascar and Comoros islands although, Guatemala is known to be the leading exporter of *C. citratus* products (Directorate 2012).

**Cultivars**
In Southern Africa for instance, there are no registered cultivars for lemongrass. But there are two known types of *C. flexuosus*. The red-stemmed grass, which is known as the true lemon grass and is recommended for higher oil yields, and the white-stemmed (*C. flexuosus* var. albiscens) yields oil of lower citral content (Directorate 2012).

## Part 2: Cultivation Practices

**Propagation**
Although propagation can be achieved by the seeds, but for better quality and yield of oil, growing lemon grass by the slips obtained from divided, well-grown clumps, is recommended. Tops of clumps are cut off within 20–25 cm of the root and the latter divided into slip and the lower brown sheath removed to expose young roots (Directorate 2012).

**Soil Preparation**
In order to obtain product yields of high quality that is in demand globally, *C. citratus* and essential oil crops are to be grown on natural soils (Directorate 2012; Ekpenyong et al. 2014). Thus, soil fertility, pH, balanced mineral status are to be ascertained.

**Planting**
Directorate (2012) recommended a row spacing of 20 cm with a row width of 40 cm, that will give a total of 125,000 plants per ha in a high rainfall area or under irrigation. In areas with lower rainfall, 60,000 plants per ha is an advisable density. An initial

high planting rate can be used and as the plants mature, every second plant can be taken out and divided again for new slips. The space created in this way can then be filled with remaining plants as they mature (Directorate 2012).

**Fertilization**

Application of fertilizer should be dependent on soil analysis (Directorate 2012), which should indicate any nutrient, compound or element that is lacking and needs to be replenished.

**Irrigation**

Lemon grass has a high water requirement. Where annual rainfall exceeds 650 mm, irrigation is not necessary. *C. citratus* is more drought tolerant than *C. flexuosus*. Overhead, flood and drip irrigation can be used. Where rust is a problem, overhead irrigation might not be a good option (Directorate 2012).

**Weed Control**

Hand-weeding and hoeing are very important as weeds affect the yield and quality of the oil (Directorate 2012). About 2–3 weedings are required in a year. Distillation waste of this crop applied as organic mulch at 3000 kg/ha is highly effective for controlling weeds on *C. citratus* farm. Another very good method of controlling weed is by excluding the sunlight through planting in such a way that the plant should form canopy quickly (Directorate 2012).

**Pest Control**

Various pests like stem-boring caterpillar and nematodes can result to enormous damage and reduction in quality and yield. In the former, burning of the dry stubble during the off-season in summer would destroy the caterpillars lurking inside the stubble but if mild, the affected shoots are pulled out and destroyed. In severe attacks, chemical control is needed (Directorate 2012). In the later situation, soil solarisation is effective but as a last resort. Due to toxicity, nematicide becomes necessary (Directorate 2012) however, with caution.

**Disease Control**

Recognizing the first symptoms of disease is quickly followed with the removal of diseased plants and continuous monitoring of fields for signs of pathogen recurrence or spread of disease is very helpful. Some major diseases of *C. citratus* are Long smut, Red leaf spot, Leaf blight, and Rust. Spraying with a recommended fungicide before flowering in all cases is necessary (Directorate 2012).

**Harvesting**

First harvest of the plant can be done within the first 6–9 months of planting the slips then on a monthly basis afterwards (Directorate 2012; Ekpenyong et al. 2014) as the plant actively grows. Frequent cutting actually stimulates vegetative growth. Moreover, oil yield is reduced if the plant is allowed to grow very large. Early in the morning harvest is recommended, which is done mechanically or by hand (Directorate 2012). Cutting too low and splitting of cutting edges should be avoided. Three harvests can be obtained in the first year and up to 5–10 harvests during each of the 3–5 succeeding years, depending on soil moisture status, management and weather (Directorate 2012).

## Part 3: Post-harvest Handling

### Part Collected and Collection Techniques Used

*C. citratus* leaf, stem, and roots are commonly used in herbal medicine (Ekpenyong et al. 2014).

### Grading

The quality of lemon grass oil is determined by the content of citral, a terpene aldehyde, which is mainly used for the production of vitamin A (Directorate 2012).

### Packaging

Essential oils are volatile and as such have to be handled with care. Lemon grass should be packed firmly as this prevents the formation of steam channels. If the grass is too long it can be cut into smaller lengths to ensure firm packaging (Directorate 2012).

### Storage

Lemon grass oil should be stored in dark, air-tight, glass bottles and away from heat or heavy metals. Once opened, refrigeration and tight sealing of the cap helps prolong its shelf-life. Deterioration begins if the liquid is much darker or more viscous than normal. Lemon grass oil is very acidic and will destroy plastic and rubber within a relatively short time (Directorate 2012).

### Marketing

*C. citratus* oil price fluctuates on the international market due to the interplay of demand and supply that can vary widely (Directorate 2012). As such, price is dependent both on the extent of availability of the oil and exchange rate at the time of sale. Important also, is the quality of the available oil on the global market (Directorate 2012).

## Part 4: Production Schedules

While the citral content (Halabi and Sheikh 2014) which is a mixture of terpenoids, neral and geranial confers the lemony smell actually justifies the large scale commercial cultivation of *C. citratus*, agro-climatic factors like soil water, and soil salinity influences the growth, quality and yield of the oil which is the most important product from *C. citratus* (Ekpenyong et al. 2014). It has been observed that a better yield of the oil and higher citral content is obtainable from *C. citratus* harvested during the drought season (March–June) than any part of the year (Ekpenyong et al. 2014). Also, the early or delayed harvest of *C. citratus* affects essential oils and citral contents of the plant, with significantly higher oil content found in plants harvested at $6 \pm 0.5$ months of the plant's growth than those harvested much later (Ekpenyong et al. 2014).

## Part 5: Utilization

### Cosmetics

In cosmetics, its essential oils are used as fragrance in the manufacture of perfumes, soaps, detergents, and creams (Ekpenyong et al. 2014).

### Pharmaceutical and Therapeutic

It is used in herbal medicine for a wide range of applications based on its anti-carcinogenic, anti-inflammatory, antioxidant, cardioprotective, antitussive, and anti-rheumatic activities (Ekpenyong et al. 2014). It has also been used to inhibit platelet aggregation, treat diabetes, dyslipidemia, gastrointestinal disturbances, anxiety, as well as in aromatherapy (Ekpenyong et al. 2014).

### Food and Flavoring

*C. citratus* is also consumed as a tea, added to non-alcoholic beverages and baked food, and used as a flavoring and preservative in confections and cuisines (Ekpenyong et al. 2014).

### Industrial

Although, the down-side in herbal medicine use is in the lack of dosage of administration as in conventional drugs, it has been reported that a team of researchers have produced tablets of *C. citratus* containing powdered dry *C. citratus* leaves using acacia and gelatin as binders at concentrations of 2, 4 and 8% w/w (Ekpenyong et al. 2014).

### Safety Data

A number of studies have attempted to answer the safety question with conflicting and inconclusive data. Thus, while some studies observed toxic effects, no toxicity was detected by others. However, most of these studies were performed in animals, making the results difficult to apply to humans (Ekpenyong et al. 2014). However, *C. citratus* has been used over many years to make caffeine-free tea and as an herbal drink, suggesting that it may be a healthier alternative to caffeine-containing tea products (Blanco et al. 2009). Akande et al. (2011) found that, in comparison to other tea brands consumed among Nigerians, *C. citratus* tea was a good source of antioxidants such as flavonoids and therefore, a nutritionally acceptable and medicinally valuable beverage (Ekpenyong et al. 2014).

## *Meliaceae*

### *Khaya senegalensis*(Desv.) A. Juss

*Khaya senegalensis* (Desv.) A. Juss is propagated by seeds, preferably from nursery and grows well on a wide range of soil conditions. Fertilizer application during planting is recommended. Survival rates are low so, regular watering of seedlings of *Khaya senegalensis* is necessary, besides weed control which is manually done by hoeing (Nikiema and Pasternak 2008). The leaves are used as remedy for diseases like diarrhea, gynecological disturbances, digestive and nervous disorders. The stem-bark and leaves have been reported to be used to cure syphilis, pyrexia, and malaria fever (Kubmarawa et al. 2008). However, safety reports have indicated that extracts may affect cellular integrity of vital organs of the body.

*Khaya senegalensis* (http://nurserylive.com/images/)

**Botanical Name**

*Khaya senegalensis*(Desv.) A.Juss (Common and local names: Mahogany (English). In Nigeria, it is called 'Madachi' in Hausa, 'Ono' in Igbo, 'Oganwo' in Yoruba, 'Ago' in Igala, 'Wuchi' in Nupe, (Mann et al. 2003) and Dalehi-kahi in Fulani.

Part 1: General Aspects

**Morphological Description**

*Khaya senegalensis* is an evergreen tree with shinny foliage, pinnate leaves and characteristic round capsules. It can grow to a height of 30 m and a girth of 3 m with dense crown and short bole covered with dark grey scaly bark, with slash dark pink and bitter yielding gum when wounded. The leaves have 3–4 (exceptionally 5) pairs of leaflets, 5–10 cm long by 2.5–5 cm broad, more or less elliptic, round, obtuse or shortly acuminate at apex stalks of leaflets 4 mm long (Kubmarawa et al. 2008). The fruit is an upright, almost spherical woody capsule, 4–6 cm in diameter, opening by four valves from the apex. The seeds are brown, six or more per cell, broadly transverse ellipsoid to flat, about 25 × 18 mm, margins and narrowly winged (Orwa et al. 2009).

**Classification**
*Khaya senegalensis* belongs to the *Meliaceae* family (Orwa et al. 2009).

**Origin and Distribution**
*Khaya senegalensis* occurs from Mauritania and Senegal east to Northern Uganda. It is found in many African countries like Cape Verde, Tanzania, Malawi, Madagascar, Egypt and South Africa. It is also found in India, Indonesia, Vietnam, Australia Sri Lanka and tropical America. It is commonly planted as ornamental and roadside tree (Nikiema and Pasternak 2008).

**Soil Requirements**
*Khaya senegalensis* prefers deep and well-drained alluvial soils and termite mounds, but can also be found on shallow, rocky soils, where it is usually much smaller. It tolerates flooding in the rainy season (Nikiema and Pasternak 2008). *Khaya senegalensis* thrives on a wide range of soil conditions, from neutral to strongly acidic, and from very well-drained coarse sandy loam to somewhat poorly drained clay (Orwa et al. 2009). It occurs in savanna woodland, often in moist localities and along water courses in areas with 650–1300 (~1800) mm annual rainfall and a dry season of 4–7 months (Nikiema and Pasternak 2008). It thrives in areas with mean annual temperature of between 24.5 °C and 31.5 °C and mean annual rainfall 400 mm to 1750 mm (Orwa et al. 2009).

**Production Levels**
In experimental plantations in Burkina Faso the annual production of *Khaya senegalensis* is estimated at 3.7 $m^3$/ha (Nikiema and Pasternak 2008).

**Major Production Areas in Africa**
Cameroon, Central African Republic, Chad, Cote d'Ivoire, Equatorial Guinea, Gambia, Ghana, Guinea, Guinea-Bissau, Mali, Niger, Nigeria, Senegal, Sierra Leone, Sudan, Togo and Uganda (Orwa et al. 2009).

Part 2: Cultivation Practices

**Propagation**
*Khaya senegalensis* is propagated by seeds, preferably from nursery Wildings are sometimes collected for planting. It also reproduces by root suckers, while grafting and layering are also possible. However, propagation by cutting is much more difficult (Nikiema and Pasternak 2008).

**Soil Preparation**
*Khaya senegalensis* seeds are best raised in seed beds in the nursery or in pots (Nikiema and Pasternak 2008).

**Planting**
High quality *Khaya senegalensis* seeds are selected for planting, and soaking of seeds for 24 h in water improves germination. The seeds can best be sown in seed beds in the nursery or in pots. Fresh healthy seeds have a high germination rate of

90–100%, and they may retain their viability for 6–8 months. Upon sowing, seeds are covered with only a thin layer of soil, or left partially uncovered. Germination takes 10–18 days (Nikiema and Pasternak 2008).

**Fertilization**
The application of a complete fertilizer at a rate of 200 g/tree at the time of planting is recommended (Nikiema and Pasternak 2008).

**Irrigation**
For good survival rates, regular watering of *Khaya senegalensis* seedling is essential. It also occurs along water courses (Nikiema and Pasternak 2008).

**Weed Control**
The recommended method of weed control on *Khaya senegalensis* plantation is by hoeing especially at the onset of the dry season (Orwa et al. 2009).

**Pest Control**
*Khaya senegalensis* is susceptible, in its natural range to severe attack by shoot borers (*Hypsipila robusta*) which may result in mis-shaped trees with no timber value. The sapwood is susceptible to attack by long-horn beetles and powder post beetles (*Lyctus* spp.). A bacterial disease of dry-zone mahogany in the Sudan caused by *Xanthomanus khaye* results in rough, scabby leaf spots and knobby stem cankers (Orwa et al. 2009).

**Harvesting**
Logs are harvested from *Khaya senegalensis* by felling using traditional equipment, which can be quite difficult because of the dense and fairly hard wood. The bark is collected whenever needed and many large trees show signs of debarking. The leaves are also harvested for fodder (Nikiema and Pasternak 2008).

Part 3: Post-harvest Handling

**Part Collected and Collection Techniques Used**
The logs are collected for timber by felling and the bark and leaves are also collected for traditional medicine and fodder respectively. Traditional methods of collection are used (Kubmarawa et al. 2008).

**Marketing**
*Khaya senegalensis* is widely used on commercial scale, particularly in West Africa as timber (Nikiema and Pasternak 2008).

Part 4: Production Schedules

*Khaya senegalensis* is grown from the seeds which are dispersed as far as 100 m by wind. The seeds bearing fruits are developed at about 20–25 years of the tree's existence, and cultivation is successful in areas with short dry seasons and high rainfall

(Orwa et al. 2009). Wijenayake (2014) however, reported that it is a fast growing tree which can reach a height of 30–33 meters and maturing in 10–12 years, and new plants could be produced by tissue culture.

Part 5: Utilization

**Cosmetic**
The seed oil is used in cosmetics production (Nikiema and Pasternak 2008).

**Pharmaceutical and Therapeutics**
*Khaya senegalensis* is used as a remedy for diseases such as diarrhea, gynecological disturbances, digestive disorders and nervous confusions (Kubmarawa et al. 2008). Decoction and concoction of the stem-bark and leaves are used to cure mucous diarrhea, syphilis, pyrexia, and malaria fever. Dried stem-bark is used externally for the treatment of skin infections and blennorrhagra, while the fixed oil is used to treat leprosy, syphilis, dermatomes, sores and ulcer in adults. The flowers are used to treat stomach diseases and as an ingredient in anti-syphilic prescriptions. The oil from the seeds is used as emmenegogue (Kubmarawa et al. 2008). Bark decoctions or macerations are used to treat malaria fever, stomach complaints, diarrhea, dysentery and anaemia, as anodyne in cases of rheumatism, headache, and as tonic and anthelmintic. They are also used as purgative, antidote to treat syphilis, leprosy, chickenpox and angina. The bark is applied externally as disinfectant in cases of inflammations and skin diseases, rash, scabies, wounds, ulcers, boils, haemorrhoids, swellings and toothache. For veterinary purposes, the bark is also used as anthelmintic, tonic, appetizer, and to treat trypanosomiasis, liver flukes, diarrhea and ulcers (Nikiema and Pasternak 2008).

**Food and Flavouring**
The foliage is common source of fodder, though of low fodder quality which is commonly used towards the end of the dry season when better quality fodder is not available, or in mixtures with better fodders. The seed oil is used for cooking (Nikiema and Pasternak 2008).

**Industrial**
*Khaya senegalensis* wood is widely used for carpentry, joinery, furniture, cabinet work, ship building, decorative veneer and implement handle (Nikiema and Pasternak 2008; Orwa et al. 2009).

**Others**
The wood ash of *Khaya senegalensis* is added to stored grain to prevent infestation by insects. The tree is commonly planted as a roadside tree, ornamental shade tree, and sometimes for soil stabilization (Nikiema and Pasternak 2008).

**Safety Data**

A study of the sub-chronic effect of aqueous stem bark extract of *Khaya senegalensis* on some biochemical, haematological and histopathological parameters of rats showed that the extract may affect cellular integrity of vital organs of the body (Onu et al. 2013).

**Chemical Constituents**

Aqueous and ethanolic extracts of root, stem-bark and leaves of *Khaya senegalensis* contain saponins, tannins and phenols (Kubmarawa et al. 2008). Limonoids have been identified in *Khaya senegalensis* stem, bark, leaves and flowers: khayanolides D and E, khayanolides, 2,6-dihydrolissinolide. Two mexicolides named khayanone and 2-hydroxyseneganolide have also been identified. Three other mexicanolide limonoids named seneganolide A, 2-hydroxyseneganolide A and 2-acetoxyseneganolide A (Kubmarawa et al. 2008). Simlarly, Atawodi et al. (2009a, b) have reported the presence of catechin, rutin and quercetin rhamnoside in the leaf, and catechin and procyanidins in the stem bark. Saponins, flavonoids, tannins, alkaloids, glycosides and carbohydrates were detected in the ethanolic leaf extracts (Wakirwa et al. 2013).

**Biological Activities**

The aqueous extract of *Khaya senegalensis* stem-bark and the ethanolic extract of the root, stem-bark and leaves, using disc diffusion method exhibited antibacterial activity against *Staphylococcus aureus, Streptococcus spp., Eschrichia coli, Pseudomonas aeruginosa, Salmonella* spp. and *Bacillus subtilis* (Kubmarawa et al. 2008). A comparison of the antibacterial activity of the aqueous, ethanolic and acetone extracts of the bark against *Staphylococcus aureus, Streptococcus pyorogenes, Salmonella typhi, Shigella dysenteriae, Klebsiella pneumonia* and *Pseudomonas aeruginosa* showed that the acetone extracts had higher antibacterial activity. The minimum inhibitory concentration of the acetone extract was in the range of 6.5–12.5 mg/ml while the minimum bactericidal concentration ranged between 6.5 and 25.0 mg/ml. The antibacterial activity was not significantly affected by heating at 100 °C for 1 h but the activity was lost at pH 8.0 (Sale et al. 2008). The ethanolic leaf extract of *Khaya senegalensis* was examined for antimicrobial activity against some selected clinical isolates: *Staphylococcus aureus, Escherichia coli, Klebsiella pneumonia* and *Candida albicans*. Dose dependent activity was observed with zones of inhibition ranging from 15.00 to 26.00 mm for *S. aureus*, 10.00–23.50 mm for *E. coli* and 3.20–16.00 mm for *K. pneumonia.* There was no inhibition against the growth of *C. albicans*. Minimum inhibitory concentrations for the clinical isolates are 25.00 mg/ml, 12.50 mg/ml, 12.5 mg/ml and no effect respectively, while the minimum bactericidal concentrations were 50.00 mg/ml and 25.00 mg/ml for *S aureus* and *E. coli* respectively (Wakirwa et al. 2013). Earlier on, the antioxidant effect of this plant has been described (Atawodi et al. 2009a, b).

## *Moringaceae*

### *Moringa oleifera* Lam.

Moringa oleifera (http://kingraah.blogspot.com.ng)

Part 1: General Aspects

**Morphological Description**

*Moringa oleifera* Lam. is a smallish deciduous tree which possesses relatively scanty foliation. It looks more like leguminous plant. It grows up to 8 m high, and bearing branches that are often forked from near the base. The bark could be smooth, dark grey, slash thin or yellowish. Twigs and shoots are short and densely hairy. The crown is wide, open, typically umbrella shaped and usually a single stemmed plant which is often deep rooted. The wood is soft.

Leaf arrangement is alternate, the old ones get yellowish and falls off. These leaves are large (up to about 90 cm long), with opposite pinnae, spaced about 5 cm apart up the central stalk, usually with a 2nd lot of pinnae, also opposite, bearing leaflets in opposite pairs, with a slightly larger terminal leaflet. Leaflets are dark green above and pale on the under surface and are of various sizes and shape, although usually rounded-elliptic, scarcely as much as 2.5 cm long.

Flowers are produced throughout the year, in loose axillary panicles up to 15 cm long; individual flower stalks up to 12 mm long and very slender; 5 pale green sepals 12 mm long, finely hairy, 5 white petals, unequal, a little longer than the sepals; 5 stamens with anthers, 5 without; style slender, and the flowers sweet smelling.

Fruits are large and distinctive, between 30 to 90 cm long and 12 mm broad, slightly constricted at intervals, gradually tapering to a point, 3-(4-) angled, with 2 grooves on each face, and light brown in color. It splits along each angle to expose the rows of rounded blackish oily seeds, each with 3 papery wings (Orwa et al. 2009; Paliwal et al. 2011).

**Classification**

*Moringa oleifera* Lam. belongs to the family of *Moringaceae* (Paliwal et al. 2011).

**Origin and Distribution**

*Moringa oleifera* Lam. is indigenous to south Asia and has been found to grow in Pakistan, West Bengal and India. It has also been introduced to many countries, including Afghanistan, Bangladesh, Sri Lanka, Southeast Asia, West Asia, Arabian Peninsula, East and West Africa, Southern Florida, West Indies, Mexico, Peru, Paraguay and Brazil (Paliwal et al. 2011).

**Soil Requirements**

*Moringa oleifera* Lam. can grow at elevations of up to 1400 m along the rivers of its native range, on sandy or gravely alluvial soils. The soils are usually well drained and low in organic matter. Although, it can thrive in drought environment for several months, the water table should be located within the maximum rooting depth of the plant. Away from its natural origin, it grows well from sea levels up to a 1200 m elevation in most light to medium textured soils. However, it best grows on sandy loam soils (Parrotta 1993).

**Production Levels**

Under ideal conditions, *Moringa oleifera* can grow very rapidly from seed. Within the same year, they can produce vegetable pods for market. For instance, early flowering types can produce market-mature pods in 6 months compared to over 1 year for other types. After coppicing, there is rapid growth and immature pods could be harvested in half a year (Paliwal et al. 2011).

**Major Production Areas in Africa**

*Moringa oleifera* has been reported to be grown in Nigeria, Egypt, Sudan, and various parts of West, East and South Africa (Stevens et al. 2013; Baba et al. 2015).

**Cultivars**

Through genetic manipulations, many different cultivars have been reported. Some of the cultivars obtainable from different countries include TOT4100 (Taiwan), TOT4880 (USA), TOT4977 AVRDC, TOT5077 AVRDC, TOT5028 AVRDC, TOT5169 AVRDC, TOT4893 AVRDC, TOT5330 AVRDC, TOT7266 AVRDC and TOT4951 AVRDC (World Vegetable Centre, Thailand), SH Silver Hill CHM Silver Hill and Limpopo (South Africa) (Ndhlala et al. 2014).

Part 2: Cultivation Practices

**Propagation**

*Moringa oleifera* is propagated through both sexual and asexual (vegetative) means. That is, either by seeds or cuttings, respectively. Seeds are planted about an inch below the soil surface and can be sprouted all year round in a well-drained soil. Propagation by cutting is usually done using coppiced portions from the stem or more commonly from the branch (about 1–2 m long). It is inserted 1/3 portion under the soil (Paliwal et al. 2011; Foidl et al. 2016). Sprouting which occur between 7 and 30 days of planting is commonly between 60 and 90% for fresh seeds. However, viability drops drastically after about 2 months of collection or in storage. Partial shadding is favourable to germinating seeds and seedlings. *Moringa oleifera* raised from seeds have been reported to be slower at both fruit yield and lower quality fruits. However, it has been reported that trees propagated from seeds yields longer roots than those from cuttings (Parrotta 1993).

**Soil Preparation**
Whether it is seedlings from the nursery, ungerminated seeds or cuttings, the soil should be loose and well drained, and moist enough to enhance growth. With availability of water, planting can be done all year round. Planting pit is dug, watered and then filled with top soil mixed with compost or manure before planting. When cuttings are planted in the nursery, the root system develops slowly. As such, addition of phosphorus to the soil could be done in order to encourage root development (Zaku et al. 2015).

**Planting**
When planting from seedlings, they can be out-planted when they are 60–90 cm high. A hole is cut in the bottom of the sack, big enough to allow the roots to emerge. In order to enhance rapid germination, the seeds can be pre-treated either by soaking the seeds in water overnight prior to planting, cracking the shells before planting, or removing the shells and planting the kernels only (Zaku et al. 2015). For intensive moringa production, the trees are planted every 3 m in rows, 3 m apart.To ensure sufficient sunlight and airflow, it is recommended to plant in an east-west direction. When the trees are part of an alley-cropping system, there should be 10 m between the rows (Zaku et al. 2015).

**Fertilization**
Moringa plant grows generally well without fertilizer addition. For extremely poor soils however, manure or compost can be mixed with the soil used to fill the planting pits. Phosphorus can be added to encourage root development and nitrogen will encourage leaf canopy growth. In some countries like India, 15 cm-deep ring trenches are dug about 10 cm from the plant during the rainy season and filled with green leaves, manure and ash. This approach has been reported to promote higher pod yields. Research done in India has also showed that applications of 7.5 kg farmyard manure and 0.37 kg ammonium sulfate per tree can increase pod yields by threefold (Zaku et al. 2015), but in Nigeria, inorganic fertilizer is hardly used, although a small proportion of growers in northern Nigeria apply manure.

**Irrigation**
Moringa tree is a drought – resistant plant that requires no need for much watering. However, during extended and harsh drought conditions, it might be necessary to water regularly for the first 2 months and afterwards only when the tree is obviously suffering from dehydration. Moringa trees starts flowering and producing pods when sufficient water is available. If rainfall is continuous throughout the year, the

trees will have a nearly continuous yield. In arid conditions, flowering can be induced through irrigation (Zaku et al. 2015).

**Weed Control**

Spaces between trees must be free of weeds. Moringa tree root system does not compete with other crops for surface nutrients and the light shade provided by the tree will be beneficial to those vegetables which are less tolerant to direct sunlight. From the second year onwards, moringa can be inter-cropped with maize, sunflower and other field crops. Sunflower is particularly recommended for helping to control weed growth. However, Moringa trees are reported to be highly competitive with eggplant (*Solanum melongena*) and sweet corn (*Zea mays*) and can reduce their yields by as much as 50% (Zaku et al. 2015). Therefore, although practiced in Nigeria and other countries, mixed and inter-cropping with moringa should be done with caution.

**Pest and Disease Control**

Although, moringa is resistant to most pests, there are many that affects them including bark-eating caterpillar, *Indarbela quadrinotata* Wlk; the hairy caterpillar, *Eupterote mollifera* Wlk.; the green caterpillar, *Noorda blitealis* Wlk.; and the bud-worm, *N. moringae* Tams, which can cause serious defoliation, and many others. Some other insect pests that cause minor or only occasional serious damage have also been reported. In Puerto Rico, *M. oleifera* have been reported to succumb to termites' attack, while seed predation has also been identified elsewhere. But in India and sub-saharan african countries like Nigeria, various caterpillars causing defoliation are successfully being controlled by spraying synthetic and natural insecticides (Parrotta 1993; Orwa et al. 2009; Zaku et al. 2015).

**Harvesting**

In Nigeria, the leaf is used as vegetables, and hence harvested when not fully mature, but when sold for making moringa tea, it is harvested and sun-dried. In other countries where they are consumed by humans, harvest is done when the pods are still young and snap easily to prevent pods from developing a tough exterior as they age, although the white seeds and flesh remain edible until the ripening process begins. When producing seed for planting or for oil extraction, the pods are allowed to dry and turn brown on the tree. In some cases, it may be necessary to prop up a branch that holds many pods to prevent it from breaking off. Harvesting the pods before they split open and seeds fall to the ground is necessary. Seeds are stored in well-ventilated sacks in dry and shady places (Zaku et al. 2015) away from rodents and other storage pests.

Part 3: Post-harvest Handling

**Part Collected and Collection Techniques Used**

It has been said that 'every part of the moringa tree is said to have beneficial properties that can serve humanity' and hence collected, they will best serve the desired purpose. For example, the seeds are highly useful in making *M. oleifera* oil and are separated from the harvested pods which are usually collected manually.

**Grading**
Like in India and most Asian countries (Radovich 2013), there are no known formal quality standards and grading for moringa products in Nigeria.

**Packaging**
The leaves are particularly perishable and are stored under cool temperatures and high humidity to avoid excessive wilting and leaflet abscission. This is most often done by bagging in plastics and refrigerating at approximately 10 °C (50 °F), while the fruit may be canned to preserve it for later consumption (Radovich 2013). However, in Nigeria, the leaves are mostly preserved by sun-drying and packaging in polythene bags.

**Storage**
Moringa oil possesses exceptional oxidative stability and can be stored for years while maintaining quality. Leaf and pods may be stored at 10 °C (50 °F) for 5–14 days, but intact pods are less perishable than peeled or cut pods, while shelf life of the seed cake has not been established (Radovich 2013). In Nigeria, the leaf may be stored and utilized when sun-dried.

**Marketing**
Pods and especially leaves are suited for local markets, but important new potential markets include ethnic groups not traditionally using the crop, restaurants, and health food stores and the West African Sub-region. Leaves, medicinal, and value-added products have potential for shipment to North America and Asia, while fruits are not acceptable in continental U.S. because of pest concerns.

However, given the increased awareness of the high nutritional value of moringa, health food stores may be the most receptive new market in the short term. Moringa is relatively easy to grow organically and organic certification may increase consumer appeal. Strong potential for internet sales also exist, particularly for oil and supplements because of product stability and established or emerging markets. Based on superfood trends, producers of processed juices and smoothies may also be a potential high-value market for moringa growers (Radovich 2013).

## Part 4: Production Schedules

Production levels and schedules differ between different ecological zones in Nigeria, but generally, leaf fresh weight yield is 1–5 kg (2.2–11 lb) per tree/year, equivalent to 10,000–50,000 kg/ha (8900–44,534 lb./ac) per year at 1 m × 1 m (3.3 ft. × 3.3 ft) spacing. Commercial leaf yield of moringa reportedly declines from November–March to 50–100% of summer yields in the west coast of Hawaii (Radovich 2013) and many regions of Nigeria, most probably as a result of function of lower solar radiation during winter/harmattan season. At 10 cm × 10 cm (4 in × 4 in) spacing, leaf yields are 7–8 kg/m2 (1.4–1.6 lb./ft2) at the first cutting in well irrigated, drained and fertilized beds, with up to seven cuttings a year.

As per the pods, yields in India are reported at 19 kg (42 lb) pods/tree/year. This is the equivalent of 31,000 kg/ha (27,600 lb./ac) per year at 2.5 m × 2.5 m (8.2 ft. × 8.2 ft) spacing. At 230 pods per tree, pods average 80–90 g (2.8–3.2 oz) each. In Hawai'i, the Indian variety 'PKM-2′ yielded 3–8 times more immature pods than local accessions 6 months after transplanting seedlings (Radovich 2013).

Because the cropping of moringa is an emerging trend in Nigeria, information on production level are rather scanty. However in Hawai'i (Radovich 2013), it is reported to produce about 3 g (0.1 oz) of kernel per dry pod. Oil yield per hectare per year may be estimated based on the numbers above. Assuming a relatively high 20% oil yield by weight from kernels and a 0.90 specific gravity for the oil, trees might be expected to produce 250 l of oil per hectare (107 qt/ac).

## Part 5: Utilization

### Cosmetic

Moringa seeds have been found to be effective against skin-infecting bacteria *Staphylococcus aureus* and *Pseudomonas aeruginosa*. They contain the potent antibiotic and fungicide terygospermin. The alkaloid spirachin (a nerve paralysant) has been found in the roots (Orwa et al. 2009).

### Pharmaceutical and Therapeutics

Decoction of the flowers is used as a remedy for cold, while the gum is utilized as a diuretic, astringent and abortificient, as well as in managing asthma. The oil is used for hysteria, scurvy, prostate problems and bladder troubles. The roots and bark are used for cardiac and circulatory problems, as a tonic and for inflammation. The bark is an appetizer and digestive aid. The iron content of the leaves is high, and they are reportedly prescribed for anaemia in the Philippines (Orwa et al. 2009). Moringa has long been known in traditional medicine worldwide as having value both as a preventative and treatment agent of several health conditions, including the treatment of inflammation, infectious diseases, cardiovascular, gastrointestinal, haematological and hepatorenal disorders.

A number of scientific articles have indeed been published describing the profiling of the antioxidant polyphenols and antioxidant properties of moringa (Atawodi et al. 2010), which can translate to its use as an anti-ageing herb (Ndhlala et al. 2014). Other medicinal uses of the leaves include treatment of fever, ear infections, lowering of blood sugar (diabetes mellitus), blood pressure (Stevens et al. 2013) and trypanosomiasis (Atawodi and Shehu 2010).

### Food and Flavoring

The leaves, a good source of protein, vitamins A, B and C and minerals such as calcium and iron, are used as a spinach equivalent. They are an excellent source of the sulphur-containing amino acids methionine and cystine, which are often in short supply. Young plants are eaten as a tender vegetable and the tap roots may be used as an alternative for horseradish. Young pods are edible and reportedly have a taste reminiscent of asparagus. Although not very popular in Nigeria, the green peas and surrounding white material can be removed from larger pods and cooked in various ways for human consumption in Asian countries. Seeds from mature pods (which can be 40–50 cm long) can be browned in a skillet, mashed and placed in boiling water, which causes an excellent cooking or lubricating oil to float to the surface. The pleasantly flavored edible oil, resembling olive oil, is an excellent salad oil. The flowers can be eaten or used to make a tea. Leaves are mainly used for human food and not to any great extent for livestock, but branches are occasionally lopped for feeding camels and cattle (Orwa et al. 2009).

**Industrial**
The oil (which is brilliant yellow) content of de-hulled seed (kernel) is approximately 42%, and it is mainly used as a lubricant for fine machinery such as timepieces because it has little tendency to deteriorate and become rancid and sticky. It is also useful as vegetable cooking and frying oil. The oil, with free fatty acid content varying from 0.5 to 3%, and containing approximately 13% saturated fatty acids and 82% unsaturated fatty acids, is popular in the perfume industry for stabilizing scents because of its known capacity to absorb and retain volatile substances. It has a particularly high level of oleic acid (70%), as compared to other vegetable oils that contain only about 40% oleic acid (Zaku et al. 2015).

*Others*

**Water Purification**
Moringa seeds contain between 30 and 42% oil and the press cake obtained as a by-product of the oil extraction process contains a very high level of protein. Some of these proteins (approximately 1%) are active cationic poly-electrolytes having molecular weights between 7 and 17 kDa. The cationic poly-electrolytes neutralize the colloids in muddy or dirty water since the majority of these colloids have a negative electrical charge. This protein can therefore be used as a non-toxic natural polypeptide for sedimentation of mineral particles and organics in the purification of drinking water, for cleaning vegetable oil, or for sedimentation of fibers in the juice and beer industries. It thus works as a primary coagulant as natural bridges are continuously formed between the colloid particles (Zaku et al. 2015).

**Safety Data**
According to Khawaja et al. (2010), moringa is an edible and extremely safe plant.

## 2 Conclusion

Cultivation of medicinal plants in Nigeria can be considered as a rarity. However, five medicinal plants namely, *Vernonia amygdalina, Cyperus esculentus, Cymbopogon citrates, Khaya senegalensis* and *Moringa oleifera* were identified and reviewed with respect to general aspects, cultivation practices, post-harvest handling, production schedules and utilization. Available literature on the medicinal plants indicated the presence of phytochemical compounds that exerts pharmacological and biological effects. However, the fewness of comprehensive literature on cultivated medicinal plants was a clear indication of total negligence or dearth of understanding about the need, importance and even the plausibility of cultivating medicinal plants. Culturally though, plants used for medicinal purposes in most of Africa including Nigeria are usually harvested from the wild. But with evident rapidly depleting natural resources due to over exploitation, urbanization and climatic changes, the availability and supply of medicinal plants cannot be sustained in the upcoming few years. As such, cultivation is the only way to ensure availability and sustenance of supply. So far however, this work has identified a near or total lack of attention to the issue of medicinal plants cultivation and by extension, conservation in Nigeria.

**Acknowledgments** Professor S.E. Atawodi's research work on antioxidants was partly supported by Alexander von Humboldt Foundation of Germany and the German Cancer Research Centre (DKFZ), Heidelberg.

## References

Agbogidi OM, Akpomorine MO (2013) Health and nutritional benefits of bitter leaf (*Vernonia amygdalina* del.) Int J A PS BMS 2(3):164–170

Akah PA, Okafor CI (1992) Hypoglycaemic effect of *vernonia amygdalina* del, in experimental rabbits. Plant Med Res 1:6–10

Akande IS, Samuel TA, Agbazue U, Olowolagba BL (2011) Comparative proximate analysis of ethanolic and water extracts of *Cymbopogon citratus* (lemon grass) and fourtea brands. Plant Sci Res 3(4):29–35

Anderson JW, Smith BM, Gustafson NJ (1994) Health benefits and practical aspects of high fibre diets. Am J Clin Nutr 59:1242S–1247S

Anon AC (1992) Macropaedia. In: The new encyclopedia britannica, vol 3, 15th edn. Encyclopaedia Britannica, Chicago, p 185

Aregheore EM (1998) A note on the nutritive value of dry ripe plantain peels as a replacement of maize for goats. J Anim Feed Sci 7(1):55–62

Aregheore EMK, Makkar HPS, Becker K (1998) Feed value of some browse plants from the central zone of Delta State. Nig Trop Sci 38:97–104

Atawodi SE, Shehu H (2010) Anti-trypanosomal property of some extracts of different parts of *Moringa oleifera,* lam. Electron J Biol 6(1):19–24

Atawodi SE, Atawodi JC, Idakwo P, Pfundstein B, Haubner R, Wurtele G, Spiegelhalder B, Bartsch H, Owen RW (2009a) Evaluation of the polyphenol composition and antioxidant activity of African variety of *Dacryodes edulis* (G.Don) H.J Lam Fruit. J Med Food 12(6):1321–1325. doi:10.1089/jmf.2008.0215

Atawodi SE, Atawodi JC, Pala Y, Idakwo P (2009b) Assessment of the polyphenol profile and antioxidant properties of leaves, stem and root barks of *Khaya senegalensis* (Desv.) Electron J Biol 5(4):80–84

Atawodi SE, Atawodi JC, Idakwo GA, Pfundstein B, Haubner R, Wurtele G, Spiegelhalder B, Bartsch H, Owen RW (2010) Evaluation of the polyphenol content and antioxidant properties of methanol extracts of the leaves, stem and root barks of *Moringa oleifera*, lam. J Med Foods 13(3):710–716

Baba MD, Yakubu G, Yelwa JM, Haruna L (2015) Costs and returns of Moringa (*Moringa oleifera*) production in Zara local government area of Kebbi state, Nigeria. N Y Sci J 8(1):36–40

Bamishaiye EI, Bamishaiye OM (2011) Tiger nut: as a plant, its derivatives and benefits. Afr J Food Agric Nutr and Dev (AJFAND) Online, 11(5):5157 September 2011

Belewu MA, Abodunrin AO (2006) Preparation of Kunnu from unexploited rich food source: tiger nut (*Cyperus esculentus*). World J Dairy Food Sci 1:19–21

Biggelaar CD, Gold MA (1996) Development of utility and location indices for classifying agroforestry species: the case of Rwanda. Agrofor Syst 34:229–246

Blanco MM, Costa CA, Freire AO, Santos JG, Costa M (2009) Neurobehavioural effects of essential oils of *Cymbopogon citratus* in mice. Phytomedicine 16:265–270

Burkill HM (1985) The useful plants of West tropical Africa, vol 3, 2nd edn. Royal Botanical Gardens, Kew

Carianne dB (2005) Organic lemongrass, a guide for smallholders. EPOPA (Export Promotion of Organic Products from Africa), Bennekom, pp 1–27

Chweya JA, Eyzaguirre P (1999) The biodiversity of traditional leafy vegetables. IPGRI, Rome, p 182

Dansi A, Adjatin A, Adoukonou-Sagbadja H, Akpagana K (2008) Production and traditional seed conservation of leafy vegetables in Benin rural areas. Bulletin de la Recherche Agronomique du Bénin, Numéro 59:59–69

Dike MC (2010) Proximate, phytochemical and nutrient compositions of some fruits, seeds and leaves of some plant species at Umudike, Nigeria. ARPN J Agric Biol Sci 5(1):7–16

Directorate [Communication Services] (2012) Essential oil crops: production guidelines for lemon grass. Department of Agriculture, Forestry and Fisheries, Republic of South Africa. www.daff.gov.za/publications. Accessed 11 Apr 2015

Dyer AR (2009) The ecology of chufa (*Cyperus esculentus sativus*). Department of Biology & Geology University of South Carolina Aiken

Ekpenyong CE, Akpan EE, Daniel NE (2014) Phytochemical constituents, therapeutic applications and toxicological profile of *Cymbopogon citratus* Stapf (DC) leaf extract. J Pharmacogn Phytochem 3(1):133–141

Ene AC, Atawodi SE, Ameh DA, Nnamani CN, Apeh YEO (2009) Antitrypanosomal effects of petroleum ether, chloroform and methanol extracts of *Artemisia maciverae* Linn. Ind J Exp Biol 47:981–986

FAO (1988) Traditional food plants: a resource book for promoting the exploitation and consumption of food plants in arid, semi-arid and sub-humid lands of Eastern Africa, vol 42. FAO Food and Nutrition Paper, Rome, pp 239–242

FAO (1998). *Cyperus esculentus*. AGRIS. www.agris.fao.org. Accessed 26 May 2015

FAO/WHO (1985) Energy and protein requirement. report of a joint FAO/WHO/UNU expert consultation, WHO Technical Report Series No. 724, Geneva

Foidl N, Makkar HPS, Becker K (2016) Planting and care instructions – for your moringa seeds and trees Moringa potential and info. http://www.moringaseeds.co.za/Planting3

Getaneh S, Girma Z (2014) An ethnobotanical study of medicinal plants in Debre Libanos Wereda, Central Ethiopia. Afr J Plant Sci 8(7):366–379

Halabi MF, Sheikh BY (2014) Anti-proliferative effect and phytochemical analysis of *Cymbopogon citratus* extract. Bio Med Res Intl, Article ID 906239. http://dx.doi.org/10.1155/2014/906239

Hamowia AM, Saffaf AM (1994) Pharmacological studies on *Vernonia amygdalina* (Del) and *tithonia diversifolia* (Gray). Vet Med J2:91–97

Ijeh II, Akomas SC (2005) Effect of oral administration of ethanolic extract of *Vernonia amygdalina* on the exocrine functions in weaning rabbits. In: Proceedings of 30th conference Nigerian Society for Animal Production 30: 109–112

Ijeh II, Ejike CC (2011) Current perspectives on the medicinal potentials of *Vernonia amygdalina* Del. J Med Plant Res 5(7):1051–1061

Izevbigie EB (2003) Discovery of water-soluble anticancer agents (edotides) from a vegetable found in Benin City, Nigeria. Exp Biol Med 228:293–298

Karkala M, Bhushan B (2014) Review on pharmacological activity of *Cymbopogon citratus*. Int J Herb Med 1(6):5–7

Kayode J (2004) Eco-physiological and conservation studies on *Vernonia amygdalina* in Ekiti State, *Nigeria*. Pak J Sci Ind Res 47:227–230

Kemka-Evans CI, Okoli B (2013) Chromosome numbers and karyotype in three species of the genus Vernonia Schreber in Southern Nigerian. Afr J Plant Sci 7(11):538–542

Khalafalla MM, Abdellatef E, Daffalla HM, Nassrallah AA, Aboul-Enein KM, Lightfoot David A, Cocchetto A, El-Shemy HA (2009) Antileukemia activity from root cultures of *Vernonia amygdalina*. J Med Plants Res 3(8):556–562

Khawaja TM, Tahira M, Ikram UH (2010) *Moringa oleifera*: a natural gift-a review. J Pharm Sci Res 2(11):775–781

Kubmarawa D, Khan ME, Punah AM, Hassan M (2008) Phytochemical screening and antimicrobial efficacy of extracts from *Khaya senegalensis* against human pathogenic bacteria. Afr J Biotechnol 7(24):4563–4566

Liman ML, Atawodi SE (2015) Hepatoprotective and nephroprotective effects of methanol extracts of different parts of *Tamarindus Indica* linn in rats following acute and chronic carbon tetrachloride intoxication. Ann Res Rev Biol 5(2):109–123

Mann A, Gbate M, Umar AN (2003) Medicinal and economic plants of Nupeland, 1st edn. Jube-Evans Books and Publications, Bida, p 158

Mpondo ME, Dibong SD (2011) Medicinal plant knowledge of ethnic groups in Douala town, Cameroon. Am J Food Nutr 1(4):178–184

Ndhlala AR, Mulaudzi R, Ncube B, Abdelgadir HA, du Plooy CP, Staden JV (2014) Antioxidant, antimicrobial and phytochemical variations in thirteen *Moringaoleifera* lam. cultivars. Molecules 19:10480–10494

Nikiema A, Pasternak D (2008) *Khaya senegalensis* (Desr.) A. Juss. [Internet] Record from PROTA4U. http://www.prota4u.org/search.asp. Accessed 24 May 2015

Nmor EI, Odeh OH (2013) Effect of tillage on the yield of bitter leaf *(Vernonia amydalina)* in Southern Nigeria. J Agric Vet Sci 5(1):159–164

Ojiako OA, Nwanjo HU (2006) Is *Vernonia amygdalina* hepatotoxic or hepatoprotective? Response from biochemical and toxicity studies in rats. Afr J Biotechnol 5:1648–1651

Okafor JNC, Mordi JI, Ozumba AU, Solomon HM, Olatunji O (2003) Preliminary studies on the characterization of contaminants in tiger nut (Yellow variety). In: Proceedings of 27th annual Nigerian Institute of Food Science and Technology (NIFST) conference, 13–17 October, pp 210–211

Oloyede GK, Sunday FA, Charles CN (2014) Antioxidant and toxicity screening of extracts obtained from *Cyperus esculentus*. Academ Arena 6(1):77–83

Onu A, Saidu Y, Ladan MJ, Bilbis LS, Aliero AA, Sahabi SM (2013) Effect of aqueous stem bark extract of *Khaya senegalensis* on some biochemical, haematological and histopathological parameters of rats. J Toxicol. http://dx.doi.org/10.1155/2013/803835

Onwuka CFI, Akinsoyinu AO, Tewe OO (1989) Feed value of some Nigerian browse plants: chemical composition and *in vitro* digestibility. East Afr Agric For J 54:157–163

Orwa C, Mutua A, Kindt R, Jamnadars R, Anthony S (2009) Agroforestree database: a tree reference and selection guide version 4.0. http://www.worldagroforestry.org/sites/treedbs/treedatabases.asp. Accessed 5 May 2015

Osagie AU, Eka OU (1998) Nutritional quality of plant foods. Post-harvest Research Unit 246–249

Osagie AU, Okoye WI, Oluwayose BO, Dawodu OA (1986) Chemical quality parameters and fatty acid position of oils of under exploited tropical seeds. Nig J Appl Sci 4:151–162

Paliwal R, Sharma V, Pracheta V (2011) A Review on horse radish tree (*Moringa oleifera*): a multipurpose tree with high economic and commercial importance. Asian J Biotech 3(4):317–328

Parrotta JA (1993) *Moringa oleifera* Lam. – Reseda, horseradish tree, SO-ITF-SM-61, pp 1–6

Radovich T (2013) Farm and forestry production and marketing profile for moringa (*Moringaoleifera*). Specialty Crops for Pacific Island Agroforestry http://agroforestry.net/scps

Raza M, Shah SM (2011) Conservation and sustainable use of medicinal plants in Baltistan. Grace Association, Pakistan

Sale M, De N, Doughari JH, Pukuma MS (2008) In vitro assessment of antibacterial activity of bark extracts of *Khaya senegalensis*. Afr J Biotechnol 7(19):3443–3446

Soares MO, Alves RC, Pires PC, Beatriz M, Oliveira PP, Vinha AF (2013) Angolan *Cymbopogoncitratus* used for therapeutic benefits: nutritional composition and influence of solvents in phytochemicals content and antioxidant activity of leaf extracts. Food Chem Toxicol 60:413–418

Stevens GC, Baiyeri KP, Akinnnagbe O (2013) Ethno-medicinal and culinary uses of *Moringa oleifera* Lam. in Nigeria. J Med Plants Res 7(13):799–804

Tadesse A, Gebre-Hiwot A, Asres K, Djote M, Frommel D (2008) The *in vitro* activity of Vernonia. Int J Environ Res Public Health 5(5):341

Tetteh JP, Ofori E (1998) A baseline survey of tiger nut (*Cyperus esculentus)* production in the Kwahu South District of Ghana. Ghana J Agric Sci 31:211–216

Ucheck FF (2004) *Vernonia amygdalina* Delile. http://www.prota4u.org/search.asp. Accessed 1 Apr 2015

Wakirwa JH, Idris S, Madu SJ, Dibal M, Malgwi T (2013) Assessment of the in-vitro antimicrobial potential of *Khaya senegalensis* ethanol leaf extract. J Chem Pharm Res 5(7):182–186

Wijenayake T (2014) Khaya: the wonder tree for commercial cultivation. http://www.ft.lk/article/337136/Khaya. Accessed 5 May 2016

Yedjou C, Izevbigie E, Tchounwou P (2008) Preclinical assessment of *Vernonia amygdalina* leaf extracts as DNA Damaging Anti-Cancer Agent in the management of breast cancer. Int J Environ Res Public Health 5(5):337–341

Zaku SG, Emmanuel S, Tukur AA, Kabir A (2015) *Moringa oleifera*: an underutilized tree in Nigeria with amazing versatility: a review. Afr J Food Sci 9(9):456–461

# The Halophytic Genus *Zygophyllum* and *Nitraria* from North Africa: A Phytochemical and Pharmacological Overview

**Soumaya Bourgou, Wided Megdiche, and Riadh Ksouri**

**Abstract** In arid and semi-arid regions, plants are often subjected to severe environmental conditions such as salinity, drought and extreme temperatures. Halophyte species are salt-tolerant plants which are characterized by high physiological plasticity and are able to thrive under stressful conditions by developing adaptive responses including the synthesis of several bioactive molecules such as terpenoids, phenolic compounds, sterols, polysaccharides, carotenoids, vitamins and glycosides. These primary and secondary metabolites possess potent biological activities including antioxidant, antiviral, antimicrobial, anti-inflammatory, and anticancer ones and epidemiological research suggests that these compounds may help prevent several chronic diseases, such as diabetes, chronic inflammation, atherosclerosis and cardiovascular disorders. Because of their high content on these bioactive metabolites, halophytes may offer a valuable source of healthy products and could be used as functional foods and nutraceuticals. This chapter deals with the ethnopharmacological uses of two famous halohyte genus; *Zygophyllum* and *Nitraria* in traditional medicine and reviews recent investigations on their phytochemical compositions and biological activities.

**Keywords** Halophytes • *Zygophyllum* • *Nitraria* • Secondary metabolites • Biological activities

## 1 Introduction

Halophytic plants are salt-tolerant species able to thrive under stressful conditions and to survive and to reproduce in environments where salt concentration reaches or exceed sea water salinity levels. There are more than 2600 halophytic species known

S. Bourgou • W. Megdiche • R. Ksouri (✉)
Laboratoire des Plantes Aromatiques et Médicinales, Centre de Biotechnologie de Borj-Cédria (CBBC), BP 901, Hammam-Lif 2050, Tunisia
e-mail: ksouri.riadh@gmail.com



M. Neffati et al. (eds.), *Medicinal and Aromatic Plants of the World - Africa Volume 3*, Medicinal and Aromatic Plants of the World,
DOI 10.1007/978-94-024-1120-1_13

worldwide to possess salinity tolerance including edible, fodder, fuel, medicine, chemicals, and ornamentals plants (Abdelly et al. 2011). Halophytes are potentially useful for ecological applications in arid and semi-arid regions, such as soil desalinisation, dune fixation, phyto-remediation, landscaping and ornamentation (Abdelly et al. 2011).

Halophytes have been collected since ancient times for alimentation and for their medicinal qualities and until now, a great number of medicinal halophytes are used to treat various diseases, infections and aging processes particularly in the rural areas, where the traditional folk medicine remains a major source to cure minor ailments. Currently, an increasing interest is granted to halophytic species because of their high content in bioactive secondary molecules with a big economic interest (Ksouri et al. 2012).

Compared to the glycophytic species, halophytes contain much higher amounts of secondary metabolites or even restricted to halophytic species which could be used in various application fields such as in the Agri-food, pharmaceutical and cosmetic industries. The biological function of bioactive compounds might be single, or combined with others present in whole plant extracts mixtures leading to synergistic effect and depending on its chemical composition and efficacy (Subhashini et al. 2013). This contribution focuses on the ethno-pharmacological uses of two famous and interesting genus of halophytes (*Zygophyllaceae* family) in traditional medicine and reviews recent investigations on their chemical composition and biological activities.

## 2 *Zygophyllum* Genus

### *Uses in Traditional North Africa Medicine*

*Zygophyllum* genus comprises several spontaneous species distributed in arid regions of Mediterranean basin, in southern, northern and north-eastern Africa (White 1983) and widely used as natural remedies in folk medicine. *Zygophyllum* species are halophyte plants widely found along saline and gypsophile areas (Chaieb and Boukhris 1998). One of the most studied species is *Z. album* L. which is an edible succulent dwarf shrub of pharmacological interest. This halophyte has long been used in traditional Africa medicine against rheumatism, gout and asthma. The aerial parts of the plant are used as an anti-diabetic agent. It is also used as diuretic and local anaesthetic and antihistaminic. Moreover, *Z. album* has been used as herbal medicine in southern Tunisia to treat diabetes mellitus (Mnafgui et al. 2012). In Algeria, *Z. album* L. is employed against aches, diabetes, dermatitis, spasms, and dysmenorrhoea and for calm thirst as well as for wound care and the treatment of dental caries (Tigrine-Kordjani et al. 2006; Ayad et al. 2012). Some Bedouin used it as hay or added it to the dry ration (Daoud and Al-Rawi 1985). In Morocco, the endemic *Zygophyllum gaetulum* is used in folk medicine for the treatments of diabetes and stomach and liver pain. It has antispasmodic, anti-inflammatory,

**Table 1** Chemical constituent of some species of *Zygophyllum* from North Africa

| Species | Origin | Plant part | Chemical constituents | References |
|---|---|---|---|---|
| *Z. album* | Egypt | shoots | flavonol 3-O-rutinosides, isorhamnetin 3-O-glucosides | Hussein et al. (2011) |
| *Z. album* | Tunisia | shoots | malvidin 3-rhamnoside, quercetin-3-sulphate, isorhamnetin-3-O-rutinoside, triterpenes saponinssterols | Megdiche Ksouri et al. (2013) |
| *Z. album* | Algeria | shoots | essential oil | Tigrine-Kordjani et al. (2006) |
| *Z. berenicense* | | | flavonol 3-O-rutinosides | Saleh and El-Hadidi (1977) |
| *Z. corntum* | Algeria | whole plant | isorhamnetin-3-O-rutinoside β-sitosterol | Ayad et al. (2012) |
| *Z. coccineum* | Egypt | roots | flavonol 3-O-rutinosides, triterpenes saponins | Saleh and El-Hadidi (1977) and Pöllmanns et al. (1997) |
| *Z. dumosum* | Egypt | roots | triterpenes saponins | Pöllmanns et al. (1997) |
| *Z. simplex* | Egypt | shoots | flavonol 3,7-diglycosides, gentisic acid, gentisic acid 5-O-α-rhamnopyranoside | Hussein et al. (2011) |
| *Z. gaetlum* | Maroc | roots | zygophyloside N, quinovic acid glycosides | Aquino et al. (2001) |

anti-diarrhoeal and anti-eczema effects (Bellakhdar et al. 1991; Ait El-Cadi et al. 2008). In addition, *Z. fabago* roots are also used as an antispasmodic remedy as an alternative to the aerial parts in Moroccan folk medicine (Aquino et al. 2001). The leaves and the seeds of *Zygophyllum simplex* are used in Egypt as anthelmintic and applied to eyes in cases of Ophthalmia and Leucoma (Mahmoud and Gairola 2013). *Z. coccineum* is the most widespread *Zygophyllum* species in Egypt, common in the desert and the saline soils. It is used as a drug against rheumatism, gout and hypertension (Pöllmanns et al. 1997).

## Phytochemical Composition

Several studies reported the separation and identification of flavonoids from some *Zygophyllum* species (Table 1). For instance, Quercetin-3-*O*-rutinoside, isorhamnetin-3-*O*-β-glucopyranoside, isorhamnetin-3-*O*-rutinoside were detected by Saleh and El-Hadidi (1977) in Z. *decumbens*. Recently, Hussein et al. (2011) from Egypt isolated from 70% methanol extract of *Z. album* several flavonoids: quercetin-3-*O*-rutinoside [Quercetin-3-*O*-α-rhamnopyranosyl-(1 → 6)-O-β-glucopyranoside], isorhamnetin-3-*O*-β-galactopyranoside, isorhamnetin-3-*O*-β-glucopyranoside, isorhamnetin-3-*O*-rutinoside [isorhamnetin-3-*O*-α-rhamnopyranosyl-

(1/6)-*O*-β-glucopyranoside], isorhamnetin 3-*O*-robinoside [isorhamnetin-3-*O*-α-rhamnopyranosyl-(1/6)-O-β-galactopyranoside]. These compounds were identified by Rf values, colour reactions, UV spectrophotometry, 1H- and 13C-NMR spectrometery and co-chromatography. Previous investigation of *Z. album* revealed others compounds such as zygophyllin, β-sitosterol-β-D-glucopyranoside, carbohydrates, tannins, lactones, protein/amino acids, saponins, terpene and flavonoid glycosides (Hani et al. 1995; Atta and Mouneir 2004). In *Z. simplex* from Egypt, quercetin, quercetin-3,7-di-*O*-β-glucopyranoside, quercetin-3,7-di-*O*-β-glucopyranoside, isorhamnetin-3-*O*-β-glucopyranoside-7-O-α-rhamnopyranoside, gentisic acid and gentisic acid 5-O-α-rhamnopyranoside were purified by Hussein et al. (2011). Isorhamnetin-3-*O*-rutinoside was found to be a chemotaxonomic marker in the genus of *Zygophyllum*. This compound was reported as one of the major phenolic compounds in the aerial part of *Z. album* grown in Tunisia by Megdiche et al. (2013) along with two other flavonoid molecules: malvidin 3-rhamnoside and quercitin-3-sulphate. Interestingly, these authors showed among 10 compounds identified by liquid chromatography coupled to electrospray time-of-flight mass spectrometry (LC-TOF/MS) such as saponins, flavonoids and sterols, five were reported for the first time in *Z. album* shoot extracts. Isorhamnetin-3-*O*-rutinoside was reported also by Ayad et al. (2012) from *Z. cornutum* endemic to Algeria. *Zygophylum* is a saponin-rich genus (Table 1). In *Z. gaetulum* from Maroc, fractionation of methanolic extract led to the isolation of the saponin zygophyloside N [3-β-*O*-α-l-rhamnopyranosyl-(1-2)-O-α-l-arabinopyranosyl-(1->2)-β-d-glucopyranosyl)oxy]-ursan-20β,28-olide], together with quinovic acid glycosides: zygophyloside G (2), zygophyloside E and 4, 3-*O*-β-D-glucopyranosyl quinovic acid 28-β-d-glucopyranosyl ester (Aquino et al. 2001). In addition, zygophyloside G (3-*O*-[β-D-2-*O*-sulphonylglucopyranosyl]-quinovic acid-28-*O*-[β-D-glucopyranosyl] ester have been isolated from the roots of *Z. dumosum* (Ahmad et al. 1993), *Z. coccineum* (Pöllmanns et al. 1997) and *Z. album* (Megdiche Ksouri et al. 2013). Pöllmanns et al. (1997) investigated the crude methanolic saponin extracts from Egyptian *Z. coccineum* and *Z. dumosum* roots and purified triterpenes saponins zygophyloside H, 3-*0*-[β-D-glucopyranosyl]-quinovic acid-28-*0*-[β-D-glucopyranosyl] ester as well as zygophyloside K. Recently, zygophyloside K have been identified by Megdiche et al. (2013) in Tunisian *Z. album*. Recently, phytochemical analysis of *Z. coccineum* have revealed that the major secondary metabolites are a class of quinovic acid compounds belonging to the ursane-type triterpene saponins (Amin et al. 2011). Tigrine-Kordjani et al. (2011) extracted the essential oil from *Zygophyllum album* L., growing in Ouargla Sahara region (Algeria) by hydro-distillation. They reported a strong musk flavour and (*E*)-β-damascenone (11.8%) chemotype. Other major compounds of *Z. album* essential oil were characterized δ-decalactone (7.8%), α-inone (7.1%), 3-nonen-2-one (4.6%), safranal (3.4%),(*E*)-2-hexen-1-ol (3.8%), (*E*)-caryophyllene (1.6%) and *cis*-linalool oxide (1.3%). On the other hand, fractionation of the ethyl acetate extract of *Z. cornutum* resulted in the isolation of β-sitosterol (Ayad et al. 2012 ). In *Z. album* from Tunisia, Megdiche et al. (2013) identified 3-*O*-[β-D-glucopyranosyl]-β-sitosterol and β-sitosterol-β-D-glucoside by comparison of their spectral and physical data with those of known compounds. 3-*O*-[β-D-glucopyranosyl]-β-sitosterol seems to be a chemotaxonomic marker in the genus of *Zygophyllum* (Ahmad et al. 1993; Hassanean and Desoky 1992; Hassanean et al. 1989; Hassanean et al. 1989).

## *Biological Activities*

The most common complementary and alternative medicine modalities are based in food supplements and herbal therapies. In this regard, several plant extracts of *Zygophyllaceae* family and their isolated bioactive compounds were shown to be effective in prevention and treatment of some diseases.

(i) **Antioxidant capacity** A number of pharmacological and clinical studies have been reported from different *Zygophylum* species which proves the potentialities and the importance of these plants. Megdiche et al. (2013) evaluated the antioxidant activity of the hexane, dichloromethane and methanol extracts of Tunisian *Z. album* shoots *in vitro* using the ORAC assay, as well as, *ex-vivo* using a cellular based-assay using DCFH-DH, a useful indicator of reactive oxygen species (ROS). The most polar extract (methanol) had a strong antioxidant ORAC capacity (1.19 ± 0.07 μmol Trolox/mg) and the dichloromethane extract proved the utmost antioxidant activity in cell (WS1) based-assay ($IC_{50}$ = 57 μg/ml). Additionally, El Ghoul et al. (2012) have also observed that the administration of *Z. album* aqueous extract to diabetic mice induced a diminution in the lipid peroxidation in the liver and the pancreas in a dose dependent manner, which was connected to an enhancement of the enzymatic (GSH) and non-enzymatic (vitamin C) anti-oxidative scavengers.

(ii) **Anti-diabetic activity** Aqueous and ethanol extracts of *Z. album* from Tunisia exerts anti-diabetic and anti-hyper-cholesterolemic activities. Oral administration of the extracts reduced the blood glucose, total cholesterol (TC), triglycerides (TG), low-density lipoprotein (LDL) and very low-density lipoprotein (VLDL) levels in treated diabetic mice in comparison with control group. *Z. album* extracts prevents the diabetic induced MDA levels via the enhancement of the tissue glutathione reductase (GSH), blood vitamin C levels, superoxide dismutase (SOD), catalase (CAT), and glutathione peroxidase (GPx) (El Ghoul et al. 2011, 2012). The ability of *Z. gaetulum* leaves from Morocco to lower the blood glucose was studied in several patients with non-insulin-dependent diabetes mellitus (Jaouhari et al. 1999). *Z. gaetulum* infusion possesses significant oral anti-hyperglycaemic activity by lowering plasma glucose levels. The anti-hyperglycaemic effect was seen both after acute and sub-acute administration of *Z. gaetulum* in non-insulin-dependent diabetics patients. Moreover, the toxicity study shows that large dose of the extract can be administered without hazard.

(iii) **Anti-diarrhoeal activity** Ait El Cadi et al. (2008) have shown that *Z. gaetulum* infusion possess anti-diarrhoeal activity in mice. The extract was prepared according to traditional methods and the study was carried out *in vitro* using chamber. Atta and Mouneir (2004) investigated the antidiarrhoeal activity of six Egyptian medicinal plant extracts and their effect on motility of isolated rabbit's duodenum. The study showed that oral administration of methanol extract obtained from the shoots of *Z. album* grown in Egypt induced anti-diarrhoeal effect against castor oil-induced diarrhoea. The anti-diarrhoeal effect of the plants could be attributed, at least partially to their content of tannins which form protein tannate that reduces intestinal secretion.

(iv) **Anti-inflammatory ability** In order to approve the traditional use of *Z. gaetulum* in Moroccan folklore as anti-inflammatory agent, Ait El Cadi et al. (2012) compared the anti-inflammatory effect of ethanolic extract obtained in soxhlet apparatus with aqueous extract obtained by infusion, and the anti-inflammatory activity was estimated by measuring the oedema induced by carragenin. Extracts of *Z. gaetulum* reduced the increase of the paw volume which suggests that the aqueous and ethanolic extracts could be used scientifically in the treatment of inflammation. Moreover, Megdiche et al. (2013) evaluated after fractionation by successive soxhlet extraction using solvents with increasing polarity, the anti-inflammatory capacity of the extracts obtained from *Z. album* grown in Tunisia using LPS-induced RAW 264.7 macrophages. The results showed that methanol extract displayed the best inhibition of NO release, in a dose-dependent manner. Thus, inhibition was ranging from 19.4 (20 μg/ml) to 84.8 % (160 μg/ml). In comparison, L-NAME, as positive control, inhibited NO release by 26.2%. Such important anti-inflammatory activity of could be due essentially to the richness of extracts on phenolic compounds including the quercetin 3-sulphate and isorhamnetin-3-O-rutinoside as well as triterpenoid saponins.

(v) **Anticancer power** In *Z. album* from Tunisia, Megdiche et al. (2013) evaluated the anti-proliferative capacity of the methanol, ethyl acetate and water fractions obtained from shoot crude extract. The cytotoxicity of each extract was assessed against Human colon carcinoma (DLD-1) and lung carcinoma (A-549) cell lines, as well as, health skin fibroblast cell line (WS1). Result showed that among the three shoot fractions, the dichloromethane extract was significantly active against the two carcinoma cells (A-549 and DLD-1), with important $IC_{50}$ values, respectively of 37 and 48 μg/ml. In fact, *Z. album* had an influence on tumour cell viability and targeted colon and lung carcinoma cell lines which could be attributed to the powerful cytotoxic compounds in dichloromethane fraction mainly triterpenes saponins and sterols. The same extract with dichloromethane was not significantly cytotoxic against fit human skin fibroblast cell (WS1). In addition, trypan blue assay was used for the evaluation anticancer activity of ethanolic and aqueous Egyptian *Z. simplex* extracts against Ehrlich Ascites Carcinoma Cells (EACC). The results displayed that the extracts gave anticancer activity more than 60% inhibition of proliferation (Aboul-Enein et al. 2012).

## 3 *Nitraria* genus

### *Uses in Traditional North Africa Medicine*

*Nitraria* species are widely distributed around the world; they are well adapted to arid climates, and some of them grow in the deserts and salt marshes of Africa, in northern and occidental parts of the Sahara desert, and in Mauritania (*N. tridendata* and *N. retusa*). These species can also be found in Algeria, Mauritania, Senegal, and Saudi Arabia. A decoction of fresh leaves of *N. retusa* is used in Morocco in case of poisoning, upset stomach, ulcers, gastritis, enteritis, heartburn, colitis and colonic abdominal pain (Bellakhdar 1997). In Tunisia, it is widespread in central and south parts. This salt-tolerant and drought resistant shrub grows along shallow and hummocks on saline grounds near the coastal areas and produces fleshy red fruits from which a tasty and refreshing juice may be extracted. *Nitraria retusa* is known in Tunisia as "Ghardaq". Its leaves and twigs are occasionally grazed by sheep, goats and camels (Le Houérou 1980). The woods are used as fuel by the local inhabitants. The sweet drupes are edible and are used to prepare drinks (Daoud and Al-Rawi 1985; Tackholm 1974). Fruits of *Nitraria* species have been used as nutritional food and traditional herb for the treatment of hypertension and abnormal menstruation (Hadj Salem et al. 2011). The leaves serve as supplement for the tea and are used as poultice (Boubaker et al. 2011). The ashes of this species have the ability to remove fluids of infected wounds. The leaves infusion (used as cataplasm) has anti-inflammatory property (Chaieb and Boukhris 1998). The ashes of this species have the ability to remove fluids of infected wounds. In Egypt, it is known locally as Ghardaq and used by the Bedouins as a source of fuel and the sweet drupes are edible (Halim Ahmed et al. 1995).

### *Phytochemical Composition*

The survey involving *N. retusa* indicated that flavonol glycosides are predominant (Zar Kalai et al. 2014; Saleh and El-Hadidi 1977). From the leaves and young stems of *N. retusa* grown in Egypt, Halim Ahmed et al. (1995) isolated and fully characterized six flavonol glycosides: isorhamnetin 3-*O*-4-rhamgalactosylrobinobioside, isorhamnetin 3-robinobioside, isorhamnetin 3-rutinoside, isorhamnetin 3-galactoside, isorhamnetin 3-glucoside, isorhamnetin 3-xylosylrobinobioside and free isorhamnetin. Hussein et al. (2009) isolated and identified twelve flavonoid compounds from the hole plant *N. retusa* ethanol extract: (1) apigenin-6-C-arabinopyranoside-8-C-glucopyranoside, (2) apigenin-6,8-di-C-glucopyranoside, (3) luteolin-7-*O*-glucopyranoside, (4) luteolin-7-*O*-xylopyranosylglucopyranoside, (5) kaempferol-3-*O*-xylopyranosyl-(1 → 2)-[O-rhamnopyranosyl-(1 → 6)]-*O*-galactopyranoside, (6) free isorhamnetin, (7) isorhamnetin-3-O-galactopyranoside, (8) isorhamnetin-3-O-glucopyranoside, (9) isorhamnetin-3-O-rutinopyranoside,

(10) isorhamnetin-3-O-robinopyranoside, (11) isorhamnetin-3-O-$\beta$-xylopyranosyl-(1 → 2) [O-rhamnopyranosyl-(1 → 6)]-O-galactopyranoside and (12) isorhamnetin-3-O-rhamnopyranoside-7-O-glucopyranoside. Recently, Hadj Salem et al. (2011) identified by LC–ESI-$MS^n$, four major flavonoids in water and methanol extracts/fractions of the leaves of *N. retusa* collected from Tunisia: isorhamnetin, isorhamnetin-3-*O*-glucoside, isorhamnetin-3-*O*-rutinoside and isorhamnetin-3-*O*-robinobioside. The chromatographic profile of Tunisian *N. retusa* extract, based on the HPLC analysis, acquired at 254 nm displayed three main components: quercetin, isorhamnetin-3-O-glucoside, and isorhamnetin-3-O-rutinoside (Zar Kalai et al. 2013). In addition, the quantification of these phenolic compounds were: 22.34 µg/ml (quercetin), 331.42 µg/ml (isorhamnetin-3-O-glucoside), and 24.91 µg/ml (isorhamnetin-3-O-rutinoside). Using GC/MS analysis, Mohamed et al. (2014) detected eugenol, germacrene D, tetratetracontane Tetratetracontane Hentriacontane as the major compounds in petroleum ether extracts of *N. retusa* shoots collected from various regions in Egypt.

## *Biological Activities*

(i) **Antioxidant capacity** Hadj Salem et al. (2011) investigated the *in vitro* antioxidant activity of four Tunisian *N. retusa* extracts obtained by successive extractions performed with a Soxhlet apparatus with increasing polarity solvents (hexane, chloroform, ethyl acetate and methanol) as well as their fractions. The authors showed that shoot extracts exhibit high antioxidant and radical scavenging and xanthine oxidase inhibition activities which derive from the richness of extracts on isorhamnetin-based flavonoids (isorhamnetin, isorhamnetin-3-*O*-glucoside, isorhamnetin-3-*O*-rutinoside and isorhamnetin-3-*O*-robinobioside). Besides, Boubaker et al. (2010) have demonstrated an antioxidant protective role of hexane, and chloroform extracts prepared from the leaves of Tunisian *N. retusa* against DNA strand scission induced by hydroxyl radical. The addition of chloroform extract induced a dose dependant DNA protection to the damage of native super-coiled circular DNA. Whereas, Hexane extract showed a low inhibiting activity against hydroxyl radicals. The scavenging potential for hydroxyl radicals of chloroform extract was found to be correlated to its polyphenol and sterol contents. When comparing the antioxidant activity of various solvent extracts, Chaâbane et al. (2014) demonstrated that the water aqueous extract possess the best antioxidant activity exhibiting high antiradical power and remarkable inhibitory effect on lipid oxidation. In fact, the last extract displayed the highest amounts of phenolics, flavonoids and anthocyanins.

(ii) **Anti-obesity effect** Obesity is a problem of growing importance, being an essential risk factor for type-2 diabetes, when accompanied with other known complications. Therefore, it has been an important therapeutic goal to reduce the risk of type-2 diabetes through weight management. In recent study, Zar Kalai et al. (2013) investigated the anti-obesity effects of *Nitraria retusa* ethanol extract in 3T3-L1 cells using different doses (25, 50, 100, 200, and 400 $\mu$g/mL). Results showed that ethanol extract could inhibit the lipid droplet accumulation compared to untreated cells, in dose dependent manner. The triglyceride accumulation significantly decreased to 42.70 ± 2.10% at 400 $\mu$g/mL. In fact, they noticed that *N. retusa* treatment in 3T3-L1 was accompanied by modulation of cell hypertrophy rather than cell hyperplasia. Moreover, Zar Kalai et al. (2014) studied the prevention and controlling weight gain effect of Tunisian *N. retusa* ethanolic extract using high fat diet-induced obesity experiment in C57B6J/L mice. The results showed that oral administration of ethanolic *N. retusa* extract significantly suppress increases in body and fat mass weight, decreases triglycerides and LDL-cholesterol levels and enhances gene expression related to lipid homeostasis in liver showing anti-obesity actions. The findings, indicate that *N. retusa* possesses potential anti-obesity effects in BKS.Cg-Dock7m+/+ Leprdb/J model mice and may relieve obesity-related symptoms including hyperlipidemia through modulating the lipolysis–lipogenesis balance.

(iii) **Antimutagenic activity** Chloroform extract obtained from Tunisian *N. retusa* showed an effective antimutagenic effect, preventing mutations induced by various oxidants in *Salmonella typhimurium* TA102 and TA 104 strains (Boubaker et al. 2010). The authors found that this antimutagenic capacity was associated to a strong antioxidant power of the extract and could be linked to the high flavonoids content since they provide strong nucleophilic centers, which enable them to react with electrophilic mutagens and form adducts that may result in the prevention of genotoxic damages. *N. retusa* hexane extract exhibited also antimutagenic effect which has been found to be correlated to high sterol content. The authors suggest that consumption of *N. retusa* could be an alternative for reducing genotoxic damage induced by free radicals.

(iv) **Anticancer activity** Hadj Salem et al. (2011) have observed that *N. retusa* leaves extracts and their fractions, especially ethyl acetate fraction, were effective towards Caco-2 cancer cells. Moreover, a correlation between the anti-proliferative activities of extracts and their flavonoid content was established; the most concentrated extracts/fractions were the most active. In addition, when tested as references, isorhamnetin and isorhamnetin-3-*O*-rutinoside exhibited high anti-proliferative activities on Caco2 cells. According to Boubaker et al. (2011) the leaf extracts of *N. Retusa* have a significant anti-proliferative effect on the human chronic myelogenous erythroleukaemia (K562). They induced in a time-dependent manner the apoptosis in cancerous cell line. This result was confirmed by ladder DNA fragmentation profile and PARP cleavage, as well as a release in caspase-3 and caspase-8 level.

# 4 Conclusion

Halophyte species as edible and folkloric medicinal plants have not yet received much attention as sources of bioactive molecules due to limited popularity or lack of commercial applications. However, recent investigations clearly demonstrate the appreciable potentialities of these salt tolerant species that are a rich source of phytochemicals including various phenolic compounds and offer opportunities for development of value-added products for nutraceutical and food applications as health promoting factors.

**Acknowledgments** This study was supported by the Tunisian Ministry of Higher Education and Scientific Research.

# References

Abdelly C, Debez A, Smaoui A, Grignon C (2011) Halophyte-fodder species association may improve nutrient availability and biomass production of the sabkha ecosystem. In: Ztürk MÖ et al (eds) Sabkha ecosystems, tasks for vegetation science, vol 46. Springer, Berlin, pp 95–104

Aboul-Enein AM, Abu El-Ela F, Shalaby AE, El-Shemy HA (2012) Traditional medicinal plants research in Egypt: studies of antioxidant and anticancer activities. *J Med Plant Res* 6:689–703

Ahmad VU, Ghazala A, Uddin S, Ali MS (1993) Saponins from *Zygophyllum propinquum*. *Phytochemistry* 33:453–455

Ait El Cadi M, Khabbal Y, Alaoui K, Faouzi M-A, Bruno E, Mahraoui L, Cherrah Y (2008) Activité anti-diarrhéique de *Zygophyllum gaetulum*. *Phytothérapie* 6:2–4

Ait El Cadi M, Makram S, Ansar M, Khabbal Y, Alaoui K, Cherrah Y, Taoufik J (2012) Activité anti-inflammatoire des extraits aqueux et éthanolique de *Zygophyllum gaetulum*. *Ann Pharm Fr* 70:113–116

Amin E, El-Hawary SS, Fathy MM, Mohammed RA, Zulfiqar TN, Wedge DE, Becnel JJ, Khan IA (2011) Triterpenoidal saponins: bioactive secondary metabolites from *Zygophyllum coccineum*. *Planta Med* 77:488–491

Aquino R, Tortora S, Fkih-Tetouani S, Capasso A (2001) Saponins from the roots of *Zygophyllum gaetulum* and their effects on electrically-stimulated guinea-pigileum. *Phytochemistry* 56:393–398

Atta AH, Mouneir SM (2004) Antidiarrhoeal activity of some Egyptian medicinal plant extracts. *J Ethnopharmacol* 92:303–309

Ayad R, Rahai M, Azouzi S, Louaar S, Dendougui H, Akkal S, Medjroubi K (2012) Phytochemical investigation of the endemic plant *Zygophyllum cornutum*. *Chem Nat Compd* 48:313–314

Bellakhdar J (1997) La pharmacopéee marocaine traditionnelle: médecine arabe ancienne et savoirs populaires. Ibis Press, Paris, p 764

Bellakhdar J, Claisse R, Fleurentin J, Younos C (1991) Repertory of standard herbal drugs in the Moroccan pharmacopoeia. J Ethnopharmacol 35:123–143

Boubaker J, Skandrani I, Bouhlel I, Ben Sghaier M, Neffati A, Ghedira K, Chekir-Ghedira L (2010) Mutagenic, antimutagenic and antioxidant potency of leaf extracts from *Nitraria retusa*. *Food Chem Toxicol* 48:2283–2290

Boubaker J, Bhouri W, Ben Sghaier M, Bouhlel I, Skandrani I, Ghedira K, Chekir-Ghedira L (2011) Leaf extracts from *Nitraria retusa* promote cell population growth of human cancer cells by inducing apoptosis. *Cancer Cell Int* 11:37

Chaâbane M, Maktouf S, Sayari N, Zouari S, Zeghal N, Ellouze-Ghorbel R (2014) Antioxidant and antimicrobial properties of the extracts from *Nitraria retusa* fruits and their applications to meat product preservation. *Ind Crop Prod* 55:295–303

Chaieb M, Boukhris M (1998) Flore succincte et illustrée des zones arides et sahariennes de Tunisie. Associations pour la protection de la nature et de l'environnement, Sfax

Daoud HS, Al-Rawi A (1985) The flora of Kuwait: dicotyledoneae, vol 1. KPI Publishers, London

El Ghoul J, Ghanem-Boughanmi N, Ben-Attia M (2011) Biochemical study on the protective effect of ethanolic extract of *Zygophyllum album* on streptozotocin-induced oxidative stress and toxicity in mice. *Biomed Prev Nutr* 1:79–83

El Ghoul J, Smiri M, Ghrab S, Boughattas NA, Ben-Attia M (2012) Antihyperglycemic, antihyperlipidemic and antioxidant activities of traditional aqueous extract of *Zygophyllum album* in streptozotocin diabetic mice. *Pathophysiology* 19:35–42

Hadj Salem J, Chevalot I, Harscoat-Schiavo C, Paris C, Fick M, Humeau C (2011) Biological activities of flavonoids from *Nitraria retusa* (Forssk.) Asch. and their acylated derivatives. *Food Chem* 124:486–494

Halim Ahmed F, Saad Hassan-elrady A, Hashish Nadia E (1995) Flavonol glycosides from *Nitraria retusa*. *Phytochemistry* 40(349):351

Hani AME, Shaker KH, Pollmann K, Seifert K (1995) Triterpenoid saponins from *Zygophyllum* species. *Phytochemistry* 4:1233–1236

Hassanean HA, Desoky EK (1992) An acylated isorhamnetin glucoside from zygophyllum simplex. *Phytochemistry*. 31(9):3293–3294

Hassanean HA, El-Shanawany MA, Bishay DW, Franz G (1989) Saponins from *Zygophyllum album*. *Bull Pharmacol Sci* 12:117

Hussein SR, Salwa Æ, Kawashty A, Mohamed Æ, Tantawy E, Nabiel Æ, Saleh AM (2009) Chemosystematic studies of *Nitraria retusa* and selected taxa of *Zygophyllaceae* in Egypt. *Plant Syst Evol* 277:251–264

Hussein SR, Mona MM, Lamyaa FI, Salwa AK, Nabiel AMS (2011) Flavonoids of *Zygophyllum album* L. and *Zygophyllum simplex* L. (*Zygophyllaceae*). *Biochem Syst Ecol* 39:778–780

Jaouhari JT, Lazrek HB, Seddik AJ (1999) Hypoglycaemic response to *Zygophyllum gaetulum* extracts in patients with non-insulin-dependent diabetes mellitus. *J Ethnopharmacol* 64:211–217

Ksouri R, Ksouri WM, Jallali I, Debez A, Magne C, Isoda H, Abdelly C (2012) Medicinal halophytes: potent source of health promoting biomolecules with medical, nutraceutical and food applications. *Crit Rev Biotechnol* 32:289–326

Le Houérou HN (1980) Browse in North Africa. In: Le Houérou HN (ed) Browse in Africa. International Livestock Center for Africa, Addis Ababa, pp 55–82

Mahmoud T, Gairola S (2013) Traditional knowledge and use of medicinal plants in the Eastern Desert of Egypt: a case study from Wadi El-Gemal National Park. *J Med Plants Stud* 1:10–17

Megdiche Ksouri W, Medini F, Mkadmini K, Legault J, Magné C, Abdelly C, Ksouri R (2013) LC–ESI-TOF–MS identification of bioactive secondary metabolites involved in the antioxidant, anti-inflammatory and anticancer activities of the edible halophyte *Zygophyllum album* Desf. *Food Chem* 139:1073–1080

Mnafgui K, Hamden K, Salah H, Kchaou M, Nasri M, Slama S, Derbali F, Allouche N, Elfeki A (2012) Inhibitory activities of *Zygophyllum album*: a natural weight-lowering plant on key enzymes in high-fat diet-fed rats. Evid Based Complement Altern Med ID 620384:9

Mohamed AA, Ali SI, El-Baz FK, Hussein SR (2014) Comparative study of antioxidant activities of *Nitraria retusa* and quantification of its bioactive components by GC/MS. Int J Pharm Sci Rev Res 29:241–246

Pöllmanns K, Agel T, Elgamal HA, Shaker CA, Ifert K (1997) Triterpenoid saponins from the roots of *Zygophyllum* species. *Phvtochemistry* 44:485–489

Saleh NAM, El-Hadidi MN (1977) An approach to the chemosystematics of the zygophyllaceae. *Biochem Syst Ecol* 5:121–128

Subhashini P, Dilipan E, Thangaradjou T, Papenbrock J (2013) Bioactive natural products from marine angiosperms: abundance and functions. Nat Prod Bioprospect 3:129–136

Tackholm V (1974) Student flora of Egypt. Published by Cairo University printed by cooperative printing company

Tigrine-Kordjani N, Meklati BY, Chemat F (2006) Analysis by gas chromatography-mass spectrometry of the essential oil of *Zygophyllum album* L., an aromatic and medicinal plant growing in Algeria. *Int J Aromather* 16:187–191

Tigrine-Kordjani N, Meklati BY, Chemat F (2011) Contribution of microwave accelerated distillation in the extraction of the essential oil of *Zygophyllum album* L. Phytochem Anal 22(1):1–9

White F (1983) The vegetation of Africa, a descriptive memoir to accompany the UNESCO/AETFAT/UNSO vegetation map of Africa. UNESCO. *Nat Resour Res* 20:1–356

Zar Kalai F, Han J, Ksouri R, El Omri A, Abdelly C, Isoda H (2013) Antiobesity effects of an edible halophyte *Nitraria retusa* Forssk in 3T3-L1 preadipocyte differentiation and in C57B6J/L mice fed a high fat diet-induced obesity. Evid Based Complement Altern Med. ID 368658:11

Zar Kalai F, Han J, Ksouri R, Abdelly C, Isoda H (2014) Oral administration of *Nitraria retusa* ethanolic extract enhances hepatic lipid metabolism in db/db mice model 'BKS.Cg-Dock7m+/+ Leprdb/J' through the modulation of lipogenesis–lipolysis balance. Food Chem Toxicol 72:247–256

# *In vivo* Anticandida Activity of Three Traditionally Used Medicinal Plants in East Africa

**D.P. Kisangau, K.M. Hosea, H.V.M. Lyaruu, C.C. Josep, Z.H. Mbwambo, and P.J. Masimba**

**Abstract** Crude extracts of *Dracaena steudneri* bark (DSB), *Sapium ellipticum* bark (SEB) and *Capparis erythrocarpos* root (CER) were investigated for their antifungal activity in immunocompromised mice infected with *Candida albicans* in an *in vivo* mice infection model. Extracts of these plant species are commonly used to treat fungal infections in East African countries. Three groups of white albino mice were immunosuppressed with 200 mg/kg body weight of cyclophosmamide for four consecutive days after which they were administered with 0.3 ml of 0.5 McFarland standard inoculum of *C. albicans*. The groups were treated with escalating doses of 100, 200 and 400 mg/kg of body weight of dichloromethane extracts. There was substantial dose dependency in all treatments given, with mice survival to the end of the experiment correlating well to the dose levels. At a dose of 400 mg/kg, *C. erythrocarpos* was the most effective with mice survival of 60% and organ burden clearance ranging from 64.0%–99.9% ($P < 0.0001$) in all treatments. The results revealed that *C. erythrocarpos* possessed significant potential for development into antifungal drugs suitable for control of *Candida* infections. The effectiveness of the plant extracts *in vivo* was a confirmation of the value of ethnopharmacological leads in drug discovery.

D.P. Kisangau (✉)
Department of Biology, South Eastern Kenya University, P.O Box 170-90300, Kitui, Kenya
e-mail: kisangau@yahoo.com

K.M. Hosea
Department of Molecular Biology and Biotechnology, University of Dar es Salaam, Dar es Salaam, Tanzania

H.V.M. Lyaruu
Department of Botany, University of Dar es Salaam, Dar es Salaam, Tanzania

C.C. Josep
Department of Chemistry, University of Dar es Salaam, Dar es Salaam, Tanzania

Z.H. Mbwambo • P.J. Masimba
Institute of Traditional Medicine, Muhimbili University of Health and Allied Sciences, Dar es Salaam, Tanzania



M. Neffati et al. (eds.), *Medicinal and Aromatic Plants of the World - Africa Volume 3*, Medicinal and Aromatic Plants of the World,
DOI 10.1007/978-94-024-1120-1_14

**Keywords** Antifungal • Medicinal plants • Ethnopharmacology • *Dracaena steudneri* • *Capparis erythrocarpos* • *Candida albicans*

# 1 Introduction

Opportunistic fungal infections represent a significant cause of morbidity and mortality in immunocompromised patients especially those living with HIV/AIDS (Masoko and Eloff 2005; Scorzoni et al. 2007; Rekha and Vidyasagar 2013). Despite the increase in fungal infections, therapeutic options are very limited as almost all the antifungal agents currently in use have toxic side effects and are relatively expensive (Diwan 2006; Ara et al. 2009). Virtually, all synthetic antifungals are associated with serious adverse effects and for some fungal infections, there is still no effective cure. Most common fungal infections are frequently difficult to eradicate with topical preparations and may require long-term use of systemic drugs (De Toledo et al. 2011; Ernst 2001). It is therefore clear that new antifungal agents with broad spectrum fungicidal activities are needed for effective management of these infections. Hence, there is a need for renewed studies on plant natural products that could provide an option for discovery of effective, safer and less toxic antifungal agents.

The aim of the present paper was therefore to investigate *in vivo* antifungal efficacy of extracts of selected plant species used in traditional medicine in East African countries.

# 2 Materials and Methods

## *Test Animals*

Normal white albino mice aged between 5 and 6 months and weighing between 20 and 30 g were locally purchased from Vocational Education and Training Authority (VETA), Dar es Salaam, and reared in the animal house facility of the Institute of Traditional Medicine, Muhimbili University of Health and Applied Sciences (MUHAS). The experimental mice were maintained in standard laboratory conditions at ambient temperature with free access to food and water according to Diwan (2006). The experimental animals were sourced and handled in accordance to Tanzanian animal ethics and regulatory guidelines.

## *Test Plant Extracts and Microorganism*

Crude dicholoromethane extracts of *C. erythrocarpos* root (CER) *D. steudneri* bark (DSB) and *S. ellipticum* bark (SEB) were chosen for the *in vivo* efficacy studies based on their significant *in vitro* anticandida activity both as crude and semi-purified fractions reported in a previous study (Kisangau et al. 2009). During treatment, the dried extracts were dissolved in normal saline with 20% ethanol (v/v) in order to avoid mice mortality due to solvent toxicity. The test organism used in this study was *C. albicans* (ATCC 90028).

## *Inoculum and Immunosuppression Dose Studies*

Preliminary experiments were carried out to determine the inoculum size and immunosuppression dose sufficient to cause death of at least 90% or more of untreated mice (Clark et al. 1991; Dogan et al. 2007; Warn et al. 2000; Rekha and Vidyasagar 2013). The 90% lethal dose ($LD_{90}$) was defined as the inoculum size of the test organism which caused 90% mortality of untreated mice 10 days post infection.

## *Treatment*

Mice were randomized into three groups of 10 in separate cages. The three groups were immunosuppressed with 200 mg/kg body weight of cyclophosmamide for four consecutive days after which they were administered with 0.3 ml of 0.5 McFarland standard inoculum of *C. albicans*. The immunosuppression dose of 200 mg/kg and inoculum size of 0.3 ml of 0.5 McFarland standard were found to be the most sufficient in causing up to 90% mice mortality ($LD_{90}$) within 10 days of the experiment. Twenty four hours post infection, one group was orally treated with escalating doses of 100, 200 and 400 mg/kg of body weight of dichloromethane extracts using a gavage needle. The second group, taken as positive control was intraperitoneally treated with fluconazole (Pfzer Ltd. S. Africa) at 10 mg/kg body weight. Treatment of the two groups was given twice daily for six consecutive days. The third group, taken as negative control was left untreated for six consecutive days post infection.

### *Efficacy Assessment*

On day 11 post infection, all surviving mice were sacrificed. The lungs, liver, brain and kidneys were removed and transferred to 2 ml of sterile phosphate buffered saline (PBS). The organs were homogenized in a tissue grinder and then diluted to $10^{-1}$, $10^{-2}$ and $10^{-3}$. A volume of 100 μl of each of the diluted suspensions was transferred to Sabouraud dextrose agar plates and uniformly spread over the agar surface. The plates were incubated at 37 °C and examined daily for 2 days. Colony counts were recorded from all plates that showed growth. Treatments were termed effective when geometric colony count was less than $3.7 \times 10^5$ c.f.u/ml per organ homogenate, equivalent to 63.0% organ burden clearance[12]. Percentage organ burden clearance in infected mice was calculated on the assumption that the initial 0.5 McFarland inoculum injected to the mice had $10^6$ c.f.u/ml judged from the inoculum turbidity.

### *Data Analysis*

Kruskal-Wallis test was performed using GraphPad Instat® software Version 3.0 to determine differences in organ burden clearance in the treatments.

## 3 Results

### *Survival Curves of Mice in Treatment Experiments*

There was 100% survival of mice in all treatment groups at 100 mg/kg, 200 mg/kg and 400 mg/kg of *D. steudneri, S. ellipticum* and *C. erythrocarpos* extracts one day post infection. However, mice in all groups appeared very sick with a profound morbidity within 24 h of infection. They became motionless, with bristled hair and difficult breathing. After this initial period of morbidity, the condition of the mice in some treated groups gradually stabilized to the end of the experiment as shown in the survivorship curves in Figs. 1, 2 and 3. In the *D. steudneri* group, all untreated mice (control) died 4 days post infection while all those in the 100 mg/kg treatment dose died 5 days post infection (Fig. 1). Mice survival in the 200 mg/kg, 400 mg/kg and fluconazole doses was reduced to 10%, 30% and 60% respectively by the end of the 10 days post infection.

In *S. ellipticum* group, all untreated mice (control) and those treated with 100 mg/kg dose died 4 days post infection (Fig. 2). Mice survival in the 200 mg/kg dose was

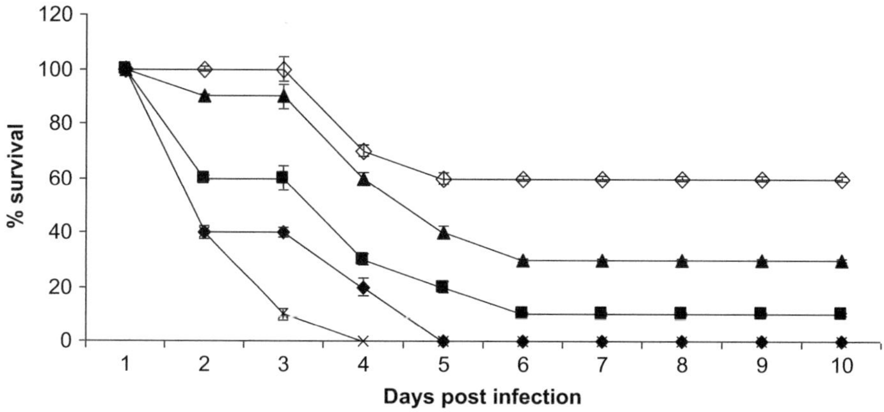


**Fig. 1** Percentage survival of mice treated with *D. steudneri* stem bark extract
**Key:** ▲ 400 mg/kg, ■ 200 mg/kg, ♦ 100 mg/kg, ◊ Fluconazole (10 mg/kg), × Untreated

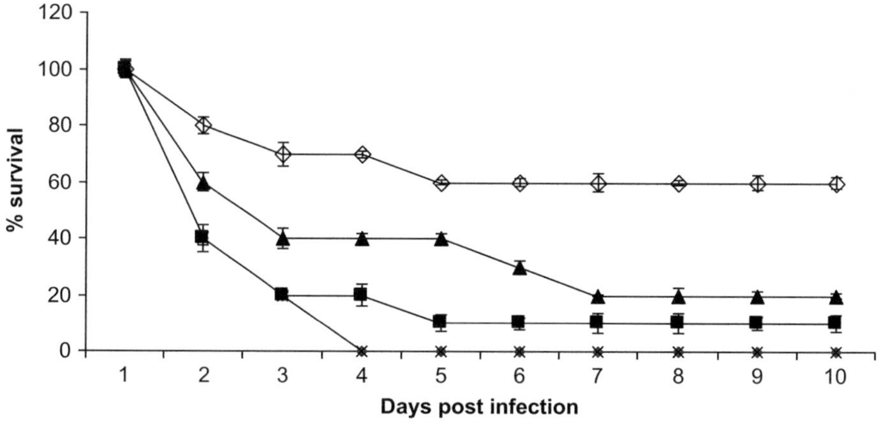


**Fig. 2** Percentage survival of mice treated with *S. ellipticum* stem bark extract
**Key:** ▲ 400 mg/kg, ■ 200 mg/kg, ♦ 100 mg/kg, ◊ Fluconazole (10 mg/kg), × Untreated

reduced to 10% by the end of the 10 days post infection while in 400 mg/kg dose, it was reduced to 20%. In the fluconazole dose (10 mg/kg), there was 60% mice survival by 10 days post infection.

In the *C. erythrocarpos* treatment group (Fig. 3), mice survival in both 100 mg/kg and 200 mg/kg doses was reduced to 20% by 10 days post infection. Mice survival in 400 mg/kg was very closely comparable to that in fluconazole dose with 60% and 70% respectively. However, all untreated mice (control) died 8 days post infection.

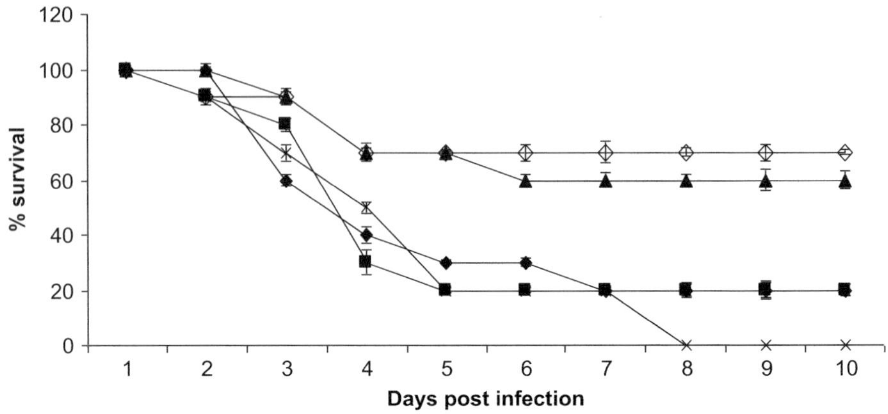


**Fig. 3** Percentage survival of mice treated with *C. erythrocarpos* root extract
**Key:** ▲ 400 mg/kg, ■ 200 mg/kg, ♦ 100 mg/kg, ◊ Fluconazole (10 mg/kg), × Untreated

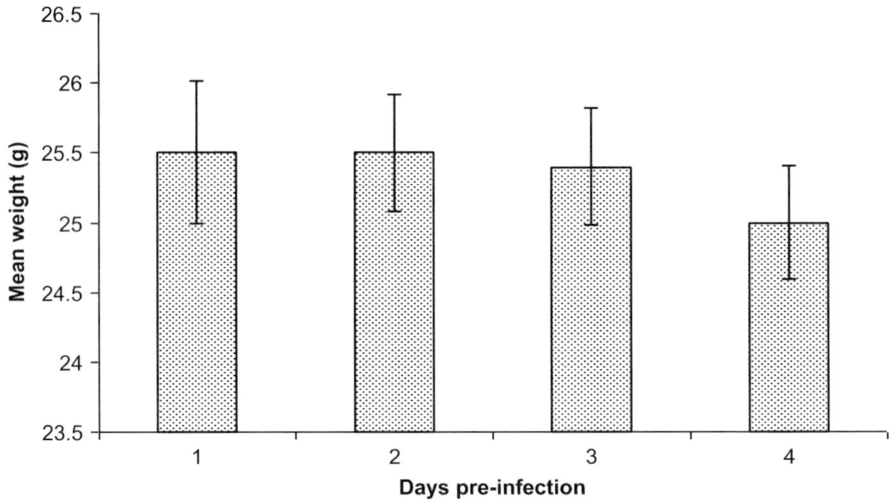


**Fig. 4** Mean weight of mice pre-infection (during immunosuppression) in *D. steudneri* treatment group

## *Mean Weights of Mice Pre- and Post-infection*

Mean weights of mice in all treatment groups were recorded for the first 10 days of the experiment. However, mean weights of mice in the untreated groups were not recorded as they were assumed to degenerate with subsequential death as observed in the treatment experiments. The mean weight of mice in the 4 days of immunosuppression for *D. steudneri, S. ellipticum* and *C. erythrocarpos* treatment groups are

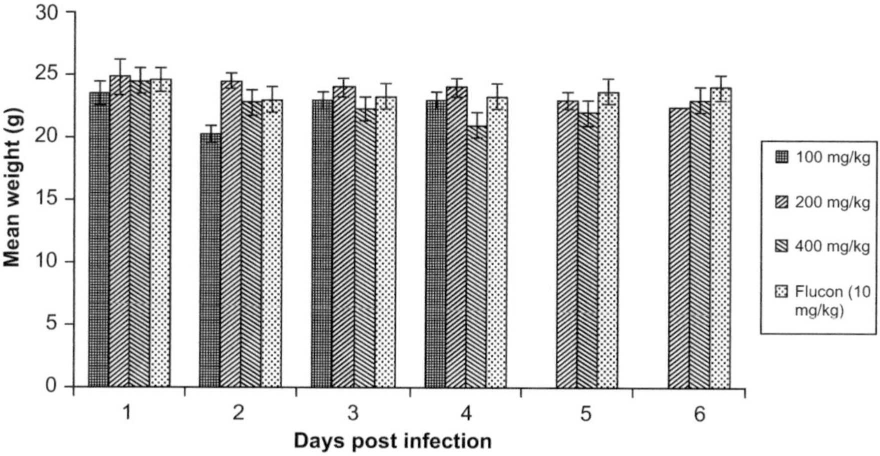


**Fig. 5** Mean weight of mice post-infection (during treatment) in *D. steudneri* treatment group

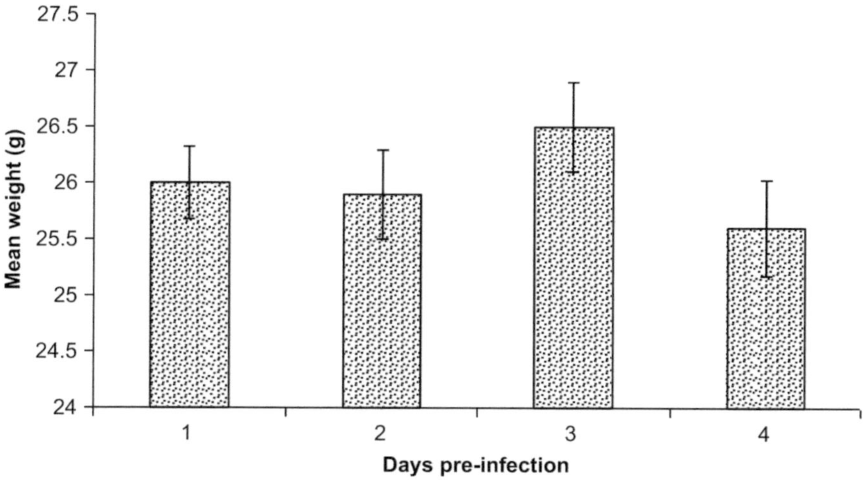


**Fig. 6** Mean weight of mice pre-infection (during immunosuppression) in *S. ellipticum* treatment group

shown in Figs. 4, 6 and 8 while the mean weight of mice post infection are shown in Figs. 5, 7 and 9. The pre-infection mean weight for *D. steudneri* treatment group ranged from 25.0 g to 25.5 g and the grand mean weight was 25.4 g for this treatment group. The pre-infection mean weight for *S. ellipticum* treatment group ranged from 25.9 g to 26.5 g with a grand mean of 26.0 g. The pre-infection mean weight for *C. erythrocarpos* treatment group ranged from 22.0 g to 22.4 g with a grand mean of 22.2 g.

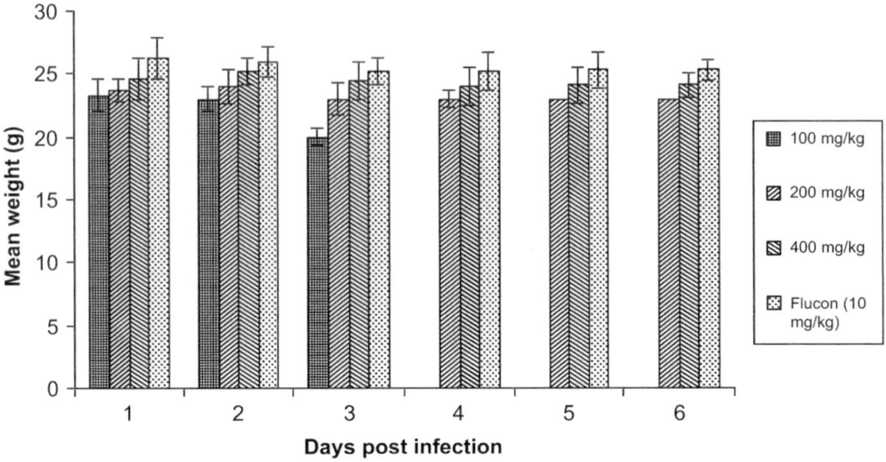


**Fig. 7** Mean weight of mice post-infection (during treatment) in *S. ellipticum* treatment group

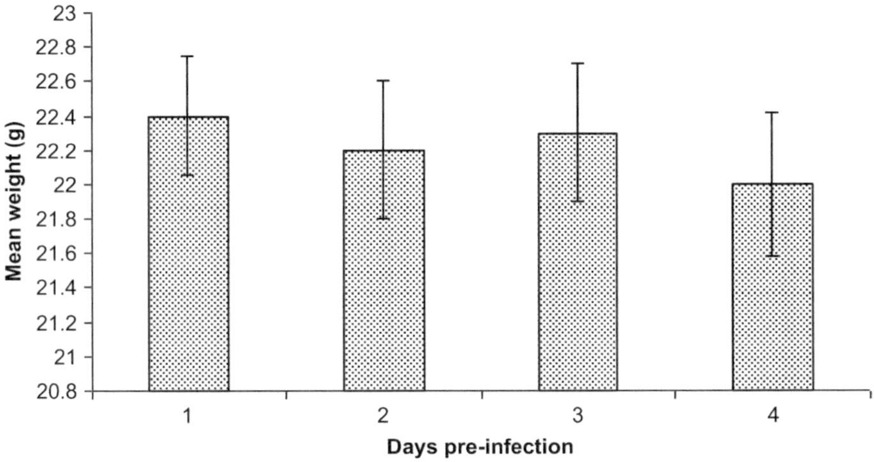


**Fig. 8** Mean weight of mice pre-infection (during immunosuppression) in *C. erythrocarpos* treatment group

In all the three treatment groups, there was a steady decline in mean weights of mice immediately after immunosuppression challenge and following infection with *C. albicans*. In the *D. steudneri* treatment group, the mean weight of mice in the 100 mg/kg dose declined up to 23.0 g on day four post infection, after which all mice in this treatment dose died (Fig. 5). Similarly, there was a steady decline in mean weight in the 200 mg/kg dose from 24.8 g on day one post infection to 22.5 g on day six post infection. In the 400 mg/kg dose, there was a mean weight decline of up to 21.0 g on day four post infection followed by a slight mean weight improvement up to 23.0 g on day six post infection. Mean weight in the fluconazole

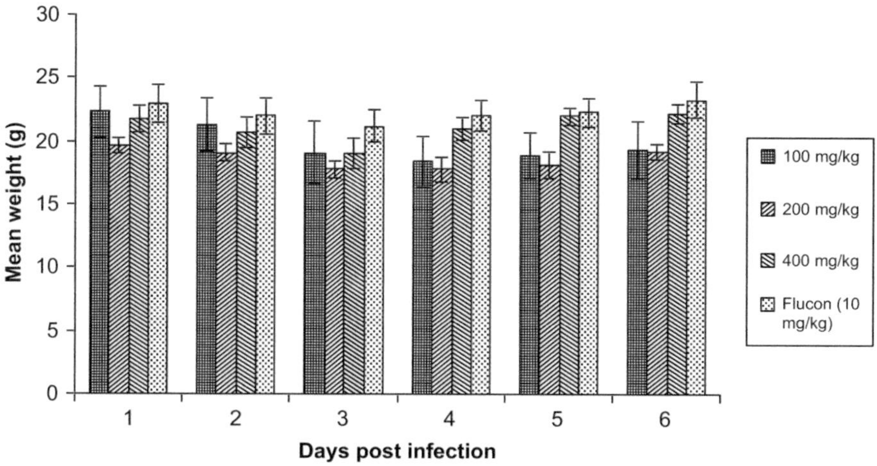


**Fig. 9** Mean weight of mice post-infection (during treatment) in *C. erythrocarpos* group

dose went down to 23.0 g day two post infection followed by a slight improvement up to 24.0 g on day six post infection.

In *S. ellipticum* treatment group, there was no mean weight improvement in 100 mg/kg and 200 mg/kg doses. There was however a slight mean weight improvement in the 400 mg/kg and fluconazole doses. In 100 mg/kg dose of *C. erythrocarpos* treatment group, mean weight declined to 18.4 g on day four post infection followed by a slight improvement up to 19.4 g on day six post infection. A similar pattern was observed in 200 mg/kg, 400 mg/kg and fluconazole doses (Fig. 9).

## *Organ Culture Counts of Mice in Autopsy*

Geometric colony counts of the brain, kidneys, liver and lungs are shown in Tables 1, 2 and 3. The geometric colony counts were determined only where distinct colony counts could be visually established in all the three dilutions ($10^{-1}$, $10^{-2}$, $10^{-3}$). However, for statistical purposes, treatments with countless colonies were calculated on the basis of 1000 colony counts as the minimum in the $10^{-3}$ dilution. In the 200 mg/kg dose of *D. steudneri* treatment group (Table 1), organ burden clearance was more effective in the kidneys than in the other organs with an average colony count of $9.3 \times 10^3$ c.f.u/ml, equivalent to 99.0% clearance. The liver was the most burdened as colony counts were well above 1000 in $10^{-1}$ and $10^{-2}$ dilutions. Also, in the 400 mg/kg dose, organ burden clearance was most effective in the kidneys with an average colony count of $3.3 \times 10^1$ c.f.u/ml (99.9% clearance), while the liver was the most burdened with colony counts above 1000. In the fluconazole group, organ burden clearance was most effective in the lungs with colony counts of $1.7 \times 10^2$ c.f.u/ml (99.8% clearance). In the overall *D. steudneri* treatment group, organ

**Table 1** Organ culture counts for *C. albicans* in *D. steudneri* stem bark extract treatment

| | | c.f.u count[a] | | | |
|---|---|---|---|---|---|
| Treatment | Dilution | Brain | Kidneys | Liver | Lungs |
| 100 mg/kg[b] | $10^{-1}$ | – | – | – | – |
| | $10^{-2}$ | – | – | – | – |
| | $10^{-3}$ | – | – | – | – |
| c.f.u average count | | – | – | – | – |
| 200 mg/kg | $10^{-1}$ | $9.0 \times 10^4$ | $2.9 \times 10^3$ | >1000 | $3.0 \times 10^5$ |
| | $10^{-2}$ | $3.5 \times 10^4$ | $2.5 \times 10^4$ | >1000 | $8.0 \times 10^2$ |
| | $10^{-3}$ | $1.0 \times 10^4$ | 0 | $1.4 \times 10^5$ | $3.0 \times 10^4$ |
| c.f.u average count | | $4.5 \times 10^4$ (95.5%) | $9.3 \times 10^3$ (99.1%) | – | $1.1 \times 10^5$ (89.0%) |
| 400 mg/kg | $10^{-1}$ | $3.6 \times 10^3$ | $1.0 \times 10^2$ | $2.0 \times 10^5$ | $2.0 \times 10^2$ |
| | $10^{-2}$ | 0 | 0 | $2.3 \times 10^5$ | $2.0 \times 10^3$ |
| | $10^{-3}$ | 0 | 0 | $3.5 \times 10^6$ | $1.0 \times 10^4$ |
| c.f.u average count | | $1.3 \times 10^3$ (99.8%) | $3.3 \times 10^1$ (99.9%) | $1.3 \times 10^6$ (–30.0%) | $4.1 \times 10^3$ (99.6%) |
| Fluconazole (10 mg/kg) | $10^{-1}$ | $1.2 \times 10^5$ | $3.5 \times 10^5$ | $1.8 \times 10^3$ | $5.0 \times 10^2$ |
| | $10^{-2}$ | $5.5 \times 10^4$ | 0 | $1.0 \times 10^3$ | 0 |
| | $10^{-3}$ | 0 | 0 | 0 | 0 |
| c.f.u average count | | $5.8 \times 10^4$ (94.2%) | $1.2 \times 10^5$ (88.0%) | $9.3 \times 10^2$ (99.9%) | $1.7 \times 10^2$ (99.9%) |

[a]All counts are expressed as c.f.u/ml of organ homogenate and percentage organ burden clearance is given in parenthesis for average colony counts only

[b]No data for 100 mg/kg dose as all mice died before the end of the experiment

burden clearance was therefore most effective in the 400 mg/kg dose in kidneys with average colony counts of $3.3 \times 10^1$ c.f.u/ml slightly higher than that of fluconazole at $1.7 \times 10^2$ c.f.u/ml in the lungs. There were significant differences in organ burden clearance in the three *D. steudneri* treatment dosages (200 mg/kg. 400 mg/kg and fluconazole) ($P < 0.05$).

In the *S. ellipticum* treatment group (Table 2), the lowest organ colony load was observed in 200 mg/kg with an average colony count of $3.4 \times 10^4$ c.f.u/ml (96.6% clearance) in the lungs. This was comparably lower than the minimum colony load of $2.2 \times 10^5$ c.f.u/ml (78.0% clearance) and $7.5 \times 10^4$ c.f.u/ml (92.5% clearance) observed in 400 mg/ml and fluconazole doses respectively. There were no significant differences in organ burden clearance in the three *S. ellipticum* treatment dosages (200 mg/kg. 400 mg/kg and fluconazole) ($P > 0.05$).

In the *C. erythrocarpos* treatment group (Table 3), percentage organ clearance in the 400 mg/kg dose was well comparable to that of the fluconazole in the brain, kidney and lungs. The lowest organ colony load was observed in the 400 mg/kg dose with an average colony count of $3.3 \times 10^1$ c.f.u/ml (99.9% clearance) in the kidneys. It was equivalent to that observed for fluconazole in the brain. The lowest colony loads in 100 mg/kg and 200 mg/kg doses were $2.0 \times 10^5$ c.f.u/ml (80.0%

**Table 2** Organ culture counts for *C. albicans* in *S. ellipticum* stem bark extract treatment.

| | | c.f.u count[a] | | | |
|---|---|---|---|---|---|
| Treatment | Dilution | Brain | Kidneys | Liver | Lungs |
| 100 mg/kg[b] | $10^{-1}$ | – | – | – | – |
| | $10^{-2}$ | – | – | – | – |
| | $10^{-3}$ | – | – | – | – |
| c.f.u average count | | – | – | – | – |
| 200 mg/kg | $10^{-1}$ | >1000 | $8.0 \times 10^4$ | $7.2 \times 10^4$ | $4.5 \times 10^4$ |
| | $10^{-2}$ | $3.8 \times 10^5$ | $2.1 \times 10^4$ | $2.2 \times 10^5$ | $3.2 \times 10^5$ |
| | $10^{-3}$ | $7.5 \times 10^5$ | 0 | $1.2 \times 10^6$ | $1.1 \times 10^6$ |
| c.f.u average count | | – | $3.4 \times 10^4$ (96.6%) | $5.0 \times 10^5$ (50.0%) | $4.8 \times 10^5$ (52.0%) |
| 400 mg/kg | $10^{-1}$ | $5.6 \times 10^4$ | $4.8 \times 10^4$ | $1.0 \times 10^5$ | $7.2 \times 10^4$ |
| | $10^{-2}$ | $4.0 \times 10^5$ | $4.0 \times 10^6$ | $6.8 \times 10^4$ | $5.2 \times 10^5$ |
| | $10^{-3}$ | $2.1 \times 10^5$ | $1.2 \times 10^6$ | $2.8 \times 10^6$ | $7.7 \times 10^5$ |
| c.f.u average count | | $2.2 \times 10^5$ (78.0%) | $5.5 \times 10^5$ (45.0%) | $9.9 \times 10^5$ (1.0%) | $4.5 \times 10^5$ (55.0%) |
| Fluconazole (10 mg/kg) | $10^{-1}$ | $1.0 \times 10^5$ | >1000 | $1.0 \times 10^5$ | $2.0 \times 10^5$ |
| | $10^{-2}$ | $9.5 \times 10^4$ | >1000 | $4.4 \times 10^5$ | $1.0 \times 10^6$ |
| | $10^{-3}$ | $3.0 \times 10^4$ | >1000 | $2.0 \times 10^4$ | $4.0 \times 10^5$ |
| c.f.u average count | | $7.5 \times 10^4$ (92.5%) | – | $1.9 \times 10^5$ (81.0%) | $5.3 \times 10^5$ (47.0%) |

[a]All counts are expressed as c.f.u/ml of organ homogenate and percentage organ burden clearance is given in parenthesis for average colony counts only
[b]No data for 100 mg/kg dose as all mice died before the end of the experiment

clearance), and $3.2 \times 10^4$ c.f.u/ml (96.8% clearance) in kidneys and lungs respectively. There were significant differences in organ burden clearance in the four *C. erythrocarpos* treatment dosages (100 mg/kg, 200 mg/kg. 400 mg/kg and fluconazole) [$P < 0.0001$].

Based on the organ burden clearance cut point of $3.7 \times 10^5$ c.f.u/ml (63.0% clearance), *D. steudneri* group had six effective treatments out of a total of 12 treatments administered to the mice, with percentage organ burden clearance ranging from 89.0% to 99.9% (Table 1). *S. ellipticum* extract had two effective treatments with percentage clearance ranging from 78.0% to 96.6% (Table 2). *C. erythrocarpos* group had six effective treatments ranging from 64.0% to 99.9% (Table 3). Most of these effective treatments of the plant extracts were quite comparable to the fluconazole treatments.

## *Percentage Organ Weights of Mice in Autopsy*

The highest percentage organ weights of mice in autopsy were recorded for liver and ranged from 5.4 to 7.5% (Table 4). Percentage organ weights for kidneys, lungs and brain ranged from 1.5 to 2.4%, 1.0 to 1.6% and 1.5 to 2.1%, respectively.

**Table 3** Organ culture counts for *C. albicans* in *C. erythrocarpos* root extract treatment

| | | c.f.u count[a] | | | |
|---|---|---|---|---|---|
| Treatment | Dilution | Brain | Kidneys | Liver | Lungs |
| 100 mg/kg | $10^{-1}$ | >1000 | $1.5 \times 10^4$ | >1000 | >1000 |
| | $10^{-2}$ | >1000 | $9.0 \times 10^4$ | $8.0 \times 10^5$ | >1000 |
| | $10^{-3}$ | $6.0 \times 10^6$ | $5.0 \times 10^5$ | $3.8 \times 10^5$ | >1000 |
| c.f.u average count | | – | $2.0 \times 10^5$ (80.0%) | – | – |
| 200 mg/kg | $10^{-1}$ | >1000 | $8.0 \times 10^4$ | $9.8 \times 10^4$ | $8.0 \times 10^4$ |
| | $10^{-2}$ | >1000 | $5.4 \times 10^5$ | $8.0 \times 10^5$ | $1.5 \times 10^4$ |
| | $10^{-3}$ | $3.6 \times 10^6$ | $9.0 \times 10^5$ | $1.9 \times 10^5$ | 0 |
| c.f.u average count | | – | $4.8 \times 10^5$ (52.0%) | $3.6 \times 10^5$ (64.0%) | $3.2 \times 10^4$ (96.8%) |
| 400 mg/kg | $10^{-1}$ | $6.0 \times 10^3$ | $1.0 \times 10^2$ | >1000 | $1.3 \times 10^3$ |
| | $10^{-2}$ | 0 | 0 | $4.5 \times 10^5$ | 0 |
| | $10^{-3}$ | 0 | 0 | $7.1 \times 10^5$ | 0 |
| c.f.u average count | | $2.0 \times 10^3$ (99.8%) | $3.3 \times 10^1$ (99.9%) | – | $4.3 \times 10^2$ (99.9%) |
| Fluconazole (10 mg/kg) | $10^{-1}$ | $1.0 \times 10^2$ | $3.0 \times 10^2$ | $9.0 \times 10^3$ | $4.8 \times 10^4$ |
| | $10^{-2}$ | 0 | $1.0 \times 10^3$ | $5.0 \times 10^3$ | $3.4 \times 10^4$ |
| | $10^{-3}$ | 0 | 0 | $1.0 \times 10^4$ | $7.0 \times 10^4$ |
| c.f.u average count | | $3.3 \times 10^1$ (99.9%) | $4.3 \times 10^2$ (99.9%) | $8.0 \times 10^3$ (99.2%) | $5.1 \times 10^4$ (94.9%) |

[a]All counts are expressed as c.f.u/ml of organ homogenate and percentage organ burden clearance is given in parenthesis for average colony counts only

## 4 Discussion

*Candida albicans* remains the most common fungal pathogen causing deep mycosis in humans (Marianne et al. 2000). Of all known opportunistic fungal infections, oral candidiasis caused by *C. albicans* is one of the most common manifestations of HIV/AIDS, and does not only influence the nutritional status of the HIV/AIDS patients, but it is a predisposing factor to the other more serious disseminated infections (Rekha and Vidyasagar 2013; Runyoro et al. 2006; Moussa et al. 2012). Among the different HIV-associated fungal infections, oral mucosal lesions caused by *Candida* species are by far the most frequent manifestations. *Candida albicans* is the most causative agent, accounting for more than 90% of cases (Blignaut 2007).

The fact that animals in all groups became very sick within 24 h of infection, showing reduced mobility was a confirmation that they had reached a state of profound neutropenia due to immunosuppression. This was also confirmed by the 100% mortality of all untreated mice during treatment experiments meaning that they were unable to withstand *Candida* challenge due to the compromised immunity. A state of immunosuppression was also confirmed by the sudden decline in mean weights of mice 24 h post infection. In a related mice infection model, Chinchilla et al. (1998), Warn et al. (2000) and Warn and Denning (1999) demon-

**Table 4** Percentage organ weight of mice in autopsy in different treatment groups

| Organ | Treatment group | Organ weight in autopsy (%) |
|---|---|---|
| Liver | DSB 200 mg/kg | 7.1 |
| | DSB 400 mg/kg | 6.4 |
| | Flucon-DSB | 6.3 |
| | SEB 200 mg/kg | 6.5 |
| | SEB 400 mg/kg | 6.1 |
| | Flucon-SEB | 6.9 |
| | CER 100 mg/kg | 7.5 |
| | CER 200 mg/kg | 5.5 |
| | CER 400 mg/kg | 5.4 |
| | Flucon-CER | 6.3 |
| Kidney | DSB 200 mg/kg | 1.9 |
| | DSB 400 mg/kg | 1.7 |
| | Flucon-DSB | 2.4 |
| | SEB 200 mg/kg | 1.7 |
| | SEB 400 mg/kg | 1.7 |
| | Flucon-SEB | 1.8 |
| | CER 100 mg/kg | 1.9 |
| | CER 200 mg/kg | 1.5 |
| | CER 400 mg/kg | 1.9 |
| | Flucon-CER | 1.6 |
| Lungs | DSB 200 mg/kg | 1.6 |
| | DSB 400 mg/kg | 1.3 |
| | Flucon-DSB | 1.6 |
| | SEB 200 mg/kg | 1.1 |
| | SEB 400 mg/kg | 1.3 |
| | Flucon-SEB | 1.3 |
| | CER 100 mg/kg | 1.1 |
| | CER 200 mg/kg | 1.1 |
| | CER 400 mg/kg | 1.2 |
| | Flucon-CER | 1.0 |
| Brain | DSB 200 mg/kg | 2.0 |
| | DSB 400 mg/kg | 1.9 |
| | Flucon-DSB | 1.8 |
| | SEB 200 mg/kg | 1.9 |
| | SEB 400 mg/kg | 1.5 |
| | Flucon-SEB | 2.1 |
| | CER 100 mg/kg | 1.7 |
| | CER 200 mg/kg | 1.7 |
| | CER 400 mg/kg | 2.0 |
| | Flucon-CER | 1.6 |

*Flucon-DSB* Fluconazole treatment in *D. steudneri* (DSB) group, *Flucon-SEB* Fluconazole treatment in *S. ellipticum* (SEB) group, *Flucon-CER* Fluconazole treatment in *C. erythrocarpos* (CER) group strated that no treatment group had 100% survival and that mice receiving no active treatment had either 90% or 100% mortality.

The present model showed a substantial dose dependency in all the treatments administered, with the mice survival to the end of the experiments correlating to the dose levels as previously reported in a different study by Warn et al. (2000, 2005), Diwan[4] and You et al. (2009). In the *C. erythrocarpos* treatments, 400 mg/kg dose which compared quite well with fluconazole in the survivorship curves was indicative of the strong potential of this extract to eliminate *Candida* infections. The ability of the 400 mg/kg dose in the *C. erythrocarpos* treatment to clear colony burden in the kidneys to $3.3 \times 10^1$ c.f.u/ml, lower than that of fluconazole ($4.3 \times 10^2$ c.f.u/ml) supports this credibility. Extract effectiveness was also portrayed in *D. steudneri* and *S. ellipticum* treatments as they were able to sustain a number of infected mice to the end of the experiment. However, it remained uncertain as to why in *C. erythrocarpos* treatments survival levels of mice in 100 mg/kg dose were higher than in 200 mg/kg in days 4–6 post infection. It was also not clear why mice survival in the *C. erythrocarpos* group was higher than in the *D. steudneri* group yet the two had equal number (six) of effective treatments. Besides, *D. steudneri* treatments had even a higher range of percentage organ clearance (89%–99.9%) than *C. erythrocarpos* treatments with 64.0%–99.9%. But overall, organ burden clearance was most effective in the *D. steudneri* and *C. erythrocarpos* treatments than in *S. ellipticum* treatments. Elsewhere, the effect of an aqueous extract of *Nigella sativa* seeds was studied on candidiasis in mice. An intravenous inoculum of *Candida albicans* produced colonies of the organism in the liver, spleen and kidneys. Treatment of mice with the plant extract produced a fivefold decrease of *Candida* in kidneys, eightfold in liver and 11-fold in spleen (Ahmad et al. 2013).

The gradual improvement in mean weights of mice in the 400 mg/kg dose in all treatment groups (*D. steudneri, S. ellipticum* and *C. erythrocarpos*) confirmed the dose dependency treatment potentials of the plant extracts as previously reported by Diwan (2006). A similar trend was observed in fluconazole treated mice. The gradual gain in mean weights thus corresponded with the recovery of mice observed in these treatments. This scenario substantiates the higher levels of mice survival to the end of the experiment observed in the 400 mg/kg dose unlike in 100 mg/kg and 200 mg/kg doses where mice survival was limited. It may also explain why there was rapid mortality in the untreated (control) groups as no mice survived by the end of the experiment.

There was a clear correlation between mice survival and colony burden clearance from organs in autopsy as previously reported in a different study by Masuoka (2004) and You et al. (2009). Mice survival to the end of the experiment in 200 mg/kg and 400 mg/kg doses of *D. steudneri* treatments could be associated with the effective colony burden clearance observed in the brain, kidney and lungs. However, mice survival in the 200 mg/kg and 400 mg/kg doses of *S. ellipticum* treatments could only be associated with colony burden clearance in the kidneys and brain, respectively. All other organs were burdened above the effective clearance level. In the *C. erythrocarpos* treatment group, mice survival in the 100 mg/kg dose could be ascribed to colony burden clearance in the kidneys while mice survival in the

200 mg/kg dose could be due to colony burden clearance in the liver and lungs. In the 400 mg/kg dose, mice survival could be due to colony burden clearance in the brain, kidneys and lungs. In a related study though on conventional antifungal, a 0.8 mg/kg dose of amphotericin B emulsion increased mice survival time compared to untreated control group and was equally effective in reducing the c.f.u counts of *C. albicans* in the kidney (Tabosa et al. 1996).

The highest percentage organ weights in autopsy were recorded in organs of mice treated with lower doses than those treated with higher doses (Table 4). For example the highest percentage organ weights for the liver were recorded in DSB 200 mg/kg (7.1%), SEB 200 mg/kg (6.5%) and CER 100 mg/kg (7.5%). A similar pattern was observed for the kidneys, lungs and brain. This could be due to inflammation of the organs caused by *Candida* infestation where colony burden clearance was not very effective as similarly reported by Dogan et al. (2007). Uncertainly, some organs of mice treated with the highest extract dose (400 mg/kg) or with fluconazole (10 mg/kg) registered high percentage organ weights. This was the case for the kidneys in 400 mg/kg and fluconazole doses with percentage organ weight of 1.9% and 2.4% respectively. It was also observed in lungs (Flucon-DSB, 1.6%) and brain (CER 400 mg/kg, 2.0%).

## 5 Conclusion

This article presents the first antifungal activity assays of the three plant species (*D. steudneri, S. ellipticum* and *C. erythrocarpos*) on animals. The results reveal that *D. steudneri* and *C. erythrocarpos* possess significant potential for development into antifungal drugs suitable for control of *Candida* infections. All the three plant species are traditionally used in treatment of candidiasis and other fungal infections in East African countries. Their effectiveness *in vivo* is a confirmation of the value of ethnopharmacological leads in drug discovery. The study presents a platform for pre-clinical trials of extracts of these plant species as potential candidates for developing therapies to manage *Candida* infections including HIV/AIDS opportunistic infections associated with *Candida spp.* With the proven *in vivo* efficacy of the extracts, standardized herbal formulations could be developed for mass production in order to meet the health needs of many people who cannot afford conventional medicines in developing countries.

**Acknowledgments** We are grateful to DAAD/NAPRECA, the Inter-University Council of East Africa Research initiative (VicRes) and the International Foundation for Science (IFS) for financial support to carry out this work. We thank Messrs. F.M. Mbago and S. Haji of the Herbarium, Botany department, University of Dar es Salaam for identifying the plant species used in this study.

# References


Ahmad A, Husain A, Mujeeb M, Khan A, Najmi AK, Siddique NA, Damanhouri ZA, Anwar F (2013) A review on therapeutic potential of *Nigella sativa*: a miracle herb. Asian Pac J Trop Biomed 3(5):337–352

Ara N, Nur MH, Amran MS, Wahid MII, Ahmed M (2009) *In vitro* antimicrobial and cytotoxic activities of leaves and flower extracts from *Lippia alba*. Pak J Biol Sci 12(1):87–90

Blignaut E (2007) Oral candidiasis and oral yeast carriage among institutionalized South African paedriatic HIV/AIDS patients. Mycopathologia 163:67–73

Chinchilla M, Guerrero OM, Abarca G, Barrios M, Castro O (1998) An *in vivo* model to study the anti-malaric capacity of plant extracts. Int J Trop Biol 46:35–39

Clark JM, Whitney RR, Olsen SJ, George RJ, Swerdel MR, Kunselman L, Bonner DP (1991) Amphotericin B Lipid Complex therapy of experimental fungal infections in mice. Antimicrob Agents Chemother 35:615–621

De Toledo CEM, Britta EA, Ceole LF, Silva ER, De Mello JCP, Filh BPD (2011) Antimicrobial and cytotoxic activities of medicinal plants of the Brazilian cerrado, using Brazilian cachac, as an extractor liquid. J Ethnopharmacol 133(2):420–425

Diwan PV (2006) *In vivo* efficacy of an antifungal fraction from *Pallavicinia lyellii*, a liverwort. Indian J Pharm 38(3):211–212

Dogan A, Ergen K, Budak F, Severcan F (2007) Evaluation of Disseminated Candidiasis on an Experimental Animal Model: A Fourier Transform Infrared Study. Appl Spectrosc 61(2):199–203

Ernst E (2001) The desktop guide to complementary and alternative medicine: an evidence-based approach. J Roy Soc Med 94:650–651

Kisangau DP, Hosea KM, Lyaruu HVM, Joseph CC, Mbwambo ZH, Masimba PJ, Gwandu CB, Bruno LN, Devkota KP, Sewald N (2009) Screening of traditionally used Tanzanian medicinal plants for antifungal activity. Int J Pharmaceut Biol 47(8):708–716

Marianne D, Ronald A, Froyen G, Frank CO, Fermin S, Roland C, Walter HMLL (2000) Single allele knock-out of *Candida albicans* CGT1 leads to unexpected resistance to hygromycin B and elevated temperature. Microbiologica 46:353–365

Masoko P, Eloff JN (2005) The diversity of antifungal compounds of six South African *Terminalia* species (Combretaceae) determined by bioautography. Afr J Biotechnol 4(12):1425–1431

Masuoka J (2004) Surface Glycans of *Candida albicans* and other pathogenic fungi: physiological roles, clinical uses, and experimental challenges. Clin Microbiol Rev 17(2):281–310

Moussa A, Noureddine D, Saad A, Abdelmalek A, Salima B (2012) The mixture of natural honey and *Peganum harmala* seeds against *Candida albicans* by well diffusion method. Asian Pac J Trop Biomed 6:1–4

Rekha S, Vidyasagar GM (2013) Anti-*Candida* activity of medicinal plants. A review. Int J Pharm Pharm Sci 5(14):9–16

Runyoro DKB, Matee MIN, Ngassapa OD, Joseph CC, Mbwambo ZH (2006) Screening of Tanzanian medicinal plants for anti-Candida activity. BMC Compl Altern Med 6:11

Scorzoni L, Benaducci T, Almeida AMF, Silva DHS, Bolzani VS, Gianinni MJSM (2007) The use of standard methodology for determination of antifungal activity of natural products against medical yeasts *Candida* sp and *Cryptococcus sp*. Braz J Microbiol 38:1–12

Tabosa Do Egito ES, Appel M, Fessi H, Barrett G, Puisieux F, Devissaguet J (1996) *In-vitro* and *in-vivo* evaluation of a new amphotericin B emulsion-based delivery system. J Antimicrob Chemother 38(3):485–497

Warn P, Denning DW (1999) Dose range evaluation of Liposomal Nystatin and comparisons with Amphotericin B and Amphotericin B Lipid Complex in temporarily neutropenic mice infected with an isolate of *Aspergillus fumigatus* with reduced susceptibility to Amphotericin B. Antimicrob Agents Chemother 43(11):2592–2599

Warn PA, Morrissey J, Moore CB, Denning DW (2000) *In-vivo* activity of amphotericin B Lipid complex in immunocompromised mice against Fluconazole-resistante or Fluconazole-susceptible *Candida tropicalis*. Antimicrob Agents Chemother 44(10):2664–2671

Warn PA, Sharp A, Morrissey G, Denning DW (2005) Activity of aminocandin (IP960) compared with amphotericin B and fluconazole in a neutropenic murine model of disseminated infection caused by a fluconazole-resistant strain of *Candida tropicalis*. Antimicrob Agents Chemother 56:590–593

You X, Li C, Yang X, Yuan M, Zhang W, Lou R, Wang Y, Li G, Chen H, Song D, Sun C, Cen S, Yu L, Zhao L, Jiang J (2009) *In vivo* antibacterial activity of vertilmicin, a new aminoglycoside antibiotic. Antimicrob Agents Chemother 53(10):4525–4528

# Wild Edible *Allium* Species from Arid Zones of Tunisia: New Perspectives of Natural Compounds

**Hanen Najjaa, Abelkarim Ben Arfa, and Mohamed Neffati**

**Abstract** *Allium* is the largest and the most important representative genus of the *Alliaceae* family. For many centuries, several *Allium* species have been used as vegetables and spices, and as folk medicines for curing various diseases. Besides the well-known garlic and onion, other species are either, widely cultivated for culinary use, such as leek, scallion, shallot and chive, or found as wild endemic species grown spontaneously.

This chapter presents an integrated view on new advances and recent trends in the nutritional value, bioactive compound and health effect of various extracts prepared from the spontaneous species, *A. roseum* (rosy garlic) and *A. ampeloprasum* (Kurrath). Chemical composition of *A. roseum* leaves was investigated. The analyses showed that *A. roseum* contained high amounts of soluble carbohydrate (32.8%), crude protein (22.7%), dietary fibre (12.3%), ash (7.2%) and fat (3.6%). The most abundant minerals were K, Ca and $SO_4^{2-}$ with 1530.5 and 712.5 mg/100 g DW, respectively. Mn, Cu and Zn were also detected in appreciable amounts ranging from 1 to 2 mg/100 g DW. Extracted leaf oil was composed of 14.93% saturated and 85.28% unsaturated fatty acids. Linolenic acid represents 53% of the unsaturated fatty acids and palmitic acid represents 13% of the saturated fatty acids. Phytochemical screening of *A. roseum* and *A. ampeloprasum* showed the presence of bioactive agents exhibit a positive effect on human health as antioxidants and antibacterial compounds such as saponins, tannins, flavonoids, coumarins, steroids, cardiac glycosides, free quinone and iridoids. The most abundant phytonutrients in *A. roseum* edible leaves included polyphenols, flavonoids, anthocyanidins, vitamin C, carotenoids and allicin which were at 736.65 mg of equivalent catechol/100 g DW, 3.37 mg of equivalent catechin/g DW, 1239.62 µg/100 g DW, 1523.35 mg/100 g DW, 242.25 µg/100 g DW and 657 mg/100 g DW, respectively. The content of polyphenols, flavonoids and tannin in *A. ampeloprasum* was 29,278 mg of equivalent gallic acid/g DW, 5.840 mg of equivalent quercetein/g DW and 5.509 mg of equivalent catechin/g DW, respectively. The results indicate that leaves of *A. roseum*

H. Najjaa (✉) • A. Ben Arfa • M. Neffati
Laboratoire d'Ecologie Pastorale, Institut des Régions Arides, 4119 Médenine, Tunisie
e-mail: hanen.najjaa@yahoo.fr

 
M. Neffati et al. (eds.), *Medicinal and Aromatic Plants of the World - Africa Volume 3*, Medicinal and Aromatic Plants of the World,
DOI 10.1007/978-94-024-1120-1_15

and *A. ampeloprasum* present a higher antioxidant capacity with a quenching DPPH value equal to 3.78 and 30.12 mg Trolox/g DW respectively. The study revealed that rosy garlic and kurrath are a rich source of many important nutrients and bioactive compounds responsible for many promising health beneficial physiological effects and they may be considered a functional food.

**Keywords** *Allium roseum* • *A. ampeloprasum* • Nutritional value • Bioactive compounds • Antioxidant activity • Antibacterial activity • Functional food

## 1 Introduction

Tunisia occupies an area of about 164.103 km$^2$. Three quarters of which are arid and desert regions (Le Houerou 1959) characterized by low rainfall, high evapotranspiration, high temperature and desiccating winds. Among 2250 species that compose its vascular flora (Le Houerou 1959), 1630 species belong to the steppic zone (Le Houerou 1959). Yet, a wide range of plants thrive in these ecosystems, which is of high economical and ecological significance. Indeed, these plants have the ability to tolerate the high temperature and drought and have a high capacity to synthesis and accumulate large amount of bioactive substances against the abiotic stresses.

In plants, unfavourable environmental conditions such as drought and high temperature increase production and accumulation of reactive oxygen species (ROS), leading to cellular damage, metabolic disorders and senescence processes. Aromatic and Medicinal Plant (AMP) of arid and desertic region are known for their ability to withstand and quench these toxic ROS, since they are equipped with a powerful antioxidant system (Khadri et al. 2010; Akrout et al. 2011; Najjaa et al. 2011b). For economical and security reasons, the identification of new sources of natural antioxidants is a promising alternative for their use in the food industry and in preventive medicine to replace synthetic antioxidants (Tadhani et al. 2007). For that, there is an increasing interest to identify among aromatic and medicinal plants (AMP) of Arid and desertic species those with high antioxidant content for their use in the agri-food industry and/or pharmaceutical and cosmetic applications.

AMP of arid and desertic regions are naturally drought-tolerant plants that contain potential medicinal/nutraceutical properties and may be potentially useful as new sources of functional compounds in dietary food. Interestingly, numerous of tunisian's AMP of arid and desertic land have been used in folklore medicines. Recently, extracts of these plants have proven activity against human, animal, and plant pathogens and more investigations have been carried out to identify the metabolites responsible for their bioactivities for interest to modern industry and medicine (Najjaa et al. 2012; Chao et al. 2013).

As for as being beneficial to human health, *Allium* plants are already well known. For example, garlic (*A. sativum*), is of particular interest owing to its prophylactic and therapeutic actions. Antioxidative and antimicrobial activities of some *Allium* species have been previously reported (Sharma and Prasad 2001; Banerjee and

Maulik 2002; Najjaa et al. 2007, 2011a, b). This ability has mainly been attributed to a variety of sulphur containing compounds, their precursors and other numerous phenolic compounds (Kim et al. 1997; Benkeblia 2007).

In this chapter we interested specially on wild *A. roseum* (rosy garlic) and *A. ampeloprasum* (kurrath) which are very polymorphous and widespread species (Cuénod 1954; Le Floc'h 1983).

The fresh young leaves and bulbs of *A. roseum* and *A. ampeloprasum* were consumed in salads and used as spice to prepare traditional recipes. Besides their culinary use, rosy garlic and Kurrath are also used in folk medicine. Le Floc'h (1983) reported that *A. roseum* is used use for the treatment of headaches and rheumatism. It is also used for the treatment of bronchitis, colds as an inhalation, fever diminution and as an appetizer. The uses of this plant as functional food ingredient and/or food supplement in the food industry and as base for the development of new drugs in phytomedecine were also reported (Najjaa et al. 2007, 2011b).

The objective of the present study was to characterize nutritional properties, bioactive components, and functional proprieties of *A. roseum* and *A.ampeloprasum* grown in arid land of Tunisia and to infer their role in human nutrition.

## 2 Nutritional Value

The nutritional proprieties of *A. roseum* has recently determined by Najjaa et al. (2012).

Compared to the carbohydrates content of aerial parts of other *Alliums* (5–12%) (Brewster 1994), the leaves of *A. roseum* can be considered as rich sources of soluble carbohydrates (32.80 g/100 g DW ± 0.21). The same other added that *A. roseum* aerial part fibres content (12.30 g/100 g DW ± 0.05) was higher than that reported for *A. cepa* bulb (1.7%), and proteins rate (22.70 g/100 g DW ± 1.51) was relatively high compared to *A. sativum* bulbs (9.3%) and *A. cepa* (1.7%) (Haciseferoğullari et al. 2005; Dini et al. 2008). Fats accounted for 0.68% of the fresh weight of *A. roseum*, making them the least abundant class of nutrients.

*A. roseum* is also characterized by high ash content (7.20 g/100 g DW ± 1.31) including macro and micro elements (Najjaa et al. 2012). The mineral element composition of *A. roseum* exhibited a higher concentration of potassium than calcium and magnesium. The high content of potassium (1530.500 mg/100 g DW ± 0.036), in *A. roseum* which is nutritionally significant, since it contributes to the control of hypertension which results in excessive excretion of potassium (Dini et al. 2008). Calcium is found at relatively high concentration in *A. roseum* (712.500 mg/100 g DW). Onion leaf calcium concentration (2540 mg/100 g fresh weight) (Boukari et al. 2001) is much higher than that of *A. roseum* leaves but bulb calcium concentration (45 mg/100 g fresh weight) (Adrian et al. 1995) is much lower. The low sodium content of *A. roseum* and consequently low Na/K ratio (0.03) is another indication that *A. roseum* consumption would reduce the incidence of hypertension (Iqbal et al.

2006). *A. roseum* leaves also contains several oligo-elements including iron, zinc, copper and manganese (Najjaa et al. 2012).

Najjaa and their collaborator show that the fatty acid composition of *A. roseum* leaves revealed twelve compounds. The unsaturated fatty acids were represented mainly by linolenic, linoleic, oleic and gadoleic, and five saturated acids (palmitic, stearic, myristic, arachidic and margaric), accounted for ~15% of the total fatty acids (Najjaa et al. 2012).

## 3 Bioactives Substance

Phytochemical screening of *A. roseum* and *A. ampeloprasum* organic and aqueous extracts using thin-layer chromatography (TLC), are reveals the presence of bioactive compounds such as saponins, tannins, flavonoids, coumarins, steroids, cardiac glycosides, free quinone and iridoids (Najjaa et al. 2011c). Such results may establish a good support to the use of this plant in herbal medicine and as base for the development of new drugs and phytomedicine.

The content of potential health-promoting substances, flavonoid, total phenolic and flavonoids content, vitamin C, tannin, anthocyanidin, carotenoids and allicin in the wild *A. roseum* growing in the arid region of Tunisia were determined recently by Najjaa et al. (2012). Total phenolic content of *A. roseum* are 736.65 ± 1.51 mg CA/100 g DW and 3.37 mg CE/g DW respectively. *Allium* species are among the richest sources of dietary flavonoids and contribute significantly to the overall intake of flavonoids (Slimestad et al. 2007). *In vitro* and *in vivo* pharmacological tests have shown that flavonoids exhibit the following variety of actions: (i) antioxidative (Boyle et al. 2000); (ii) reduction of cardiovascular disease (Janssen et al. 1998) and (iii) reduction of carcinogenic activity (Steiner 1997). The content of polyphenols, flavonoids and tannin in *A. ampeloprasum* was 29.278 mg of equivalent gallic acid/g DW, 5.840 mg of equivalent quercetein/g DW and 5.509 mg of equivalent catechin/g DW, respectively.

The high *A. roseum* vitamin C content (1523.35 mg/100 g DW) may be an important reason that it has been reputedly used as a traditional Tunisian medicine for treating rheumatism and cold.

Garlic antibacterial bioactive principal was identified as diallylthiosulphinate and was given allicin as trivial name since 1944. This bioactive substance is also detected and with the same concentration in *A. roseum* (0.0328 μg/mL) (Najjaa et al. 2012).

The global chromatographic analysis of *A. roseum* essential oil showed nineteen compounds representing 91.4% of the whole oil, with a relatively high content of organic sulphurous compounds representing about 26% of the total eluted compounds in essential oil *A. roseum* (Najjaa et al. 2007). Methional presented the highest percentage (17.1%) among all the compounds of the essential oil of *A. roseum.* As this compound is going to be mentioned by Najjaa et al. (2007) for the first time in the essential oil of *Allium* genus with such relatively high level, methional used as an important descriptor to characterize the essential oil of *A. roseum* and consequently this endemic species itself (Najjaa et al. 2007). The S-containing compounds are responsible for the appropriate smell and taste and also the health benefits of members of *Aliliaceae* family.

Thiosulfonates which are very unstable would further decompose to produce a series of degraded compounds including sulphides, according to the solvent used and extraction conditions (Lanzotti 2006). For that raison, the analysed the thiosulfonate precursors in *A. roseum* and *A. ampeloprasum* using CLPH. The results show that both studied species are characterized by the presence of the two -glutamyl peptides γ-L-glutamyl-S-(allyl)-L-cysteine (GluAISC) and γ-L-glutamyl-S-(*trans*-1-propenyl)-Lcysteine (GluPeCS). Their corresponding sulfoxide derivatives, are (+)-S-methyl-L-cysteine sulfoxide (methiin), (+)-S-(trans-1-propenyl)-L-cysteine sulfoxide (isoalliin) and (+)-S-(2-propenyl)-L-cysteine sulfoxide (alliin) in *A roseum* and isoalliin and methiin in *A. ampeloprasum.* These differences in their flavours can correlated with the vicinity of *A. roseum* with garlic and *A. ampeloprasum* with onion (Najjaa et al. 2011c).

The antioxidant activities of *A. roseum* leaf extracts were assessed and confirmed using two functional analytical methods based on the radicals (ABTS and DPPH) scavenging potential. The extracts obtained were all able to inhibit the DPPH, as well as ABTS radicals. The antioxidant potential was 3.78 mg Trolox/g DW with the DPPH method, and 3.99 mg Trolox/g DW with the ABTS (Najjaa et al. 2011b). In comparison to previous data based on the ABTS scavenging capacity, *A. roseum* leaf extracts were comparable or higher than other investigated species known to be rich in antioxidants including strawberry (25.9), raspberry (18.5), red cabbage (13.8), broccoli (6.5), and spinach (7.6) (Proteggente et al. 2002).

The assessment of antioxidant activity showed that *A. ampeloprasum* methanol extract able to scavenge DPPH (30.12 mg Trolox/g DW) and has an important iron (II) chelators with $IC_{50}$ values of 6.03 mg/ml.

The important antioxidant activity of *Allium* species collected from arid land of assessed by the different systems could be attributed to their high total polyphenolic contents; in fact, it has been found that polyphenols are one of the most effective antioxidative constituents in the plant (Velioglu et al. 1998). Moreover, the high yield of the different phenolic compounds found in these plants might contribute to the potent antioxidant activity of the extracts, since a positive correlation between phenolic composition and antioxidant activity was proved (Que et al. 2006).

# 4 Health Effects and Future Trends

The importance of functional foods, nutraceuticals and other natural health products has been well recognized in connection with health promotion, disease risk reduction and diminution in health care costs (Shahidi and Marian 2003). The expression of nutraceutical was firstly mentioned 20 years ago to describe a union between nutrition and pharmaceutics, both key contributors to human wellness. In the last years, several research publications were devoted to so-called "functional food" and "nutraceuticals" (Bernal et al. 2011). The boundary between nutraceutical and functional food is not always clear, but the main difference is the format in which they are consumed: nutraceuticals are consumed as capsules, pills, and tablets, whereas functional foods are always consumed as ordinary foods. The results of the screening assays in these recent publications, indicating a synergistic of functional properties and composition of the extracts activities, justify the use of the investigated plant in the Tunisian food and ethnomedicine. Thus, their possible use as natural additives emerged from a growing tendency to replace synthetic substances by naturally derived ones. From this point of view, our study may be considered as an important report based on antioxidative potential of a wild *Allium* species growing in the arid region of Tunisia and could be evaluated as a starting point for further investigations with other species from their bioclimatic stage.

## 5 Conclusions

This study revealed that *A. roseum* gowning in Tunisia had a high soluble carbohydrates, crude protein and dietary fibre contents, comparatively to other *Alliums*. Its mineral content was high in potassium, and calcium. The mineral composition of 'rosy garlic' is sufficient in Ca, P, K, Cu, Fe, Zn and Mg so that it can meet many macronutrient and micronutrient requirements of the human diets. Furthermore, edible part oil included 15% saturated and 85% unsaturated fatty acids. Linolenic acid and palmitic acid were the most abundant unsaturated and saturated fatty acids, respectively. This fatty composition confers to the *A. roseum* oil considerable nutritional value, acting on physiological functions and reducing cardiovascular, cancer and arthroscleroses diseases occurrence risk. The most abundant phytonutrients found in *A. roseum* and *A. ampeloprasum* (polyphenolic compounds, flavonoids, anthyacinidins, vitamin C and allicin) exhibit a positive effect on human health as antioxidants compounds. Since *A. roseum* and *A. ampeloprasum* are a rich source of many important nutrients and bioactive compounds responsible for many promising health beneficial physiological effects, it may be considered a nutraceutical that serves as a natural source of necessary components to help fulfil our daily nutritional needs and as a functional food as well as in ethnomedecine.

*A. roseum* and *A. ampeloprasum* extracts exhibit interesting antioxidant properties correlated with their phytochemical composition, explaining, at least partially, their usefulness in traditional medicine. The in use of these plants and their derivatives for the primary purpose of flavouring and preserving foods will be of interest for further study.

## References

Adrian J, Potus J, Frangne R (1995) La science alimentaire de a à z. Lavoisier, 2ème Edition, Paris, 457 p

Akrout A, Alarcon Gonzalez L, El Jani H, Campra Madrid P (2011) Antioxidant and antitumor activities of *Artemisia campestris* and *Thymelaea hirsuta* from southern Tunisia. Food Chem Toxicol 49:342–347

Banerjee SK, Maulik SK (2002) Effect of garlic cardiovascular disorders. Nutrition 1:1–14

Benkeblia N (2007) Antimicrobial activity of essential oil extracts of various onions (*Allium cepa*) and garlic (*Allium sativum*). Lebensm-Wiss Technol 37:263–268

Bernal J, Mendiola JA, Ibanez E, Cifuentes A (2011) Advanced analysis of nutraceuticals. J Pharm Biomed Anal 55(4):758–744

Boukari I, Shier NW, Xinia EFR, Frisch J, Bruce AW, Pawloski L, Fly AD (2001) Calcium analysis of selected Western African foods. J Food Compos Anal 14:37–42

Boyle SP, Dobson VL, Duthie SJ, Kyle JAM, Collins AR (2000) Absorption and DNA protective effects of flavonoid glycosides from onion meal. Eur J Nutr 39:213–223

Brewster JL (1994) Onions and other vegetable *Alliums*. CAB International, Wallingford

Chao HC, Najjaa H, Villareal OM, Ksouri R, Junkyu H, Neffati M, Hiroko I (2013) *Arthrophytum scoparium* inhibits melanogenesis through the down-regulation of tyrosinase and melanogenic gene expressions in B16 melanoma cells. Exp Dermatol 22:131–136

Cuénod A (1954) Flore analytique et synoptique de la Tunisie. Cryptogames vasculaires, Gymnospermes et Monocotylédones. S.E.F.A.N, Tunis

Dini I, Tenore GC, Dini A (2008) Chemical composition, nutritional value and antioxidant properties of Allium caepa L. Var. tropeana (red onion) seeds. Food Chem 107:613–621

Haciseferoğullari H, Özcan M, Demir F, Çalişir S (2005) Some nutritional and technological properties of garlic (*Allium sativum* L.) J Food Eng 68:463–469

Iqbal A, Khalil IA, Ateeq N, Khan MS (2006) Nutritional quality of important food legumes. Food Chem 97:331–335

Janssen K, Mensink RP, Cox FJ, Harryvan JL, Hovenier R, Hollman PC, Katan MB (1998) Effects of the flavonoids quercetin and apigenin on hemostasis in healthy volunteers: results from an *in vitro* and a dietary supplemented study. Am J Clin Nutr 67:255–262

Khadri A, Neffati M, Smiti S, Falé P, Lino ARL, Serralheiro MLM, Araújo EMM (2010) Antioxidant, antiacetylcholinesterase and antimicrobial activities of *Cymbopogon schoenanthus* L. Spreng (lemon grass) from Tunisia. Food Sci Technol 43(2):331–336

Kim SM, Kubota K, Kabayashi A (1997) Antioxidative activity of sulphur-containing flavour compounds in garlic. Biosci Biotechnol Biochem 61:1482–1485

Lanzotti V (2006) The analysis of onion and garlic. J Chromatogr A 1112:3–22

Le Floc'h E (1983) Contribution à une étude ethnobotanique de la flore tunisienne. Programme flore et végétation tunisienne. Ministère de l'Enseignement Supérieur et de la Recherche Scientifique, Tunis

Le Houerou HN (1959) Recherches écologiques et floristiques sur la végétation de la Tunisie méridionale. Institut de Recherches sahariennes de l'Université d'Alger 2:510p

Najjaa H, Neffati M, Zouari S, Ammar E (2007) Essential oil composition and antibacterial activity of different extracts of a North African endemic species. C R Chim 10:820–826

Najjaa H, Zerria K, Fattouch S, Ammar E, Neffati M (2011a) Antioxidant and antimicrobial activities of *Allium roseum* "Lazoul", a wild edible endemic species in North Africa. Int J Food Prop 14:371–380

Najjaa H, Zouari S, Ammar E, Neffati M (2011b) Phytochemical screening and antibacterial properties of *Allium roseum* L. a wild edible species in North Africa. J Food Biochem 35:699–714

Najjaa H, Zouari S, Ingrid A, Auger J, Ammar E, Neffati M (2011c) Différences et similitudes des métabolites secondaires chez deux espèces du genre *Allium, Allium roseum* L. et *Allium ampeloprasum* L. Acta Botanica Gallica 158(1):111–123

Najjaa H, Fattouch S, Ammar E, Neffati M (2012) *Allium* species, ancient health food for future. Scient Health Soc Aspect Food Chem 17:343–354 (edition by Benamin Valdez)

Proteggente AR, Pannala AS, Pagana G, Van Buren L, Wagner E, Wiseman S (2002) The antioxidant activity of regularly consumed fruit and vegetables reflects their phenolic and vitamin C composition. Free Radic Res 36:217–233

Que F, Mao L, Pan X, (2006) Antioxidant activities of five Chinese rice wines and involvement of phenolic compounds. Food Research International 581–587

Shahidi F, Marian N (2003) Phenolics in food and neutraceuticals, vol 1. CRC Press LLC, Boca Raton, pp 144–150

Sharma GP, Prasad S (2001) Drying of garlic (*Allium sativum* L.) cloves by microwave-hotair combination. J Food Eng 50:99–105

Slimestad R, Fossen T, Vagen IM (2007) Onions: a source of unique dietary flavonoids. J Agric Food Chem 55:10067–10080

Steiner M (1997) The role of flavonoids and garlic in cancer prevention. In: Ohigashi H (ed) Food Factors for Cancer Prevention. Springer, Tokyo, pp 222–225

Tadhani MB, Patel VH, Subhah R (2007) *In vitro* antioxidant activities of *Stevia rebaudian* leaves and callus. J Food Compos Anal 20:323–329

Velioglu YS, Mazza G, Gao L, Oomah BD (1998) Antioxidant activity and total polyphenolics in selected fruits, vegetables, and grain products. J Agric Food Chem 46:4113–4117

# Utilization of Pteridophytes as Herbal Medicines in Sub-Saharan Africa

**Alfred Maroyi**

**Abstract** Use of pteridophytes for medicinal purposes is entrenched in cultural practices throughout sub-Saharan Africa. The current study is aimed at answering the following questions through a study of the literature: (i) Which pteridophyte species are used for medicinal purposes in sub-Saharan Africa? (ii) What are their therapeutic uses, preparation and administration? and (iii) What are the geospatial patterns of medicinal pteridophytes in sub-Saharan Africa? This investigation is based on review of the literature published in scientific journals, books, conference papers, floras, theses, reports from national, regional and international organizations obtained from libraries and electronic search of Google Scholar, ISI Web of Science and Scopus. A total of 99 pteridophyte species are used for medicinal purposes in sub-Saharan Africa, corresponding to 8.4% of the total pteridophyte species of the African continent. A total of 42 medical conditions are treated using herbal remedies made from pteridophyte species. Ethnopharmacological studies are recommended for the documented pteridophytes species as some of them have already shown interesting applications in primary healthcare.

**Keywords** Pteridophytes • Medicinal • Cultural and ritualistic uses • Traditional medicine • Sub-Saharan Africa

## 1 Introduction

Traditional medicines have been used to treat human and animal diseases in sub-Saharan Africa for many generations before the introduction of orthodox medicine. Use of traditional medicines in sub-Saharan Africa is widely practiced and this practice is generally regarded as part of the African culture. Traditional medicines therapy, although still an unwritten science, is well established in some cultures and traditions (Shale et al. 1999), with the World Health Organization estimating that up

A. Maroyi (✉)
Department of Botany, Faculty of Science and Agriculture, University of Fort Hare, Private Bag X1314, Alice 5700, South Africa
e-mail: amaroyi@ufh.ac.za



M. Neffati et al. (eds.), *Medicinal and Aromatic Plants of the World - Africa Volume 3*, Medicinal and Aromatic Plants of the World,
DOI 10.1007/978-94-024-1120-1_16

**Fig. 1** Young fronds of *Pteridium aquilinum* (L.) Kuhn developing after the start of the rains

to 80% of the people in developing countries still depend on herbal medicines to fulfil their primary healthcare needs (WHO 2002). About 25–50% of all current pharmaceutical drugs are derived from plants (Harnett et al. 2005). Of the 252 drugs considered as basic and essential by the World Health Organization (WHO), 11% are exclusively of plant origin and a significant number are synthetic drugs obtained from natural precursors (Rates 2001). It is not so much the case that traditional medicines offer an alternative form of medical care, but rather it is in fact the only source of treatment available for many people, particularly those living in undeveloped rural and marginalized areas in sub-Saharan Africa (WHO 2002).

The medicinal qualities of pteridophytes were recognized as early as 300 BC by the Greek philosopher Theophrastus (Puri 1970). Pteridophytes existing today represent an ancient plant group which appeared about 300 million years ago in the late Devonian period (Fernández et al. 2011). Pteridophyte species (Fig. 1), including ferns and fern allies (Maroyi 2014a) play an important role as sources of traditional medicines and other functions important for people's livelihoods in the tropics and subtropics. Isolated studies conducted in Lesotho (Moteetee and van Wyk 2011), Nigeria (Nwosu 2002), South Africa and Swaziland (Roux 2003) showed that herbal medicines derived from pteridophytes are used to cure many human and animal diseases and disorders. Published literature on pteridophytes used in African traditional medicine usually lack details on therapeutic uses, preparation and administration of medicinal pteridophytes. A significant number of biologically active compounds have been isolated from medicinal pteridophyte species in the past, based on their medicinal applications and uses (Ho et al. 2010). Therefore, pteridophytes used as traditional medicines in sub-Saharan Africa may represent one of the untapped sources of biologically active compounds that need to be exploited as possible sources of pharmaceutical drugs. The current study is aimed at answering the following questions through a study of the literature:

1. Which pteridophyte species are used for medicinal purposes in sub-Saharan Africa?
2. What are their therapeutic uses, preparation and administration?
3. And finally, what are the geospatial patterns of medicinal pteridophytes in sub-Saharan Africa?

## 2 Research Methods

Pteridophytes documented in literature as having medicinal uses in sub-Saharan Africa are recorded in this study. Most of the available ethnobotanical information on pteridophytes characterized by medicinal uses is scattered in various publications such as books, floras, scientific journals, theses, reports from natitional, regional and international organizations. The total number of scientific papers, books and other literature sources used in this study is 56. Literature was searched on international online databases such as ISI Web of Science, Scopus and Google Scholar using specific search terms such as "medicinal ferns or pteridophytes". References were also identified by searching the library collections of the University of Fort Hare, South Africa. All plant scientific names, plant families and plant authorities were verified using internet sources such as the International Plant Name Index (www.ipni.org), the Missouri Botanical Garden's Tropicos Nomenclatural database (www.tropicos.org) and the Royal Botanic Garden and Missouri Botanic Garden plant name database (www.theplantlist.org). For each species, data was also collected from literature on countries in which the species are utilized, therapeutic use(s), mode of administration and preparation of the species. Disease and ailment categories were classified according to ailment categories proposed by Cook (1995).

## 3 Results and Discussion

### *Pteridophyte Diversity*

A total of 99 pteridophyte species are used for medicinal purposes in sub-Saharan Africa (Table 1), corresponding to 8.4% of the total pteridophyte species of the African continent. About 627 pteridophyte species are known to grow in Africa and an additional 557 species have been recorded in Madagascar (Aldasoro et al. 2004; Kornaś 1993). These species belong to 24 families and 41 genera (Table 1). The medicinal use category includes pteridophytespecies used for treating human and

**Table 1** Pteridophytes used for medicinal purposes in sub-Saharan Africa

| Species and family | Country | Use(s) | Reference(s) |
|---|---|---|---|
| *Acrostichum aureum* L., Pteridaceae | Tanzania | Tea made from dry leaves drunk as remedy for epilepsy | Matthews (1993) |
| *Adiantum aethiopicum* L., Adiantaceae | Nigeria | Decoction of whole plant used as tonic against bruises, gangrenous wounds, expectorant and diuretic disorders | Nwosu (2002) |
| | South Africa | Leaves smoked for head and chest colds | van Wyk et al. (2013) |
| *Adiantum capillus-veneris* L., Adiantaceae | Lesotho | Leaves smoked for head and chest colds | Moteetee and van Wyk (2011) |
| | Nigeria | Leaf infusion drunk against infertility and vaginal discharge; decoction of roots used for washing hair to remove dandruff; powdered rhizome mixed with camphor and dissolved in palm-kernel oil applied externally to hair infested by lice | Nwosu (2002) |
| | South Africa | Tea made from the plant and leaves smoked for head and chest colds | van Wyk et al. (2013) |
| | Swaziland | Leaves smoked for respiratory tract disorders and used for hair and scalp problems | Roux (2003) and Long (2005) |
| *Adiantum caudatum* L., Adiantaceae | Nigeria | Decoction used for treating jaundice in animals and scabies; potherb taken internally against abdominal pain and constipation | Nwosu (2002) |
| *Adiantum incisum* Forsk., Adiantaceae | Nigeria | Leaf infusion used against malaria and bronchial diseases | Nwosu (2002) |
| *Adiantum poiretii* Wikstr., Adiantaceae | South Africa | Leaves smoked for head and chest colds | Roux (2003) |
| | Swaziland | Leaves smoked for respiratory tract disorders and used for hair and scalp problems | Long (2005) |

(continued)

**Table 1** (continued)

| Species and family | Country | Use(s) | Reference(s) |
|---|---|---|---|
| *Adiantum raddianum* C. Presl, Adiantaceae | Swaziland | Leaves smoked for respiratory tract disorders and used for hair and scalp problems | Long (2005) |
| *Adiantum venustum* G. Don, Adiantaceae | Nigeria | Extract of rhizome and leaves drunk against diabetes, liver problems and used as diuretic; infusion taken internally for circulatory disorders | Nwosu (2002) |
| *Adiantum vogelii* Mett. ex Keys, Adiantaceae | Gabon | Leaves used as herbal medicine | Sassen and Wan (2006) |
| | Ivory coast | Plant used for faints, treatment of debility, edema and skin lesions of leprosy | Bouquet (1974) |
| *Aleuriopteris albomarginata* (C.B. Clarke) Ching, Pteridaceae | Nigeria | Decoction of whole plant taken against peptic ulcers; powdered leaves mixed with *Zingiber officinale* Roscoe. rhizomes used against infertility in women and infusion taken against dysentery | Nwosu (2002) |
| *Ampelopteris prolifera* (Retz.) Copel., Thelypteridaceae | Tanzania | Leaf sap mixed with leaves of *Hypoestes aristata* (Vahl) Sol. ex Roem. & Schult. and drunk to treat meningitis and encephalitis | Nwosu (2002) |
| *Anemia vestita* (Baker) Christenh., Anemiaceae | Swaziland | Used to rid children of worms and mixed with fat for burns | Long (2005) |
| *Arthromeris wallichiana* (Spr.) Ching, Polypodiaceae | Nigeria | Leaf paste applied externally for sprains, paste mixed with egg albumen applied to fractured bone and used for calcinosis | Nwosu (2002) |
| *Arthropteris palisotii* (Desv.) Alston, Tectariaceae | Ivory Coast | Leaf juice used for treatment of faints | Bouquet (1974) |
| *Asplenium achilleifolium* (M. Martens & Galeotti) Liebm., Aspleniaceae | Nigeria | Decoction of leaves and roots drunk as health tea and purge; root powder mixed with *Elaeis guineense* Jacq. oil taken internally for stomachache and intestinal worms | Nwosu (2002) |

(continued)

**Table 1** (continued)

| Species and family | Country | Use(s) | Reference(s) |
|---|---|---|---|
| *Asplenium adiantum-nigrum* L., Aspleniaceae | Nigeria | Leaf extract mixed with *Ricinus communis* L. oil and used as purge to remove intestinal worms; leaf decoction taken against urinary disorders; tonic used as antiseptic after ear-piercing | Nwosu (2002) |
| | South Africa | Anthelmintic and emetic | Watt and Brandwijk (1962) |
| *Asplenium africanum* Desv., Aspleniaceae | DRC | Leaf infusion used as enema for spleen disease | Terashima et al. (1991) |
| | Gabon | Whole plant used as herbal medicine | Sassen and Wan (2006) |
| *Asplenium blastophorum* Hieron., Aspleniaceae | Tanzania | Pounded leaves mixed with cold water applied to stiffy neck | Matthews (1993) |
| *Asplenium bulbiferum* G. Forst., Aspleniaceae | Nigeria | Leaf paste applied externally for hemorrhoids; leaf decoction taken for liver problems | Nwosu (2002) |
| *Asplenium cuneatum* Lam., Aspleniaceae | South Africa | Emetic | Watt and Brandwijk (1962) |
| *Asplenium monanthes* L., Aspleniaceae | Lesotho | Leaves smoked for head and chest colds | Moteetee and van Wyk (2011) |
| | South Africa | Leaves smoked for head and chest colds and also as diaphoretic | Hutchings (1992) |
| *Asplenium trichomanes* L., Aspleniaceae | Lesotho | Leaves smoked for head and chest colds | Moteetee and van Wyk (2011) |
| | South Africa | Leaves smoked for head and chest colds | Hutchings (1992) |
| *Blotiella glabra* (Bory) R.M. Tryon, Dennsteadtiaceae | DRC | Leaves used for menstruation pain | Termote (2012) |
| *Botrychium lanuginosum* Wall. ex Hook et Grev., *Ophioglossaceae* | Nigeria | Boiled rhizomes used for pneumonia, catarrh and expectorant | Nwosu (2002) |
| *Cheilanthes albomarginata* C.B. Clarke, Pteridaceae | Nigeria | Leaf extract used externally for abscess and decoction of rhizomes administered internally against diarrhoea | Nwosu (2002) |

(continued)

**Table 1** (continued)

| Species and family | Country | Use(s) | Reference(s) |
|---|---|---|---|
| *Cheilanthes eckloniana Mett.*, Pteridaceae | Lesotho | Leaves smoked for head and chest colds | Moteetee and van Wyk (2011) |
| | South Africa | Leaves smoked for head and chest colds and ashes from burnt leaves used on sores | van Wyk et al. (2013) |
| | *Swaziland* | Leaves smoked for head and chest colds and ashes from burnt leaves used for sores | Long (2005) |
| *Cheilanthes farinosa* (Forssk.) Kaulf., Pteridaceae | *Ethiopia* | Leaf powder crushed with leaves of *Rhamnus prinoides* L'Hér. and the whole plant of *Crepis rueppellii* Sch. Bip. and brushed over the body as remedy for hepatitis | Yineger et al. (2007) |
| | *Uganda* | Decoction of leaves, roots and stems used for headache, paludism and gastritis | Tabuti et al. (2012) |
| *Cheilanthes hirta* Sw., Pteridaceae | Lesotho | Decoction drunk for colds and sore throat | Moteetee and van Wyk (2011) |
| | South Africa | Used as anthelmintic, sore throats, head and chest colds. Burned with *Mohria caffrorum* (L.) Desv. as an inhalation for a restless child and rhizomes used as cancer cure | Roux (2003) and van Wyk et al. (2013) |
| | Swaziland | Decoction used for sore throat, colds and asthma | Roux (2003) and Long (2005) |
| *Cheilanthes involuta* (Sw.) Schelpe & N.C.Anthony, Pteridaceae | Lesotho | Leaves smoked for head and chest colds | Moteetee and van Wyk (2011) |
| | South Africa | Leaves smoked for head and chest cold | van Wyk et al. (2013) |

(continued)

**Table 1** (continued)

| Species and family | Country | Use(s) | Reference(s) |
|---|---|---|---|
| *Cheilanthes quadripinnata* (Forssk.) Kuhn, Pteridaceae | Lesotho | Rhizome mixed with milk drunk for diarrhoea and spider bites | Moteetee and van Wyk (2011) |
| | South Africa | Rhizome decoction injected into the vagina of a cow if it does not get rid of the placenta after calving. Crushed leaves mixed with milk and used for diarrhoea and spider bites | Watt and Brandwijk (1962) and Roux (2003) |
| | Swaziland | Rhizome mixed with milk used for diarrhoea | Long (2005) |
| *Cheilanthes spp.*, Pteridaceae | Lesotho | Whole plant slightly roasted and powdered material applied to wounds | Shale et al. (1999) |
| *Cheilanthes viridis* (Forssk.) Swart, Pteridaceae | *Madagascar* | Aerial parts used as diuretics | Jenkins (1987) |
| | South Africa | Paste made from dried ground leaves applied to burns, sores and other skin complaints | Hutchings et al. (1996) |
| | Swaziland | Paste made from dried ground leaves applied to burns, sores and other skin complaints | Roux (2003) and Long (2005) |
| *Coniogramme caudata* Ching, Adiantaceae | Nigeria | Leaf extract taken as remedy for urinary disorders, backache and gall stones | Nwosu (2002) |
| *Cyathea dregei* Kunze, Cyatheaceae | South Africa | Infusions taken regularly during pregnancy to ensure easy childbirth and dried roots used as anthelmintics | Hutchings et al. (1996) |
| | Swaziland | Tonic taken in pregnancy to ensure easy birth; roots taken for toothache and anthelmintic | Long (2005) |
| *Cyathea manniana* Hook., Cyatheaceae | DRC | Leaves used for menstruation pain | Termote (2012) |
| | Ethiopia | Leaf decoction used for sexually transmitted diseases | Tesfaye and Sebsebe (2009) |
| | South Africa | Leaf decoction used as worm remedy | Watt and Brandwijk (1962) |

(continued)

**Table 1** (continued)

| Species and family | Country | Use(s) | Reference(s) |
|---|---|---|---|
| *Cystopteris fragilis* Bernh., Woodsieaceae | Lesotho | Rhizome decoction used as enema for deworming and constipation | Moteetee and van Wyk (2011) |
| *Dicranopteris linearis* (Burm. f.) Underw., Gleichenaceae | Nigeria | Extract of whole plant administered as remedy for convulsions in children | Nwosu (2002) |
| *Diplazium esculentum* (Retz.) Sw., Athyriaceae | Nigeria | Leaf infusion used for toothache and by pregnant women as protection against difficult childbirth | Nwosu (2002) |
| *Diplazium sammatii* (Kuhn) C.Chr., Athyraceae | DRC | Roots used for menstruation pain | Termote (2012) |
| *Dryopteris athamantica* (Kunze) Kuntze, Dryopteridaceae | South Africa | Used as general anthelmintic and rhizome decoction used for retained placenta in cows | Hutchings et al. (1996) |
| | Swaziland | Rhizome used as a general anthelmintic | Long (2005) |
| *Dryopteris filix-mas* (L.) Schott, Dryopteridaceae | Nigeria | Leaves used as vermifuge and leaf infusion used as aphrodisiac | Nwosu (2002) |
| *Dryopteris inaequalis* (Schlechtd.) Kuntze, Dryopteridaceae | Ethioipia | Root decoction used as ethnoveterinary medicine | Yineger et al. (2007) |
| | Kenya | Herbal medicine, uses not specified | Timberlake (1994) |
| | South Africa | Rhizome taken as anthelmintic | Hutchings et al. (1996) |
| *Dryopteris lewalleana* Pic. Serm., Dryopteridaceae | South Africa | Rhizome used as an anthelmintic | Roux (2003) |
| | Swaziland | Rhizome used as a general anthelmintic | Long (2005) |
| *Dryopteris pentheri* (Krasser) C. Chr., Dryopteridaceae | Burundi | Used against tenia and parasites | Baerts and Lehmann (1996) |
| | Swaziland | Rhizome used as a general anthelmintic | Long (2005) |
| *Dryopteris wallichiana* (Spreng) Hyl., Dryopteridaceae | Nigeria | Tonic from roots drunk against rheumatism | Nwosu (2002) |
| *Elaphoglossum petiolatum* Bonap., *Lomariopsidaceae* | South Africa | Root decoction used for sore throat | Watt and Brandwijk (1962) |
| *Equisetum diffusum* D. Don, Equisetaceae | Nigeria | Root decoction used to treat psychosis and dried leaf powder mixed with water used as insect repellant | Nwosu (2002) |

(continued)

**Table 1** (continued)

| Species and family | Country | Use(s) | Reference(s) |
|---|---|---|---|
| *Equisetum ramosissimum* Desf., Equisetaceae | Lesotho | Rhizome decoction drunk for infertility in women, colds and colic in infants | Moteetee and van Wyk (2011) |
| | South Africa | Plant sap used to relieve toothache and applied to wound after extraction. Powdered stem and water or milk infusions administered as enemas to children with abdominal upsets. Rhizome decoctions used for fertility, colds, venereal diseases, diarrhoea, septic inflammations, glandular swellings, earache and diuretics. Roots crushed and smoked as remedy for asthma | Hutchings et al. (1996) and Roux (2003) |
| | Swaziland | Cooked rhizome enhances fertility in women | Long (2005) |
| *Glaphyropteridopsis erubescens* (Wall ex Hook.) Ching, Thelypteridaceae | Nigeria | Leaf decoction used against indigestion; leaf paste applied externally for rheumatism; root powder mixed with water and used as antidote for scorpion bite | Nwosu (2002) |
| *Lycopodiella cerma* (L.) Pic. Serm., Lycopodiaceae | DRC | Flea repellant | Lemmens (2008) |
| | Madagascar | Decoction used as tonic and a mixture with *Tristemma mauritianum* J.F. Gmel. used to treat neuralgia and hypertension; whole plant used to prepare tea that is drunk for stomach ulcers | Lemmens (2008) |
| | Rwanda | Whole plant crushed and applied as dressing to wounds | Lemmens (2008) |
| *Lycopodium cernuum* L., Lycopodiaceae | DRC | Whole plant used as insect repellant | Lemmens (2008) |

(continued)

**Table 1** (continued)

| Species and family | Country | Use(s) | Reference(s) |
|---|---|---|---|
| *Lycopodium clavatum* L., Lycopodiaceae | DRC | Whole plant placed on the bed like a mattress to keep insects away | Yamada (1999) |
| | Lesotho | Leaves dried and smoked for headache, colds in the head and chest. Whole plant dried and smoked with *Selaginella caffrorum* (Milde.) Hieron. to cure head and chest colds | Zimudzi and Bosch (2008) and Moteetee and van Wyk (2011) |
| | Madagascar | Used as neuralgia; whole plant is grilled with sugarcane and banana skins and applied to cracked lips to promote healing. | Jenkins (1987) and Zimudzi and Bosch (2008) |
| | Nigeria | Leaf decoction applied externally to wounds or ulcer; pulverized leaves mixed with unripe *Musa acuminata* Colla mixed with milk given in small doses to infants suffering from diarrhoea and dysentery | Nwosu (2002) |
| | Rwanda | Extract of aerial parts drunk as cure for dysentery and malaria | Zimudzi and Bosch (2008) |
| | South Africa | Whole plant smoked for headaches and used as decoction diuretic and anti-spasmatic | Hutchings et al. (1996) |
| | Swaziland | Used to increase urine flow and relieve spasms | Long (2005) |
| *Lycopodium selago* L., Lycopodiaceae | Nigeria | Spores and root powder mixed with *Elaeis guineense* Jacq. oil and applied externally to treat eczema, cuts; leaf extract drunk against snake bite and as antidote | Nwosu (2002) |
| *Lygodium flexuosum* (L.) Sw., Lygodiaceae | Nigeria | Leaf infusion used for treating female infertility; paste applied to fix fractured bones and against hemorrhoids | Nwosu (2002) |
| *Lygodium microphyllum* (Cav.) R. Br., Lygodiaceae | Ivory Coast | Crushed leaves used to cure hiccough | Bouquet (1974) |

(continued)

**Table 1** (continued)

| Species and family | Country | Use(s) | Reference(s) |
|---|---|---|---|
| *Lygodium palmatum* (Bernh.) Sw., Lygodiaceae | Nigeria | Leaf extract applied externally to boils, abscess, swellings; used as astringent and emollient | Nwosu (2002) |
| *Marattia fraxinea* Sm., *Marattiaceae* | DRC | Root infusion used as enema for pregnant women near delivery and for those who suffer from abdominal troubles | Terashima et al. (1991) |
| | Tanzania | Remedy for ankylostomiasis | Watt and Brandwijk (1962) |
| *Marsilea minuta* L., Marsileaceae | Nigeria | Plant extract used as aphrodisiac and for increased fertility | van der Burg (2004a) |
| *Marsilea quadrifolia* L., Marsileaceae | Nigeria | Whole plant extract used as aphrodisiac and for increased fertility by married couple | Nwosu (2002) |
| *Microgramma lycopodioides* (L.) Copel., Polypodiaceae | South Africa | Whole plants used against public lice | Hutchings et al. (1996) |
| | Tanzania | Crushed leaves mixed with porridge used as remedy for anaemia and stomach pains | Matthews (1993) |
| *Microgramma mauritiana* (Willd.) Tardieu, Polypodiaceae | South Africa | Whole plant used against lice on humans | Hutchings et al. (1996) |
| | Swaziland | Whole plant used against lice on humans | Roux (2003) and Long (2005) |
| *Microgramma owariensis* (Desv.) Alston, Polypodiaceae | DRC | Women put this fern on their body for 3 days as remedy for menorrhagia and used for abdominal disorder | Yamada (1999) |
| *Microsorum punctatum* (L.) Copel, Polypodiaceae | DRC | Leaf decoction used as enema | Yamada (1999) |
| | Ivory Coast | Leaves and juice purgative, diuretic and wound healing | Bouquet (1974) |
| *Mohria caffrorum* (L.) Desv., Anemiaceae | South Africa | Smoke of burnt leaf is inhaled by children for nightmares, dried leaf ointment used as cooling ointment for burns and leaves smoked for head and chest colds. Burned with *Cheilanthes hirta*Sw. as an inhalation for a restless child | Hutchings et al. (1996) and van Wyk et al. (2013) |

(continued)

**Table 1** (continued)

| Species and family | Country | Use(s) | Reference(s) |
|---|---|---|---|
| *Mohria vistata* Baker, Anemiaceae | South Africa | Plant said to rid children of worms and the ground leaves are mixed with fat and used as an ointment against burns | Roux (2003) |
| *Nephrolepsis acutifolia* (Desv.) Christ, Davalliaceae | Gabon | Whole plant used as herbal medicine | Sassen and Wan (2006) |
| *Nephrolepsis biserrata* (Sw.) Schott, Davalliaceae | Cameroon | Leaf decoction used for abdominal pains | Jiofack et al. (2008) |
| | DRC | Leaves used to remove thorns in the body | Termote (2012) |
| | Ivory Coast | Leaves used as enema or therapeutic meal during pregnancy and wounds | Malan and Neuba (2011) |
| | Liberia | Snake bites and urinary complaints | Harley (1941) |
| *Nephrolepsis cordifolia* (L.) Presl., Davalliaceae | Nigeria | Leaf infusion administered for amnesia | Nwosu (2002) |
| *Nephrolepsis undulata* (Afzel. ex Sw.) J. Sm., Davalliaceae | Gabon | Whole plant used as herbal medicine | Sassen and Wan (2006) |
| *Ophioglossum capense* Sw., Ophioglossaceae | South Africa | Rhizome decoction used as remedy for boils | Watt and Brandwijk (1962) |
| *Ophioglossum costatum* R.Br., Ophioglossaceae | Madagascar | Rhizome decoction taken internally to treat lung and heart diseases; dried and pulverized rhizome applied to sores, wounds and burns | van der Burg (2004b) |
| *Ophioglossum reticulatum* L., Ophioglossaceae | Gabon | Whole plant used as herbal medicine | Sassen and Wan (2006) |
| | Lesotho | Rhizome decoction used topically on boils | van der Burg (2004b) |
| | Nigeria | Rhizome extract used as antidote for snake bite | van der Burg (2004b) |
| | Tanzania | Leaf juice drunk against spasms of the heart | van der Burg (2004b) |
| *Ophioglossum vulgatum* L., Ophioglossaceae | Lesotho | Rhizome decoction used to bathe boils | Moteetee and van Wyk (2011) |
| | Nigeria | Rhizome decoction taken internally to treat pulmonary, bronchial and heart diseases; dried pulverized rhizome externally applied to ulcerated sores, burns and wounds | Nwosu (2002) |

(continued)

**Table 1** (continued)

| Species and family | Country | Use(s) | Reference(s) |
|---|---|---|---|
| *Osmunda regalis* L., Osmundaceae | Nigeria | Whole plant extract taken internally for psychosis; infusion of roots used against malaria and jaundice | Nwosu (2002) |
| *Pellaea calomelanos* (*Sw.*) *Link*., Pteridaceae | Kenya | Herbal medicine, uses not specified | Timberlake (1994) |
| | Lesotho | Cooked rhizome decoction used to get rid of afterbirth in cows and colds in humans | Moteetee and van Wyk (2011) |
| | Malawi | Whole plant burnt and smoke directed into vagina as remedy for painful uterus | Gelfand et al. (1985) |
| | South Africa | Smoke from burnt leaves inhaled for headaches, asthma, head colds and leaf decoction taken for diarrhoea, coughs and colds. Rhizome decoctions taken or applied to boils and mouth or nasal ulcers while boiled roots and rhizomes are used for tuberculosis and as poultices for abscesses. Milk decoctions of rhizomes used for internal sores and parasites | Hutchings et al. (1996) and Semenya and Maroyi (2013) |
| | Swaziland | Rhizome decoction used for sores and boils | Roux (2003) and Long (2005) |
| | Zimbabwe | Powder of whole plant taken in porridge to prevent abortion or smoke directed into vagina to prevent abortion. Infusions of whole plant taken by mouth and plant also burnt and smoke inhaled as remedy for convulsions. Ointment of whole plant applied to depressed fontanelle. Whole plant mixed with roots of *Peltophorum africanum* Sond. and the powder taken in porridge as remedy for abdominal pains | Gelfand et al. (1985) |

(continued)

**Table 1** (continued)

| Species and family | Country | Use(s) | Reference(s) |
|---|---|---|---|
| *Pellaea involuta* (Sw.) Baker, Pteridaceae | South Africa | Rhizome used for spider bite and diarrhoea | Watt and Brandwijk (1962) |
| *Pellaea longipilosa* Bonap., Pteridaceae | Kenya | Herbal medicine, uses not specified | Timberlake (1994) |
| *Pellaea rufa* A.F. Tryon, Pteridaceae | South Africa | Roots used for sores and skin complaints | Hutchings et al. (1996) |
| *Pellaea spp.*, Pteridaceae | Zimbabwe | Fronds and roots burnt and smoke inhaled as remedy for chest pains | Maroyi (2013) |
| *Polypodiodes microrhizoma* (C. B. Clarke ex Baker) Ching, Polypodiaceae | Nigeria | Leaf and rhizome decoction used for backache, gastrointestinal disorders and jaundice; leaf paste applied externally to fissures on hand and wounds; paste mixed with palm-kernel oil applied externally to sheep or cattle | Nwosu (2002) |
| *Polystichum pungens* (Kaulf.) C.Presl, Dryopteridaceae | Lesotho | Rhizome decoction used as vermifuge | Moteetee and van Wyk (2011) |
| | South Africa | Rhizome decoctions administered as enemas for intestinal worms and wounds | Hutchings et al. (1996) |
| *Pteridium aquilinum* (L.) Kuhn, Pteridaceae | Angola | Leaves and stems used for toothache and madness | Hutchings et al. (1996) |
| | DRC | Leaves used for eye injuries and young buds to remove thorns in feet; leaf infusion given by enema as a purgative for cleaning bowels. Shoot pounded to a paste and applied to swellings. | Terashima et al. (1991) and Termote (2012) |
| | Ivory Coast | Pulp of cooked crosiers used as enema to overcome sterility in women. Rhizome mixed with rhizome of *Zingiber officinale* Roscoe pounded and juice drunk as aphrodisiac | van der Burg (2004c) |
| | Nigeria | Rhizome decoction drunk as herbal tea and purge | Nwosu (2002) |

(continued)

**Table 1** (continued)

| Species and family | Country | Use(s) | Reference(s) |
|---|---|---|---|
| | South Africa | Root decoction used as vermifuge and mixed with rootstock of *Gymnanthemum corymbosum* (Thunb.) H. Rob. for menstrual irregularities and as abortifacient. Rhizome sap used for chronic septic sores. Rhizome used as anthmelmintic for livestock and for stomachache and diarrhoea in humans. Stem sap used for insect bites, stings and rashes. Decoction of roots taken orally to facilitate trouble-free childbirth and during painful and extended labour | Hutchings et al. (1996) and Roux (2003) |
| | Swaziland | Roots make teas for rickets, stomach cramps, diarrhoea, worms and male sexual impotence | Long (2005) |
| | Tanzania | Leaf decoction used for oral candidiasis and tuberculosis | Kisangau et al. (2011) |
| *Pteris prolifera* Hieron., Pteridaceae | DRC | Whole plant used as an insect repellant | Yamada (1999) |
| *Pteris quadriaurita* Retz., Pteridaceae | Nigeria | Crushed rhizome used as astringent and emollient and extract of leaves mixed with *Elaeis guineense* Jacq. used as insect repellant | Nwosu (2002) |
| *Selaginella caffrorum (Milde.) Hieron.*, Selaginellaceae | Lesotho | Whole plant dried and smoked with *Lycopodium clavatum* L. to cure head and chest colds | Moteetee and van Wyk (2011) |
| | South Africa | Whole plant mixed with *Lycopodium clavatum* L. or *Selaginella sellowii* Hieron. smoked for headaches; decoction diuretic and anti-spasmatic | Hutchings et al. (1996) |

(continued)

**Table 1** (continued)

| Species and family | Country | Use(s) | Reference(s) |
|---|---|---|---|
| *Selaginella cinerascens* A. A. Eaton, Selaginellaceae | Nigeria | Dried rhizomes and spores mixed with boiled water used as douche and emollient; leaf infusion drunk to relieve gastrointestinal disorders such as heart burns and atrophy | Nwosu (2002) |
| *Selaginella dregei* (C. Presl) Hieron., Selaginellaceae | South Africa | Whole plant burnt and smoke directed into mouth for toothache | van Wyk and Gericke (2007) |
| *Selaginella fissidentodes* (Hook. & Grev.) Spring, Selaginellaceae | Madagascar | Leaf decoction used for cough | de Winter and Jansen (2003) |
| *Selaginella myosurus* Alston, Selaginellaceae | Cameroon | Whole plant maceration used for headache | Jiofack et al. (2008) |
| *Selaginella pallescens* (C. Presl) Spring, Selaginellaceae | Nigeria | Leaf paste applied externally on joints to treat rheumatism; rhizome infusion drunk against insomnia; young leaves steamed mixed with *Vitellaria paradoxa* C. F. Gaertn. and used for hemorrhoids as ointment. It is applied on fistulas | Nwosu (2002) |
| *Selaginella sellowii* Hieron., Selaginellaceae | South Africa | Whole plant smoked for headaches; decoction diuretic and anti-spasmatic | Hutchings et al. (1996) |
| *Selaginella* spp., Selaginellaceae | DRC | Dried leaves of *Afromomum laurentii* (De Wild. & T. Durand) K. Schum. and the resin of *Canarium schweinfurtii* Engl. are used in baths for measles. The plant is used for snake bites | Terashima et al. (1991) |
| *Selaginella tenuissima* Fée, Selaginellaceae | Nigeria | Leaf decoction drunk against liver problems, hepatoma; leaf paste mixed with ashes of *Elaeis guineense* Jacq. applied externally to tattooed skin as an antiseptic and cosmetics | Nwosu (2002) |
| *Selaginella vogelii* Mett., Selaginellaceae | Cameroon | Whole plant maceration used for kidney problems, skin and cutaneous diseases | Jiofack et al. (2008) |

(continued)

**Table 1** (continued)

| Species and family | Country | Use(s) | Reference(s) |
|---|---|---|---|
| *Stenochlaena tenuifolia* (Desv.) T.Moore, *Blechnaceae* | Congo (Brazzaville) | Sap taken with ripe banana as aphrodisiac | van der Burg (2004d) |
| *Tectaria coadunata* (Wall. ex Hook. & Grev.) C. Chr., Dryopteridaceae | Nigeria | Leaf powder mixed with *Ricinus communis* L. oil and given to goats and sheep for running stomach; young leaves chewed by cow after delivery to accelerate the expulsion of the placenta; root paste applied externally to stop foot itching caused by fungal growth and root infusion drunk against gonorrhea | Nwosu (2002) |
| *Thelypteris dentata* (Forssk.) E. P. St. John, Thelypteridaceae | Tanzania | Leaves crushed and applied over body parts against bacterial and fungal infections | Chhabra et al. (1987) |
| *Vittaria elongata* Sw., Pteridaceae | Tanzania | Leaves cooked in water then rubbed over patient as remedy for fever | Matthews (1993) |
| *Woodwardia unigemmata* (Makino) Nakai, Blechnaceae | Nigeria | Infusion of spores and rhizomes administered internally for ringworm; leaf decoction used as tea as well as bath for infertility in women, abdominal pain, constipation and sore throat | Nwosu (2002) |

animal ailments and diseases. Aspleniaceae, Polypodiaceae, Pteridaceae and Selaginellaceae families have at least five species used for medicinal purposes (Fig. 2a). Members of these pteridophyte families are also used in other regions of the world as herbal medicines, for example, Aspleniaceae, Polypodiaceae and Selaginellaceae are among taxa employed as herbal medicines in northeast Brazil (Agra et al. 2008). Genera with the highest number of species used for medicinal purposes are *Asplenium* and *Selaginella*, represented by at least six species each (Fig. 2b). The medicinal value of the genus *Asplenium* was early recognized by taxonomists, as the name "*Asplenium*" comes from the Greek word "splen", for "spleen", denoting the ancient medicinal use of the genus in treating diseases of the spleen (Yarborough and Powell 2002). Members of the genus *Selaginella* have high pharmaceutical value due to the presence of a variety of secondary metabolites such as alkaloids, phenol, terpenoids and some species demonstrated anti-inflammatory, antibacterial, antifungal, antiviral, antimicrobial and antioxidant activities

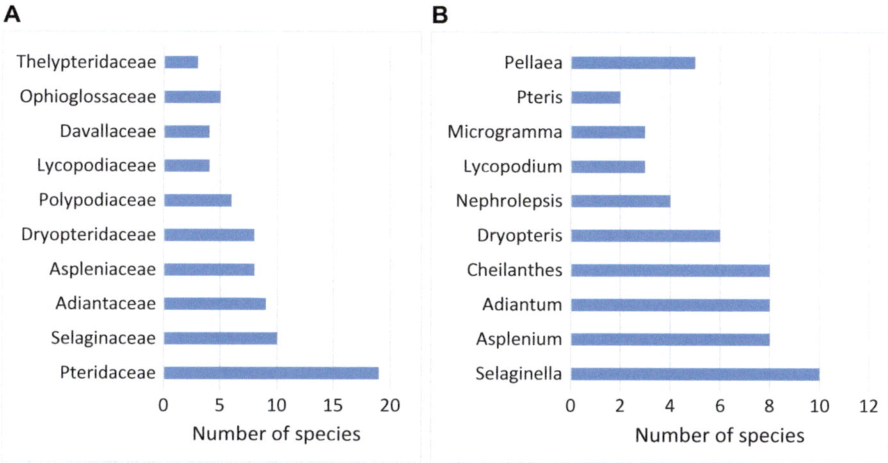


**Fig. 2** Pteridophyte families (**a**) and genera (**b**) with the largest number of species used for medicinal purposes in sub-Saharan Africa

(Setyawan 2009). Other important genera include *Adiantum, Cheilanthes, Dryopteris, Lycopodium, Microgamma, Nephrolepsis, Pallaea* and *Pteris* (Fig. 2b).

Pteridophytes used for medicinal purposes in sub-Saharan Africa have been recorded from 20 countries (Fig. 3). Most of the ethnobotanical data on medicinal plants have been reported in DRC, Gabon, Ivory Coast, Lesotho, Madagascar, Nigeria, South Africa, Swaziland and Tanzania (Table 1), all these countries have miombo woodland or mountainous areas characterized by small pockets of afromontane forests or tropical rain forests, the natural habitats of pteridophytes (Kornaś 1993). Low pteridophyte diversity of sub-Saharan Africa is mainly the result of the paucity of the rain forest flora (Aldasoro et al. 2004). In Nigeria, 37 pteridophyte species are used for medicinal purposes, followed by South Africa where 33 species are used, Lesotho and Swaziland with 16 species each, DRC with 14 species and the rest of the other countries haveless than 10 species each (Fig. 3). Popular pteridophyte species used for medicinal purposes in at least three countries include *Adiantum capillus-veneris* (Lesotho, Nigeria, South Africa, Swaziland), *Cheilanthes eckloniana* (Lesotho, South Africa, Swaziland), *Cheilanthes hirta* (Lesotho, South Africa, Swaziland), *Cheilanthes quadripinnata* (Lesotho, South Africa, Swaziland), *Cheilanthes viridis* (Forssk.) Swart (Madagascar, South Africa, Swaziland), *Cyathea manniana* Hook. (DRC, Ethiopia, South Africa), *Dryopteris inaequalis* (Ethiopia, Kenya, South Africa), *Equisetum ramosissimum* (Lesotho, South Africa, Swaziland), *Lycopodiella cerma* (L.) Pic. Serm. (DRC, Madagascar, Rwanda), *Lycopodium clavatum* L. (DRC, Lesotho, Madagascar, Nigeria, Rwanda, South Africa, Swaziland), *Nephrolepsis biserrata* (Sw.) Schott (Cameroon, DRC, Ivory Coast, Liberia), *Ophioglossum reticulatum* L. (Gabon, Lesotho, Nigeria, Tanzania), *Pellaea calomelanos* (Kenya, Lesotho, Malawi, South Africa, Swaziland, Zimbabwe) and *Pteridium aquilinum* (L.) Kuhn (Angola, DRC, Ivory Coast, Nigeria, South Africa,

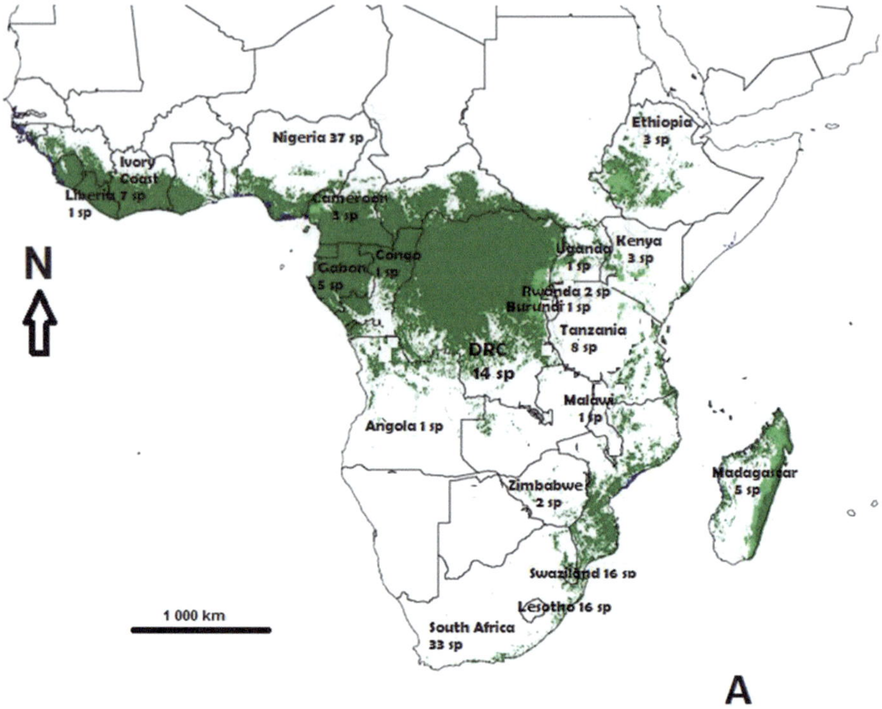


**Fig. 3** Pteridophyte species used for medicinal purposes in sub-Saharan Africa with forests in green

Swaziland, Tanzania) (Table 1). Only two pteridophyte species are used for cultural and ritualistic purposes in at least three countries: *Equisetum ramosissimum* (Lesotho, South Africa, Swaziland) and *Pteridium aquilinum* (DRC, Ivory Coast, Uganda) (Table 1).

About 9.4% of the pteridophyte species documented in this study are sold by traditional medicine sellers in sub-Saharan Africa (Table 2). The commercialization of medicinal plants is a strategy that is employed by poor people in developing countries as a means of generating income (Cunningham 1993). Research has shown that millions of people from sub-Saharan Africa derive a significant part of their income from gathered wild plants (Williams et al. 2001). Although, the contribution of medicinal plants to household income is widely recognized (Cunningham 1993), there are few studies that have examined the significance of the trade in pteridophytes to the livelihoods of poor and marginalized people as well as their conservation status throughout their geographical range. Knowing what species are traded commercially is the foundation for identifying taxa threatened by the trade and comparisons with other regional and national medicinal plant markets (Williams et al. 2001). Conservation assessments made in southern Africa (Bingham and Smith 2002; Mapaura and Timberlake 2002; Talukdar 2002) revealed that over-

**Table 2** Pteridophytes sold by traditional medicine sellers in sub-Saharan Africa

| Species | Plant part(s) | Country | Reference(s) |
|---|---|---|---|
| *Asplenium rutifolium* | Whole plant | South Africa | Cunningham (1993) |
| *Cheilanhes hirta* | Leaves, roots | South Africa | Williams et al. (2001) |
| *Cyathea dregei* | Roots, stem | South Africa | Cunningham (1993) |
| *Dryopteris athamantica* | Roots | South Africa | Cunningham (1993) |
| *Dryopteris inaequalis* | Roots | South Africa | Williams et al. (2001) |
| *Equisetum ramosissimum* | Roots | South Africa | Williams et al. (2001) |
| *Lycopodium clavatum* | Whole plant | South Africa, Swaziland | Cunningham (1993) and Williams et al. (2001) |
| *Microgramma lycopodioides* | Whole plant | South Africa | Cunningham (1993) |
| *Pellaea calomelanos* | Leaves, roots, whole plant | Zimbabwe | Cunningham (1993) and Williams et al. (2001) |
| *Pellaea rufa* | Roots | South Africa | Cunningham (1993) and Williams et al. (2001) |
| *Selaginella imbricata* | Whole plant | Zambia, Zimbabwe | Cunningham (1993) |

exploitation of some of the documented pteridophyte species as herbal medicines, limited distribution, habitat loss and modification are major causes of threats. *Acrostichum aureum* is critically endangered in Zimbabwe due to severely fragmented habitat and continuing population decline (Mapaura and Timberlake 2002). *Cyathea dregei* is also critically endangered in Lesotho due mainly to habitat degradation (Talukdar 2002). *Selaginella imbricata* is vulnerable in Zambia due to a very restricted area of occupancy and therefore prone to anthropogenic disturbances and stochastic events (Bingham and Smith 2002). Although *Pellaea rufa* is not listed in the Southern African Plant Red Data List, it is endemic to the Western Cape Province, South Africa (Burrows 1990). Baseline data on collection of pteridophyte species for medicinal value play an important role in drafting conservation plans for such species in the regions where they occur.

## Ailments and Diseases Treated by Pteridophyte Species

A total of 42 medical conditions are treated using remedies made from pteridophyte species (Table 1). Cold, cough and sore throat, injuries, parasitic worms, gastro-intestinal system, gynaecological, general body pains, inflammations, urinary system, respiratory system, ethno-veterinary medicine, insect repellent, pregnancy, birth and puerperium, snake and spider antidote, headache, circulatory system, dermatological, fainting and fits, fever, hair lice and dendrufs, odontological and sexual dysfunctional are treated with the highest number of medicinal pteridophytes in

**Table 3** Major ailments and disease categories and pteridophyte species reported

| Ailment category | Number of pteridophyte species |
|---|---|
| Cold, cough and sore throat | 24 |
| Injuries | 23 |
| Parasitic worms | 19 |
| Gastro-intestinal system | 16 |
| Gynaecological | 13 |
| General body pains | 11 |
| Inflammations | 11 |
| Urinary system | 9 |
| Respiratory system | 8 |
| Ethnoveterinary medicine | 8 |
| Insect repellent | 7 |
| Pregnancy, birth and puerperium | 7 |
| Snake and spider antidote | 7 |
| Headache | 6 |
| Circulatory system | 5 |
| Dermatological | 5 |
| Fainting and fits | 5 |
| Fever | 5 |
| Hair lice and dendrufs | 5 |
| Odontological | 5 |
| Sexual dysfunctional | 7 |

Most species are reported in more than one ailment and disease category

sub-Saharan Africa (Table 3). For hundreds of years, indigenous people in southern Africa, particularly South Africa have used a variety of pteridophytes species as herbal medicines for coughs, colds and respiratory infections and gastro-intestinal problems (Watt and Brandwijk 1962; Hutchings et al. 1996; van Wyk et al. 2013). Injuries and wound infections are one of the most common diseases in developing countries mainly because of poor hygienic conditions (Ayyanar and Ignacimuthu 2009). Leaf decoction of *Pteridium aquilinum*is used to treat and manage oral candidiasis in Tanzania (Kisangau et al. 2011), which is one of the prevalent HIV/AIDS opportunistic infections in sub-Saharan Africa (Maroyi 2014b). Almost all plant parts are used to prepare different remedies: leaves, rhizomes, roots, sap and spores. Plant parts most frequently used in preparing herbal medicines are leaves (44.4%), whole plant (39.3%) and rhizome and/or roots (36.8%) (Table 1). Monotherapy preparations made from a single pteridophyte species are the most dominant (84.6%), while 12.8% and 2.6% of the herbal concortions are prepared from a combination of two and three species respectively (Table 1). Plant remedies are often utilized in the form of decoction, infusion or maceration (47.9%), sap or paste (33.3%) and smoked or smoke inhaled (19.7%) (Table 1). Combination of at least two herbal medicines (Table 1) is believed to increase efficacy of the herbal concortions due to the additive or synergistic effects that herbal medicines have during disease or ailment treatment (Bussmann and Sharon 2006).

## 4 Conclusion

The annotated checklist in Table 1 provides baseline data on pteridophytespecies used for medicinal purposes in sub-Saharan Africa. Published accounts of this category of pteridophytes are available from Angola, Burundi, Cameroon, Congo, DRC, Ethiopia, Gabon, Ivory Coast, Kenya, Lesotho, Liberia, Madagascar, Malawi, Nigeria, Rwanda, South Africa, Swaziland, Tanzania, Uganda, Zambia and Zimbabwe only. Although the results from these 20 countries are diverse in nature, they may not apply directly to 48 countries in the sub-Saharan African region. Based on the results of this study, it appears that the use of pteridophytes for medicinal purposes is entrenched in cultural practices throughout the sub-Saharan African region. Documentation of how local communities use pteridophytes and other plant resources as herbal medicines is critical to safeguard indigenous knowledge and conserve plant species which are rapidly being lost (Bingham and Smith 2002; Mapaura and Timberlake 2002; Talukdar 2002). More importantly, documentation of pteridophytes used for medicinal, cultural and ritualistic purposes can facilitate future research on the safety and efficacy of these plant resources. Poor health facilities and limited access to drugs have perpetuated and increased the use of traditional medicines particularly in rural and marginalized areas (Kisangau et al. 2011; Maroyi 2013). Studies like the current investigation can help stimulate confidence in the use of traditional medicines among local communities and appreciate the value of plant resources. Sharing of traditional knowledge on medicinal plants is important for maintaining options for the use of traditional medicines, particularly as use of alternative medicine is growing because of its moderate costs and increasing faith in herbal medicines. As demand for medicinal plants continue to accelerate, awareness creation should be made among local communities to ensure sustainable use and conservation of medicinal plants.

This inventory is a crucial starting point in trying to demonstrate that traditional knowledge on pteridophytes used for medicinal purposes represent not only an important heritage developed over centuries but it is also an important mass of data which should be exploited in order to assist future workers with a repository of herbal plants to evaluate for phytochemical safety and pharmaceutical efficacy. Further ethnopharmacological studies are recommended for the documented pteridophytes species as some of these species have already shown interesting applications in primary healthcare in several sub-Saharan African countries (Table 1). There is need therefore, to assess phytochemical and pharmacological properties of popular pteridophyte species used for medicinal purposes such as *Adiantum capillus-veneris*, *Cheilanthes eckloniana*, *Cheilanthes hirta*, *Cheilanthes quadripinnata*, *Cheilanthes viridis*, *Cyathea manniana*, *Dryopteris inaequalis*, *Equisetum ramosissimum*, *Lycopodiella cerma*, *Lycopodium clavatum*, *Nephrolepsis biserrata*, *Ophioglossum reticulatum*, *Pellaea calomelanos* and *Pteridium aquilinum* with the aim of identifying active ingredients contained by such plants. Historically, active compounds have been isolated from medicinal plants, and a landmark discovery includes the discovery of quinine from the bark of the cinchona (quina-quina) tree

(Achan et al. 2011). Further investigations on phytochemical constituents and subsequent screening are needed for opening new opportunities to develop pharmaceutical drugs based on medicinal applications of pteridophytes in sub-Saharan African.

**Acknowledgments** The author is thankful to the Govan Mbeki Research and Development Centre (GMRDC), University of Fort Hare, South Africa for funding the research project.

## References


Achan J, Talisuna AO, Erhart A, Yeka A, Tibenderana JK, Baliraine FN, Rosenthal PJ, D'Alessandro U (2011) Quinine, an old anti-malarial drug in a modern world: role in the treatment of malaria. Malar J 10:144

Agra MF, Silva KN, Basílio IJLD, de Freitas PF, Barbosa-Filho JM (2008) Survey of medicinal plants used in the region Northeast of Brazil. Braz J Pharmacog 18(3):472–508

Aldasoro JJ, Cabezas F, Aedo C (2004) Diversity and distribution of ferns in sub-Saharan Africa, Madagascar and some islands of the South Atlantic. J Biogeogr 31(10):1579–1604

Ayyanar M, Ignacimuthu S (2009) Herbal medicines for wound healing among tribal people in southern India: ethnobotanical and scientific evidences. Int J Appl Res Nat Prod 2(3):29–42

Baerts M, Lehmann J (1996) Plantes médicinales à effet placebo au Burundi. In: Schröder E, Balansard G, Cabalion P, Fleurentin J, Mazars G (eds) Médicaments et aliments: Approche ethnopharmacologique. Colloque Européen d'Ethnopharmacologie; Conférence Internationale d'Ethnomédecine, Colloques et Séminaires, 24–27 March 1993, Heidelberg, pp 343–352

Bingham MG, Smith PP (2002) Zambia. In: Golding JS (ed) Southern African plant red data lists. Southern African Botanical Diversity Network (SABONET) report no. 14, Pretoria, pp 135–156

Bouquet A (1974) Plantes medicinales de la Côte d'Ivoire. Oratom, Paris

Burrows JE (1990) Southern African ferns and fern allies. Frandsen Publishers, Johannesburg

Bussmann RW, Sharon D (2006) Traditional medicinal plant use in northern Peru: tracking two thousand years of healing culture. J Ethnobiol Ethnomed 2:47

Chhabra SC, Mahunnah RLA, Mshiu EN (1987) Plants used in traditional medicine in eastern Tanzania 1. Pteridophytes and angiosperms (Acanthaceae to Canellaceae). J Ethnopharmacol 21(3):253–277

Cook F (1995) Economic botany data collection standard. Royal Botanic Gardens, Kew

Cunningham AB (1993) African medicinal plants: setting priorities at the interface between conservation and primary healthcare. People and Plants Working Paper no. 1, UNESCO, Paris

de Winter WP, Jnasen PCM (2003) *Selaginella* Pal. Beauv. In: de Winter WP, Amoroso VB (eds) Plant resources of south-east Asia 15(2) cryptogams: ferns and fern allies. Backhuys Publishers, Leiden, pp 261–297

Fernández H, Kumar A, Revilla MA (2011) Working with ferns: issues and applications. Springer, New York

Gelfand M, Drummnd RB, Mavi S, Ndemera B (1985) The traditional medical practitioner in Zimbabwe: his principles of practice and pharmacopoeia. Mambo Press, Gweru

Harley GW (1941) Native African medicine: with special reference to its practice in the Mano tribe of Liberia. Cambridge, MA

Harnett SM, Oosthuizen V, van de Venter M (2005) Anti-HIV activities of organic and aqueous extracts of *Sutherlandia frutescens* and Lobostemon trigonus. J Ethnopharmacol 96(1–2):113–119

Ho R, Teai T, Bianchini J-P, Lafont R, Raharivelomanana P (2010) Ferns: from traditional uses to pharmaceutical development, chemical identification of active principles. In: Fernández H, Kumar A, Revilla MA (eds) Working with ferns: issues and applications. Springer-Verlag, Berlin, pp 321–346

Hutchings A (1992) Plants used for some stress-related ailments in traditional Zulu, Zhosa and Sotho medicine. MSc dissertation, University of Natal, Pietermaritzburg
Hutchings A, Scott AH, Lewis G, Cunningham AB (1996) Zulu medicinal plants: an inventory. University of Natal Press, Scottsville
Jenkins MD (1987) Madagascar: an environmental profile. IUCN, Gland
Jiofack T, Fokunang C, Kemeuze V, Fongnzossie E, Tsabang N, Nkuinkeu R, Mapongmetsem PM, Nkongmeneck BA (2008) Ethnobotany and phytopharmacopoea of the south-west ethnoecological region of Cameroon. J Med Pl Res 2(8):197–206
Kisangau DP, Herrmann TM, Lyaruu HVM, Hosea KM, Joseph CC, Mbwambo ZH, Masimba PJ (2011) Traditional knowledge, use practices and conservation of medicinal plants for HIV/AIDS care in rural Tanzania. Ethnobot Res Appl 9:43–57
Kornaś J (1993) The significance of historical factors and ecological preference in the distribution of African pteridophytes. J Biogeogr 20(3):281–286
Lemmens RMHJ (2008) *Lycopodiella cerma* (L.) Pic. Serm. In: Schmelzer GH, Gurib-Fakim A (eds) Plant resources of tropical Africa 11: medicinal plants 1. PROTA Foundation, Leiden, pp 365–366
Long C (2005) Swaziland's flora: siSwati names and uses. http://www.sntc.org.sz/flora/clbotalpha.asp?l=C&pg=10. Downloaded on 2 August 2014
Malan DF, Neuba DFR (2011) Traditional practices and medicinal plant use during pregnancy by Anyi-Ndenye women (Eastern Côte d'Ivoire). Afr J Reprod Health 15(1):85–93
Mapaura A, Timberlake JR (2002) Zimbabwe. In: Golding JS (ed) Southern African plant red data lists. Southern African Botanical Diversity Network (SABONET) report no. 14, Pretoria, pp 157–186
Maroyi A (2013) Traditional use of medicinal plants in south-central Zimbabwe: review and perspectives. J Ethnobiol Ethnomed 9:31
Maroyi A (2014a) Not just minor wild edible forest products: consumption of pteridophytes in sub-Saharan Africa. J Ethnobiol Ethnomed 10:78
Maroyi A (2014b) Alternative medicines for HIV/AIDS in resource-poor settings: insight from traditional medicines use in sub-Saharan Africa. Trop J Pharm Res 13(9):1527–1536
Matthews P (1993) Medicinal plants of the Tanzanian coastal forest: a list of species with local names and applications. The Society for Environmental Exploration, London
Moteetee A, van Wyk BE (2011) The medical ethnobotany of Lesotho: a review. Bothalia 41(1):209–228
Nwosu MO (2002) Ethnobotanical studies on some pteridophytes of southern Nigeria. Econ Bot 56(3):255–259
Puri HS (1970) Indian pteridophytes used in folk remedies. Amer Fern J 60(4):137–143
Rates SMK (2001) Plants as sources of drugs. Toxicon 39(5):603–613
Roux JP (2003) Swaziland ferns and fern allies. SABONET report 19, National Botanical Institute, Pretoria
Sassen M, Wan M (2006) Biodiversity and local priorities in a community near the Ivindo national park, Makokou, Gabon. Project IRET/CENAREST and CIFOR, Makokou.
Semenya SS, Maroyi A (2013) Medicinal plants used for the treatment of tuberculosis by Bapedi traditional healers in the Limpopo Province, South Africa. Afr J Tradit Compl Altern Med 10(2):316–323
Setyawan AD (2009) Traditionally utilization of *Selaginella*: field research and literature review. Nusantara Biosci 1(3):146–158
Shale TL, Stirk WA, van Staden J (1999) Screening of medicinal plants used in Lesotho for anti-bacterial and anti-inflammatory activity. J Ethnopharmacol 67(3):347–354
Tabuti JRS, Kukunda CB, Kaweesi D, Kasib OMJ (2012) Herbal medicine use in the districts of Nakapiripirit, Pallisa, Kanungu and Mukono in Uganda. J Ethnobiol Ethnomed 8:35
Talukdar S (2002) Lesotho. In: Golding JS (ed) Southern African plant red data lists. Southern African Botanical Diversity Network (SABONET) report no. 14, Pretoria, pp 21–30
Terashima H, Kalaya S, Malasi N (1991) Ethnobotany of the Lega in the tropical rainforest of eastern Zaire: part one, zone de Mwenga. Afr Stud Monogr 15(suppl.):1–61

Termote C (2012) Wild edible plant us in Tshopo District, Democratic Republic of Congo. Ph.D.Thesis. Faculty of Bioscience Engineering, University of Ghent, Ghent

Tesfaye A, Sebsebe D (2009) Ethnobotanical study of medicinal plants in Kafficho people, south-western Ethiopia. In: Svein E, Harald A, Birhanu T, Shiferaw B (eds) Proceedings of the 16th international conference of Ethiopian studies. NTNU-Trykk Press, Trondheim, pp 711–726

Timberlake J (1994) Vernacular names and uses of plants in northern Kenya. J East Afr Nat Hist 83(1):31–69

van der Burg WJ (2004a) *Marsilea minuta* L. In: Grubben GJH, Denton OA (eds) Plant resources of tropical Africa 2: vegetables. PROTA Foundation, Leiden, pp 379–380

van der Burg WJ (2004b) *Ophioglossum reticulatum* L. In: Grubben GJH, Denton OA (eds) Plant resources of tropical Africa 2: vegetables. PROTA Foundation, Leiden, pp 404–405

van der Burg WJ (2004c) *Pteridium aquilinum* (L.) Kuhn. In: Grubben GJH, Denton OA (eds) Plant resources of tropical Africa 2: vegetables. PROTA Foundation, Leiden, pp 439–441

van der Burg WJ (2004d) *Stenochlaena tenuifolia* (Desv.) T. Moore. In: Grubben GJH, Denton OA (eds) Plant resources of tropical Africa 2: vegetables. PROTA Foundation, Leiden, pp 515–516

van Wyk B-E, Gericke N (2007) Peoples plants: a guide to useful plants of southern Africa. Briza Publications, Cape Town

van Wyk B-E, van Oudtshoorn B, Gericke N (2013) Medicinal plants of South Africa. Briza Publications, Cape Town

Watt JM, Brandwijk MG (1962) Themedicinal and poisonous plants of southern and eastern Africa. E and S Livingstone, London

Williams VL, Balkwill K, Witkowski ETF (2001) A lexicon of plants traded in the Witwatersrand umuthi shops, South Africa. Bothalia 31(1):71–98

World Health Organization (WHO) (2002) Traditional medicine strategy 2002–2005. WHO, Geneva

Yamada T (1999) A report on the ethnobotany of the Nyindu in the eastern part of the former Zaire. Afr Stud Monogr 20(1):1–72

Yarborough SC, Powell AM (2002) Ferns and fern allies of the Trans-Pecos and adjacent areas. Texas Tech University Press, Texas

Yineger H, Kelbessa E, Bekele T, Lulekal E (2007) Ethnoveterinary medicinal plants at Bale Mountains National Park, *Ethiopia*. J Ethnopharmacol 112(1):55–70

Zimudzi C, Bosch CH (2008) *Lycopodium clavatum* L. In: Schmelzer GH, Gurib-Fakim A (eds) Plant resources of tropical Africa 11: medicinal plants 1. PROTA Foundation, Leiden, pp 366–369

# Index





M. Neffati et al. (eds.), *Medicinal and Aromatic Plants of the World - Africa Volume 3*, Medicinal and Aromatic Plants of the World,
DOI 10.1007/978-94-024-1120-1

MIX
Papier aus verantwortungsvollen Quellen
Paper from responsible sources
FSC® C105338

If you have any concerns about our products,
you can contact us on
**ProductSafety@springernature.com**

In case Publisher is established outside the EU,
the EU authorized representative is:
**Springer Nature Customer Service Center GmbH**
**Europaplatz 3, 69115 Heidelberg, Germany**

Printed by Libri Plureos GmbH
in Hamburg, Germany